Larry L. Barton · Martin Mandl
Alexander Loy

Editors

Geomicrobiology: Molecular and Environmental Perspective

 Springer

Editors
Larry L. Barton
Department of Biology
MSCO3 2020
University of New Mexico
Albuquerque, New Mexico 87131-0001
USA
lbarton@unm.edu

Martin Mandl
Masaryk University
Faculty of Science
Department of Biochemistry
Kotlarska 2
611 37 Brno
Czech Republic
mandl@chemi.muni.cz

Alexander Loy
Department of Microbial Ecology
Faculty of Life Sciences
University of Vienna
Althanstrasse 14
A-1090 Wien
Austria
loy@microbial-ecology.net

ISBN 978-90-481-9203-8   e-ISBN 978-90-481-9204-5
DOI 10.1007/978-90-481-9204-5
Springer Dordrecht Heidelberg London New York

Library of Congress Control Number: 2010931683

Cover illustration: Outflow channel of Alvord Hot Springs, Oregon. Microbial biofilms are the catalysts for several geomicrobiological phenomena visible here, including deposition of carbonates (white), and precipitation of iron- and arsenic-bearing minerals (red). A clear transition of biofilm appearance in the channel correlates with increased arsenic oxidation. Plant and cyanobacterial denizens also contribute to biogeochemical reactions in this habitat.

Printed on acid-free paper

Springer is part of Springer Science+Business Media (www.springer.com)

Preface

The interaction of microorganisms with geological activities results in processes influencing development of the Earth's geo- and biospheres. In assessing these microbial functions, scientists have explored short- and longterm geological changes attributed to microorganisms and developed new approaches to evaluate the physiology of microbes including microbial interaction with the geological environment. As the field of geomicrobiology developed, it has become highly interdisciplinary and this book provides a review of the recent developments in a cross section of topics including origin of life, microbial-mineral interactions and microbial processes functioning in marine as well as terrestrial environments. A major component of this book addresses molecular techniques to evaluate microbial evolution and assess relationships of microbes in complex, natural communities. Recent developments in so-called 'omics' technologies, including (meta) genomics and (meta)proteomics, and isotope labeling methods allow new insights into the function of microbial community members and their possible geological impact. While this book summarizes current knowledge in various areas, it also reveals unresolved questions that require future investigations. Information in these chapters enhances our fundamental knowledge of geomicrobiology that contributes to the exploitation of microbial functions in mineral and environmental biotechnology applications. It is our hope that this book will stimulate interest in the general field of geomicrobiology and encourage others to explore microbial processes as applied to the Earth. Not only have authors provided skillful reviews but also they have outlined unique perspectives on environmental microorganisms and their related processes. This book should be of interest to scientists of various fields including biogeochemistry, geology, microbiology, microbial ecology, evolution, and environmental sciences.

We appreciate the efforts of the authors who have generously contributed their time to prepare chapters for this book and we encourage readers to consult them for further advancements in their research areas.

Larry L. Barton
Martin Mandl
Alexander Loy

Contents

List of Authors

Daniel R. Bond
Department of Microbiology and Biotechnology Institute,
University of Minnesota, 1479 Gortner Ave, St Paul, MN 55105, USA
dbond@umn.edu

George T. Bonheyo
Pacific Northwest Laboratory, Marine Sciences Laboratory,
1529 W. Sequim Bay Rd, Sequim, WA 98382, USA
Department of Geology, University of Illinois at Urbana-Champaign,
1301 West Green Street, Urbana, IL 61801-2938, USA
George.Bonheyo@pnl.gov

Violaine Bonnefoy
Institut de Microbiologie de la Méditerranée, Laboratoire de Chimie Bactérienne,
C.N.R.S., Université de la Méditerranée, 31 chemin Joseph Aiguier,
13402 Marseille Cedex 20, France
bonnefoy@ifr88.cnrs-mrs.fr

Charles Seaton Cockell
CEPSAR, Open University, Milton Keynes, MK7 6AA, UK
c.s.cockell@open.ac.uk

Katerina Demnerova
Department of Biochemistry and Microbiology, Faculty of Food and Biochemical
Technology, ICT Prague, Technicka 3, 16628, Prague, Czech Republic
katerina.demnerova@vscht.cz

Giovanni Emiliani
Tree and Timber Institute – National Research Council, Via Biasi 75 I-38010,
San Michele all'Adige (TN), Italy
giovanni.emiliani@unifi.it

Annette Summers Engel
Department of Geology & Geophysics, Louisiana State University,
E235 Howe-Russell Geoscience Complex, Baton Rouge, Louisiana 70803, USA
aengel@lsu.edu

Renato Fani
Laboratory of Microbial and Molecular Evolution, Department of Evolutionary
Biology, University of Florence, Via Romana 17–19, I-50125 Florence, Italy
renato.fani@unifi.it

Marco Fondi
Laboratory of Microbial and Molecular Evolution,
Department of Evolutionary Biology, University of Florence,
Via Romana 17–19, I-50125 Florence, Italy
marco.fondi@unifi.it

Bruce W. Fouke
Department of Geology, University of Illinois at Urbana-Champaign,
1301 West Green Street, Urbana, IL 61801-2938, USA
Department of Microbiology, University of Illinois Urbana-Champaign,
601 S. Goodwin Avenue, Urbana, IL 61801, USA
University of Illinois Urbana-Champaign, Institute for Genomic Biology,
1206 W. Gregory Drive, Urbana, IL 61801, USA
fouke@illinois.edu

Kati Geszvain
Division of Environmental and Biomolecular Systems, Oregon Health & Science
University, 20000 NW Walker Road, Beaverton, OR 97006, USA
geszvaik@ebs.ogi.edu

Nigel Goldenfeld
Department of Physics, University of Illinois at Urbana-Champaign,
1110 West Green Street, Urbana, IL 61801-3080, USA
University of Illinois Urbana-Champaign, Institute for Genomic Biology,
1206 W. Gregory Drive, Urbana, IL 61801, USA
nigel@uiuc.edu

Zhili He
Department of Botany and Microbiology,
Institute for Environmental Genomics, University of Oklahoma,
101 David L. Boren Blvd, Norman, OK 73019, USA
zhili.he@ou.edu

David Barrie Johnson
School of Biological Sciences, Bangor University, Deiniol Road,
Bangor LL57 2UW, UK
d.b.johnson@bangor.ac.uk

Andreas Kappler
Center for Applied Geosciences, University of Tübingen,
Sigwartstrasse 10, D-72076 Tübingen, Germany
andreas.kappler@uni-tuebingen.de

Inga Koehler
Center for Applied Geosciences, University of Tübingen,
Sigwartstrasse 10, D-72076 Tübingen, Germany
inga.koehler@uni-tuebingen.de

Kurt Konhauser
Department of Earth and Atmospheric Sciences, University of Alberta, Edmonton,
Alberta, T6G 2E3, Canada
kurtk@ualberta.ca

Rhesa N. Ledbetter
Department of Biological Sciences, Idaho State University, P.O. Box 8007,
Pocatello, ID 83209, USA
ledbrhes@hotmail.com

Sung-Woo Lee
Division of Environmental and Biomolecular Systems, Oregon Health & Science
University, 20000 NW Walker Road, Beaverton, OR 97006, USA
lees@ebs.ogi.edu

Pietro Liò
Computer Laboratory, University of Cambridge, 15 JJ Thompson Avenue,
cb03fd, Cambridge, UK
pl219@cam.ac.uk

Petra Lovecka
Department of Biochemistry and Microbiology, Faculty of Food and Biochemical
Technology, ICT Prague, Technicka 3, 16628 Prague, Czech Republic
petra.lovecka@vscht.cz

Alexander Loy
Department of Microbial Ecology, Faculty of Life Sciences, University of Vienna,
Althanstrasse 14, A-1090 Wien, Austria
loy@microbial-ecology.net

Tomas Macek
Department of Biochemistry and Microbiology, Faculty of Food and Biochemical
Technology, ICT Prague, Technicka 3, 16628 Prague, Czech Republic
Institute of Organic Chemistry and Biochemistry CAS, Joint Laboratory of ICT
Prague and IOCB, Flemingovo n. 2, 16610 Prague, Czech Republic
tom.macek@uochb.cas.cz

Martina Mackova
Department of Biochemistry and Microbiology, Faculty of Food and Biochemical
Technology, ICT Prague, Technicka 3, 16628 Prague, Czech Republic
Institute of Organic Chemistry and Biochemistry CAS, Joint Laboratory of ICT
Prague and IOCB, Flemingovo n. 2, 16610 Prague, Czech Republic
martina.mackova@vscht.cz

Timothy S. Magnuson
Department of Biological Sciences, Idaho State University, P.O. Box 8007,
Pocatello, ID 83209, USA
magntimo@isu.edu

Héctor García Martín
Department of Physics, University of Illinois at Urbana-Champaign,
1110 West Green Street, Urbana, IL 61801-3080, USA
Joint BioEnergy Institute, Emeryville, CA 94608, USA
Physical Biosciences Division, Lawrence Berkeley National Laboratory,
Berkeley, CA 94710, USA
hectorg@gmail.com

Laura E. McAllister
Lancaster Environment Centre, Lancaster University,
Lancaster LA1 4YQ, UK
l.mcallister@lancaster.ac.uk

Kentaro Nakamura
Precambrian Ecosystem Laboratory,
Japan Agency for Marine-Earth Science and Technology (JAMSTEC),
2-15 Natsushima-cho, Yokosuka 237-0061, Japan

Joy D.Van Nostrand
Department of Botany and Microbiology,
Institute for Environmental Genomics, University of Oklahoma,
101 David L. Boren Blvd, Norman, OK 73019, USA
joy.vannostrand@ou.edu

Martina Novakova
Department of Biochemistry and Microbiology,
Faculty of Food and Biochemical Technology, ICT Prague,
Technicka 3, 16628 Prague, Czech Republic
Institute of Organic Chemistry and Biochemistry CAS,
Joint Laboratory of ICT Prague and IOCB, Flemingovo n. 2,
16610 Prague, Czech Republic
martina1.novakova@vscht.cz

Jörg Overmann
Bereich Mikrobiologie, Ludwig-Maximilians-Universität München,
Großhadernerstrasse 2-4, 82152, Planegg-Martinsried, Germany
j.overmann@LRZ.uni-muenchen.de

Michael Pester
Department for Microbial Ecology, Faculty of Life Sciences,
University of Vienna, Althanstrasse 14, A-1090 Wien, Austria
pester@microbial-ecology.net

Kathrin Riedel
Department of Microbiology, Institute of Plant Biology,
University of Zürich, Winterthurerstrasse 190, CH-8057 Zurich, Switzerland
kriedel@botinst.uzh.ch

Thomas Schneider
Department of Microbiology, Institute of Plant Biology,
University of Zürich, Winterthurerstrasse 190, CH-8057 Zurich, Switzerland
Thomas.Schneider@botinst.uzh.ch

Kirk T. Semple
Lancaster Environment Centre, Lancaster University,
Lancaster LA1 4YQ, UK
k.semple@lancaster.ac.uk

Michel Sylvestre
Institut National de la Recherche Scientifique, INRS-Institut Armand-Frappier,
531, boul. des Prairies, Laval (Quebec) H7V 1B7, Canada
Michel.Sylvestre@iaf.inrs.ca

Ken Takai
Subsurface Geobiology Advanced Research (SUGAR) Project/Precambrian
Ecosystem Laboratory, Japan Agency for Marine-Earth Science and Technology
(JAMSTEC), 2-15 Natsushima-cho, Yokosuka 237-0061, Japan
kent@jamstec.go.jp

Bradley M. Tebo
Division of Environmental and Biomolecular Systems, Oregon Health & Science
University, 20000 NW Walker Road, Beaverton, OR 97006, USA
tebo@ebs.ogi.edu

Ondrej Uhlik
Department of Biochemistry and Microbiology, Faculty of Food and Biochemical
Technology, ICT Prague, Technicka 3, 16628 Prague, Czech Republic
Institute of Organic Chemistry and Biochemistry CAS, Joint Laboratory of ICT
Prague and IOCB, Flemingovo n. 2, 16610 Prague, Czech Republic
ondrej.uhlik@gmail.com

John Veysey
Department of Physics, University of Illinois at Urbana-Champaign,
1110 West Green Street, Urbana, IL 61801-3080, USA
veysey@gmail.com

Jitka Viktorova
Department of Biochemistry and Microbiology, Faculty of Food and Biochemical
Technology, ICT Prague, Technicka 3, 16628 Prague, Czech Republic
Institute of Organic Chemistry and Biochemistry CAS, Joint Laboratory of ICT
Prague and IOCB, Flemingovo n. 2, 16610 Prague, Czech Republic
jitka.viktorova@vscht.cz

Günter Wächtershäuser
D-80333, Munich, Weinstr. 8, Germany
gwmunich@yahoo.com

Huifang Xu
Department of Geoscience and NASA Astrobiology Institute, University of
Wisconsin-Madison, 1215 West Dayton Street, Madison, WI 53706, USA
hfxu@geology.wisc.edu

Jizhong Zhou
Department of Botany and Microbiology, Institute for Environmental Genomics,
University of Oklahoma, 101 David L. Boren Blvd, Norman, OK 73019, USA
jzhou@ou.edu

Chapter 1
Chemoautotrophic Origin of Life: The Iron–Sulfur World Hypothesis

Günter Wächtershäuser

Introduction

The study of the origin of life is an immature science. If we apply the strictures of Immanuel Kant it may not be considered a mature science until it can be said to have embarked on a course of orderly progress. Indeed, if we review the development of research into the origin of life, we have to admit that it is still far from presenting the image of progress. It may be best characterized as an exercise of randomly groping around – and doing so at a number of different levels.

At the philosophical level we still are faced with the conflict between mechanistic explanations and teleological judgments. Biochemistry is providing ever more refined mechanistic explanations of the chemistry of life, down to the finest molecular details. A theory of biology, by contrast, would have to treat organisms, i.e. organized beings, as integrated wholes. Biochemistry is reductionistic and mechanistic while biology is holistic and teleological. This provides us with our first question. What would count as a solution of the problem of the origin of life – a molecular reaction mechanism or a primordial organism? Kant of course held that a natural science would be impossible without mechanistic explanations (Kant 1790). Applying this requirement he suggested that a full replacement of all teleological notions of "natural purposes" (end-means relations or functions within a whole organism or between an organism and its environment) by mechanisms would be impossible. Kant did see a "ray of hope" that a mechanistic theory of evolution, i.e. of transformations from one type of organism to another might one day be achievable. He was, however, convinced that a scientific theory of the origin of life, defined as replacement of all teleological judgments by mechanistic explanations, would be impossible. It would require postulating an ultimate ancestral organism, which would have to be endowed with the

G. Wächtershäuser (✉)
D-80333 Munich, Weinstr. 8, Germany
e-mail: gwmunich@bellsouth.net

L.L. Barton et al. (eds.), *Geomicrobiology: Molecular and Environmental Perspective*,
DOI 10.1007/978-90-481-9204-5_1, © Springer Science+Business Media B.V. 2010

means toward the ends of all future organisms, and so it would remain inescapably teleological. This then is our problem: to postulate a primordial organism – here termed "pioneer organism", which is at the same time mechanistic and organizational. Central to this effort will be the notion of a "synthetic autocatalysis", the chemical equivalent of biological reproduction, which is a chemical reaction mechanism and at the same time a functional whole, and by being synthetic it is endowed from the start with the primary vector of complexity increase.

The problem of the origin of life is situated within three major scientific disciplines: biology, geology and chemistry. Biology and geology are both sciences of natural history, interconnected in a multitude of ways. Geological processes provide the habitats of life and products of organismic activities leave their mark in the compositions of the atmosphere, the hydrosphere and the lithosphere of the Earth; and in the bio-geo-chemical cycles: the carbon cycle, the nitrogen cycle and the sulfur cycle. In these cycles geochemical and biochemical reaction segments are integrated. Therefore, broadly speaking, the problem of the origin of life is the problem of the origin and early evolution of the bio-geo-chemical connection.

In principle we may think of two different possible sources of evidence for the origin and early evolution of life: geological evidence locked in the rocks of the extant lithosphere and biochemical evidence conserved in the extant biosphere. The global process of plate tectonics seems to have obliterated any rock formation old enough for bearing witness to the origin and earliest evolution of life on Earth. This leaves us with the second source of evidence, the biochemical record. The most astounding fact of this record is its complexity in terms of both molecular structures and reaction networks. In view of this complexity the best we could hope for would be an initial theory that points in the right general direction and a program of theory evolution towards ever increasing organismic and evolutionary comprehensiveness and ever finer molecular detail. For establishing such an initial theory we need a heuristic that correlates known facts of extant biochemistry with the unknown chemistry of the pioneer organism.

The conventional heuristic of "backward projection" establishes *one-to-one* relations between extant biochemical features and chemical features of the origin of life (Fig. 1.1a). The repeated application of this heuristic leads to a recipe for life's beginning that comprises a large number of compounds each identical with or similar to an extant bio-compound. These are all projected into an aqueous concoction, widely known by the names "prebiotic broth" or "primordial soup". Its birthmark is obscurity. Nobody has ever been able to specify a chemically convincing mechanism, by which the first organism could have appeared in a prebiotic broth.

By contrast, the heuristic of "biochemical retrodiction" (Wächtershäuser 1997) generates *many-to-one* relationships, whereby several different extant biochemical features are correlated to one simpler functional precursor feature. By employing this heuristic to include more and more extant features we move to ever deeper, fewer and simpler precursor features. In this manner the heuristic of biochemical retrodiction generates an overall pattern of backward convergence (Fig. 1.1b). It is immediately apparent that this heuristic does not lead into the chaos of a prebiotic broth. It rather focuses on one specific, coherent chemical entity, the "pioneer organism" of life.

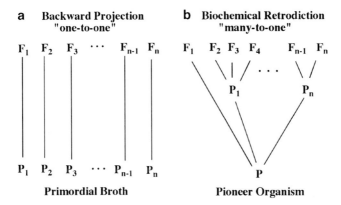

Fig. 1.1 Schematic comparison of (**a**) the conventional "backward projection" and (**b**) "biochemical retrodiction" (F_1, F_2 … extant functions; P, P_1, P_2 … precursor functions)

When we view these correlations not in the backward heuristic direction, but in the forward evolutionary direction, we speak of an explanatory connection between the hypothetical pioneer organism and extant biochemistry. In any event, the heuristic of biochemical retrodiction remains methodological and its results remain problematic and in need of empirical evaluation and improvement. Chemical testing provides the best means for an empirical evaluation. In spite of its historic nature a theory of the origin of life may readily satisfy the logic of empirical testing. For example, if a theory of the origin of life requires a certain chemical reaction and if we accept experiments that show this reaction to be impossible, we are forced to admit the falsity of that theory of the origin of life. Theories on the origin of life, therefore, have the same logical status as all other scientific theories.

At the very origin of life there must be a phase transition between the pioneer organism and its aqueous environment. Recent findings point to the availability of liquid water on Earth as early as 4.4 billion years ago and the origin of life may date back to this time (Wilde et al. 2001; Peck et al. 2001; Mojzsis et al. 2001). Most notions of geological uniformitarianism break down when we enter this early hadean period of Earth's history. Therefore, all geochemical clues concerning the parameters of the origin of life are highly problematic. Moreover, the pioneer organism must be seen as a coherent entity in a spatially and temporally coherent situation. This require-ment of coherence puts severe constraints not only on the possible sets of reaction parameters, but also on the possible sites of habitats of the pioneer organism.

Retrodicting the Origin from the Chemical Elements of Life

We are long used to it by now, but it should still fill us with a sense of wonder that by the universal laws of nature and by a few natural constants a total of 80 stable elements are enabled, each a discrete chemical individual with discrete chemical properties.

All these chemical individuals are organized in the Periodic System of elements. When we rank the elements in the order of decreasing abundance in our Solar System (Lodders 2003), we note that the most abundant elements of the Solar System comprise also the most important elements of life as shown below (bold letters indicating the bio-elements; numerals in parentheses indicating the abundance as numbers of atoms relative to an arbitrary numerical value of 10^6 for the abundance of Si):

H(2.43×10^{10})>He>**O**(1.413×10^7)>**C**(7.079×10^6)>Ne>**N**(1.95×10^6)>**Mg**(1.02×10^6)>Si$(\equiv 10^6)$>**Fe**(8.38×10^5)>**S**(4.449×10^5)>Ar>Al>**Ca**(6.287×10^4)>**Na**>**Ni**(4.93×10^4)>Cr>**Mn**$(9,168)$>**P**$(8,373)$>Cl>**K**>Ti>**Co**$(2,323)$>**Zn**$(1,226)$>F>**Cu**(527)>**V**(288.4)>Ge>**Se**(65.79)>Li>Kr>Ga>Sc>Sr>B>Zr>Br>Rb>As>Te>Y>Xe>Ba>Sn>Pb>**Mo**(2.601)>Ru>Cd>Pd>Pt>Ce>I>Nd>Nb>Be>Os>Ir>Ag>La>Hg>Dy>Rh>Cs>Gd>Sb>Er>Sm>Yb>Au>Tl>In>Pr>Hf>Bi>**W**(0.1277)>Eu>Ho>Tb>Re>Tm>Lu>Ta.

The 20 bio-elements fall into three chemically distinct clusters in the periodic system. The seven main group non-metal bio-elements hydrogen (H), carbon (C), nitrogen (N), oxygen (O), phosphorus (P), sulfur (S) and selenium (Se) form inorganic, small molecule compounds that are gaseous or volatile and escape from the interior of the Earth as volcanic exhalations, notably as dihydrogen (H_2), methane (CH_4), dinitrogen (N_2), ammonia (NH_3), steam (H_2O), hydrogen sulfide (H_2S), hydrogen selenide (H_2Se), carbon monoxide (CO), carbon dioxide (CO_2), carbon oxysulfide (COS), carbon oxyselenide (COSe), hydrogen cyanide (HCN), methyl mercaptan (CH_3SH), sulfur dioxide (SO_2), selenium dioxide (SeO_2), phosphorus pentoxide (P_4O_{10}). These small molecules constitute the ultimate nutrients for the metabolism that forms the bulk organic compounds of all extant organisms. Steam (H_2O) constitutes the dominant compound in volcanic gases and it condenses to liquid water, which is the universal medium of life. Its properties are uniquely suited for transporting nutrients, facilitating redox and acid–base reactions, and providing the medium for the phase separations that form the basis of the organism-environment dichotomy of life.

The nine transition metal bio-elements vanadium (V), molybdenum (Mo), tungsten (W), manganese (Mn), iron (Fe), cobalt (Co), nickel (Ni), copper (Cu) and zinc (Zn) are located in minerals of the crust of the Earth. Their function in biochemistry is mainly catalytic owing to their abilities to cycle through different redox states and to acquire diverse ligand coordination geometries. They form the catalyst centers of metallo-enzymes. The four main group metal bio-elements sodium (Na), magnesium (Mg), potassium (K), calcium (Ca) are also located in minerals in the crust of the Earth. The alkali metals Na and K are subject to facile leaching as mono-cations from these minerals. The alkaline earth metals Mg, Ca are di-cationic and remain essentially fixed in minerals. They operate mainly as counter ions and as structure-forming bridges for anionic groups in the organic constituents of life. The absence of abundant chromium (Cr) from the set of bio-metals may be explained by its tendency to form weak bonds with sulfur, which is not true for any one of the transition metal bio-elements. The presence of the ultra-rare transition metal tungsten (W) in the set of bio-metals may be explained by its increased abundance in volcanic/hydrothermal settings and the presence of molybdenum (Mo) may be due to a relatively late replacement of tungsten (W).

The distinction between main group non-metal bio-elements and transition metal bio-elements constitutes perhaps the deepest functional dichotomy in life. Carbon and the other main group non-metal elements are pre-ordained to form the required range of molecular structures owing to the ability of carbon to form the C–C bonds in diverse molecular skeletons, that are stabilized by having their valences occupied by hydrogen. Moreover, the ability of carbon to form relatively stable bonds to O, N, S, Se is responsible for the formation of the functional groups of the bioorganic compounds that provide reactivity and form the basis for their ability to aggregate to larger structures. The transition metal bio-elements form lumped coordination structures with a transition metal center that is surrounded by a ligand sphere. They are pre-ordained to function as catalysts, notably in the fixation of volcanic gases. *Thus, the main group non-metal bio-elements are mainly structural, while the transition metal bio-elements are mainly catalytic.*

The fundamental dichotomy of the bio-elements has been overlooked for the longest time. This is partly due to the misconception that the most abundant bulk material in extant cells must also be the most ancient material of life. Already Carl von Nägeli in his groundbreaking theory on the origin of life assumed that the bulk material of extant organisms, i.e. proteins, constituted the original material of life (Nägeli 1884). His proposal resurfaced 40 years later, when a young Russian student, A. I. Oparin, after returning from his studies in Kossel's laboratory in Heidelberg, Germany, published a theory on the origin of life (Oparin 1924) and variants thereof continue to be proposed to this day. After it became clear in the 1940s that the nucleic acids are the molecules of inheritance in extant organisms, it was suggested that life began with replicating nucleic acids (in a prebiotic broth), known today as "RNA-world" theories (Woese 1967; Orgel 1968).

All these conventional theories associate the origin of life with the onset of replicative polycondensation of low molecular weight monomers to high molecular weight polymers or aggregates thereof. The chemical synthesis of the monomers is relegated to an obscure chemical evolution in the "prebiotic broth", wherein the monomers are supposed to have accumulated and become activated over thousands or millions of years. All "prebiotic broth" theories suffer from the same paradox: they have to postulate liquid water as the universal medium of life and of the prebiotic broth. Yet it is precisely this medium that tends to counteract the formation and accumulation of polycondensation products (activated monomers or polycondensation agents). The RNA world theories have the added problem of an intramolecular catalysis of RNA hydrolysis due to the 2'-OH groups of the ribose rings. This aggravates the hydrolysis problem considerably, notably at elevated temperatures and under alkaline conditions. The assumption of RNA molecules as carriers of genetic information, notably over longer geologic periods of time, is therefore chemically unreasonable. Contrary to common perceptions there is not a single example for an RNA genome under cellular conditions. In RNA viruses the RNA genome is packaged under stabilizing neutral or even water-free conditions within an impervious casing. As soon as the viral RNA genome enters a host cell is rapidly replicate or its information is incorporated into the host DNA genome. For these reasons all the popular RNA-world theories are chemically unreasonable and symbolic for the distance that exists between origin-of-life research and level-headed chemical knowledge.

The above difficulties seem to be insurmountable and render any "prebiotic broth" theory at the level of the formation of polymers from monomers a lost cause. Wouldn't it then be about time to seriously consider a radically new approach: To see whether it would not be possible to locate the onset of the pioneer organism at the level of the formation of low molecular-weight organic compounds from small inorganic molecules as they are found in volcanic exhalations. Such a theory has been proposed in 1988 for a volcanic flow setting, and it has evolved over the subsequent years (Wächtershäuser 1988a, b, 1990, 1992, 1997, 1998a, 2001, 2003, 2006, 2007) toward greater detail and greater comprehensiveness, which now goes up to the origin of the domains Bacteria, Archaea and Eukarya. The theory departs from conventional notions on several points:

1. The origin of life is associated with the synthesis of low-molecular weight organic compounds from inorganic nutrients.
2. The origin of life is identified with a seemingly instantaneous induction of synthetic, autocatalytic carbon fixation (termed "chemoautotrophic origin").
3. The origin of life is dependent on redox reactions, for which the presence of liquid water is a benefit rather than a detriment. Redox reactions are chemically "creative" with thoroughgoing changes of the electron configurations, while polycondensation reactions are merely organizational, leaving the electron configurations of the monomers essentially unchanged.
4. The origin of life is critically dependent on transition metal catalysis.
5. The origin of life is associated with a dynamic inheritance of "analogue" information in the form of autocatalytic feedback effects while static inheritance of "digital" sequence information is seen as the result of evolution.
6. The origin of life is associated with the redox potential of kinetically inhibited, quenched (non-equilibrium) volcanic exhalations.

On the Minimal Organization of the Pioneer Organism

The notion of a pioneer organism based on the onset of autocatalytic carbon fixation requires a sufficient compositional coherence for preventing a rapid decrease of product activity by diffusion, while permitting uninhibited access of nutrients. This requirement leads us to postulate a minimal organization in the form of a composite structure (Fig. 1.2). It consists of an organic superstructure attached to an inorganic substructure, which may be amorphous or crystalline; compact, porous or layered; aggregated or agglomerated; suspended or packed.

The superstructure consists of low-molecular weight organic compounds, which are formed by reductive autocatalytic carbon fixation pathways that are catalyzed by the transition metal centres in or on the inorganic substructure. The organic compounds come to be bonded *in statu nascendi* as ligands onto the transition metal centers, whereafter they engage in a surface metabolism. This minimally organized

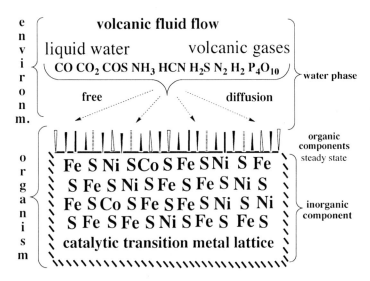

Fig. 1.2 Minimum organization of the pioneer organism

entity with the totality of the organic molecules that are bonded at any given time to outer (or inner) surfaces of the inorganic substructure (plus the proximal regions of the inorganic substructure) is defined as the pioneer organism. The pioneer organism is indefinite in terms of the profiles and lateral extensions of the surfaces of the substructure. In the direction normal to the outer or inner surfaces the organic superstructure has a definite extension and a vectorial orientation. The water phase constitutes the environment and the source of nutrients.

The so-called iron group elements Fe, Co, Ni are the most effective and versatile catalysts of life. Among these Fe is most outstanding and also of greatest geochemical abundance. Moreover, the most stable compounds of the iron group elements and notable of Fe under anaerobic conditions are the sulfides, while in a metabolic context the most prominent forms are clusters with sulfur ligands. Therefore, the theory of a chemoautotrophic pioneer organism under volcanic conditions has been dubbed "iron-sulfur world theory".

The organic superstructure is a dynamic entity. It has a steady state composition determined by the rates of synthesis and by the rates of detachment from the substructure and subsequent irreversible dissolution in the vast expanses of the ocean. The inorganic substructure should also be viewed as undergoing dynamic changes in accordance with the "Ostwald-Vollmer step rule": *The withdrawal of energy from a chemical system that can exist in several states of density will not go directly, but rather stepwise, into the state of greatest density, which is typically, but not always, also the state of greatest thermodynamic stability.* In this sense the initial substructure may be a highly energetic, amorphous hydrated structure. It subsequently becomes denser and stabilizes by dehydration and crystallization. In fact, the initial substructure may be interpreted as an extended polynuclear, polymodal, heteroleptic complex. Its ligands may at first be mainly hydroxy and aquo ligands.

The later development is characterized by a competitive input of polymerizing bidentate ligands (sulfidation or cyanidation) and of depolymerising ligands (carbonylation or ammination). Sufficient sulfide activity will ultimately drive this stabilization cascade to the formation of transition metal sulfide crystals. Notably iron, the most abundant of all transition metals may form at first ferrous hydroxide, which undergoes sulfidation to ferrous sulfide as poorly ordered crystals. These will subsequently go down a cascade of stabilizations with oxidative pyrite formation as a source of reducing power at the end stage,

$$FeS + H_2S \rightarrow FeS_2 + 2H^+ + 2e^-. \tag{1.1}$$

Catalytic minerals have been invoked practically from day 1 in conjunction with a prebiotic broth. Nägeli suggested for the formation of reproducing protein micells (Nägeli 1884): "Probably it does not occur in a free body of water, but rather in the wetted surface layer of a fine porous substance (clay, sand), where the molecular forces of the solid, liquid and gaseous bodies cooperate" (translation by the author). Bernal reinvented this idea later (Bernal 1951) and clay minerals have been invoked extensively ever since, mainly for catalyzing polycondensation reactions. Subsequently, Cairns-Smith (1982) took the bold step of replacing one-dimensional nucleic acid templating by two-dimensional layered clay templating, thus postulating inorganic clay organism. Arrhenius introduced divalent-trivalent metal hydroxide minerals, notably of iron, with a layered structure for the absorptive concentration and selective synthetic conversion of organic compounds in the prebiotic broth (Kuma et al. 1989). Volcanic sites were invoked early on, but merely as a source for the prebiotic broth (Mukhin 1974). Hydrothermal vents began to be suggested as sites for the origin of life soon after their discovery (Corliss et al. 1981), but without recognition of the need for volcanic quenching. Several authors engaged in a formal analysis of an origin by autocatalysis of small organic molecules, but still remained victim to the notion of a heterotrophic origin in a "prebiotic broth" (Ycas 1955; King 1977).

Metabolic Reproduction and Evolution of the Pioneer Organism

The most characteristic feature of the iron–sulfur world theory resides in the need for a mechanism of metabolic reproduction and evolution, by which the products of the synthetic reactions may exhibit autocatalytic feedback into the synthetic reactions themselves. Extant metabolisms comprise two aspects that must have evolved along two distinctive tracks: (1) Evolution of catalysts beginning with inorganic transition metal centers and proceeding through ligand variation to metallo-peptides and later to metallo-enzymes; (2) Evolution of biochemical pathways to more and more extended and integrated pathway systems. These two tracks are here assumed to have always been interconnected in the following manner (Fig. 1.3). A transition metal center (Me_1) catalyzes a pathway for the synthesis of

Fig. 1.3 Metabolic evolution by ligand feedback

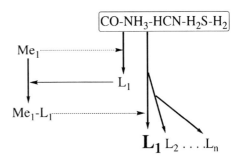

organic compounds from volcanic gases, e.g. CO, NH_3, HCN, H_2S. Some products of this pathway become in turn ligands to generate a modified metallo-catalyst. Now, it is well established that certain ligands bonded to catalytic transition metal centers may increase their catalytic activity by factors of more than 10^3 in an idiosyncratic, theoretically unpredictable manner (Berrisford et al. 1995). Therefore, some of the ligands (L_1) are expected to promote the catalytic activity of the transition metal centers. The new metallo-catalyst (Me-L_1) augments the concentration of ligand L1 (autocatalysis). This is a positive feedback effect, which in biological parlance is termed *reproduction*. Incidentally, from this point of view racemates of the organic product ligands mean a higher number of potential ligand structures and thus a higher likelihood of a positive feedback. This means that at the early stage of life a lack of enantioselectivity of the synthetic reactions is a benefit rather than a detriment.

Now, we have to recognize that under primordial conditions of low catalytic specificity the positive feedback of a certain ligand (L1) will promote not only the formation of that ligand, but also the formation of a set of other reaction products. This means that the steady state concentrations of some other reaction products (L_2 ... L_n) come up to levels, at which further feedback effects are induced. In this manner a single positive feedback effect will usher in further feedback effects. These in turn elicit new metallo-catalysts, new metabolic expansions, new ligands, and in turn new metallo-catalysts and so forth. In this manner the metabolism of the pioneer organism is subject to an avalanche of self-expansions.

Of all the products of the metabolism some will bond to transition metal centers with more or less long residence times (or retention times under flow conditions) and others may not essentially bond at all. The longer the residence time or retention time of a product on the surface the higher is the likelihood that the product is protected from destruction by hydrolysis or other reactions. In this sense the products of the pioneer metabolism are *self-selective*.

In addition to the metabolic self-expansion effect, we also see an expansion of the range of catalytic metals (e.g. $Me_1 \Rightarrow Me_1$ to Me_n) due to ligands from the expanded set of reaction products (L_1 to L_n), which give rise to an expanded set of catalysts and to new metabolic reactions with new reaction products (L_x to L_z). These in turn will recruit further transition metals into the metabolic system and so

Fig. 1.4 Catalytic self-expansion by recruitment of additional transition metals

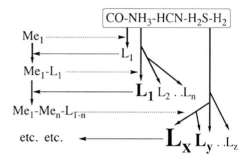

Fig. 1.5 Recruitment of additional nutrients

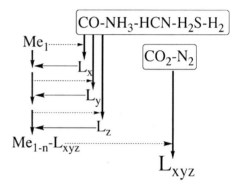

forth (Fig. 1.4). Finally, the picture is completed by adding the recruitment of additional volcanic nutrients for the metabolism (Fig. 1.5). Some of these expansions may occur only after the emergence of sophisticated ligands in the form of coded metallo-proteins. At a certain stage in this process of reproduction/evolution a primitive form of the autocatalytic reductive citric acid cycle will begin on the metabolic track (Wächtershäuser 1990).

The above-discussed hypothetical mechanism of evolution exhibits a form of autocatalysis by ligand feedback effect. At the same time and more importantly it also exhibits an avalanche self-expansion, which is not a feedback effect, because it leads beyond the established products of synthesis. Therefore, it is here termed "feed-forward effect".

All chemical reactions are dependent on the reaction conditions. This dependence, however, is lessened more and more by more and more evolved metallo-catalysts compared to the de novo transition metal catalysts without the promotion by ligands. Therefore, every instance of a feedback or feed-forward effect means an incremental increase of independence from the environment, i.e. a deepening autonomy. The metabolism acquires an increasing propensity to run even under conditions, which no longer would allow its de novo initiation. This is how the metabolism becomes precarious, life becomes mortal, and chemistry becomes historic. Each additional feedback or feed-forward effect constitutes an instance of inheritance or memory and the process of evolution is seen largely as a concatenation of such memory effects.

Whenever the self-expanding metabolism brings forth a new branch pathway this will weaken to some extent the pre-existing metabolism, whence it branches.

Table 1.1 Mechanisms of evolution of Fe–S world and RNA world theories

Fe–S World	RNA World
Transition metals are essential	Transition metals have no function
Catalytic inheritance (dynamic)	Structural inheritance (static)
Analogue feedback information	Digital sequence information
Persistence until successor feedback	Long-time persistence essential
Direct mechanism	Indirect mechanism
Order out of order out of order	Order out of chaos
Mutation = new metallo-catalyst	Mutation = new RNA sequence coding for new protein ligand for new metallo-enzyme
Selection not required	Selection required
Rapid evolution by avalanche of metabolic expansions	Evolution by slow accumulation of RNA sequence mutations
Chemical reaction type	Genotype-phenotype
Mutations: chemically pre-ordained	Mutations: stochastic, accidental
Transition metal catalysts insensitive to pH and temperature allow for an origin in a hot, alkaline volcanic flow	RNA extremely sensitive to pH and temperature requires an origin in cold, neutral ocean

If a product of the new branch pathway feeds back autocatalytically into this very same branch pathway, this weakening effect is amplified ("egotistic feedback"). However, if the product of the new branch pathway exhibits in addition to the egotistic feedback an "altruistic feedback", i.e. a positive feedback effect or a feed-forward effect for the pre-existent metabolism, this compensates for the egotistic feedback. A reaction product with such a double feedback effect is termed "vitalizer". Examples for vitalizers in extant metabolism are: coenzymes, ribosomes, protein translocases and nucleic acid polymerases.

The main features of the proposed mechanism of metabolic evolution are summarized in Table 1.1 in comparison to an RNA-implemented mechanism of evolution. A most significant difference is seen in the required persistence of a mutation. According to the RNA world theory any mutated RNA sequence must be stable enough to persist as a digital information storage medium over long periods of time. According to the iron–sulfur world theory a ligand feedback is not required to persist any longer than it takes for the next successor ligand feedback to emerge. The better ligand will wipe out its less efficient predecessor and so there is no need for persistence over long periods of time. This important point is sometimes not understood owing to an RNA world mindset (Blackmond 2009).

Volcanic Flow Setting of the Pioneer Organism

The Solar System started about 4.567 billion years ago. The Earth may have been cool enough for supporting liquid water as early as 4.4 billion years ago. At that time the partially parallel processes of accretion, core-mantel differentiation and crust segregation were still significantly incomplete (Kleine et al. 2002; Yin et al. 2002; Jacobsen 2003). This means that the juvenile Earth had a very hot magma and a relatively thin, hot, highly reducing crust. The crust suffered heavy bombard-ment and impact cratering with intense fracturing and deposition of debris. As a consequence a large number of volcanic flow sites were available for the origin of life (Cockell 2006). These sites were characterized by chemically diverse flow channels for volcanic exhalations in combination with hydrothermal flows.

The volcanic flow setting may be described as follows: At very high tempera-tures the gas mixture in the volcanic exhalations is in chemical equilibrium. If the volcanic exhalations are cooled slowly, the gas mixture remains in chemical equi-librium. However, if the volcanic exhalations are quenched rapidly enough, the mixture of volcanic gases is "frozen" into a non-equilibrium state, i.e. it acquires a chemical potential. Quenching may occur in a variety of ways: Conduction cool-ing by contact with cold solids, internal cooling by mixing with cold water, melt-cooling by contact with ice and/or evaporation cooling by pressure decrease.

In view of the highly reduced state of the crust the volcanic exhalations must have also been highly reducing. High molar ratios of CO/CO_2 would have existed in equilibrium at high temperatures. For example, under conditions of saturation with graphite the equilibrium molar ratio of CO/CO_2 of the system C–H–O increases with temperature and decreases with pressure and it is about 1:1 at 1,200°C and 2 kbar or at 900°C and 0.1 kbar (Holloway and Blank 1994). With decreasing temperature the water gas equilibrium,

$$CO + H_2O \rightarrow CO_2 + H_2, \qquad (1.2)$$

shifts in favour of CO_2. If quenching along the flow path is fast enough compared to the rate of the water gas shift reaction, a disequilibrium CO concentration arises providing not only a highly reactive carbon source, but also reducing potential for the pioneer organism,

$$CO + H_2O \rightarrow CO_2 + 2H^+ + 2e^-, \qquad (1.3)$$

akin to the carbon monoxide dehydrogenase reaction in biochemistry. In the metabolism of all extant autotrophs carbon dioxide is the dominant carbon source. This is the reason, why it was proposed as the primordial carbon source of life (Wächtershäuser 1988a, 1990, 1992) and this proposal was confirmed in conjunction with pyrite formation (Heinen and Lauwers 1996), even though chemical properties of CO_2 (insolubility of carbonates, thermodynamic stability of CO_2) seem to point to a later entry into the metabolism.

The water gas shift reaction generates dihydrogen (H_2), one molecule for each CO molecule that is oxidized to CO_2. This means that the water gas shift reaction replaces the reductant CO by the reductant dihydrogen (H_2). Dihydrogen is a prominent component of volcanic gases and it is produced by reaction of FeS and H_2S at the site of the pioneer organism or upstream thereof (Taylor et al. 1979; Wächtershäuser 1988c; Drobner et al. 1990). Dihydrogen has been suggested as possible reductant in a chemoautotrophic origin of life (Wächtershäuser 1988b, 1998a). It may generate electrons as in the hydrogenase reaction:

$$H_2 \rightarrow 2H^+ + 2e^-. \qquad (1.4)$$

In a volcanic flow system, HCN may be formed by reaction of CO with NH_3. This is suggested by industrial chemistry. The conversion is favoured thermodynamically by increasing temperature (Owen 1961). Ammonia in turn forms by nitrogen fixation, which is favoured thermodynamically by decreasing temperature:

$$CO + NH_3 \rightarrow HCN + H_2O, \tag{1.5}$$

$$N_2 + 3H_2 \rightarrow 2NH_3. \tag{1.6}$$

Therefore, the equilibrium concentration of HCN at low temperature (e.g. 100°C) and low pressure would be quite low. However, aside from quenching effects the equilibrium may be shifted in favour of cyanide formation, if sequestration of cyanide by transition metals like Ni^{2+} occurs. This reaction would form stable complex bonds, e.g. Ni–CN or Ni–CN–Ni. The Ni-centers are available as catalytic centers and the cyano ligands are available as C-source and N-source for the synthetic processes.

Volcanic exhalations may also comprise COS (Corazza 1986), which is known to form by reaction of CO with H_2S (Fukuda et al. 1977):

$$CO + H_2S \rightarrow COS + 2H^+ + 2e^-. \tag{1.7}$$

COS may serve as source for group activation in the pioneer organism, but it hydrolyzes readily:

$$COS + H_2O \rightarrow CO_2 + H_2S. \tag{1.8}$$

COS may also serve as carbon source, thereby generating activated thioacid groups (–CO–SH). COS also functions as intermediate for converting $CO + H_2 + H_2S$ into methylmercaptan (Barrault et al. 1987), which is also found in volcanic gases:

$$COS + 6H^+ + 6e^- \rightarrow CH_3 - SH + H_2O. \tag{1.9}$$

A hadean volcanic flow system may be conceptually subdivided into three flow regions:

1. A deep upstream flow region with a temperature above the critical temperature of water may be likened to a flow reactor with a stationary solid medium and a mobile gas phase in equilibrium.
2. A medium flow region with a temperature below the critical temperature of water may be likened to a trickle-bed reactor with a stationary solid phase, a volcanic gas phase and a mobile liquid water phase with dissolved volcanic exhalations. The water phase condenses progressively in gas flow direction. For its vital functions in nutrient conversions a thin transient film of water that merely wets the solid surfaces may be sufficient.
3. An uppermost flow region with a still lower temperature may be viewed as a chromatographic reactor with a stationary packed bed of solid debris or particles and a mobile liquid water phase.

Such a volcanic flow system exhibits temperature and pressure gradients as well as chemical composition gradients in flow direction. The third flow region may exhibit chemical zoning (e.g. pH zoning or sulfidation zoning). Moreover, it may exhibit chromatographic characteristics for the organic products in the superstructure of the pioneer organism. This means that the constituents of the surface metabolism may have been selected by differential retention times in view of the state of volcanic flow. Reactive and chromatographic processes would have interacted along the flow path with best surface bonders being the slowest travellers. This renders the chemoautotrophic origin of life a spatially and temporally coherent, local affair as opposed to the global affair of the "prebiotic broth theory".

A volcanic flow condition is of significance for the location of the onset of the pioneer organism by ligand-based feedback and fed-forward effects. We note that strong-bonding ligands have a long retention time and migrate slowly in flow direction. This means that they tend to remain concentrated in a relatively early and slow flow zone. This means in turn that stronger and stronger bonding ligands occur in ever more retarded flow zones, termed "ligand zones". Thereby the pioneer organism becomes localized in spite of the volcanic flow condition.

Hydrolysis of orthosilicates in komatiite rock material is seen as the starting point for the formation of the inorganic substructure of the pioneer organism according to the following simplified, notional reaction formulae:

$$(Mg,Ca)_2 SiO_4 + H_2O \rightarrow (Mg,Ca)(OH)_2 + (Mg,Ca)SiO_3, \quad (1.10)$$

$$(Fe,Ni)_2 SiO_4 + H_2O \rightarrow (Fe,Ni)(OH)_2 + (Fe,Ni)SiO_3. \quad (1.11)$$

The basic minerals $(Mg,Ca)(OH)_2$ serve as pH-buffer against acidic volcanic gases and acidic reaction products of the pioneer metabolism. Reaction with CO_2 generates $CaCO_3$ and basic $MgCO_3$. $(Fe,Ni)(OH)_2$ undergoes subsequent alterations by dehydration, cyanidation, carbonylation, sulfidation and ultimately pyritization. Carbonylation products may mobilize transition metals as carbonyls (Cody et al. 2000) for subsequent transport through volcanic flow channels and bonding into or onto the substructure. Progressive sulfidation of $(Fe,Ni)(OH)_2$ generates $(Fe,Ni)S$. FeS appears first as "amorphous" FeS, which ultimately undergoes pyritization (1.1). An alternative source of electrons in the substructure of the pioneer organism is provided by the oxidation of ferrous hydroxide:

$$3Fe(OH)_2 \rightarrow Fe_3O_4 + 2H_2O + 2H^+ + 2e^-. \quad (1.12)$$

By the above conversions the inorganic substructure offers not only a rich set of possible polynuclear, polymodal, heteroleptic structures exposed as catalytic centers in its surfaces, but also reducing power, e.g. for nitrogen fixation as demonstrated experimentally with $^{15}N_2$ (Dörr et al. 2003):

$$N_2 + 3FeS + 3H_2S \rightarrow 3FeS_2 + 2NH_3; \quad (1.13)$$

or for the formation of methylmercaptan from CO_2 with FeS/H_2S (Heinen and Lauwers 1996):

$$CO_2 + 3FeS + 4H_2S \rightarrow CH_3SH + 3FeS_2 + 2H_2O. \quad (1.14)$$

Experimental Synthetic Reactions

The above considerations give us a grasp of the possible energy sources, nutrients, reductants and physical conditions for a volcanic origin of life. Based on these insights we now turn to the problem of reconstructing the pioneer metabolism. This problem can only be solved by experiments and the theoretical-experimental research program has as its ultimate goal the reconstitution of an evolvable pioneer organism. This contrasts with the experimental program of the "prebiotic broth theories" that merely aims at a rationalization of some "prebiotic broth reactions" of an otherwise intractable process of early evolution. We chose the following aqueous, volcanic/hydrothermal reaction conditions:

Temperature: $100 \pm 20°C$, the temperature range of hyperthermophilic organisms.

Pressure: Autogenous pressure developed in a closed batch reactor. This pressure may be too low. The lack of a higher pressure for the acceleration of reactions may be compensated by increase of the concentrations of the reactants.

Catalytic transition metals: iron group transition metals: Fe^{2+}, Co^{2+}, Ni^{2+}.

Volcanic reactants: CO, COS, HCN, NH_3, H_2S, CH_3SH.

Reductant (electron source): Water gas shift reaction (1.3), pyrite formation (1.1).

pH-buffer: $(Mg,Ca)(OH)_2$, $(Mg,Ca)CO_3$.

Activated Acetic Acid Thioester

In extant acetyl-CoA synthase a Fe–Ni–S cluster catalyzes the formation of acetyl-CoA (CH_3–CO–S–CoA) from Ni-methyl, carbon monoxide and coenzyme A. It was retrodicted that activated acetic acid (CH_3–CO–SH or CH_3–CO–S–CH_3), may form from CH_3SH as methyl source and CO (Wächtershäuser 1990). This retrodiction was confirmed experimentally at 100°C for FeS, CoS, NiS, (Ni,Fe)S or $Ni(OH)_2$ as catalysts for a spectrum of slightly acidic to alkaline pH-values. If CH_3–SH is replaced by H_2S, the reactions proceed as well, whereby CH_3–SH is formed as intermediate from CO and H_2S.

These synthetic reactions provide a C2-carbon skeleton and at the same time a portion of the chemical energy of the system is conserved in an activated carboxylic group (–CO–S–H or –CO–S–CH_3). The reaction must be seen as an organo-metal reaction with metal-carbon bonds (Huber and Wächtershäuser 1997). These group activations are available for further reactions and for transformation into phosphorylation energy (Wächtershäuser 1998a).

Pathways to α-Hydroxy Acids and α-Amino Acids

A series of ordered pairs of α-hydroxy acids and α-amino acids with the general formula R–CHA–COOH (A = OH, NH_2; R = H, $HOCH_2$, CH_3, C_2H_5, C_3H_7) was formed by reaction of a Ni^{2+}-catalyst system with cyano ligands (from ^{13}C–KCN) attached to Ni–centers (as carbon and nitrogen source), with CO (as carbon and electron source) in the presence of $(Mg,Ca)(OH)_2$ as hydroxy ligand source and as buffer. The presence of Fe^{2+}, HS^-, CH_3S^- ions does not disturb the reaction. It is noteworthy that α-hydroxy acids and α-amino acids are excellent ligands for transition metals. Moreover, pyruvate is formed as intermediate as well as glycol, perhaps via glycol aldehyde, the simplest sugar. Acetic, propionic and butyric acids are also found. Overall the product spectrum indicates a tendency of incremental chain extensions towards long-chain lipids, which perhaps is broken off early due to low pressure (Huber and Wächtershäuser 2006).

Activation of α-Amino Acids and Peptide Cycle

α-Amino acid activation and peptide formation were discovered when α-amino acids were reacted with CO plus H_2S (or CH_3SH) in the presence of NiS under alkaline conditions. Nascent COS appears to be the activating intermediate since its use instead of CO and H_2S (or CH_3SH) also produces peptides (Huber and Wächtershäuser 1998). Under the same conditions a peptide cycle was discovered (Huber et al. 2003). Peptides were found to react N-terminally with CO (or COS) to acquire an N-terminal hydantoin ring, which subsequently hydrolyzes in two stages, first to an N-terminal urea group and subsequently by urea cleavage. This reaction sequence constitutes N-terminal peptide elongation in competition with N-terminal chain degradation, each by one amino acid unit at a time.

In the first step of the peptide cycle an amino acid (aa) is apparently converted to the highly reactive aminoacyl N-carboxyanhydride (c). This means that the redox energy of reaction (c) is coupled to amino acid activation via COS, which may well be the simplest pioneer energy coupling of life. The free amino group of the dipeptide may react with another molecule of aminoacyl N-carboxyanhydride to form a tripeptide, which again may react by N-terminal chain extension to a tetrapeptide and so forth. A hypothetical mechanism of the entire peptide cycle is in shown in Fig. 1.6.

Up to the stage of the urea derivative the peptide cycle is synthetic (anabolic). The subsequent hydrolysis is catabolic and reminiscent of the extant Ni-enzyme urease. The peptide cycle may be seen as a catalytic cycle for the conversion of CO to CO_2 akin to extant carbon monoxide dehydrogenase (CODH). The peptide cycle will run as long as the redox energy supply of CO lasts. The hydrolytically sensitive COS is merely an intermediate and its accumulation is not required. With additional α-amino acids a variety of peptides (and their hydantoins and ureas) are formed to coexist in mixture it constitutes a dynamic library, in which the peptides (hydantoins, ureas) come and go with steady state concentrations corresponding to

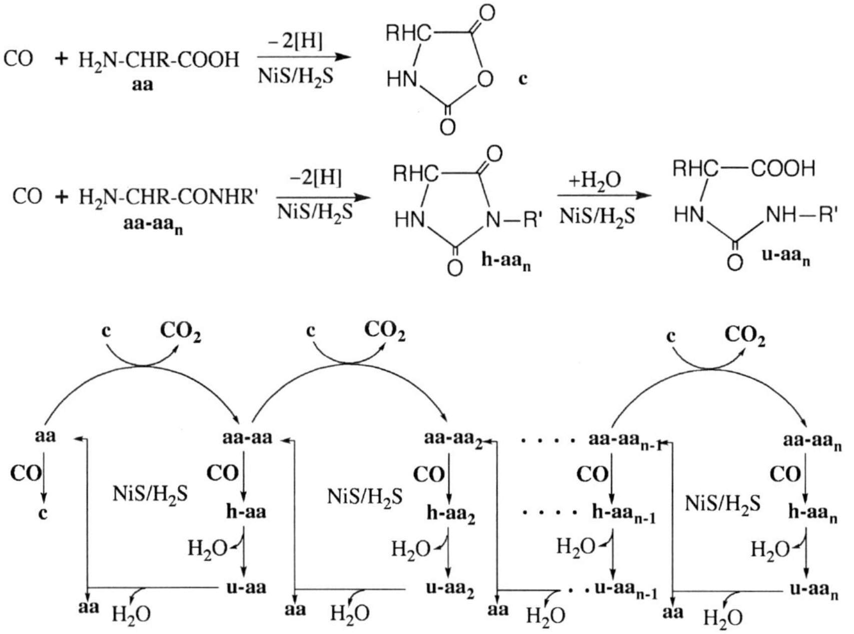

Fig. 1.6 Peptide cycle (aa = amino acid unit; h = hydantoin ring; u = urea group; NHR' = –(NH–CHR–CO)$_n$–OH)

the difference between the rates of formation and degradation. Importantly, the dynamic library of structures is also self-selecting, because a selective bonding of a certain structure as ligand to a metal centre of the substructure causes its stabilization against hydrolysis and thus an increase of its overall steady state concentration. It means a self-selection of stable metallo-peptides.

Emergence of the Genetic Machinery and Enzymatization of the Metabolism

According to the above proposal the earliest mechanism of evolution is based on the bootstrapping effect of mutual promotion of catalyst evolution and pathway evolution. The catalyst evolution leads to ever more efficient metallo-catalysts with ligands that result from the pathway evolution. Among the sets of ligands figures most prominently the compound class of peptides and peptide derivatives. Therefore, any catalyst that augments the synthesis of peptides is an immediate advantage for the earliest organisms of life, because the peptide function is already pre-established. This brings us to the simple conclusion that one of the earliest functions of the incipient genetic machinery must have been the promotion of peptide synthesis, which is actually the functional core of the ribosome. It operates by base pair-assisted anchimeric positioning of a peptide to be elongated C-terminally and an elongating amino acid. From then on the evolution of the genetic machinery must have been a co-evolution of replication and translation of nucleic acid sequences.

How would the first functional nucleic acids (polynucleotides) have arisen? From the vantage point of ligand accelerated catalysis the answer is trivially simple. All components of nucleic acids: sugars, phosphorylated sugars, bases, nucleosides, nucleotides and oligo-nucleotides are excellent potential ligands for transition metals. We therefore simply assume that they first entered the pioneer organism as ligands for accelerating transition metal catalysis. But how could these components of nucleic acids have arisen in the first place? Clues for an answer may be gleaned from the above-reported results of synthetic carbon fixation reactions.

1. We detected glycol, a carbon fixation product, as a regular companion of amino and hydroxy acid synthesis. It is believed to be the result of a reduction of glycol aldehyde, the simplest sugar and suitable for condensation reactions.
2. The required phosphorylation energy for oligonucleotide formation may be the result of hydrolysis of volcanic phosphorus pentoxide (P_4O_{10}). Phosphorylation energy may alternatively be a derivative of the chemical energy of acetyl thioester (Wächtershäuser 1998a). Moreover, group activation energy may also be derived from the surmised aminoacyl N-carboxy-anhydrides that are formed as activated amino acids with CO/H_2S or COS (Wächtershäuser 2003).
3. The evolution of de novo biosynthesis of purine bases may be traced to the peptide cycle reported above. We first note that in the peptide cycle hydantoin rings are formed at the N-terminal ends of peptides by carbon fixation. Such hydantoin rings have a close structural similarity to the imidazole portion of the purine bases in extant nucleic acids. The purines are to this day biosynthesized by a pathway of carbon fixation. This experimental result brings the two ancient syntheses of peptides and purines into a close synthetic connection. Pyrimidine bases would have infiltrated the nucleic acids much later.

In the protracted coevolution of the genetic machinery each new increment of added sequence fidelity of translation to peptides would have served as basis for a subsequent increment of added sequence fidelity of template-directed nucleic acid synthesis (bootstrapping). By such evolution of nucleic acid-implemented synthesis of peptides the earliest direct peptide feedback would have become first supplemented and later increasingly substituted by a secondary, indirect, genetic mechanism of evolution, whereby variations of replicating nucleic acid sequences led indirectly to variations of peptide sequences. The transformation led to a step-by-step replacement of primitive catalytic transition metal centers by more and more complex coded metallo-proteins. It was the basis for the admission of more diversified amino acids into the system and for the establishment and expansion of the genetic code.

Finally, let us briefly consider the place of this crucial development within the flow system of the volcanic setting of early life. Initially, the flow channel would have been alkaline, because the primordial alkaline rock material must have been ultramafic. Subsequently, however, neutralization would have taken place by acidic volcanic gases, notably carbon dioxide; and also by the carboxylic acid products of CO-fixation. This means that a neutralization front would have advanced in flow direction. This neutralization was a decisive precondition for the emergence of a genetic machinery involving RNA, for which alkalinity is intolerable. Similarly, a volcanic flow segment with a sufficiently reduced temperature of the volcanic fluids would have been crucial for the emergence of a genetic machinery involving temperature-sensitive RNA.

Finally, let us turn to the question of how folded proteins could have come about under the hot conditions of the volcanic flow system. The earliest functional peptides bond as ligands to metals, thus forming metallo-peptides wherein the bonding to the metal determines the overall peptide conformation. With the advent of cysteine units in peptides by exergonic sulfidation of serine, covalent metal-sulfur bonds will be particularly strong folding determinants. In this fashion the emergent folded metallo-proteins acquired from the start the hyperthermostability that is needed in the high temperature volcanic environment of early life. A few strong covalent bonds were sufficient as discrete, localized folding determinants and overall sequence fidelity was not critical. Only after a sufficient, coordinated increase of replication fidelity *and* translation fidelity could proteins arise that owed their folding stability not so much to a few strong, covalent bonds to transition metal centres, but rather to the cooperation of a multitude of weak, non-covalent group interactions. This would have happened while the organisms were still hyperthermophiles. Only much later could the covalent metal-sulfur folding determinants disappear opportunistically one by one, in an irreversible, polyphyletic evolution down the temperature scale and out of the volcanic region (Wächtershäuser 2001).

Cellularization

The conversion from an early, direct mechanism of evolution to a later, indirect genetic mechanism of evolution is truly an evolution of the mechanism of evolution. It surely must have occurred in multiple stages and in parallel with a protracted multi-stage process of cellularization, which we consider now.

Inorganic Cells?

Hans Kuhn has worked out a detailed theory (Kuhn 1972) of an origin of life within interconnected open pores of pre-existent rocks, like porous limestone (Kelley et al. 2001). The theory is based on the understanding that such pore structures would restrict the diffusion of high molecular-weight nucleic acids while permitting a free diffusion of the small molecules of low molecular-weight organic compounds.

In a highly imaginative shift from Kuhn's theory it has been proposed that life originated within open pore structures due to a semipermeable FeS membrane that is postulated to precipitate at the boundary between mildly acidic ocean water (laden with Fe^{2+} and CO_2) and alkaline hydrothermal fluids (laden with H_2 and HS^-). Repeated bursting and re-precipitation of these membranes is supposed to generate a growing mound of FeS-cells at the bottom of the ocean. A proton-motive force across the membrane separating external acidic from internal alkaline fluid is proposed to have somehow driven a chemoautotrophic metabolism between ocean-side

CO_2 and vent-side H_2. Eventually, full-blown Bacteria and Archaea are supposed to have broken loose from the top of the mound (Martin and Russell 2003).

The basic assumptions of this elaborate proposal are simple enough that empirical evaluation by precipitation experiments would seem to be mandatory. Such experiments have indeed been reported (Russell and Hall 1997). A discrete glob of FeS is obtained, if highly alkaline 0.5 M sodium sulfide solution is injected into a highly acidic 0.5 M $FeCl_2$ solution. A photomicrograph of a section through the glob showed a myriad of open pores (about 20 μm diameter). The micrograph was widely publicized. Unfortunately, however, the micrograph is an artefact due to sample preparation by freeze-drying, which unavoidably generates a foamy froth, because freezing generates ice crystals and the evaporation of ice crystals leaves behind pores. The size of the pores is determined by the rate of cooling. This has been acknowledged by one of the authors, who now argues that cell formation by freeze-drying is an acceleration model for the geochemical theory (Russell 2007).

There are of course no hadean rocks that could support such an FeS membrane theory. Moreover, the notion of semipermeable FeS-membranes capable of supporting a protonmotive force has been experimentally refuted (Filtness et al. 2003). Finally, RNA cannot coexist with the alkalinity required for the interior of the envisioned cells. We therefore leave this topic behind and turn to cellularization by the formation lipid membrane structures.

Lipid Synthesis

It has been proposed that cellularization of life occurred in several stages by autotrophically generated lipids (Wächtershäuser 1988a, 1988b, 1988c, 2003). For addressing the problem we recall the experimentally demonstrated formation of acetylthioester as evolutionary precursor of acetyl-CoA. In extant organisms the biosynthesis of fatty acids and of isoprenoid compounds by the mevalonate pathway begins with a Claissen condensation of acetyl-CoA to β-keto butyryl-CoA. Analogous acetyl thioester condensations may have figured in an early synthesis of lipids (Wächtershäuser 1992) as shown in Fig. 1.7. At the level of 3-keto-butyryl thioester the hypothetical reaction system splits into two sibling pathways: to fatty acid lipids and to isprenoid acid lipids. The latter pathway proceeds via 3-hydroxy-3-methyl-glutaroyl thioester (HMGT) and 3-methyl-glutaconyl thioester (MGAT) to long-chain isoprenoid acid lipids. MGAT is a vinylogue of malonyl thioester. Its condensation is expected to lead to chain elongations by four carbon atoms. Later in evolution this pathway must have been replaced by the well known mevalonate pathway to isoprenoid alcohol phosphate lipids in Archaea and Eukarya and by the so-called non-mevalonate pathway to isoprenoid alcohol phosphates in Bacteria (Gräwert et al. 2004) as shown in Fig. 1.8. It is also possible that carboxylic acid lipids formed from C1-units by organo-metal reactions under pressure ($12 \times C1 = C12$). In this regard it is noteworthy that the reaction system for forming α-hydroxy and α-amino acids proceeds in an iterative manner with C1-units and may well be an early source of lipids.

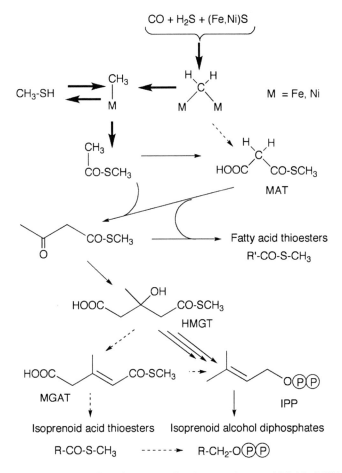

Fig. 1.7 Schematic representation of a system of pathways to isoprenoid lipids (MAT = malonyl thioester; HMGT = 3-hydroxy-3-methyl-glutaryl thioester; MGAT = 3-methylglutaconyl thioester; IPP = isopentenyl pyrophosphate)

Surface Lipophilization

Fatty acids, isoprenoid acids, α-hydroxy acids or α-amino acids with sufficiently long carbon chains, synthesized by chemo-autotrophic carbon fixation would have functioned as primitive lipids in the iron-sulfur world context. They have a low solubility in water and therefore must have accumulated at the interface between the inorganic substructure and the water phase, becoming in fact part of the organic superstructure. This led to a progressive lipophilization of the surfaces of the inorganic substructure, i.e. to a two-dimensional solvent of sorts. Lipophilization in turn has the effect of lowering the activities of H_2O and H_3O^+ near the surfaces, which disfavours hydrolytic reactions while it favours all condensation reactions (altruistic feedback) thus ushering

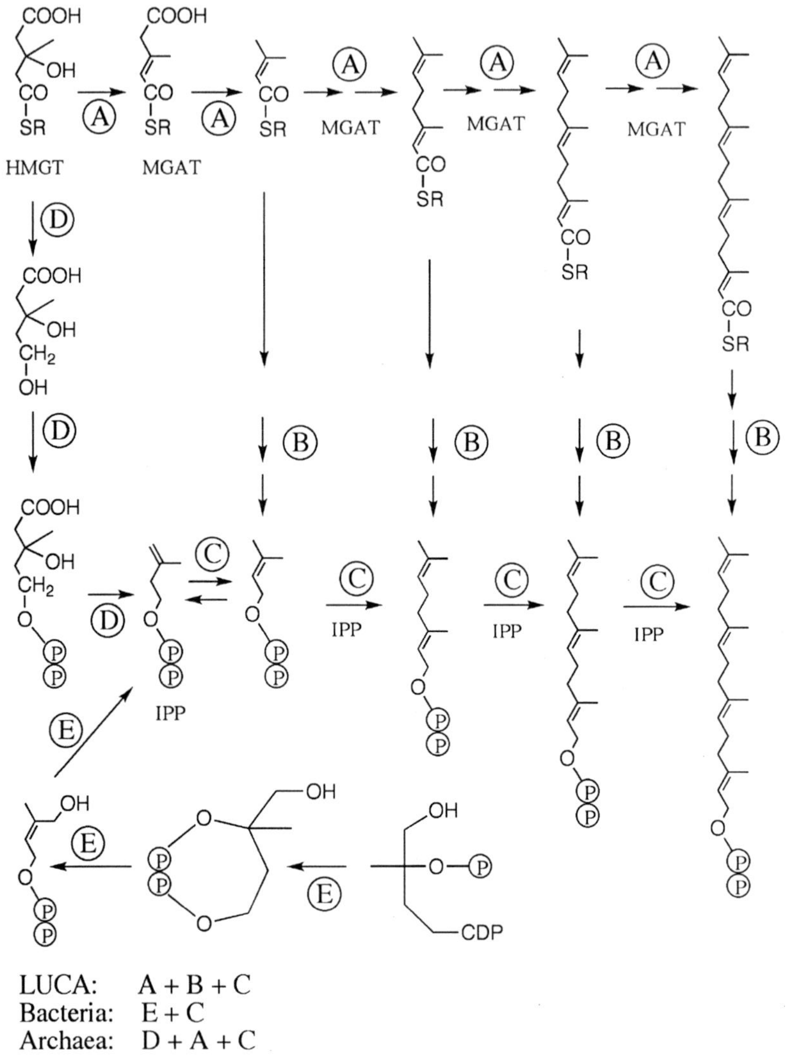
HMGT
MGAT
MGAT
MGAT
MGAT
IPP
IPP
IPP
IPP
CDP
LUCA: A + B + C
Bacteria: E + C
Archaea: D + A + C

Fig. 1.8 Lipid pathway evolution to extant lipid pathways

in greatly expanded metabolic possibilities. Of course, the formation of fatty or isoprenoid acids is itself based on condensation reactions, which also benefit from surface lipophilization (egotistic feedback). This form of double feedback is not based on the effect of single ligand molecules, but rather on the collective effect of an accumulation of lipid molecules on the surface.

Such autocatalytic lipid synthesis has the consequence of ever longer and ever more accumulated lipid molecules. At a certain surface concentration a self-organization of a multitude of lipid molecules into bi-layer membranes set in by physico-chemical forces. First the membranes formed discrete surface-supported patches, which later coalesced into large coherent surface-supported lipid membranes.

Semi-cellular Structures

The surface-supported membrane would invariably cover surface areas with tiny defects in the form of crevices or cavities. For energetic reasons a bi-layer lipid membrane would span such cavities rather than line them thereby forming embryonic caverns, which would subsequently enlarge by further lipid accumulation. This mechanism created a "semi-cellular structure", which is defined as a volume of aqueous solution (cytosol), bounded partly by inorganic substructure and partly by lipid membrane.

The semi-cellular structures constitute the beginning of individuation and of an inside-outside dichotomy with an isolation of the interior cytosol from adverse chemical conditions in the exterior water phase. The small inorganic nutrients (CO_2, CO, H_2S, NH_3) would pass easily through the membrane as long as the pH allows for a significant equilibrium proportion of non-charged neutral species. Therefore, the formation of the semi-cellular structure would not create a nutritional impasse.

Within the semi-cellular structures the inorganic surfaces are still available for catalysis. Carbon fixation by these catalysts leads to the formation polar, ionic or zwitter-ionic organic compounds, which remain inside the semi-cellular structure. This increases the osmolality of the cytosol, which in turn leads to an osmotic pressure as driving force for further expansion of the semi-cellular structure. The cytosol is available for an unfolding cytosol metabolism with catabolic pathways beginning as internal salvage pathways by energy coupling. It is this stage that begins to accommodate multi-component genetic machineries. A new class of metabolic processes would be anchored to the lipid membrane. Chemiosmosis is the most prominent membrane-metabolic process and its origin shall be discussed next.

Origin of Chemiosmosis

Chemiosmosis must begin soon after the formation of semi-cellular structures, because chemical gradients would develop automatically across the lipid membrane. Given the development of a pH gradient along the volcanic flow path, we may assume that semi-cellular structures form at a stage where the external medium may be slightly alkaline to neutral. By the same token, carbon fixation would lead by necessity to internal acid formation, i.e. to a pH difference across the membrane.

The next step would consist of a machinery for coupling exergonic proton translocation across the membrane from inside to outside with endergonic, membrane-bound redox reactions. Once established, this machinery would also operate readily in reverse, whereby certain exergonic, membrane-bound redox reactions would drive endergonic proton translocation against the proton gradient from outside to inside, thus increasing the proton potential. At the same time other endergonic, membrane-bound redox reactions would continue to be coupled to exergonic proton translocation from inside to outside. Overall this means that the proton potential would mediate an energy coupling between endergonic and exergonic redox reactions. Next, membrane-bound unit for pyrophosphate formation driven by proton flow from inside to outside would become established, thus causing the proton potential to mediate energy coupling between redox energy and group activation energy. Finally, membrane-bound pyrophosphate formation would be first supplemented and later substituted by ATP formation.

Such a chemiosmotic energy coupling would immediately benefit from any improvement of the membrane in terms of stability and resistance against idle proton leakage. This makes lipid evolution the most beneficial part of early evolution, leading eventually to stable membranes comprised of wedge-shaped lipids with two long-chain lipophilic rests attached to one small hydrophilic head group, e.g. of phosphatidyl lipids. With such wedge-shaped lipids the lipid membranes become self-supporting cell membranes. This ushers in yet another stage of cellularization, the formation of pre-cells, to which we turn next.

Pre-cells and the Dawn of Speciation

The discovery of the domain Archaea by Carl Woese and George Fox was tantamount to the discovery of the problem of the last universal common ancestor (LUCA) at the deepest node in the tree of life, and they surmised that it was a primitive state, a "progenote", still "in the throes of evolving the genotype-phenotype" (Woese and Fox 1977). Woese suggested that the disjoint genes of LUCA would "have existed in high copy numbers" and suffered genetic mixing that created "a state of genetic communion" giving rise to a high rate of evolution (Woese 1982, 1987). Subsequently, Otto Kandler recognized that the common ancestor at the eve of speciation must have had a highly sophisticated cell type, termed "pre-cell". It would have occupied a whole biosphere with diverse habitats by multi-phenotypical pre-cells, which were uniform in terms of the genetic machinery, and underwent rapid genetic exchange and thus rapid evolution (Kandler 1994a, b, 1998). This means that perhaps hundreds of million years of evolution may have separated the origin of life from the eve of speciation.

A characterization of the pre-cells at the dawn of speciation, is in principle possible by comparison of bacterial and archaeal genomes. It was found (Wächtershäuser 1998b) that in all sequenced bacterial and archaeal genomes a number of relatively short conserved gene clusters, mainly for transcription and translation, had different lengths from phyla to phyla and could be fitted into overlapping gene cluster alignments, whereby astonishingly long segments of ancestral genome segments can be reconstructed (Fig. 1.9) for the last bacterial common ancestor (LBCA) (46 genes) and the last archaeal common ancestor (LACA) (53 genes). Previously, only individual short clusters (≤11 genes) had been recognized as ancestral (Siefert et al. 1977).

These reconstructed genome segments are amazingly similar and even the few deviations are again constant throughout the domain Bacteria or throughout the domain Archaea, respectively. Lateral cross-phyla transfers of gene clusters cannot

```
LBCA|.... SecE NusG L11P L1P  L10P L12P  —          RpoB RpoC L30P  —      S12P S7P  EF-G
LACA|.... SecE NusG L11P L1P  L10P L12P  ‖RpoH RpoB RpoC      L30P  NusA   S12P S7P  EF-G
LUCA|.... SecE NusG L11P L1P  L10P L12P  —          RpoB RpoC L30P  —      S12P S7P  EF-G

LBCA|EF-Tu S10P L3P  L4P L23P L2P S19P L22P S3P  L16  L29P  —      S17P L14P L24P
LACA|EF-Tu S10P ‖L3P L4P L23P L2P S19P L22P S3P  —    L29P  Rpp29  S17P L14P L24P
LUCA|EF-Tu S10P L3P  L4P L23P L2P S19P L22P S3P  —    L29P  —      S17P L14P L24P

LBCA|—   L5P S14P S8P L6P  —    —    L18P S5P L30P L15P SecY Adk Map IF-1
LACA|S4E L5P S14P S8P L6P  L32E L19E L18P S5P L30P L15P SecY Adk —   —
LUCA|—   L5P S14P S8P L6P  —    —    L18P S5P L30P L15P SecY Adk —   —

LBCA|L36 —       S13P —    S11P S4P  RpoD L17 L13P S9P  —     ‖S2P EF-Ts. . .
LACA|—   7 genes S13P S4P  S11P —    RpoD L17 L13P S9P  RpoN  S2P  X     . . .
LUCA|—   —       S13P S4P? S11P S4P? RpoD L17 L13P S9P  —     S2P  —     . . .
```

Fig. 1.9 Reconstruction of ancestral genome segments (Designated by gene products)

be invoked for explaining the extreme conformity within and between the domains of Bacteria and Archaea. Lateral transfer between distantly related groups of organisms of such large clusters of genes that code for plural components of vital and complex genetic machineries would generate a high proportion of defunct hybrid machineries. This would be lethal. It means that the pre-cells at the eve of speciation (LUCA) had already a genome with an extremely conserved cluster of more than 40 canonical genes mainly for transcription and translation.

The reconstructed genome segments reveal much about the LUCA pre-cells on the eve of speciation:

1. LUCA had a complex genetic machinery with plasmid-sized circular chromosomes and a highly conserved order of genes for the machinery.
2. Absence of genes for energy and intermediary metabolism from the reconstructed genome segment of LUCA suggests separate metabolic chromosomes as required by Kandler's proposal of multi-phenotypical LUCA pre-cells.
3. The presence of the gene for anti-termination factor NusG suggests rolling circle transcription generating long polycistronic transcripts of varying lengths.
4. Rolling circle replication would have followed rolling circle transcription seamlessly and without change of the polymerase. This would have required nothing more than an activation of a ribonucleotide reductase function for the conversion of ribonucleotides to deoxyribonucleotides, since an early polymerase could not have discriminated between ribonucleotides and deoxyribonucleotides. This readily explains why RNA primers are needed in DNA replication to this day. With subsequent decay of the ribonucleotide reductase function the system would have returned to the transcription mode. Initiating nicks for rolling circle replication would have occurred in both strands, which means roughly equal numbers of copies of both strands. These would have undergone annealing to linear, double stranded daughter chromosomes with sticky ends for ring formation (Fig. 1.10).

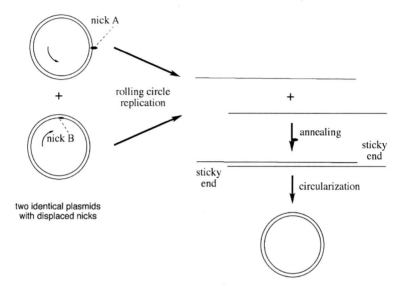

Fig. 1.10 *Rolling circle* replication of two identical plasmids with displaced nicks and a linear intermediate with sticky end segments

5. Presence of the genes for SecE and SecY, the essential sub-units of membrane-embedded protein translocase means that LUCA had a stable bi-layer lipid membrane.
6. Comparisons of Archaea and Bacteria suggest an absence of cell walls from LUCA pre-cells, which as a consequence would have undergone facile fusions with wholesale genome combinations and a stepwise increase of polyploidy. Cell fusions are still found in extant *Thermococcus coalescens* cells that grow too fast for cell wall formation to keep up (Kuwabara et al. 2005). Such mechanism of pre-cell fusion constitutes a highly promiscuous quasi-sexuality generating a huge gene pool with a massively parallel, very rapid evolution of the pre-cell population. It enabled large-scale reshuffling and differential assorting of the genetic endowments for the metabolism of multi-phenotypical pre-cells in accordance with the chemical conditions of diverse habitats.
7. It has been argued compellingly that the last bacterial common ancestor (LBCA), the last archaeal common ancestor (LACA) and the last universal common ancestor (LUCA) were hyperthermophiles (Di Giulio 2003). We can now add that pre-cell fusions across the spectrum of phenotypes are only possible among essentially isothermal pre-cells, which means hyperthermophily for all LUCA pre-cells, and also for all LBCA cells and all LACA cells.

The situation of the biosphere and of the evolution of life changes decisively with the segregation of the pre-cells into the two domains Bacteria and Archaea. To this topic we turn next.

Divergence of the Domains Bacteria and Archaea

Kandler's pre-cell theory is a significant departure from previous phylogeny, a
Gestalt switch of sorts. Conventionally, phylogenetic trees have been represented as
bifurcations, whereby both branches of a bifurcation are topologically equivalent.
Kandler breaks with this pattern regarding the two deepest nodes in Woese's tree of
life and replaces it by a pattern of divergence of founder populations of the domains
Bacteria and Archaea from a straight trunk of a rapidly evolving population of pre-
cells, the Bacteria branching off first at a less advanced stage and the Archaea
branching off later at a more advanced stage (Fig. 1.11).

We adopt this proposal and suggest a causal mechanism for the divergence of the
domains Bacteria and Archaea that is based on well-established or readily testable
physical chemistry of lipids. We begin with the problem of the structure of LUCA
lipids. The core lipids of the domain Bacteria and of the domain Archaea differ in
three respects (Fig. 1.12): fatty versus isoprenoid rests, ester versus ether bridges,
G-1-P versus G-3-P enantiomers of the glycerol unit. As a most simple solution of
the problem of the ancestral lipids in terms of both chemistry and evolution it has
been suggested that LUCA had a membrane consisting of a lipid racemate, i.e. a
1:1 mixture of enantiomers G-1-P and G-3-P (Fig. 1.13).

We now consider that heterochiral membranes of racemic lipids, while capable
of forming stable vesicle membranes, will undergo spontaneous segregation

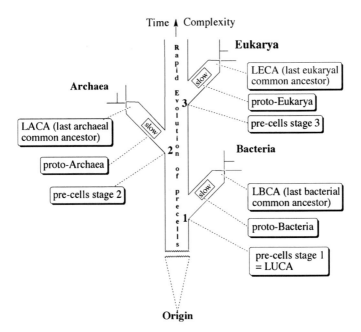

Fig. 1.11 Kandler's pre-cell phylogeny

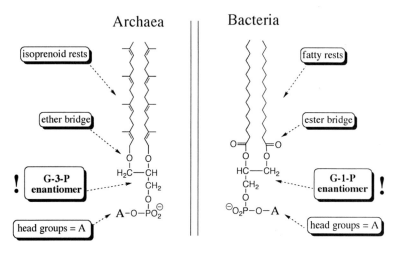

Fig. 1.12 Core lipids of the domains bacteria and archaea

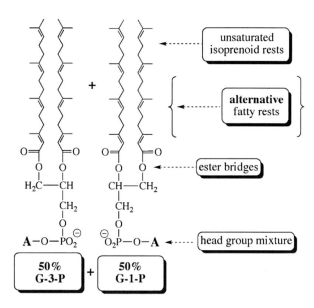

Fig. 1.13 Proposed racemic core lipid of LUCA

(symmetry breaking) into a pattern of homochiral membrane patches ("membrane domains") within an overall heterochiral pre-cell membrane envelope. Now, by frequent fusions and divisions the pre-cell population will segregate into two sub-populations with a predominance of one lipid enantiomer or the other. This segregation is maintained continuously through the process of pre-cell multiplication, even though the metabolism will continue to add racemic lipid material. The segregation

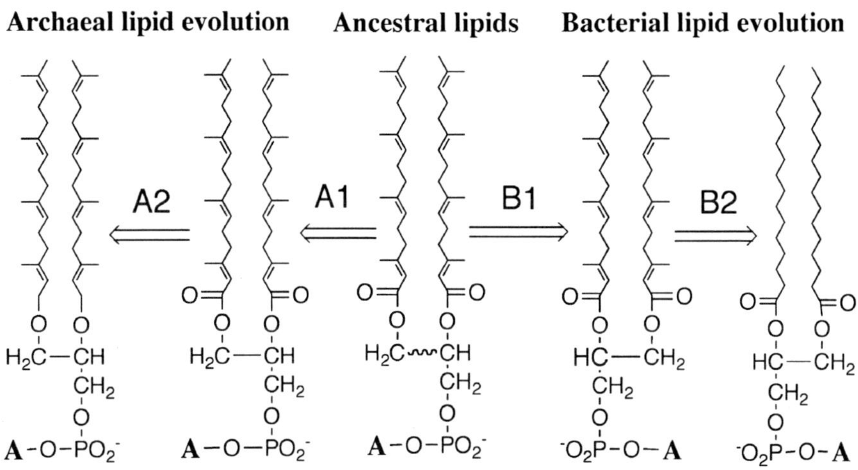

Archaeal lipid evolution   Ancestral lipids   Bacterial lipid evolution

Fig. 1.14 Bacterial and archaeal lipid evolution (A1, B1 = enantioselective enzymes; A2 = conversion from esters to ethers; B2 = conversion from isoprenoid rests to fatty rests)

occurs strictly by forces of physical chemistry (line tension). It generates placeholders for the future divergence of the domains Bacteria and Archaea.

For the divergence of the domains Bacteria and Archaea from the trunk evolution of pre-cells it is merely required that new enzymes for enantiospecific lipid synthesis emerge, one for the Bacteria and the other for the Archaea. These would be internally selected as adaptation to the dominant lipid enantiomer in one or the other placeholder subpopulation. According to this proposal the phylogenetic domains Bacteria and Archaea are pre-ordained by the membrane domains of racemic lipids. This means that two and only two domains could diverge by chiral lipid segregation, here on Earth or anywhere in the Universe.

After the divergence of the domains Bacteria and Archaea their unitary populations embarked on their own distinct paths of evolution exhibiting a progressive lipid incompatibility by a progressive increase of the degree of chirality of the lipid molecules (Fig. 1.14).

In the course of the bacterial and archaeal evolution the genetic machineries changed independently, yet analogously by the combination of multiple plasmid-type chromosomes into one large chromosome, thereby facilitating linkage of replication with cell division. DNA-replication and transcription became separated. In the course of evolution of the proto-Bacteria and in the course of evolution of the pre-cells up to pre-cell stage 2 non-homologous enzymes for dDNA replication with coordinated leading strand and lagging strand synthesis departed from the much simpler rolling circle replication of the pre-cells of stage 1 (LUCA). In addition, the mechanisms of transcription and translation were refined independently (Woese 1998). The proto-Bacteria and proto-Archaea continued to evolve as unitary populations until the emergence of fusion-inhibiting means (e.g. cell walls), which triggered the branching into bacterial phyla and into archaeal phyla.

Divergence of the Domain Eukarya

Numerous theories postulate an origin of the Eukarya by endosymbiosis between cells of advanced species of the domains Archaea (e.g. methanogens) and Bacteria (e.g. α-proteobacteria). Such theories entail an intermediate formation of heterochiral hybrid membranes with a mixture of bacterial lipids and archaeal lipids, which would be of reduced stability and therefore energetically counterselective.

This problem can be avoided by a theory, which begins with the assumption of a metabolic syntrophy between the sub-population of stage 3 pre-cells with a predominance of the bacteria-type lipid enantiomer G-1-P and a population of primitive, wall-less bacteria. Soon facultative syntrophy must have turned into obligatory syntrophy. Eventually, a number of bacteria surrounding a pre-cell would have undergone fusion to a large host cell containing a stage 3 pre-cell guest surrounded by a double membrane having numerous transport units. A cytoskeleton of stage 3 ppre-cells would have prevented fusion of the pre-cells with the bacterial cells. The stage 3 pre-cell guest evolved into the eukaryal nucleus.

This proposal delivers a number of explanations:

1. The peculiar relationship of Eukarya to the Archaea in terms of the genetic machinery is explained by the order of divergence, whereby the Eukarya diverged after the Archaea.
2. The double membrane trapped the polyploidal pre-cell chromosomes so that their replication could not become linked with the outer cell membrane during cell division. Thus, there was no selective advantage in their unification into one large circular chromosome. This explains the polychromosomal status of the nuclear genome of extant Eukarya.
3. The endoplasmatic reticulum is explained as consequence of a high rate of lipid biosynthesis for the outer nuclear membrane.
4. The linear structure of the nuclear chromosomes is explained by the linear intermediates of rolling circle replication of pre-cell plasmids.
5. Closed mitosis (prior to open mitosis) is explained by the pre-cell cytoskeleton.
6. The short monocistronic eukaryal mRNAs are explained by the need to transport the mRNAs through nuclear pores, which is prohibitive for long mRNA operons.
7. The cytoplasmic location of translation is explained by the need to move the production of enzymes into the outer cytosol where they are functional for the metabolism and where the nutrients arrive.
8. The roundabout biosynthesis of eukaryal ribosomes (rRNA production inside nucleus; production of ribosomal proteins outside nucleus; import of most ribosomal proteins into the nucleus for assembly with matured rRNAs; export of incomplete ribosomal sub-units out of the nucleus; completion of ribosomal subunits outside nucleus) is explained by a multi-stage pre-cell guest evolution of the nucleus:

    (Stage a) Enzymes generated in the pre-cell guest are exported through pores into the host cytoplasm while one by one host genes are deleted or relocated into the guest genome (pre-cell dominance).

(Stage b) Guest ribosomal subunits (and tRNAs) are exported into the cyto-plasm with the selection advantage of generating metabolic enzymes where the nutrients enter, which in turn drives progressive conversion of polycis-tronic mRNAs into monocistronic mRNAs with caps and polyA tails and abandoment of operons (pre-nuclear stage).

(Stage c) Ribosomal proteins as well as polymerases etc. are imported into the nucleus to complete nuclear-metabolic segregation (proto-nuclear stage).

9. The infestation of the nuclear genes by introns is explained not as a driving force (Martin and Koonin 2006), but as an opportunistic result of nucleariza-tion, which provided from the start a safe compartment for mRNA splicing well isolated from the site of translation.
10. Mitochondria appear at the proto-nuclear stage (at or prior to origin of LECA). The problem of ribosomal chimerism (Jékely 2008) does not arise, because the genes for most mitochondrial ribosomal proteins remain fixed for a long time in the mitochondrial genome as evidenced by an extended mitochondrial gene cluster of *Reclinomonas americana* (Lang et al. 1997) with stunning similarity to the gene cluster of LBCA (Fig. 1.9).

Natural-Historic Considerations

The theory of Wallace and Darwin is based on the assumption of stochastic micro-mutations, which suggests to most biologists that there is no inner directionality in the process of evolution. The prebiotic broth theories and notably the RNA world theories extend this stochasticity down to the very beginning of life. By contrast, the theory of a chemoautotrophic origin of life forces us to view the overall process of evolution in a fundamentally novel manner, recognizing aspects of both direc-tionality and uniqueness in the early process of evolution. Let me explain.

The pioneer organism arises in a vectorial flow system. All essential outer parameters, like temperature, pressure, pH, chemical activities have gradients in flow direction. These gradients determine the interactions between inorganic substructure and organic superstructure, which is from the beginning vectorially oriented between the inorganic substructure and the water phase. In addition to this spatial directionality we see a chemical directionality that is intrinsic to the pro-cesses of autotrophic syntheses from C1 compounds and chemically necessary.

By the same token we note that metabolic evolution is subject to strict internal limitations, which force uniqueness. These limitations are a result of the pre-ordained bonding possibilities of carbon atoms. The number of structural possibilities increases with an increase of the molecular size. It is however restricted by the limi-tations of the volcanic reaction system in terms of the starting materials, reaction parameters, available metal catalysis, and idiosyncratic ligand-accelerated catalysis. These factors jointly amount to a "chemical determinism" that is bound to hold in the farthest reaches of the Universe as on Earth. As we have seen this chemical determinism was still at work at the level of the divergence of the three domains of life. With the advent of the chemistry of the genetic machinery we see an explosion of possibilities of structures and reactions. Now life progresses from the narrow confines of a canyon of synthetic chemistry to the wide equi-probability plane of combinatorial polymer chemistry.

So, we are led here to recognize a chemically pre-ordained primary track of evolution that begins with the origin of the pioneer organism. It is not only directional and unique, but also direct. Mutations consist of the direct emergence of new metallo-catalysts by feedback of new ligands (direct inheritance). Later this primary, track of evolution gives rise to a secondary, track of evolution with accidental, stochastic and non-directional DNA sequence mutations. These engage the metabolism indirectly via the products of transcription/translation that turn into metallo-enzymes. Inheritance of these mutations requires a fixation in the whole population by natural selection. This secondary track superposes the primary track and appears to dominate it more and more. Teleologically speaking, the effects of the mechanism of selection make themselves felt as adaptations, but not only as adaptations to the environmental conditions of life, but rather first and foremost as adaptations to the unique, chemically pre-ordained virtual pathways in the direct, primary track of evolution.

With these results our inquiry has arrived at two distinct prospects: (1) The chemical predetermination of the virtual pathways raises in us the hope that we might ultimately succeed in our efforts to reconstruct the pioneer organism in the laboratory and to watch it evolve. (2) The apparent chemical predetermination of the divergence of the three domains of life generates in us the suspicion that other salient changes throughout the subsequent evolution of the three domains of life may also be predetermined on a chemical trajectory of evolution, rather than the accidental result of a random walk through sequence space.

References

Barrault J, Boulinguiez M, Forquy C, Maurel R (1987) Synthesis of methyl mercaptan from carbon oxides and H_2S with tungsten-alumina catalysts. Appl Catal 33:309–330

Bernal JD (1951) The physical basis of life. Routledge/Kegan Paul, London

Berrisford DJ, Bolm C, Sharpless KB (1995) Ligand accelerated catalysis. Angew Chem Int Ed 34:1059–1070

Blackmond DG (2009) An examination of the role of autocatalytic cycles in the chemistry of proposed primordial reactions. Angew Chem Int Ed 48:386–390

Cairns-Smith AG (1982) Genetic takeover. Cambridge University Press, London

Cockell CS (2006) The origin and emergence of life under impact bombardment. Phil Trans R Soc B 361:1845–1875

Cody GD, Boctor NZ, Filley TR, Hazen RM, Scott JH, Sharma A, Yoder HS Jr (2000) Primordial carbonylated iron–sulfur compounds and the synthesis of pyruvate. Science 289:1337–1340

Corazza E (1986) Field workshop on volcanic gases, Volcano (Italy), 1982, General Report. Geothermics 15:197–200

Corliss JB, Baross JA, Hoffman SE (1981) An hypothesis concerning the relationship between submarine hot springs and the origin of life on Earth. Oceanol Acta SP:59–69

Di Giulio M (2003) The universal ancestor and the ancestor of bacteria were hyperthermophiles. J Mol Evol 57:721–730

Dörr M, Käßbohrer J, Grunert R, Kreisel G, Brand WA, Werner RA, Geilmann H, Apfel C, Robl C, Weigand W (2003) A possible prebiotic formation of ammonia from dinitrogen on iron-sulfide surfaces. Angew Chem Int Ed 42:1540–1543

Drobner E, Huber H, Wächtershäuser G, Rose D, Stetter KO (1990) Pyrite formation linked with hydrogen evolution under anaerobic conditions. Nature 346:742–744

Filtness MJ, Butler IB, Rickard D (2003) The origin of life: the properties of iron sulphide membranes. Trans Inst Min Metall Sect B 112:171–172

Fukuda F, Dokiya M, Kameyama T, Kotera Y (1977) Catalytic activity of metal sulfides for the reaction, $H_2S + CO = H_2 + COS$. J Catal 49:379–382

Gräwert T, Kaiser J, Zepeck F et al (2004) IspH Protein of *Escherichia coli*: Studies on iron-sulfur cluster implementation and catalysis. J Am Chem Soc 126:12847–12855

Heinen W, Lauwers AM (1996) Organic sulfur compounds resulting from the interaction of iron sulfide, hydrogen sulfide and carbon dioxide in an anaerobic aqueous environment. Orig Life Evol Biosph 26:131–150

Huber C, Eisenreich W, Hecht S, Wächtershäuser G (2003) A possible primordial peptide cycle. Science 301:938–940

Huber C, Wächtershäuser G (1997) Activated acetic acid by carbon fixation on (Fe, Ni)S unnder primordial conditions. Science 276:245–247

Huber C, Wächtershäuser G (1998) Peptides by activation of amino acids on (Fe, Ni)S surfaces: Implications for the origin of life. Science 281:670–672

Huber C, Wächtershäuser G (2006) α-Hydroxy and α-amino acids under possible hadean, volcanic origin-of-life conditions. Science 324:630–632

Holloway JR, Blank JG (1994) Application of experimental results to C–O–H species in natural melts. Rev Mineralog 30:187–230

Jacobsen SB (2003) How old is planet Earth? Science 300:1513–1514

Jékely G (2008) Origin of the nucleus and Ran-dependent transport to safeguard ribosome biogenesis in a chimeric cell. Biol Direct 3:31–45

Kandler O (1994a) Cell wall biochemistry in Archaea and its phylogenetic implications. J Biol Phys 20:165–169

Kandler O (1994) The early diversification of life. In: Bengtson S (ed) Early life on earth: Nobel Symposium No. 84. Columbia University Press, New York, p 152

Kandler O (1998) The early diversification of life and the origin of the three domains: a proposal. In: Wiegel J, Adams MWW (eds) Thermophiles: the keys to molecular evolution and the origin of life. Taylor & Francis, London, pp 19–28

Kant I (1790) Krtik der Urteilskraft, Translation by Meredith JC. 1952. The critique of judgment. Clarendon, Oxford, pp 81, 82

Kelley DS, Karon JA, Blackman DA et al (2001) An off-axis hydrothermal vent field near the Mid-Atlantic Ridge at 30 degrees N. Nature 412:145–149

King GAM (1977) Symbiosis and the origin of life. Orig Life 8:39–53

Kleine T, Münker C, Mezger K, Palme H (2002) Rapid accretion and early core formation on asteroids and on terrestrial planets from Hf-W chronometry. Nature 418:952–955

Kuhn H (1972) Selbstorganisation molekularer systeme und die evolution des genetischen apparats. Angew Chem 84:838–862

Kuma K, Paplawsky W, Gedulin B, Arrhenius G (1989) Mixed-valence hydroxides as bioorganic host minerals. Orig Life Evol Biosph 19:573–582

Kuwabara T, Minaba M, Ogi N, Kammekura M (2005) *Thermococcus coalescens* sp. nov., a cell-fusing hyperthermophilic archaeon from Suiyo Seamount. Int J Syst Evol Microbiol 55:2507–2514

Lang BF, Burger G, O'Kelly CJ, Cedergren R, Golding GB, Lemieux C, Sankoff D, Turmel M, Gray MW (1997) An ancestral mitochondrial DNA resembling a eubacterial genome in miniature. Nature 387:493–497

Lodders K (2003) Solar system abundances and condensation temperatures of the elements. Astrophys J 591:1220–1247

Martin W, Koonin EV (2006) Introns and the origin of nucleus-cytosol compartmentalization. Nature 440:41–45

Martin W, Russell MJ (2003) On the origin of cells: an hypothesis for the evolutionary transitions from abiotic geochemistry to chemoautotrophic prokaryotes, and from prokaryotes to nucleated cells. Phil Trans R Soc B 358:27–85

Mojzsis SJ, Harrison TM, Pidgeon RT (2001) Oxygen-isotope evidence from ancient zircons for liquid water at the Earth's surface 4.300 Myr ago. Nature 409:178–181

Mukhin L (1974) Evolution of organic compounds in volcanic regions. Nature 251:50–51

Nägeli C (1884) Mechanisch-physiologische Theorie der Abstammungslehre. Oldenbourg, München, pp 83–101

Orgel LE (1968) Evolution of the genetic apparatus. J Mol Evol 38:381–393

Oparin AI (1924) Proiskhozhdenie zhizny. Moscow. Izd. Mosk. Rabochii. English translation by Synge A (1967). In: Bernal JD (ed) The origin of life. Weidenfeld & Nicolson, London, pp 199–234

Owen AJ (1961) Calcium cyanamide synthesis. Part 1. – Thermodynamic studies. Trans Faraday Soc 57:670–677

Peck WH, Valley JW, Wilde SA, Graham CM (2001) Oxygen isotope ratios and rare earth elements in 3.3 to 4.4 Ga zircons: Ion microprobe evidence for high $\delta^{18}O$ continental crust and oceans in the early archaean. Geochim Cosmochim Acta 65:4215–4229

Russell MJ, Hall AJ (1997) The emergence of life from iron monosulphide bubbles at a submarine hydrothermal redox and pH front. J Geol Soc 154:377–402

Russell MJ (2007) The alkaline solution to the emergence of life: energy, entropy and early evolution. Acta Biotheor 55:133–179

Siefert JL, Martin KA, Abdi F, Widger WR, Fox GE (1977) Conserved gene clusters in bacterial genomes provide further support for the primacy of RNA. J Mol Evol 45:467–472

Taylor P, Rummery TE, Owen DG (1979) Reactions of iron mono-sulfide solids with aqueous hydrogen sulfide up to 160°C. J Inorg Nucl Chem 41:1683–1687

Wächtershäuser G (1988a) Before enzymes and templates: theory of surface metabolism. Microbiol Rev 52:452–484

Wächtershäuser G (1988b) German Patent Application P 38 12 158.1, filed April 4, 1988 and published November 3, 1988. p 9

Wächtershäuser G (1988c) Pyrite formation, the first energy source for life: a hypothesis. Syst Appl Microbiol 10:207–210

Wächtershäuser G (1990) Evolution of the first metabolic cycles. Proc Natl Acad Sci USA 87:200–204

Wächtershäuser G (1992) Groundworks for an evolutionary biochemistry: the iron–sulphur world. Prog Biophys Mol Biol 58:85–201

Wächtershäuser G (1997) The origin of life and its methodological challenge. J Theor Biol 187:483–494

Wächtershäuser G (1998a) The case for a hyperthermophilic, chemolithoautotrophic origin of life in an iron–sulfur world. In: Wiegel J, Adams MWW (eds) Thermophiles: the keys to molecular evolution and the origin of life. Taylor & Francis, London, pp 47–57

Wächtershäuser G (1998b) Towards a reconstruction of ancestral genomes by gene cluster alignment. Syst Appl Microbiol 21:473–477

Wächtershäuser G (2001) RNA world vs. autocatalytic anabolism. In: Dworkin M (ed) The prokaryotes, an evolving electronic resource for the microbial community. Springer, New York

Wächtershäuser G (2003) From pre-cells to Eukarya – a tale of two lipids. Mol Microbiol 47:13–22

Wächtershäuser G (2006) From volcanic origins of chemoautotrophic life to bacteria, archaea and eukarya. Phil Trans R Soc B London 361:1787–1808

Wächtershäuser G (2007) On the chemistry and evolution of the pioneer organism. Chem Biodivers 4:584–602

Wilde SA, Valley JW, Peck WH, Graham CM (2001) Evidence from detrital zircons for the existence of continental crust and oceans on the Earth 4.4 Gyr ago. Nature 409:175–178

Woese CR (1967) The genetic code: the molecular basis for genetic expression. Harper and Row, New York
Woese CR, Fox GE (1977) The concept of cellular evolution. J Mol Evol 10:1–6
Woese CR (1982) Archaebacteria and cellular origins: an overview. Zbl Bakt Hyg, I Abt Orig C3:1–17
Woese CR (1987) Bacterial evolution. Microbiol Rev 51:221–271
Woese CR (1998) The universal ancestor. Proc Natl Acad Sci USA 95:6854–6859
Ycas M (1955) A note on the origin of life. Proc Natl Acad Sci USA 41:714–716
Yin Q, Jacobsen SB, Yamashita K, Blichert-Toft J, Télouk P, Albarède F (2002) A short timescale for terrestrial planet formation from Hf-W chronometry of meteorites. Nature 418:949–952

Chapter 2
Evolution of Metabolic Pathways and Evolution of Genomes

Giovanni Emiliani, Marco Fondi, Pietro Liò, and Renato Fani

The Microbial Role in Geochemistry

Bacteria can be considered as the interface between geochemical cycles and the superior forms of life. Therefore, how the origin of life has been constructing metabolic complexity from earth geochemistry and how bacterial evolution is continuously modifying it represent major issues cross-linking both geochemical and evolutionary viewpoints.

In this chapter the current theories about the origin and evolution of metabolic pathways will be reviewed.

The present day Earth ecosystems are the result of 4.5 billion years evolutionary history and have been shaped by the combined effect of tectonic, photochemical and biological metabolic processes. Since the arisal of primordial life, that very likely took place around 3.8–3.5 billion years ago (Lazcano and Miller 1996), nearly all elements have been used and altered by microbial metabolic activities. From a geochemical perspective, the energy and mineral element (especially H, C, N, O and S) flows fueling the inner working of life and biogeochemistry processes can be considered thermodynamically constrained redox

G. Emiliani
Trees and Timber Institute – National Research Council, Via Biasi 75 I-38010, San Michele all'Adige (TN), Italy

M. Fondi and R. Fani (✉)
Laboratory of Microbial and Molecular Evolution, Department of Evolutionary Biology, University of Florence, Via Romana 17–19, I-50125 Florence, Italy
e-mail: renato.fani@unifi.it

P. Liò
Computer Laboratory, University of Cambridge, 15 JJ Thompson Avenue, cb03fd Cambridge, UK

L.L. Barton et al. (eds.), *Geomicrobiology: Molecular and Environmental Perspective*, DOI 10.1007/978-90-481-9204-5_2, © Springer Science+Business Media B.V. 2010 reactions mainly catalyzed by microbial metabolic pathways (Falkowski et al. 2008). Hence, the biogeochemical cycles might be interpreted as the interconnection existing between abiotically driven acid–base reactions, acting on geological time scale to resupply the system with elements through volcanism and rock weathering and biotically redox reactions. Such microbial reactions transformed the redox state of the planet to an oxidative environment with the development of oxygenic photosynthesis that appeared in cyanobacteria about 3–2.7 billion years ago (Canfield 2005), leading to an increasing oxygen concentration in the atmosphere between 2.4 and 2.2 billion year ago (Bekker et al. 2004). Depending on the original atmosphere composition and redox state (Kasting and Siefert 2002), several models can be proposed for the evolution of abiotic and biotic elements cycles (metabolic pathways) as well as the microbial lineages able to drive them (Canfield et al. 2006; Capone et al. 2006; Navarro-Gonzalez et al. 2001; Wächtershauser 2007), but it is clear that oxygenation led to drastic changes in elements availability.

Nitrogen compounds redox state and availability is a case in point, being biunivocally related to microbial evolutionary scenarios and metabolic pathways origin and evolution. The present day reconstruction of the nitrogen cycle offers a scenario much more complex than previously thought (Jetten 2008). Up to 20 years ago three steps in N cycle were recognized: nitrogen fixation, nitrification (oxidation) of ammonia to nitrite and nitrate and denitrification (Fig. 2.1). The evolutionary order of appearance of these three steps is still under debate (Klotz and Stein 2008) and resides mainly on geological evidences (chemical composition and isotope signatures); for example it has been postulated that denitrification (Falkowski 1997; Falkowski et al. 2008) evolved after nitrification as a consequence of the assembly of oxygenic photosynthesis. This idea relies on the finding that there is no nitrate without oxygen. The late evolution of denitrification coupled with an early origin of N_2 fixation would have led to a "nitrogen crisis" (Klotz and Stein 2008). Such observations, along with some molecular data led to the proposal of a late emergence of N_2 fixation and an early emergence of denitrification (Canfield et al. 2006; Capone et al. 2006) when N fluxes were still mainly driven by abiotic processes.

From a molecular perspective, these scenarios must also consider that the origin and evolution of enzymes involved in N cycle metabolic pathways required the availability in the Archaean and Proterozoic atmosphere (whose composition is still under debate) of metal cofactors (Klotz and Stein 2008). The isotopic signatures suggest the presence of iron, nickel and molybdenum in the Archaean era, whereas copper, zinc and cadmium were probably rare and became unavailable in the Proterozoic for their precipitation in oceanic sulfidic minerals (Canfield 1998).

Regardless of the relative importance of abiotic or biologically driven nitrogen cycle and the timing and order of appearance of its parts in the early Earth system, the present day nitrogen cycle is based on an interconnected network of metabolic pathways working as coupled reactions performed by a microbial group or by spatially and/or temporarily diversified microbial communities

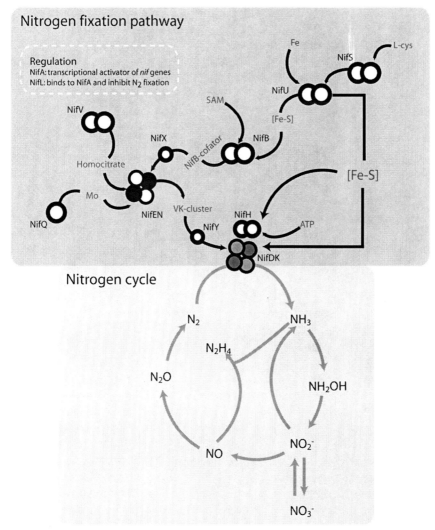

Fig. 2.1 Schematic representation of the nitrogen fixation process together with the whole nitrogen cycle

inhabiting different ecological niches (Fig. 2.1). The extant N cycle can be interpreted as the outcome of the appearance and further evolution of different metabolic capabilities. The present day metabolic pathways are based on the presence of highly complex protein systems that are often highly conserved, for example those involved in energy flows, but such diversity of metabolic abilities in extant prokaryotes did not evolve instantaneously and simultaneously. Nevertheless, since the onset of life, co-evolution of organisms led to interconnected element and energy cycles in which the outcome or an intermediate compound of a metabolic pathway is the entry point for another one, both at the single organism and

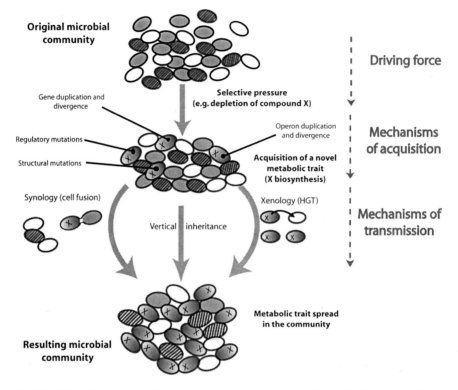

Fig. 2.2 Schematic representation of an ancestral cell community with a selective pressure allowing for the acquisition and the spreading of a new metabolic trait (from Fondi et al. 2009)

ecosystem level such as the nitrogen cycle, mediated by complex multi-species microbial communities.

Interestingly, similar metabolic pathways can be used to perform forward assimilative reductive (energy consuming) reactions and reverse, oxidative, dissimilatory ATP producing processes (Falkowski et al. 2008) thus building network linking different biologically driven elements cycles. The gaining and/or evolution of metabolic pathways in Archean cells very likely took place by modification of existing ones by molecular processes such as gene duplication followed by evolutionary divergence, enabling vertical inheritance of new capabilities and/or *via* horizontal gene transfer (HGT) of genes or operons (possibly coding for entire metabolic pathways) with spreading of such abilities to closely or distantly (but sharing the same environment) related organisms (Fig. 2.2).

From an evolutionary perspective it is interesting that even if a specific taxonomic unit or clade is eliminated by selection its metabolic abilities can be "saved" from extinction by vertical inheritance and/or HGT in a spatially and temporarily heterogeneous selective environment resulting in the evolutionary "success" and maintenance of the metabolic processes and related gene *core*.

Origin and Evolution of Metabolic Pathways

From Ancestral to Extant Genomes

It is commonly assumed that early organisms inhabited an environment rich in organic compounds spontaneously formed in the prebiotic world. This idea, commonly referred to as the Oparin-Haldane theory (Lazcano and Miller 1996) posits that life originated in a "prebiotic soup" containing different organic molecules, probably formed spontaneously under reducing conditions during the Earth's first billion years and/or delivered by extraterrestrial sources (Bada and Lazcano 2003). This soup of nutrients was available for the early heterotrophic organisms, so they had to do only a minimum of biosynthesis. As an alternative to the heterotrophic theory, autotrophic scenarios on origin of life have been proposed. A main reason for such alternative hypothesis is based on the CO_2-rich model of the primitive Earth's atmosphere (Kasting and Siefert 2002). In the presence of high CO_2 concentration, the reducing conditions are no more met and therefore it would be essential for the first organisms to synthesize by themselves the organic compounds, or use the organic compounds brought in by comets and meteorites.

The first autotrophic model for the origin of life was proposed by Wächtershauser (1988, 2006) (see also Chapter 1). In such scenario it is suggested that a primitive metabolism evolved at the surface of pyrite minerals with the reduction of carbon dioxide using a FeS/H_2S combination. A hot origin of life is thus proposed, an assumption supported by the discovery of hyperthermophiles and hydrothermal vents ecological niches. In such conditions a high temperature would have favored chemical reactions on the surface and/or inside minerals.

Despite the controversy between the heterotrophic and autotrophic theories, there is a general agreement that minerals played an important role in cell origin and early evolution, catalyzing reactions with metal sulfides providing reducing potential (Bada and Lazcano 2002). It is also widely accepted that reactions occurring in hydrothermal vents and volcanic environments – important for the formation of phosphoric compounds (Schwartz 2006) – may have contributed to prebiotic synthesis. A pivotal step was represented by the development of an energy metabolism for the evolution of more complex cells. Ferry and House (2006) proposed the synthesis of phosphorylated compounds from energy obtained by geothermal fluxes.

Independently from their origin, the community of the first living cells evolved into a smaller number of more complex cell types, which ultimately developed into the ancestor(s) of all the extant life domains usually referred to as Last Universal Common Ancestor (LUCA) (Fig. 2.3).

The increasing number of available sequences from organisms belonging to the three domains of life (Bacteria, Archaea and Eukarya) has allowed estimating the minimal gene content of LUCA whose genome was probably composed by about 1,000–1,500 genes (Ouzounis et al. 2006). However, despite this small gene content, ancestral genomes were probably fairly complex, similar to those of the extant free-living prokaryotes and included a variety of functional capabilities including

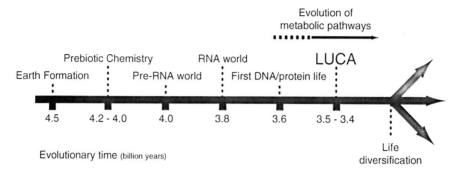

Fig. 2.3 Evolutionary time line from the origin of Earth to the diversification of life (Fani and Fondi 2009)

metabolic transformation, information processing, membrane/transport proteins and complex regulation (Ouzounis et al. 2006). Hence, starting from a common pool of highly conserved genetic information, still shared by all the extant life forms, genomes have been shaped to a considerable extent during evolution, leading to the great diversification of life (and genomes) that we observe nowadays. This raises the intriguing question of how both genome complexity and size could have been increased during evolution.

The Primordial Metabolism

All living (micro)organisms possess an intricate network of biosynthetic and catabolic routes. How these pathways originated and evolved is still under debate. Assuming that life arose in a prebiotic soup containing most, if not all, of the necessary small molecules, then a large potential availability of nutrients in the primitive Earth can be surmised, providing both the growth and energy supply for a large number of ancestral organisms. If this scenario is correct, **why** did heterotrophic primordial cells expand their metabolic abilities and genomes? The answer to this question is rather intuitive. Indeed, the increasing number of early cells thriving on primordial environment would have led to the depletion of essential nutrients that might have imposed a progressively stronger selective pressure that, in turn, favoured those (micro)organisms that have become able to synthesize the nutrients whose concentration was decreasing in the primordial soup. Thus, the origin and the evolution of basic metabolic routes represented a crucial step in molecular and cellular evolution, since it rendered the primordial cells less dependent on the external source of nutrients.

Since ancestral cells probably contained small chromosomes and consequently had limited coding capabilities, it is likely that their metabolism was based on a limited number of enzymes. Thus, **how** did the early cells expand their gene content?

The next session will focus on the molecular mechanisms that guided the expansion and the refinement of ancestral metabolic routes.

The Role of Duplication and Fusion of DNA Sequences in the Evolution of Metabolic Pathways in the Early Cells

The Starter Types

It has been recognized that most genetic information is not essential for cell growth and division. Therefore, it is possible that the early cells possessed a genome containing about 200–300 genes most of which arising by duplication of a limited number of "starter type genes". This term, coined by Lazcano and Miller (1994) refers to the original ancestral genes that underwent (many) duplications and gave rise to the extant paralogous gene families. The origin of the starter types is still unclear, but their number has been estimated to range between 20 and 100 (Lazcano and Miller 1996). It is quite possible that once a starter type gene encoding a functional catalyst (or structural protein) appeared, it will undergo duplications at a rate surprisingly fast on the geological timescale. Hence, the basic biosynthetic routes might have been assembled in a short geological timescale (Peretó et al. 1997). Concerning the timing, even though it is not possible to assign a precise chronology to the development of biochemical pathways, most of them were likely assembled in a DNA/protein world.

The Explosive Expansion of Metabolism in the Early Cells

Different molecular mechanisms may have been responsible for the expansion of early genomes and metabolic abilities. However, data obtained in the last decade indicate that a very large proportion of the gene set of different (micro)organisms is the outcome of more or less ancient gene duplications predating or following the appearance of LUCA (Ohta 2000). Thus, duplication and divergence of DNA sequences of different size represents one of the most important forces driving the evolution of genes and genomes. Indeed, this process may allow the formation of new genes from pre-existing ones. However, there are a number of additional mechanisms that could have increased the rate of metabolic evolution, including the modular assembly of new proteins, gene fusion events, and HGT (see below).

Gene Duplication

In principle, a DNA duplication may involve (a) part of gene, (b) a whole gene, (c) entire operons, (d) entire chromosomes, and (e) a whole genome. Two structures or sequences that evolved from a single ancestral structure or sequence, after a duplication event, are referred to as "homologs" that, in turn, can be classified as orthologs or paralogs. Orthologous structures (or sequences) in two organisms are homologs that evolved from the same feature in their last common ancestor. Therefore, the evolution of orthologs reflects organism evolution. Homologs whose evolution reflects gene duplication events are called paralogs. Consequently, orthologs usually perform the same function in different organisms, whereas paralogous proteins often catalyze different, although similar, reactions. Two paralogous genes may also undergo successive and differential duplication events involving one or both of them giving rise to paralogous gene families.

One of mechanisms for the rapid expansion of genomes and gaining of metabolic abilities is the duplication of entire clusters of genes involved in the same metabolic pathways, i.e. entire operons or part thereof. Indeed, if an operon A, responsible for the biosynthesis of aminoacid A, duplicates giving rise to a pair of paralogous operons, one of the copies (B) may diverge from the other and evolve in such a way that the encoded enzymes catalyze reactions leading to a different amino acid, B. Once acquired, metabolic innovations might spread rapidly between (micro)organisms through horizontal gene transfer mechanisms.

Fate of Duplicated Genes

The structural and/or functional fate of duplicated genes is an intriguing issue that has led to the proposal of different evolutionary models accounting for the possible scenarios emerging after the appearance of a paralogous pair.

Structural fate. Duplication events can generate genes arranged in-tandem, which are often the result of an unequal crossing-over between two DNA molecules; however, other processes, such as replication slippage, may explain the existence of tandemly arranged paralogous genes. In addition, duplication by recombination involving different DNA molecules or transposition can generate a copy of a DNA sequence at a different location within the genome (Fani et al. 2000; Li and Graur 1991).

If an in-tandem duplication occurs, at least two different scenarios for the structural evolution of the two copies can be depicted:

1. The two genes undergo an evolutionary divergence becoming paralogs
2. The two genes fuse doubling their original size generating an elongated gene

If the two copies are not tandemly arranged:

1. They may become paralogous genes
2. One copy may fuse to an adjacent gene, with a different function, giving rise to a mosaic or chimeric gene that potentially may evolve to perform other(s) metabolic role(s)

Functional fate. The functional fate of the two (initially) identical gene copies originated from a duplication event depends on the further modifications (evolutionary divergence) that one (or both) of the two redundant copies accumulates during evolution. It can be surmised, in fact, that after a gene duplicates, one of the two copies becomes dispensable and can undergo several types of mutational events, mainly substitutions, that can lead to the appearance of a new gene harboring a different function in respect to the ancestral coding sequence. On the contrary, duplicated genes can also maintain the same function in the course of evolution, thereby enabling the production of a large quantity of RNAs or proteins (gene dosage effect).

Neo-functionalization. The classical model of gene duplication (neofunctional-ization) predicts that in most cases one duplicate may become functionless, whereas the other copy will retain the original function (Li and Graur 1991; Liò et al. 2007; Ohno 1972). At least in the early stages following the duplication event, the two copies maintain the same function. Then, it is likely that one of the redundant copies will be lost, due to the occurrence of mutation(s) negatively affecting its original function that will be preserved by the other copy. However, although less probably, an advantageous mutation may change the function of one duplicate and both copies may be maintained.

Sub-functionalization. The sub-functionalization model is based on the observa-tion that a gene contains other "accessorial" components, i.e. promoter regions, different functional and/or structural domains of the protein it encodes and so on. These elements can be considered as a sub-functional module for a gene or protein, each of wich contributing to the global function of that gene or protein. Starting from this idea, Lynch and Force (2000) proposed that multiple sub-functions of the original gene may play an important role in the preservation of gene duplicates. After the duplication event, deleterious mutations can reduce the number of active sub-functions of one or both the duplicates, but the sum of the sub-functions of the duplicates will be equal to the number of original functions before duplication (i.e. the original functions have been partitioned among the two duplicates). The sub-functionalization differs from the classical model because the preservation of both gene copies mainly depends on the partitioning of sub-functions between dupli-cates, rather than the occurrence of advantageous mutations.

Sub-neofunctionalization. More recently He and Zhang (2006) proposed the sub-neofunctionalization model, according to which the sub-functionalization is a rapid process, while the neo-functionalization requires more time and continues even long after duplication.

Gene Fusion

Another major route of gene evolution is the fusion of independent cistrons leading to bi- or multifunctional proteins (Brilli and Fani 2004; Xie et al. 2003). Gene fusions provide a mechanism for the physical association of different catalytic domains or of catalytic and regulatory structures. Fusions frequently involve genes encoding proteins functioning in a concerted manner, such as enzymes catalysing sequential steps within a metabolic pathway. Fusion of such catalytic centres likely promotes the channelling of intermediates that may be unstable and/or in low con-centration; this, in turn, requires that enzymes catalysing sequential reactions are co-localized within cell and may (transiently) interact to form complexes that are termed metabolons (Srere 1987). The high fitness of gene fusions can also rely on the tight regulation of the expression of the fused domains.

Hypotheses on the Origin and Evolution of Metabolic Pathways

As discussed in the previous sections, the emergence and refinement of basic biosynthetic pathways allowed primitive organisms to become increasingly less dependent on exogenous sources of essential compounds accumulated in the primitive environment as a result of prebiotic syntheses. But how did these metabolic pathways originate and evolve? Then, which is the role that gene duplication and/or fusion played in the assembly of metabolic routes?

How the major metabolic pathways actually originated is still an open question, but several different theories have been suggested to account for the establishment of metabolic routes (Fani and Fondi 2009). Two of them are discussed below.

The Retrograde Hypothesis

The first attempt to explain in detail the origin of metabolic pathways was made by Horowitz (1945), who suggested that biosynthetic enzymes had been acquired via gene duplication that took place in the reverse order found in current pathways. This idea, known as the Retrograde Hypothesis, states that if the contemporary biosynthesis of compound "A" requires the sequential transformations of precursors "D", "C" and "B" *via* the corresponding enzymes, the final product "A" of a given metabolic route was the first compound used by the primordial heterotrophs (Fig. 2.4). In other words, if a compound A was essential for the survival of primordial cells, when A became depleted from the primitive soup, this should have imposed a selective pressure allowing the survival and reproduction of those cells that were become able to perform the transformation of a chemically related compound "B" into "A" catalyzed by enzyme "a" that would have lead to a simple, one-step pathway. The selection of variants having a mutant "b" enzyme related to "a" via a duplication event and capable of mediating the transformation of molecule "C" chemically related into "B", would lead into an increasingly complex route, a process that would continue until the entire pathway was established in a backward

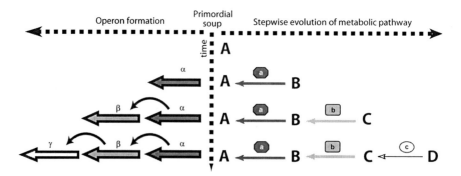

Fig. 2.4 Schematic representation of the Horowitz hypothesis on the origin and evolution of metabolic pathways and operons (modified from Fani and Fondi 2009)

fashion, starting with the synthesis of the final product, then the penultimate pathway intermediate, and so on down the pathway to the initial precursor. Twenty years later, the discovery of operons prompted Horowitz to restate his model, arguing that it was supported also by the clustering of genes that could be explained by a series of early tandem duplications of an ancestral gene; thus, genes belonging to the same operon should have formed a paralogous gene family.

The retrograde hypothesis establishes an important evolutionary connection between prebiotic chemistry and the development of metabolic pathways, but requires special environmental conditions in which useful organic compounds and potential precursors accumulated. Although these conditions might have existed at the dawn of life, they must have become less common as life forms became more complex and depleted the environment of ready-made useful compounds (Copley 2000). Furthermore, many anabolic routes involve *many unstable intermediates* whose synthesis and accumulation in both the prebiotic and extant environments is difficult to explain. The Horowitz hypothesis also fails to account for the origin of *catabolic pathway regulatory mechanisms*, and for the development of biosynthetic routes involving dissimilar reactions. Lastly, if the enzymes catalyzing successive steps in a given metabolic pathway resulted from a series of gene duplication events, then they must share structural similarities. However, the list of known examples confirmed by sequence comparisons is small.

The Patchwork Hypothesis

The so-called "patchwork" hypothesis (Jensen 1976; Ycas 1974) states that metabolic pathways may have been assembled through the recruitment of primitive enzymes that could react with a wide range of chemically related substrates. Such relatively slow, un-specific enzymes may have enabled primitive cells containing small genomes to overcome their limited coding capabilities. Figure 2.5 shows a schematic three-step model of the patchwork hypothesis; (a) an ancestral enzyme E0 endowed with low substrate specificity is able to bind to three substrates (S1, S2 and S3) and catalyze three different, but similar reactions; (b) a duplication of the gene encoding E0 and the divergence of one of the two copies leads to the appearance of enzyme E2 with an increased and narrowed specificity; (c) a further gene duplication event, followed by evolutionary divergence, leads to E3. In this way the ancestral enzyme E0, belonging to a given metabolic route is "recruited" to serve other novel pathways. In this way primordial cells might have expanded their metabolic capabilities.

The patchwork hypothesis is supported by several lines of evidence. The broad substrate specificity of some enzymes means they can catalyse different chemical reactions. As demonstrated by whole genome sequence comparisons, there is a significant percentage of metabolic genes that are the outcome of paralogous duplications. The recruitment of enzymes belonging to different metabolic pathways to serve novel biosynthetic routes is also well documented by the so-called "directed evolution experiments", in which microbial populations are subjected to a strong selective pressure leading to the appearance of phenotypes capable of using new substrates (Fani and Fondi 2009; Mortlock and Gallo 1992).

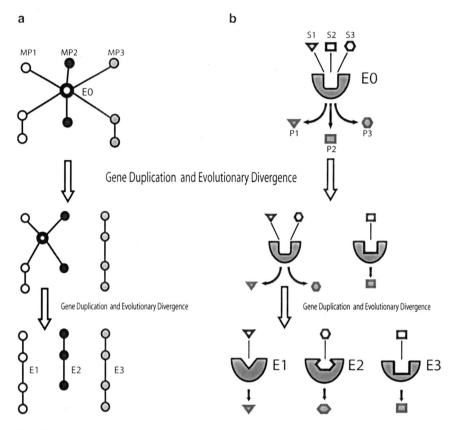

Fig. 2.5 The patchwork hypothesis on the origin and evolution of metabolic pathways (**a**) Hypothetical overall structure of the metabolic pathways (MP) in which enzymes (E0, E1, E2, E3) are involved (**b**) The origin of enzymes with narrowed specificity from an ancestor unspecific one (modified from Fani and Fondi 2009)

The Role of Horizontal Gene Transfer in the Evolution of Genomes and Spreading of Metabolic Functions

The Darwinian view of organism evolution predicts that such process can be interpreted and represented by a "tree of life" metaphor. Any functionally significant (phenotypic) and so selectable evolutionary "invention", arising from gene or genome level molecular processes (point mutations, gene duplication, etc.) is vertically transmitted – if not lethal. Nevertheless, there are exceptions to the tree of life paradigm, although providing a still valid framework: evolutionary landmark events of cellular and genome evolution mediated by symbiosis (i.e. chloroplast and mitochondria) defines an example of non-linear evolution. Such processes define a different model of evolution - the reticulate one (Gogarten and Townsend 2005) – that eventually took place along with the "classical" vertical transmission. Thus, a single bifurcating tree is insufficient to describe the microbial evolutionary process (that is furthermore problematical for the difficulty to define species boundaries in prokaryotes) as "only 0.1–1% of each genome fits the metaphor of a tree of life" (Dagan and Martin 2006). Indeed, the phylogenomic and comparative genomic approaches based on the availability of a large number of completely sequenced genomes has highlighted the importance of non-vertical transmission in shaping genomes and evolution processes. Incongruence existing in the phylogenetic recon-structions using different genes is considered as a proof of HGT events (Gribaldo and Brochier-Armanet 2006; Ochman et al. 2005), some of which probably (very) ancient (Brown 2003; Huang and Gogarten 2006). The extent of HGT events occurred during evolution is still under debate (Dagan and Martin 2006, 2007; Koonin et al. 2001) and is especially intriguing in the light of early evolution elu-cidation as well as the notion of a communal ancestor (Koonin 2003). It has been in fact proposed that HGT dominated during the early stages of cellular evolution and was more frequent than in modern systems (Woese 1998, 2000, 2002).

The emergence of a "horizontal genomics" well explains the interest in the role of HGT processes in genome and species evolution. From a molecular perspective HGT is carried out by different mechanisms and is mediated by a mobile gene pool (the so called "mobilome") comprising plasmids, transposons and bacteriophages (Frost et al. 2005). HGT can involve single genes or longer DNA fragment contain-ing entire operons and thus the genetic determinants for entire metabolic pathways conferring to the recipient cell new capabilities. It has been hypothesized that HGT does not involve equally genes belonging to different functional categories. Genes responsible for informational processes (transcription, translation, etc.) are likely less prone to HGT than operational genes (Shi and Falkowski 2008), even though the HGT of ribosomal operon has been described (Gogarten et al. 2002). This latter finding and the observation that only a 40% of the genes are shared by three *Escherichia coli* strains (Martin 1999) raise the question of the stability of bacterial genomes (Itoh et al. 1999; Mushegian and Koonin 1996). It is therefore important for phylogenetic and evolutionary analysis to individuate the "stable core" and the "variable shell" in prokaryotic genomes (Shi and Falkowski 2008).

It is also quite possible that, in addition to HGT (xenology), the early cells might have exchanged (or shared) their genetic information through cell fusion (sinol-ogy). The latter mechanism might have been facilitated by the absence of a cell wall in the Archaean cells and might have been responsible for large genetic rearrange-ments and rapid expansion of genomes and metabolic activities.

The Nitrogen Cycle

On the planetary scale the biogeochemical N cycle has suffered major anthropo-genic alterations in the last decades shifting the priorities from boosting food production to control large scale environmental changes (Galloway et al. 2008). Half of the fixed nitrogen entering Earth ecosystems is produced via the Haber-Bosch process and cultivation of nitrogen fixing crops. Furthermore, reactive nitrogen is also produced by fossil and bio-fuels combustion. These inputs of reactive nitrogen might alter the terrestrial and marine N cycles (Deutsch et al. 2007; Houlton et al. 2008) as well as interconnected biogeochemical cycles, such as those related to carbon and phosphorus (Gruber and Galloway 2008).

In the absence of human perturbations, the nitrogen cycle is the result of geological time-scale abiotic processes including NH_4^+ production from N_2 (Wächtershauser 2007), combustion of N_2 to nitrate (Mancinelli and McKay 1988; Navarro-Gonzalez et al. 2001), mineralization (McLain and Martens 2005) and biologically driven metabolic reactions. The abiotic production of fixed nitrogen, which is mainly due to lightening discharge, is ten-fold lower than microbial production (Falkowski 1997). It has been postulated that abiotic fixed nitrogen was limiting in the early Earth (Kasting and Siefert 2001), a condition that might have favored an early appearance of microbial N_2 fixation (Raymond et al. 2004).

Schematically, the microbial driven nitrogen cycle comprises three steps (Fig. 2.1):

1. The fixation of the atmospheric N_2 to ammonia (NH_4^+)
2. The stepwise oxidation of ammonia to nitrite and of nitrite to nitrate
3. The denitrification of nitrite and nitrate to gaseous dinitrogen through anaerobic respiration in anoxic environment (complete denitrification) or the detoxifying reduction of nitrite to NO in aerobic environment (incomplete or nitrifier denitrification)

Nevertheless, a complete picture of the microbial nitrogen cycle must take into account other relevant processes and the list of prokaryotic players in the biogeochemical N fluxes is continuously increasing.

Nitrification

Nitrification, the stepwise oxidation of ammonia to nitrite (NO_2^-) via hydroxylamine and the successive oxidation of NO_2^- to NO_3^- (nitrate) is a catabolic O_2 dependent process that evolved after the oxygenation of the atmosphere and it is considered as the last step of nitrogen cycle appeared on Earth (Klotz and Stein 2008). Such process, enabling chemolithoautotrophic growth (even though several heterotrophic bacteria can perform the same reaction) is performed by different players of the "nitrifying community" (Arp et al. 2007). The ammonia oxidizing bacteria use ammonia as an energy source for carbon assimilation.

Nitrite oxidizers bacteria catalyze the second step of nitrification and are so far restricted to five bacterial genera (Alawi et al. 2007). These microorganisms are able to catalyze the oxidation of nitrite in the reaction $NO_2^- + H_2O \rightarrow NO_3^- + 2H+ 2e^-$ with the activity of nitrite oxidoreductase (NXR).

Denitrification

The denitrification process, the dissimilatory reduction of nitrate and nitrite to gaseous nitrogen, proceeds stepwise following the reactions $NO_3^- \rightarrow NO_2^- \rightarrow NO \rightarrow N_2O \rightarrow N_2$ and is an anaerobic or microaerophilic process performed by denitrifying (facultative) heterotrophic soil and water bacteria using organic carbon source and nitrate as electron acceptor.

Anaerobic Ammonia Oxidation (ANAMMOX)

The recent discovery of anaerobic ammonia oxidation has been regarded as one of main advancement in the comprehension of nitrogen cycle (Jetten 2008). Microorganisms with this metabolic ability are able to couple nitrification (oxidation of ammonia) and denitrification (until N_2 production) in anaerobic environments. The exact enzymology and genetic inventory of such process are still unsettled (Strous et al. 2006).

The importance of ANAMMOX in the global nitrogen cycle is striking (Jetten 2008) and its evolutionary origin intriguing. It is in fact proposed that ANAMMOX evolved soon after the incomplete denitrification pathway (in absence of the copper dependent NOS) (Strous et al. 2006) and provided the first metabolic pathway to resupply the atmospheric N_2 pool, performing this role until the evolutionary origin of the full denitrification pathway.

Ammonification

Ammonification, the dissimilatory electrogenic reduction of nitrate to ammonia via formate or H_2 in oxygen limited conditions, is performed by many facultative and obligate chemolithotrophic proteobacteria (Simon 2002). Interestingly, since this process does not require oxygen and needs iron it is proposed that this pathway evolved very early and was responsible for fixed nitrogen resupply from abiotic formed NO_2^- before the advent of N_2 fixation (Mancinelli and McKay 1988).

Nitrogen Fixation: A Paradigm for the Evolution of Metabolic Pathways

Nitrogen fixation, the biological conversion of atmospheric dinitrogen to ammonia, represents an excellent model for studying the evolutionary interconnections linking different pathways and the functional divergence of paralogs (Fig. 2.1).

Table 2.1 List of the 20 genes involved in nitrogen fixation in *Klebsiella pneumoniae*

Gene name	Poduct function	Source of reference
nifH	structural dinitrogenase reductase Fe protein	Mevarech et al. 1980
nifY	involved in nitrogenase maturation	Homer et al. 1993
nifT	involved in nitrogenase maturation	Simon et al. 1996
nifD	structural component of dinitrogenase (FeMo protein)	Lammers and Haselkorn 1984
nifK	structural component of dinitrogenase (FeMo protein)	Mazur and Chui 1982
nifU	required for the activation of Fe and FeMo proteins	Jacobson et al. 1989; Dos Snatos et al. 2004
nifS	required for the activation of Fe and FeMo proteins	Jacobson et al. 1989; Dos Snatos et al. 2004
nifM	required for accumulation of active FeMo protein	Jacobson et al. 1989; Howard et al. 1986; Paul and Merrick 1989
nifZ	acts as a chaperone in the assembly of the FeMo protein	Hu et al. 2004
nifW	protect the MoFe protein from oxygen damage	Kim et al.1996
nifN	scaffold for the FeMo and FeVn cofactor biosynthesis	Roll et al. 1 995
nifE	scaffold for the FeMo and FeVn cofactor biosynthesis	Roll et al. 1995
nifO	involved in the biosynthesis of FeMo cofactor	Rodriguez-Quignones et al. 1993; Shah et al. 1994
nifQ	involved in the biosynthesis of FeMo cofactor	Rodriguez-Quignones et al. 1993; Shah et al. 1994
nifX	involved in FeMo-co biosynthesis (able to transfer NifB-co to nifEN)	Shah et al. 1999; Hernandez et al. 2006
nifB	crucial for FeMo cofactor biosynthesis	Bishop and Joergerp (1990)
nifV	(homocitrate synthase) involved in the biosynthesis of FeMo cofactor	Filler et al. 1986
nifF	electron transport to the structural components	Hill and Kavanagh 1980
nifJ	electron transport to the structural components	Hill and Kavanagh 1980
nifA	(together with rpoN) activates transcription of all nitrogenase promoters	Dixon et al. 1980; Merrick 1983
nifL	Modulates the activity of the transcriptional activator NifA	Hill et al, 1981; Merrick et al. 1982; Blanco et al. 1993; Sidoti et al. 1993

Nitrogen fixation is the most important input of biologically available nitrogen in Earth ecosystems. It is a metabolic ability possessed only by some *Bacteria* (green sulfur bacteria, *Firmicutes, Actinomycetes, Cyanobacteria* and *Proteobacteria*) and *Archaea*, where it is mainly present in methanogens (Dixon and Kahn 2004). Nitrogen fixation is compatible with different microbial lifestyles: aerobic, anaerobic and facultative anaerobic heterotrophs, anoxygenic and oxygenic photosynthetic bacteria and chemolithotrophs. Diazotrophs inhabit many ecological niches, marine and terrestrial environments as free living or plant symbiotic or endophytic microorganisms (Raymond 2005). The correlation between nitrogen fixation – that is poisoned by O_2 – and oxygen rich environment or oxygenic (photosynthetic) metabolism is particularly intriguing from an evolutionary viewpoint (see below).

Nitrogen fixation is a complex process with a high energetic cost, requiring the activity of several genes (Fig. 2.1). In the free-living diazotroph *Klebsiella pneumoniae* 20 genes involved in nitrogen fixation (*nif* genes) have been identified (Table 2.1). The enzyme responsible for nitrogen fixation, the nitrogenase, shows high degree of conservation of structure, function and amino acid sequence across wide phylogenetic ranges (Fani et al. 2000). Nitrogenase contains an unusual metal clusters, the Iron-Molybdenum cofactor (FeMo-co), that is considered to be the site of dinitrogen reduction, and whose biosynthesis requires the products of *nifE, nifN* and several other *nif* genes (Fig. 2.1). All known Mo-nitrogenases consist of two components, component I (dinitrogenase, or Fe–Mo protein), an $\alpha_2\beta_2$ tetramer encoded by *nifD* and *nifK*, and component II (dinitrogenase reductase, or Fe-protein) a homodimer encoded by *nifH*.

In the last years some light has been shed on the molecular mechanisms responsible for the evolution of *nif* genes and the interconnections of nitrogen fixation with other metabolic pathways, such as bacteriochlorophyll biosynthesis (Xiong et al. 2000). In spite of this, many questions remain still open:

1. Is nitrogen fixation an ancestral character, arising prior to the appearance of LUCA?
2. How many genes were involved in the ancestral nitrogen fixation process?
3. How did the *nif* genes originate and evolve?
4. How and at what extent was nitrogen fixation correlated to other metabolic processes in the earliest cells?
5. Which were the molecular mechanisms involved in the origin, evolution and spreading of nitrogen fixation?

Is Nitrogen Fixation an Ancestral Character?

The time and order of appearance of nitrogen fixation in relation to the other nitrogen related metabolic pathways is still under debate. However, it is generally thought that N_2 fixation represents an early invention of evolution since the biological importance of the elements and the rapid depletion of abiotically fixed nitrogen in the primordial metabolism (Falkowski et al. 2008). Such model is consistent with both geological evidence, for example the availability of molybdenum and iron in the Archaean (Canfield et al. 2006), and phylogenomics analyses (Raymond et al. 2004).

Nevertheless, since *nif* genes can be organized in (compact) operons that are prone to HGT, the presence of *nif* genes in *Archaea* and *Bacteria* is not considered a straight-forward demonstration of the antiquity of the metabolic pathway (Raymond et al. 2004; Shi and Falkowski 2008). Moreover Mancinelli and McKay (1988), basing on the complexity of the pathway, the high energy costs of fixation, and the absence in eukaryotic organelles, suggested that these findings are not compatible with an early origin of N_2 fixation that they proposed evolved after denitrification when fixed nitrogen was available for early metabolism by abiotic reactions or ammonification. This model is in agreement with the lack of supporting data for a depletion of atmospheric N_2 in presence of coupling of early nitrogen fixation and absence of denitrification (Capone and Knapp 2007). However, this scenario has some pitfalls (Klotz and Stein 2008), such as the absence, in the Archean and Proterozoic eras, of nitrous oxide reductase (NOS), an enzyme possessed by extant denitrifiers for the lack of its copper cofactor and the low concentration of nitrite that could had formed only in limited amounts by combustion in the early neutral to mildly reducing CO_2 depleting Archean atmosphere (Navarro-Gonzalez et al. 2001).

How Many Genes were Involved in the Ancestral Nitrogen Fixation?

Recently (Fondi et al., unpublished data) the phylogenetic distribution of *nif* genes was checked in completely sequenced prokaryotes. The analysis performed by probing 842 prokaryotic genomes (52 archaea and 790 bacteria) for the presence of *nifH* genes revealed that 124 possessed it. All these genomes were scanned for the presence of genes homologous to each of the 20 *K. pneumoniae nif* genes. As shown in Fig. 2.6, only six *nif* genes (*nifHDKENB*), involved in the synthesis of nitrogenase, nitrogenase reductase and Fe–Mo Cofactor biosynthesis, were present in almost all the genomes. All the other *nif* genes have a patchy phylogenetic distribution revealing a complex evolutionary history.

This finding strongly suggests that if nitrogen fixation is an ancestral metabolic trait possessed by LUCA, it is quite possible that only *nifHDKENB* genes were present in the genome of the LUCA community. Thus, if nitrogen fixation required other enzymes, their function might have been performed by enzymes with low substrate specificity (in agreement with the Jensen hypothesis on the origin and evolution of metabolic pathways). According to this idea, the *nifHDKENB* might represent a "universal core" for nitrogen fixation, whereas the other genes might have been differentially acquired during evolution in the different phylogenetic lineages.

How Did the nif Genes Originate and Evolve?

In/out – paralogs of nif genes. The hypothesis proposed in the previous paragraph implies that during evolution some genes might have been recruited from other metabolic pathways through duplication and divergence of genes coding for enzymes with a low substrate specificity.

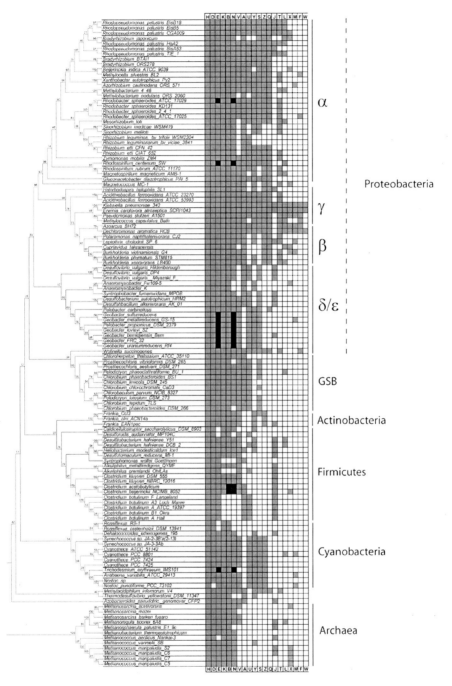

Fig. 2.6 The distribution of *nif* genes within 124 diazotrophic Bacteria and Archaea (whose genomes were completely sequenced and available on NCBI). *White* and *light grey boxes* represent the absence or presence of the corresponding genes, respectively. *Dark grey boxes* represent fusions of the corresponding genes

This idea is supported by the finding that most of *nif* genes have in- paralogs (i.e. paralogs involved in the same pathway) and/or out-paralogs (i.e. paralogs involved in different pathways) as pointed out by Fondi et al. (unpublished data) using a Psi-blast analysis using each Nif protein as query (Fig. 2.7). The analysis did not retrieve any known paralogs for *nifW* (*nifO*), *nifT* (*fixU*), *nifQ* and *nifZ* which are also missing from a large fraction of diazotrophs genomes. Eight *nif* genes (*nifAFHJLMSU*) are related at a different extent to proteins involved in other metabolic pathways (out-paralogs). NifS is related to a number of paralogs mainly involved in amino acid and carbon metabolisms. NifJ, a multidomain pyruvate:ferredoxin (flavodoxin) oxi-doreductase, exhibited a large number of paralogs. Several of the proteins involved in Fe–Mo cofactor biosynthesis have paralogs in other cofactor biosyntheses.

Eight Nif proteins share a significant degree of sequence similarity with proteins involved in other metabolic routes, and also with other *nif* genes products; this group can be further separated into two different clusters, the first of which includes *nifDKEN*, and the second being composed by *nifBXY* and *nifV*. Actually, NifBXY are related through a common domain of about 90 aminoacids; moreover, *nifB* has an additional domain belonging to the S:-adenosylmethionine (SAM) family, found in proteins that catalyze diverse reactions, including unusual methylations, isomeri-sation, sulphur insertion, ring formation, anaerobic oxidation and protein radical formation. Evidence exists that these proteins generate a radical species by reduc-tive cleavage of SAM through an unusual Fe-S centre. The genes *nifV* and *nifB* are not directly linked and their connection is due to multidomain proteins sharing homology with NifV and NifB in different domains.

As expected, NifDKEN showed sequence similarity with Bch proteins involved in bacterial photosynthesis.

Nitrogen fixation and bacterial photosynthesis: an ancestral interconnections through a cascade of gene and operon duplication. The two gene pairs *nifD-nifK* and *nifE-nifN*, coding for nitrogenase and the tetrameric complex Nif N_2E_2, form a paralogous gene family, and arose through duplications of an ancestral gene, by a two-step model in which an ancestor gene underwent an in-tandem duplication event giving rise to a bicistronic operon; this, in turn, duplicated leading to the ancestors of the present-day *nifDK* and *nifEN* operons (Fani et al. 2000).

The model proposed is in agreement with the Retrograde Hypothesis but also fits the Jensen's hypothesis of the metabolic pathways assembly. Accordingly, the ancestor of the *nif* gene family encoded a protein which might assemble to give a homotetra-meric (or a homomultimeric) complex with a low substrate specificity able to catalyse more than one enzymatic reactions (Fani et al. 2000). By assuming that the ability to fix nitrogen was a primordial property dating back to LUCA (Woese 1998; Zillig et al. 1992), then the duplication events leading to the two operons predated the appearance of LUCA and the function(s) performed by this primordial enzyme might have depended on the composition of the early atmosphere. There is no agreement on the composition of the primitive atmosphere, but it is generally accepted that O_2 was absent and this represents an essential prerequisite for the appearance of (an ancestral) nitrogenase, which is inactivated by free oxygen (Fay 1992). The appearance of nitro-

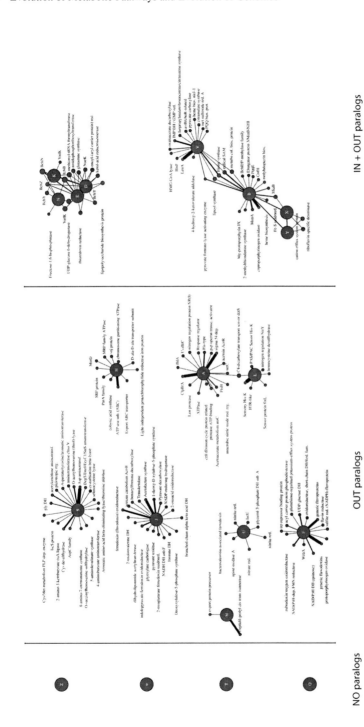

Fig. 2.7 In- and Out-paralogs network of *nif* genes. Nodes represent protein, links represent similarity values

genase on the primitive Earth would have represented a necessary event for the first cells, living in a planet whose atmosphere was neutral, containing dinitrogen, but not ammonia (first scenario). In fact, if ammonia was required by the primitive micro-organisms for their syntheses, then its absence must have imposed a selective pressure favouring those cells that had evolved a system to synthesise ammonia from atmospheric dinitrogen. Therefore, according to this scenario, the function of the ancestral enzyme might have been that of a "nitrogenase", slow, inefficient and with low substrate specificity able to react with a wide range of compounds with a triple bond.

According to a second theory, the early atmosphere was a reducing one and contained free ammonia (Fig. 2.8). In those conditions, the evolution of a nitrogen fixation system was not a prerequisite because of the abundance of abiotically produced ammonia. Hence, why a nitrogenase in those days? The answer to this question relies in the catalytic properties of nitrogenase. In fact the enzyme is able to reduce also other molecules such as acetylene, hydrogen azide, hydrogen cyanide, or nitrous oxide, all of which contain a triple bond. Therefore, according to this second scenario (Fig. 2.8), the primitive enzyme encoded by the ancestor gene, would have been a *detoxyase*, an enzyme involved in detoxifying cyanides and other chemicals present in the primitive reducing atmosphere (Silver and Postgate 1973). This scenario implies that the progressive exhaustion of combined nitrogen would have imposed the refinement of the enzyme specificity which very likely modified and adapted to another triple-bond substrate, dinitrogen, and was selected for, and retained by some bacterial and archaeal lineages to enable survival in nitrogen-deficient environments. Finally, the decreasing of free ammonia and cyanides in the atmosphere triggered the evolution of the detoxyase toward nitrogenase, that might have been a common feature of all microbial life until photosynthesising cyanobacteria largely increased the oxygen concentration and burned cyanides.

Particularly intriguing is the finding that genes coding for nitrogenase (*nifDK*) and nitrogenase reductase (*nifH*) are evolutionary related to the genes involved in bacteriochlorophyll biosynthesis (see below). Chlorophyll (Chl) and bacteriochlorophyll (Bchl) are the photochemically active reaction centre pigments for most of the extant photosynthetic organisms. During the synthesis of both Chl and Bchl, reduction of the tetrapyrrole ring system converts protochlorophyllide (Pchlide), into a chlorin. A second reduction that is unique to the synthesis of Bchl converts the chlorin into a bacteriochlorin. There are two mechanisms for reducing the double bond in the fourth ring of protochlorophyllide. One enzyme complex functions irrespective of the presence or absence of light and is thus termed "light-independent protochlorophyllide reductase". The second is a light-dependent reaction that utilizes the enzyme NADPH-protochlorophyllide oxidoreductase (Suzuki et al. 1997). In *Rhodobacter capsulatus*, the products of three genes are required for each reduction: *bchL*, *bchN*, and *bchB* for the Pchlide reductase and *bchX*, *bchY*, and *bchZ* for the chlorin reductase (Burke et al. 1993b). Both enzymes are three-subunit complexes. Burke et al. (1993a, b) detected a significant degree of sequence similarity between BchlL, BchN, BchB, and BchX, BchY and BchZ, respectively, suggesting that the six genes represent two triads of paralogs and that the two enzymes are derived from a common three-subunit ancestral reductase.

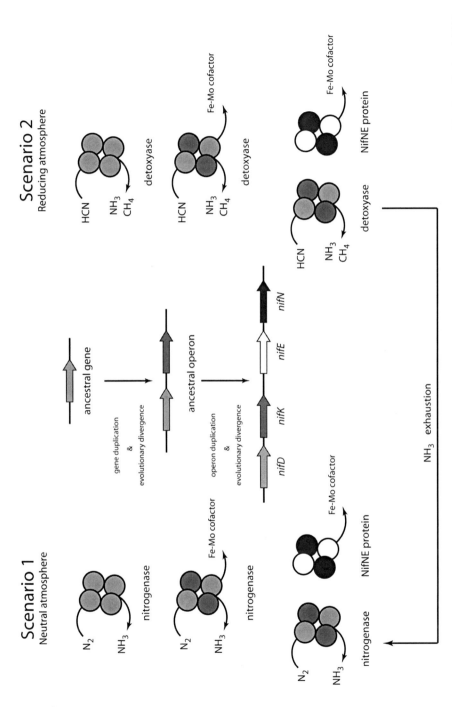

Fig. 2.8 Two possible scenarios depicted for the original function performed by the *nifDKEN* genes and their ancestor(s) gene(s) (modified from Fani et al. 2000)

It was also found that the so-called "chlorophyll iron protein" subunits encoded by *bchX*, *bchL*, and *chlL* shared a remarkable sequence similarity with the nitrogenase Fe proteins (Burke et al. 1993a). Burke et al. (1993a) suggested that genes involved in bacteriochlorophyll biosynthesis and nitrogen fixation were related mechanistically, structurally and evolutionarily. Similarly to NifH protein, which serves as the unique electron donor for the nitrogenase complex, the products of *bchL* and *bchX* could serve as the unique electron donor into their respective catalytic subunits (BchB-BchN and BchY-BchZ). The idea of a common ancestry of *nifH*, *bchL* and *chlL* genes (Burke et al. 1993b; Fujita et al. 1993) has had an elegant experimental support by Cheng et al. (2005) who demonstrated in the photosynthetic eukaryote *Chlamydomonas reinhardtii* that NifH is able to partially complement the function of ChlL in the dark-dependent chlorophyll biosynthesis pathway.

Nitrogenases and carboxylases might have represented bacterial preadaptations, multigenic traits that were retained because of new selective advantages in altered environments. As abiotically produced organic matter became depleted, competition for the organic prerequisites for reproduction ensued. As the carboxylation and nitrogen-fixing functions were achieved, a new, abundant, and direct source of carbon and nitrogen for organic synthesis became available-the atmosphere. The ability to take up atmospheric carbon and nitrogen would be of great selective advantage (Margulis 1993). It is possible to propose a model (Fig. 2.9) for the origin and evolution of nitrogen fixation and bacterial photosynthesis based on multiple and

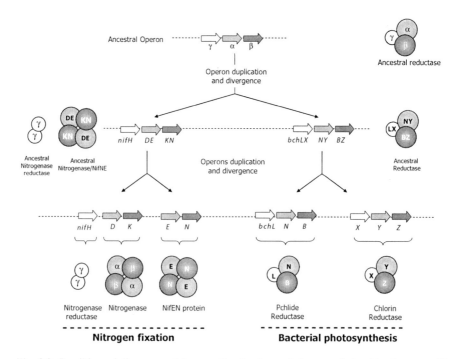

Fig. 2.9 Possible evolutionary model accounting for the evolutionary relationships between *nif* and *bch* genes

successive paralogous duplications of an ancestral operon encoding an ancient reductase. The eight genes (*nifDKEN* and *bchYZNB*) are members of the same paralogous gene family, in that that all of them are the descendant of a single ancestral gene. The model proposed posits the existence of an ancestral three-cistronic operon (Fig. 2.9) coding for an unspecific reductase. One might assume that this complex was (eventually) able to perform both carboxylation and nitrogen fixation. The following evolutionary steps might have been the duplication of the ancestral operon followed by an evolutionary divergence that led to the appearance of the ancestor of *nifH*, *nifDE*, and *nifKN* on one side, and *bchLX*, *bchNY* and *bchBZ* on the other one. In this way the two reductases narrowed their substrate specificity with one of them channelled toward nitrogen fixation and the other one toward photosynthesis.

However, each of the two multicomplex proteins was able to perform at least two different reactions:

1. The ancestor of *nifDKEN*, was likely able to carry out the reduction of dinitrogen to ammonia and the synthesis of Fe-Mo cofactor (Fani et al. 2000)
2. The ancestor of protochlorophyllide reductase and chlorin reductase performed both of the reactions that in the extant photosynthetic bacteria are carried out by two triads (BchN and BchLX, respectively)

The complete diversification of the function of the two heteromeric complexes was likely achieved thorough duplication of *nifDE nifKN* ancestors and by the duplication of the three-cistronic operon *bch(LX)(NY)(NZ)* followed by evolutionary divergence (Fig. 2.9). In our opinion, this idea may perfectly fit the Jensen's hypothesis.

Concerning the timing of the above reported evolutionary events (Fani et al. 2000) the two paralogous duplication events leading to *nifDK* and *nifEN* likely predated the appearance of the LUCA. Conversely, other authors (Raymond et al. 2004) have proposed a different scenario, according to which nitrogen fixation *per se* was invented by methanogenic Archaea and subsequently transferred, in at least three separate events, into bacterial lineages.

Differently from nitrogen fixation, tetrapyrrole-based photosynthesis occurs only in bacteria and bacterially derived chloroplasts, therefore it can be surmised that the appearance of photosynthesis should have not predated the divergence of *Archaea* and *Bacteria*.

Which were the Molecular Mechanisms Involved in the Spreading of Nitrogen Fixation?

The phylogenetic analysis performed using a concatenation of NifHDKEN proteins (Fig. 2.10) may help to shed light on the main evolutionary steps leading to the extant distribution of nitrogen fixation in prokaryotes.

As shown in Fig. 2.10, a group of bacteria (including representatives from green sulphur bacteria (GSB) δ-*Proteobacteria* and *Chloroflexi*) are strongly supported as sister groups of a cluster embedding *Methanosarcina* (*Euryarchaea*). Similarly, some *Firmicutes* (mainly *Clostridium* species) cluster as a sister clade with the euryarchaeote *Methanoregula boonei*. Their position in the phylogenetic tree

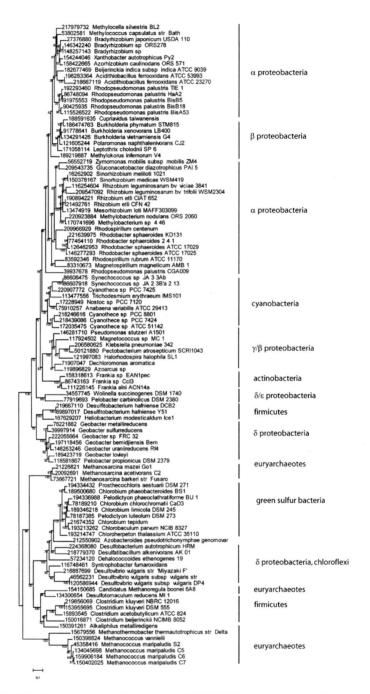

Fig. 2.10 ML phylogenetic tree of concatenated NifHDKEN sequences from 105 representative microorganisms

suggests that these bacteria might have acquired nitrogen fixation via HGT from an archaeon. It is worth of noticing that all the microorganisms embedded in this clade are frequently found among syntophic consortia in anaerobic environment, providing a viable environment for gene sharing (Garcia et al. 2000).

All the other bacterial sequences are embedded in a single monophyletic group. Interestingly, the sequences from *Cyanobacteria*, *Firmicutes* and *Actinobacteria* form three monophyletic clades that emerge as sister groups of α-, δ- and γ-*Proteobacteria*, respectively. The monophyly of the three groups that are surrounded by proteobacteria, points toward a later acquisition of nitrogen fixation in these bacteria from a proteobacterium; hence, HGT appears to have played a key role in spreading nitrogen fixation within the different bacterial lineages. The phylogenetic analyses also suggested that the ancestor of extant proteobacteria was a diazotroph.

An evolutionary model for origin and spreading of nitrogen fixation is shown in Fig. 2.11. The available data do not permit to discern whether LUCA was a diazotroph or not. If we assume that LUCA already possessed the set of genes necessary for nitrogen fixation (the LCA hypothesis, Fig. 2.11a) then gene loss should have played a major role in the evolution of nitrogen fixation pathway. Conversely, if we assume that nitrogen fixation was not present in LUCA but was later "invented" by methanogenic Archaea (Raymond et al. 2004), extensive HGT must be invoked to account for the distribution and the phylogeny that we observe in present-day prokaryotes (Fig 2.11b).

Finally, phylogenetic data suggest that, once appeared in bacteria, *nif* genes flowed through the ancestral prokaryotic communities by vertical inheritance and HGT events.

Conclusions

Metabolic pathways of the earliest heterotrophic organisms arose during the exhaustion of the prebiotic compounds present in the primordial soup and it is likely that the first biosynthetic pathways were partially or wholly non-enzymatic.

In the course of molecular and cellular evolution different mechanisms and different forces might have concurred in the arisal of new metabolic abilities and shaping of metabolic routes.

Several data confirm that duplication of DNA regions represents a major force of gene and genome evolution. The evidence for gene elongation, gene duplication and operon duplication events suggests in fact that the ancestral forms of life might have expanded their coding abilities and their genomes by "simply" duplicating a small number of mini-genes (the starter types) via a cascade of duplication events, involving DNA sequences of different size.

In addition to this, gene fusion also played an important role in the construction and assembly of chimeric genes.

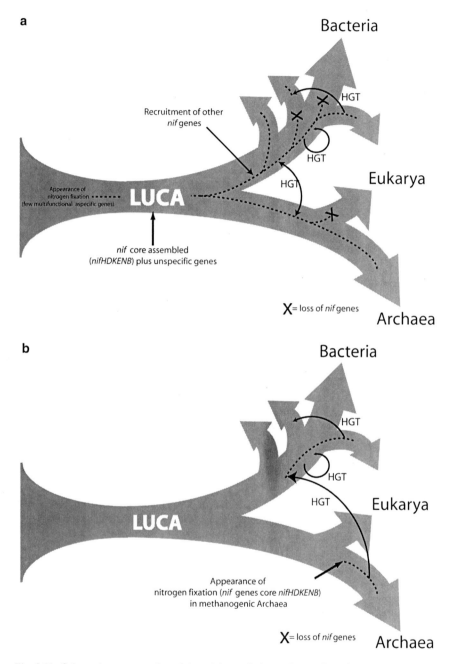

Fig. 2.11 Schematic representation of the origin, evolution and spreading of *nif* genes in Bacteria and Archaea assuming (**a**) the presence of a core of *nif* gene in LUCA or (**b**) the appearance of Nitrogen Fixation in methanogenic Archaea

The dissemination of metabolic routes between micro-organisms might be facilitated by horizontal transfer events. The horizontal transfer of entire metabolic pathways or part thereof might have had a special role during the early stages of cellular evolution when, according to Woese (1998), the "genetic temperature" was high.

Many different schemes can be proposed for the emergence and evolution of metabolic pathways depending on the available prebiotic compounds and the available enzymes previously evolved. Even though most of data coming from the analysis of completely sequenced genomes and directed-evolution experiments strongly support the patchwork hypothesis, we do not think that all the metabolic pathways arose in the same manner. In our opinion the different schemes might not be mutually exclusive. Thus, some of the earliest pathways may have arisen from the Horowitz scheme, some from the semi-enzymatic proposal, and later ones from Jensen's enzyme recruitment. However, other ancient pathways, such nitrogen fixation might be assembled using (at least) two different schemes (Horowitz and Jensen). Recent data speak toward a pivotal role played by HGT events in the evolution and the spreading on nitrogen fixation genes within the microbial world.

The investigation of the origin of life and early molecular evolution will help the understanding of the interactive dynamics between geochemical cycles and expansion of the variety of the life forms. We are confident that this research direction will be actively pursued in the future by researchers in both life and earth sciences.

References

Alawi M, Lipski A, Sanders T, Pfeiffer E-M, Spieck E (2007) Cultivation of a novel cold-adapted nitrite oxidizing betaproteobacterium from the Siberian Arctic. ISME J 1:256–264

Arp DJ, Chain PSG, Klotz MG (2007) The impact of genome analyses on our understanding of ammonia-oxidizing bacteria. Ann Rev Microbiol 61:21–58

Bada J, Lazcano A (2002) Origin of life. Some like it hot, but not the first biomolecules. Science 296:1982–1983

Bada J, Lazcano A (2003) Perceptions of science. Prebiotic soup-revisiting the Miller experiment. Science 300:745–746

Bekker A, Holland HD, Wang PL, Rumble D, Stein HJ, Hannah JL, Coetzee LL, Beukes NJ (2004) Dating the rise of atmospheric oxygen. Nature 427:117–120

Brilli M, Fani R (2004) The origin and evolution of eucaryal HIS7 genes: from metabolon to bifunctional proteins? Gene 339:149–160

Brown JR (2003) Ancient horizontal gene transfer. Nat Rev Genet 4:121–132

Burke D, Alberti M, Hearst J (1993a) The *Rhodobacter capsulatus* chlorin reductase-encoding locus, *bchA*, consists of three genes, *bchX*, *bchY*, and *bchZ*. J Bacteriol 175:2407–2413

Burke D, Hearst J, Sidow A (1993b) Early evolution of photosynthesis: clues from nitrogenase and chlorophyll iron proteins. Proc Natl Acad Sci USA 90:7134–7138

Canfield DE (2005) The early history of atmospheric oxygen: homage to Robert M. Garrels. Annu Rev Earth Planet Sci 33:1–36

Canfield DE (1998) A new model for Proterozoic ocean chemistry. Nature 396:450–453

Canfield DE, Rosing MT, Bjerrum C (2006) Early anaerobic metabolisms. Philos Trans R Soc Lond B Biol Sci 361:1819–1836

Capone DG, Knapp AN (2007) Oceanography: a marine nitrogen cycle fix? Nature 445:159–160

Capone DG, Popa R, Flood B, Nealson KH (2006) Geochemistry: follow the nitrogen. Science 312:708–709

Cheng Q, Day A, Dowson-Day M, Shen G-F, Dixon R (2005) The *Klebsiella pneumoniae* nitrogenase Fe protein gene (*nifH*) functionally substitutes for the *chlL* gene in *Chlamydomonas reinhardtii*. Biochem Biophys Res Commun 329:966–975

Copley SD (2000) Evolution of a metabolic pathway for degradation of a toxic xenobiotic: the patchwork approach. Trends Biochem Sci 25:261–265

Dagan T, Martin W (2007) Ancestral genome sizes specify the minimum rate of lateral gene transfer during prokaryote evolution. Proc Natl Acad Sci USA 104:870–875

Dagan T, Martin W (2006) The tree of one percent. Genome Biol 7:118

Deutsch C, Sarmiento JL, Sigman DM, Gruber N, Dunne JP (2007) Spatial coupling of nitrogen inputs and losses in the ocean. Nature 445:163–167

Dixon R, Kahn D (2004) Genetic regulation of biological nitrogen fixation. Nat Rev Microbiol 2:621–631

Falkowski PG (1997) Evolution on the nitrogen cycle and its influence on the biological sequestration of CO_2 in the ocean. Nature 387:272–275

Falkowski PG, Fenchel T, Delong EF (2008) The microbial engines that drive Earth's biogeochemical cycles. Science 320:1034–1039

Fani R, Fondi M (2009) Origin and evolution of metabolic pathways. Phys Life Rev 6:23–52

Fani R, Gallo R, Liò P (2000) Molecular evolution of nitrogen fixation: the evolutionary history of the *nifD*, *nifK*, *nifE*, and *nifN* genes. J Mol Evol 51:1–11

Fay P (1992) Oxygen relations of nitrogen fixation in cyanobacteria. Microbiol Rev 56:340–373

Ferry J, House C (2006) The stepwise evolution of early life driven by energy conservation. Mol Biol Evol 23:1286–1292

Fondi M, Emiliani G, Fani R (2009) Origin and evolution of operons and metabolic pathways. Res in Mic 160:502–512

Frost LS, Leplae R, Summers AO, Toussaint A (2005) Mobile genetic elements: the agents of open source evolution. Nat Rev Microbiol 3:722–732

Fujita Y, Matsumoto H, Takahashi Y, Matsubara H (1993) Identification of a *nifDK*-like gene (ORF467) involved in the biosynthesis of chlorophyll in the cyanobacterium *Plectonema boryanum*. Plant Cell Physiol 34:305–314

Galloway JN, Townsend AR, Erisman JW, Bekunda M, Cai Z, Freney JR, Martinelli LA, Seitzinger SP, Sutton MA (2008) Transformation of the nitrogen cycle: recent trends, questions, and potential solutions. Science 320:889–892

Garcia JL, Patel BKC, Ollivier B (2000) Taxonomic phylogenetic and ecological diversity of methanogenic Archaea. Anaerobe 6:205–226

Gogarten JP, Doolittle WF, Lawrence JG (2002) Prokaryotic evolution in light of gene transfer. Mol Biol Evol 19:2226–2238

Gogarten JP, Townsend JP (2005) Horizontal gene transfer, genome innovation and evolution. Nat Rev Microbiol 3:679–687

Gribaldo S, Brochier-Armanet C (2006) The origin and evolution of Archaea: a state of the art. Philos Trans R Soc Lond B Biol Sci 361:1007–1022

Gruber N, Galloway JN (2008) An Earth-system perspective of the global nitrogen cycle. Nature 451:293–296

He X, Zhang J (2006) Transcriptional reprogramming and backup between duplicate genes: is it a genome wide phenomenon? Genetics 172:1363–1367

Horowitz NH (1945) On the evolution of biochemical syntheses. Proc Natl Acad Sci USA 31:153–157

Houlton BZ, Wang YP, Vitousek PM, Field CB (2008) A unifying framework for dinitrogen fixation in the terrestrial biosphere. Nature 454:327–330

Huang J, Gogarten JP (2006) Ancient horizontal gene transfer can benefit phylogenetic reconstruction. Trends Genet 22:361–366

Itoh T, Takemoto K, Mori H, Gojobori T (1999) Evolutionary instability of operon structures disclosed by sequence comparisons of complete microbial genomes. Mol Biol Evol 16:332–346

Jensen RA (1976) Enzyme recruitment in evolution of new function. Annu Rev Microbiol 30:409–425

Jetten MS (2008) The microbial nitrogen cycle. Environ Microbiol 10:2903–2909

Kasting JF, Siefert JL (2001) Biogeochemistry: the nitrogen fix. Nature 412:26–27

Kasting JF, Siefert JL (2002) Life and the evolution of Earth's atmosphere. Science 296:1066–1068

Klotz MG, Stein LY (2008) Nitrifier genomics and evolution of the nitrogen cycle. FEMS Microbiol Lett 278:146–156

Koonin EV (2003) Comparative genomics, minimal gene-sets and the last universal common ancestor. Nat Rev Microbiol 1:127–136

Koonin EV, Makarova KS, Aravind L (2001) Horizontal gene transfer in prokaryotes: quantification and classification. Annu Rev Microbiol 55:709–742

Lazcano A, Miller S (1994) How long did it take for life to begin and evolve to cyanobacteria? J Mol Evol 39:546–554

Lazcano A, Miller SL (1996) The origin and early evolution of life: prebiotic chemistry, the pre-RNA world, and time. Cell 85:793–798

Li WH, Graur D (1991) Fundamentals of molecular evolution. Sinauer Associates, Sunderland, Massachussets

Liò P, Brilli M, Fani R (2007) Phylogenetics and computational biology of multigene families molecules, networks, populations. In: Bastolla U, Porto M, Roman HE, Vendruscolo M (eds) Structural approaches to sequence evolution. Springer, Berlin/Heidelberg, pp 191–205

Lynch M, Force A (2000) The probability of duplicate gene preservation by subfunctionalization. Genetics 154:459–473

Mancinelli RL, McKay CP (1988) The evolution of nitrogen cycling. Orig Life Evol Biosph 18:311–325

Margulis L (1993) Symbiosis in cell evolution: microbial communities in the archean and proterozoic eons. WH Freeman and Company, New York

Martin W (1999) Mosaic bacterial chromosomes: a challenge en route to a tree of genomes. Bioessays 21:99–104

McLain JET, Martens DA (2005) Nitrous oxide flux from soil amino acid mineralization. Soil Biol Biochem 37:289–299

Mortlock R, Gallo MA (1992) Experiments in the evolution of catabolic pathways using modern bacteria. In: Mortlock R, Gallo MA (eds) The evolution of metabolic functions. CRC Press, Boca Raton, FL, pp 1–13

Mushegian AR, Koonin EV (1996) Gene order is not conserved in bacterial evolution. Trends Genet 12:289–290

Navarro-Gonzalez R, McKay CP, Mvondo DN (2001) A possible nitrogen crisis for Archaean life due to reduced nitrogen fixation by lightning. Nature 412:61–64

Ochman H, Lerat E, Daubin V (2005) Examining bacterial species under the specter of gene transfer and exchange. Proc Natl Acad Sci USA 102:6595–6599

Ohno S (1972) Simplicity of mammalian regulatory systems. Dev Biol 27:131–136

Ohta T (2000) Evolution of gene families. Gene 259:45–52

Ouzounis CA, Kunin V, Darzentas N, Goldovsky L (2006) A minimal estimate for the gene content of the last universal common ancestor-exobiology from a terrestrial perspective. Res Microbiol 157:57–68

Peretó J, Fani R, Leguina J, Lazcano A (1997) Enzyme evolution and the development of metabolic pathways. In: Cornish-Bowden A (ed) New beer in an old bottle: Eduard Buchner and the growth of biochemical knowledge. Universitat de Valencia, Valencia, pp 173–198

Raymond J (2005) The evolution of biological carbon and nitrogen cycling -a genomic perspective. Rev Mineral Geochem 59:211–231

Raymond J, Siefert JL, Staples CR, Blankenship RE (2004) The natural history of nitrogen fixation. Mol Biol Evol 21:541–554

Schwartz A (2006) Phosphorus in prebiotic chemistry. Philos Trans R Soc Lond Ser B 361:1743–1749

Shi T, Falkowski P (2008) Genome evolutuon in cyanobacteria: the stable core and the variable shell. Proc Natl Acad Sci USA 107:2510–2515

Silver VS, Postgate JR (1973) Evolution of asymbiotic nitrogen fixation. J Theor Biol 56:340–373

Simon J (2002) Enzymology and bioenergetics of respiratory nitrite ammonification. FEMS Microbiol Rev 26:285–309

Srere PA (1987) Complexes of sequential metabolic enzymes. Annu Rev Biochem 56:89–124

Strous M, Pelletier E, Mangenot S et al (2006) Deciphering the evolution and metabolism of an anammox bacterium from a community genome. Nature 440:790–794

Suzuki J, Bollivar D, Bauer C (1997) Genetic analysis of chlorophyll biosynthesis. Annu Rev Genet 31:61–89

Wächtershauser G (1988) Before enzymes and templates: theory of surface metabolism. Microbiol Rev 52:452–484

Wächtershauser G (2006) From volcanic origins of chemoautotrophic life to bacteria, archaea and eukarya. Philos Trans R Soc Lond Ser B 361:1787–1806

Wächtershauser G (2007) On the chemistry and evolution of the pioneer organism. Chem Biodivers 4:584–602

Woese C (2000) Interpreting the universal phylogenetic tree. Proc Natl Acad Sci USA 97:8392–8396

Woese C (2002) On the evolution of cells. Proc Natl Acad Sci USA 99:8742–8747

Woese C (1998) The universal ancestor. Proc Natl Acad Sci USA 95:6854–6859

Xie G, Keyhani NO, Bonner CA, Jensen RA (2003) Ancient origin of the tryptophan operon and the dynamics of evolutionary change. Microbiol Mol Biol Rev 67:303–342

Xiong J, Fischer WM, Inoue K, Nakahara M, Bauer CE (2000) Molecular evidence for the early evolution of photosynthesis. Science 289:1724–1730

Ycas M (1974) On earlier states of the biochemical system. J Theor Biol 44:145–160

Zillig W, Palm P, Klenk HP (1992) A model for the early evolution of organisms: the arisal of the three domains of life from a common ancestor. In: Hartman H, Matsuno K (eds) The origin and evolution of the cell. World Scientific, Singapore, pp 163–182

Chapter 3
Novel Cultivation Strategies for Environmentally Important Microorganisms

Jörg Overmann

The Significance of Culture-Based Approaches

Prokaryotes dominate the biosphere (Whitman et al. 1998). Although bacteria were already discovered three centuries ago by Antonie van Leeuwenhoek (1632–1723), they continue to represent the least understood group of organisms since analyses of their diversity and biogeochemical significance are impeded by the persisting difficulties of culturing relevant bacteria from environmental samples.

Many of the techniques currently employed for the isolation of bacteria date back to the nineteenth century. After microscopic techniques enabled the direct enumeration of bacteria after the 1930s, it became apparent that only 0.001–1% of all bacterial cells in natural samples typically grow on standard cultivation media. Yet, up to 91% of the cells were demonstrated to be metabolically active in situ (Ouverney and Fuhrman 1999). For more than half a century this so-called "great plate count anomaly" (Staley and Konopka 1985) continued to remain a major obstacle for the study of environmentally relevant bacteria and led to a shift towards novel culture-independent molecular approaches for the study of the diversity and ecology of microorganisms.

Broad-scale analyses of bacterial 16S rRNA gene sequences uncovered a vast diversity of different types of microorganisms in the environment. Today, the estimated number of bacterial species ranges from one million to one billion (Gans et al. 2005), whereas only 8,400 species (as of May 2010) have been validly described (Deutsche Sammlung von Mikororganismen und Zellkulturen; http://www.dsmz.de/). These studies also revealed that 16S rRNA gene sequences of strains isolated from natural bacterial communities frequently do not match the

J. Overmann (✉)
Mikrobielle Ökologie und Diversitätsforschung, Leibniz-Institut DSMZ-Deutsche Sammlung von Mikroorganismen und Zellkulturen, Inhoffenstraße 7B, 38124 Braunschweig, Germany
email: joerg.overmann@dsmz.de

L.L. Barton et al. (eds.), *Geomicrobiology: Molecular and Environmental Perspective*,
DOI 10.1007/978-90-481-9204-5_3, © Springer Science+Business Media B.V. 2010

numerically dominant sequence types. Meanwhile, the number of bacterial phyla consisting entirely of not-yet-cultured sequence types outnumber those phyla which encompass the known bacteria. Obviously, many of the relevant bacteria so far have escaped isolation and characterization. Evidently, numerous novel capacities with relevance to biogeochemical cycles, biotechnology and medicine are awaiting their discovery.

Parallel to the rapid advances of culture independent molecular techniques, a considerable number of novel cultivation approaches have been established over the past two decades. As a result, the rate of discovery of novel bacterial species has increased steadily over the past 10 years and a growing number of environmentally significant bacteria could be enriched or even isolated in pure culture, like novel *Acidobacteria*, the marine mesophilic Crenarchaeon *Nitrosopumilus marinus*, or symbionts of phototrophic consortia, to name just a few. State-of-the-art molecular and culture-based approaches are now seen as complementary and of equal significance for the understanding of the ecology of microbial communities. A combination of the two offers the most promising way for the retrieval of key microbial players and the study of their specific adaptations, regulation mechanisms and evolution in the environment.

The successful enrichment, isolation and cultivation of prokaryotes critically depends on an educated choice of appropriate inoculum, growth media and incubation conditions. Existing methods for cultivation are introduced and discussed in the following paragraphs. Without doubt, however, another important prerequisite for cultivation success will always be the persistence of the individual researcher.

Basic Requirements of the Bacterial Cell

For a rational design of cultivation-based studies of bacterial communities, the nutritional and energetic needs of the particular target bacteria have to be considered. Essential macroelements and trace elements must be included in the media and an appropriate amount of energy-yielding substrates or, in the case of phototrophic bacteria, light of suitable wavelengths must be provided.

Macroelements (listed in Overmann 2002) are constituents of structural components and enzymes, or represent major ionic constituents of the cytoplasma and therefore are required at relatively high concentrations. The molar ratio of macroelements necessary for a balanced growth can be calculated from the average composition of the bacterial cell

$$C_4H_{6.4}O_{1.5}NP_{0.09}S_{0.024}.$$

By comparing the ratio of macroelements in a medium recipe with that in bacterial biomass, it can be determined which of the macroelements will limit the growth of the cells in a culture.

Essential trace elements (Overmann 2002) usually are added at concentrations between 0.1 and 1 μM since they are only minor constituents of the biomass but exert toxic effects at higher concentrations. While Mn, Co, Cu, Mo, Zn, Ni and V are routinely included in cultivation media, strict anaerobes often require selenium and tungsten. Lately, the requirement for boron has been recognized as being rather widespread, since many bacteria employ the AI-2 autoinducer, a cyclic borate diester.

In addition, certain bacteria depend on specific organic compounds for growth. These growth factors may be amino acids, purins/pyrimidins, or vitamins and are supplied at small concentrations of 0.1–1 μM. Fewer bacteria require porphyrins, short branched or straight chain fatty acids (Balch and Wolfe 1976), cholesterol, betaine or polyamine. If the growth factor requirements of a target bacterium are not known, complex substrates like digested yeast extract or fermented manure extract can be added instead of a defined mixture of a large number of potentially essential compounds (Widdel 1983). Yeast extract is often employed as a convenient source of most vitamins and also amino acids. Freshly prepared yeast autolysate has been found to be superior to commercially available dried preparations (Leadbetter et al. 1999; Vogl et al. 2006). Sterilized rumen extract contains a number of volatile, particularly short branched, fatty acids, but also vitamins and hemine. However, yeast extract or casamino acids inhibit the growth of many photolithoautotrophic and chemolithoautotrophic bacteria even at low concentrations of 0.01% (w/v) (Overmann and Pfennig 1989). Sludge supernatant is another complex source of supplines but has been applied less frequently. The polyol Tween 80 (polyoxyethylenesorbitan monooleate) is used as a water soluble source of the long chain fatty acid oleate (*cis*-9-octadecenoic acid) and also contains linoleate, palmitate and stearate as minor compounds. Finally, cold soil extract has been observed to stimulate growth of certain soil bacteria and humic substances are utilized as electron-donating substrates by various *Proteobacteria*. The quinone groups of humic substances, or their quinone analogs (anthroquinone-2,6-disulfonate), are redox active compounds and hence can serve as electron shuttles between the cells of dissimilatory iron reducing bacteria and their insoluble electron acceptors like $Fe(III)$ oxides. Organic ligands are used for the mobilization of $Fe(III)$. Many of these supplements are heat-labile and are added to the basal medium after sterilization by filtration to avoid deterioration.

The macroelements N and S occur at various redox states, but many bacteria can assimilate only particular compounds. A number of obligately anaerobic bacteria, like about half of the known strains of anoxygenic phototrophic bacteria or the methanogenic archaea, are not capable of assimilatory sulfate reduction and thus require reduced sulfur compounds (mostly sulfide, in some cases alternatively cysteine) to satisfy their requirements for growth. Similarly, bacteria lacking the enzymes for assimilatory nitrate reduction need to be cultivated with ammonia or an organic nitrogen source.

Bacterial growth often results in the production or consumption of acidic compounds, in particular during fermentative growth. At low pH, weak organic acids (like benzoic acid, propionic acid and sorbic acid) influence the cytoplasmic pH in a direct manner since they are lipophilic in their undissociated form and rapidly diffuse through cell membranes. This results in a transport of hydrogen ions along the transmembrane gradient and a decrease of the intracellular pH. In high nutrient complex media, the various acidic and basic groups of the organic constituents provide sufficient buffering capacity to maintain the desired external pH. However, in many mineral media and low nutrient strength complex media, it is mandatory to maintain a suitable pH value during the growth of the cells by adding suitable buffers. To this end, K- or Na-phosphate are frequently employed at concentrations of 10–30 mM. Yet, such high concentrations of inorganic phosphate have been found to inhibit the growth of freshwater aquatic bacteria (Bartscht et al. 1999; Bast 2001). HEPES (4-(2-hydroxyethyl)-1-piperazineethanesulfonic acid) has been documented to be superior for the cultivation of fastidious bacteria from natural samples. A second disadvantage of phosphate is its precipitation with Ca^{2+}, Mg^{2+} and Fe^{3+} ions even at concentrations <10 mM phosphate. If phosphate is used as buffer, the Ca^{2+} concentration therefore has to be decreased to 200 μM which still permits a sufficiently high cell yield for most bacterial strains.

Principles of the Selective Enrichment

In selective enrichments, culture conditions are tailored to favor microbes with a particular metabolic activity. The target bacteria grow faster than accompanying bacteria and reach a numerical dominance even if the original cell numbers were very low in the original natural sample.

In each novel enrichment trial, the first task is to determine whether the energetics of the envisioned metabolism is favorable for bacterial growth, i.e. whether the free energy change is negative. It is generally accepted that an organism can exploit a reaction for growth, if the Gibbs free energy is ≥ 20 kJ (mol substrate)$^{-1}$ which is sufficient for the translocation of one proton across the cytoplasmic membrane (Schink 1991). Most of the ATP-yielding metabolic pathways involve redox processes such that the standard free energy change can be calculated from the difference between the standard redox potential of the electron donor (E_0'(don)) and that of the electron acceptor (E_0'(acc)) according to:

$$\Delta G_0' = -n \cdot F \cdot \Delta E_0' = -n \cdot F \cdot [E_0'(acc) - E_0'(don)] \tag{3.1}$$

where n is the number of electrons transferred and F is the Faraday constant (96.5 kJ·V^{-1}). For all types of reactions, the standard free energy change can be calculated from the tabulated free energy of formation (G_f^0) of the products and reactants (Thauer et al. 1977):

$$\Delta G_0' = \sum G_f^0(products) - \sum G_f^0(reactants) \tag{3.2}$$

However, since biogeochemical transformations in nature rarely proceed under standard conditions, the feasibility of a potential reaction (hence the free energy change $\Delta G'$) must be assessed based on the actual (natural) concentrations of reactants A and B and the products C and D according to

$$aA + bB \leftrightarrow cC + dD \tag{3.3}$$

$$\Delta G' = \Delta G_0' + RT \ln \frac{[C]^c \cdot [D]^d}{[A]^a \cdot [B]^b} \tag{3.4}$$

where R is the universal gas constant (8.314 J mol^{-1} K^{-1}) and T the temperature in K. This means that reactions which are endergonic under standard conditions can actually provide energy for growth in the natural habitat. A well documented example is the oxidation of short-chain fatty acids which involves the liberation of molecular hydrogen. Thus, if the hydrogen partial pressure is kept at very low values (<10 Pa), fatty acids like propionate or butyrate can be oxidized to acetate, CO_2 and H_2 despite a positive value of $\Delta G_0'$. This represents one prominent example of interspecies interactions between different types of bacteria which needs to be considered for their cultivation (see below).

When novel substrate combinations were selected based on the thermodynamic considerations outlined above, previously unknown types of bacteria could be cultivated and isolated. These include bacteria oxidizing Fe(II) with oxygen at circumneutral pH, bacteria reducing arsenate or oxidizing arsenite, or even strains exploiting phosphite as electron donors to reduce sulfate (Schink and Friedrich 2000). The continuing success in the retrieval of novel metabolic types suggests that many ecological niches of microorganisms so far have been overlooked.

However, numerous not-yet-cultured bacteria seem to utilize the same substrates as known bacteria, rather than depending on novel exotic compounds. Aquatic microbial communities in particular are dominated by oligotrophic bacteria which are adapted to low substrate concentrations and reach only low maximum growth rates. In contrast, copiotrophic bacteria exhibit a low affinity towards the limiting growth substrate but attain high maximum specific growth rates at substrate saturation. Consequently, copiotrophs overgrow the oligotrophs rapidly when high substrate concentrations are used for enrichment.

Media with significantly lowered substrate concentrations have been found to provide a limited selective advantage for oligotrophic forms. The cultivation success can be increased further and significantly by physical separation of target bacteria from the competing cells. A well established protocol is the separation by Percoll gradient centrifugation. Percoll is superior to other agents like sucrose in that it is osmotically inert and therefore exerts only marginal effect on the viability of the bacterial cells. However, density gradient centrifugation can only be used for the separation of bacterial cells with sufficiently different buoyant densities. If target bacteria dominate the natural microbial community, the dilution-to-extinction method provides a convenient means for separation. To this end, serial 1:10 or 1:5 dilutions are prepared in test tubes or microtiter dishes. The method requires a sufficient number of dilution steps such that at least the last dilution statistically receives a single bacterial cell. The chance to obtain a pure culture or at least a culture consisting of very few types of bacteria can be increased by preparing several dilution series in parallel.

It has to be kept in mind, however, that the dilution to extinction methods not only relies on a dominance of the target bacteria in the original sample but also requires that a high fraction of these cells are indeed capable of growing in the laboratory medium. Even in improved cultivation media, however, the fraction of dividing cells of certain target species is very low. Accordingly, high-throughput cultivation methods have been introduced which permit the generation of hundreds or thousands of cultures, each inoculated with one or few bacterial cells (Bruns et al. 2003a; Connon and Giovannoni 2002) (Fig. 3.1a). Some of these methods require only short periods of time (minutes) for the inoculation of hundreds of cultures in microtiter plates. While the detection of growth can be automated using microtiter plate readers, screening such a high number of cultures for particular target bacteria is a demanding task and can be accomplished by PCR amplification with specific primers, PCR amplification combined with dot blot hybridization using specific oligonucleotide probes, or by fluorescent in situ hybridization (FISH).

Improved Classical and Advanced Cultivation Methods

Preconditions for a successful enrichment and growth of target bacteria are (1) to obtain sufficient information on the potential ecological niche, particularly for stenoec microorganisms (e.g., methylotrophs or psychrophiles), and (2) to control the growth of undesired competitors, either by specific inhibition or by exploiting the actual (rather than the potential) ecological niche of the target bacteria. An example for the latter case is the specific enrichment strategy applied for green sulfur bacteria which are capable of growing at higher light intensities but compete successfully with accompanying phototrophs like purple sulfur bacteria only at low light intensities.

Determining Potential Growth Substrates

Theoretically, enrichment trials could be performed by tedious testing of a large number of different substrate combinations at different concentrations. However, state-of-the-art molecular methods are now available which can provide essential information about the metabolism of target bacterial species in a much more rapid way.

Monitoring the uptake of radiolabeled substrates by microautoradiography can be combined with FISH, such that potential growth substrates can be determined in a culture-independent manner (Ouverney and Fuhrman 1999). The recently developed nano-scale secondary-ion mass spectroscopy (nanoSIMS) can be used to monitor the uptake of ^{13}C- or ^{15}N-labeled compounds at single-cell resolution (Musat et al. 2008). An alternative, non-microscopic approach is stable isotope probing which is based on the detection of the incorporation of ^{13}C-labeled

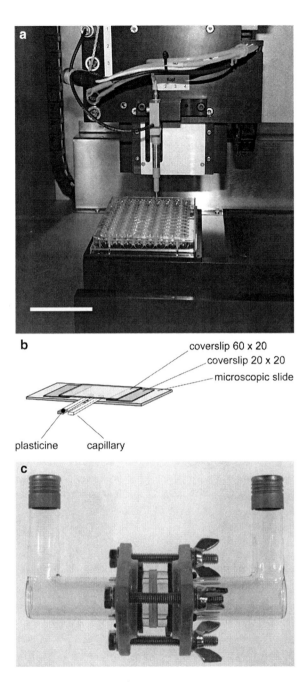

Fig. 3.1 (a) Photograph of the MicroDrop® AutoDrop microdispenser for automated inoculation of 170-pl aliquots of a bacterial suspension into a 96-well microtiter plate. In the center of the pipeting robot, the glass capillary (volume, 25 µl) surrounded by a piezo actuator can be seen. *Bar* denotes 5 cm. (b) Scheme of a laboratory-made microscopic chamber for chemotaxis assays. Two sides and the back of the chamber are sealed with paraffin/mineral oil (not shown). Flat glass capillaries are loaded with substrate solutions, sealed at one end with plasticine and inserted into the chamber through the opening in front. (c) All glass vessel for the cocultivation of two cell suspensions separated by a dialysis membrane. The latter (diameter, 25 mm) is fixed between the flanges of the two side arms

substrates into target 16S rRNA gene sequences (Radajewski et al. 2000). A third technique to identify potential growth substrates is based on the incorporation of bromodeoxyuridine (BrdU) into genomic DNA by metabolically active cells which are subsequently detected by immunological methods (Pernthaler et al. 2002). An advantage of this latter method is that a nested testing scheme can be performed easily since mixtures of substrates can be added to the assay and compounds of activating mixtures can be tested individually in a subsequent step. Potential polymeric substrates can be tested by co-localization of key extracellular enzymes with particular types of microorganisms using the enzyme labeled fluorescence (ELF) technique (Nedoma and Vrba 2006).

A principally different, yet also culture-independent method is the chemotaxis assay which can be applied if the target bacteria are motile (Overmann 2005). Flat rectangular glass capillaries with a length of 50 mm, an inside diameter of 0.1×1.0 mm and a capacity of 5 μl (Vitrocom, Mountain Lakes, N.J., USA) are filled with sterile filtered water from the natural environment, to which test substrates have been added. These capillaries are inserted into chemotaxis chambers consisting of microscope slides (Fig. 3.1b) or bottles, or can even be exposed directly in the aquatic habitat of the target bacteria. After exposure of the chemotaxis assay, capillaries are recovered and examined directly by phase contrast microscopy. Bacterial cells from positive enrichments can be identified based on their 16S rRNA gene sequences. Since substrates mixtures as well as single substrates can be tested rapidly (i.e. within hours), the spectrum of potential substrates can be determined in a very efficient manner.

Finally, the membrane filter microcultivation technique has been used to rapidly determine the growth requirements of bacteria. For this purpose, bacterial cells are filtered onto a polycarbonate membrane and the membrane positioned on a soil slurry or another growth medium to be tested. After a time period of 7–10 days, the membrane is removed from the substrate and the bacteria capable of growing are identified as microcolonies by FISH using group-specific oligonucleotide probes (Ferrari et al. 2005).

Mimicking the Chemical Composition in the Natural Environment

In many instances, cultivation strategies must be modified to represent the environmental conditions of the target microorganisms more correctly and to account for their specific growth kinetics. Synthetic media have been developed which mimic the ionic composition of freshwater (Bartscht et al. 1999), seawater (Coolen and Overmann 2000), or that typically found in soil solution (Angle et al. 1991). These media represent an improvement as compared to other types of media employed previously. Special attention should be paid not only to the composition (see above), but also to the overall concentrations of organic carbon substrates.

Some fastidious bacteria like the *Planctomycetales* are long known to require low concentrations of organic carbon substrates (yeast extract, peptone or casamino acids) in the range of 0.25 g·l^{-1}. Growth experiments on solid media with diluted nutrient broth (80 mg·l^{-1}) resulted in significantly increased colony-forming units of *Alphaproteobacteria*, *Firmicutes*, *Acidobacteria*, *Actinobacteria* and *Verrucomicrobia* from soil samples (Janssen et al. 2002). For planktonic bacteria, the apparent cultivation success (the ratio of most probable numbers relative to the total cell count) can be increased significantly from 0.1% to values up to 20–60% by decreasing the organic carbon content further to ≤5 mg·l^{-1} or using just straight filtered autoclaved seawater (Button et al. 1993; Connon and Giovannoni 2002). In the pelagic and deep ocean, organic carbon concentrations is present at concentrations of 30–200 μM, corresponding to 0.36–2.4 mgC·l^{-1} (Jannasch et al. 1996). Consequently, not more than 1–10 mg·l^{-1} of bacto yeast extract (Difco) must be added to the respective cultivation media in order to simulate the natural organic carbon concentrations. These observations indicate that the majority of heterotrophic prokaryotic cells in different aquatic and soil environments are adapted to lower concentrations for growth as routinely applied. Oligotrophic bacteria are defined as those which, on first cultivation, develop on media containing 1–15 mg C l^{-1} but cannot grow above 350 mg organic carbon l^{-1} (Overmann 2002). Cell division of typical oligotrophs like the marine Alphaproteobacteria *Sphingomonas alaskensis* (Schut et al. 1993, 1997) and *Pelagibacter ubique* (Rappé et al. 2002), or the *Gammaproteobacteria* of the marine OMG group (Cho and Giovannoni 2004) is already inhibited at nutrient concentrations above 5–200 mg organic carbon·l^{-1} and some of these bacteria reach maximum growth rates at nutrient concentrations as low as 1 mg organic carbon l^{-1}.

The filtration-acclimatization method (Hahn et al. 2004) targets planktonic ultramicrobacteria which are inhibited by high nutrient concentrations only during initial cultivation, but can be gradually adapted to higher nutrient concentrations. In order to avoid that these bacteria are outcompeted by fast growing, larger bacteria, the latter are first removed by filtration through 0.2 μm pore size membrane filters, which reduces cell numbers by 90% of total cell numbers. During the subsequent acclimatization procedure, complex carbon substrates are added in increasing doses, until concentrations of organic carbon reach those typically employed in standard microbial media. This method has been used successfully for the isolation of typical freshwater *Actinobacteria*.

The cultivation success of planktonic bacteria can also be increased by reducing nitrogen and phosphorus concentrations to 50 μM N and 1.5 μM P (Eilers et al. 2001). The use of dilute growth media, in some instances nothing more than sterilized seawater amended with low amounts of phosphate (0.1 μM) and ammonium (1 μM), permits the growth of typical planktonic bacteria like the marine chemoheteroph *Pelagibacter ubique* or planktonic cyanobacteria from oligotrophic environments (Rappé et al. 2002). Evidently, phosphate cannot be used for buffering these low-nutrient concentration media and has to be replaced by non-ionic compounds like HEPES (see above). Reduction of phosphate concentrations to 5.2 nM has been very successful for the isolation of pelagic cyanobacteria like *Trichodesmium thiebautii* and *T. erythraeum* (Waterbury 1991).

Many chemoorganotrophic bacteria require carbon dioxide in small amounts (~6% of total cell carbon synthesized) for anaplerotic reactions, like the synthesis of oxaolacetate by phosphoenolpyruvate (PEP) carboxylase, PEP carboxykinase, PEP carboxytransphosphorylase, and by pyruvate carboxylase (Wood-Werkman reactions). Since CO_2 is produced during catabolism of organic compounds, it does not normally become a limiting nutritional factor. However, some pathogenic bacteria, such as *Neisseria* and *Brucella*, require elevated concentrations of carbon dioxide of up to 10% in the atmosphere for good growth in organic media. Similarly, elevated concentrations of CO_2 prevail in soil. Here, values between 0.07% and 0.23% are common, but can increase to 12 vol % (per vol air-filled space). Certain aerobic bacteria, like the nitrifyers, but also many obligate anaerobic bacteria are adapted to CO_2 levels greater than those prevalent in the atmosphere. Correspondingly, incubations at elevated CO_2 concentrations resulted in a significantly increased cultivation success of soil *Acidobacteria* (Stevenson et al. 2004).

Three-electrode poised-potential amperometric culture systems consist of a platinum counter electrode, a platinum or graphite working electrode and a AgCl–Ag reference electrode connected to a potentiostat. These systems offer the possibility to grow anaerobic prokaryotes at a precisely controlled and constant redox potential despite continuing redox reactions and a flow of electrons to the working electrode (Bond et al. 2002). In fact, a graphite electrode can function as sole electron acceptor if it is inserted into an anoxic sediment. It functions as an anode by connection in an electrical circuit via a fixed resistor to a cathode in the overlying aerobic water. These so-called fuel cells not only can be used to harvest electric energy (currents of 20–30 mA·[m² electrode surface]⁻¹), but also yield enrichments of Fe(III)-reducing *Deltaproteobacteria* when incubated in freshwater or marine anoxic sediments. *Geobacter*, *Geothrix* and *Pelobacter* spp. are the dominant bacteria enriched in freshwater sediments, while *Desulfuromonas* has been obtained in marine sediments. Vice versa, a potentiostat-poised graphite electrode (at −500 mV versus Ag/AgCl) has been shown to promote nitrate reduction or fumarate respiration in sediment enrichments as the sole electron donor. In contrast to *Geobacter* spp., numerous bacteria, including *E. coli* and *Proteus vulgaris*, require mediator compounds like neutral red, thionin, methyl viologen, phenazine ethosulfate or anthraquinone 2,6-disulfonate (AQDS) as electron shuttles.

Finally, viable cell counts of freshwater or marine planktonic bacteria reached on agar-solidified media are usually orders of magnitude smaller than those obtained as most probable numbers in liquid extinction dilution series (see below for methods of purification of agar). Many bacteria, especially from planktonic samples, do not appear to grow on solid media and have to be isolated in liquid dilution series. In contrast, the cultivation success of soil bacteria on improved solid media has been reported to surpass that in liquid cultures at least in one case (Janssen et al. 2002).

Effect of Cyclic Adenosine Monophosphate (cAMP)

The transcription factor σ^S is involved in the transition to stationary phase. The transcription of the corresponding gene (*rpoS*) is negatively regulated by cAMP in a complex with the cAMP receptor protein (CRP-cAMP). Addition of extracellular cAMP exerts the same effect. In addition, cAMP is known to regulate scavenging transporters. It has therefore been suggested to add extracellular cAMP to enrichments of bacteria from nutrient-limited habitats in order to maintain bacterial cells in a nutrient scavenging state and prevent them to enter the protective stationary phase response. In the absence of other stress factors, this could facilitate cultivation.

If a growth limiting substrate (e.g., glycerol, glucose, ribose, phosphate or ammonia) is added to cells previously starved for the same substrate, growth is inhibited. This so-called "substrate-accelerated death" has been described for several laboratory strains like *Klebsiella*, *Escherichia*, *Streptococcus*, *Azotobacter*, *Arthrobacter* and *Mycobacterium*, but its significance for the not-yet-cultured majority of the bacteria remains unknown. In laboratory cultures, substrate-accelerated death can be suppressed by the addition of extracellular cAMP or the addition of Mg^{2+}. Correspondingly, it has been reported that the addition of small amounts of cAMP significantly increases the cultivation success for bacteria from natural samples (Bruns et al. 2003b).

Mimicking the Physical Structure and Heterogeneity of the Natural Environment: Polymer Matrices, Solid Surfaces and Defined Laboratory Gradient Systems

Although solid surfaces may lead to stimulation of cell division and growth of starved bacteria, the capability of attachment has rarely been exploited as a selective factor for the enrichment and isolation of novel types of prokaryotes. Recent, systematic analyses demonstrated that selective enrichments on solid surfaces with a specific chemical composition yielded a variety of previously non-cultured types of freshwater bacteria (Frederic Gich, Melanie König, Jörg Overmann; unpublished results). Gliding bacteria, like the green sulfur bacterium *Chloroherpeton thalassium*, require a solid matrix for growth. For initial isolation, washed soft agar at a final concentration of 0.8% has been found to be most suitable matrix which permits some spreading of the filaments during growth and leads to a fluffy appearance of the colonies. However, a gelling agent needs to be added even to the pure cultures for growth. Good results have been obtained by adding Gelrite (4% w/v stock solution) to a final concentration of 0.025% which produces a visible increase in viscosity of the liquid medium and significantly stimulates cell division of the gliding bacteria. The filamentous gliding multicellular sulfate-reducing *Desulfonema* spp. can also be maintained on insoluble aluminum phosphate precipitate which can readily be precipitated in the anoxic medium by the addition of $AlCl_3$ and Na_2CO_3 solutions (Widdel et al. 1983).

Colorless chemolithoautotrophic sulfur bacteria *Beggiatoa*, *Thioploca*, *Thiovulum* or *Thiomargarita*, or magnetoctacti cocci, occur in highly stratified environments like the chemocline of stratified lakes or marine sediments. These bacteria are adapted to a defined and very narrow concentration regime of sulfide and oxygen. Appropriate conditions can be established in gradient tubes, in which sulfide-containing and higher strength agar is overlaid with low concentration oxic agar containing growth medium. Subsequently, an agar-stabilized artificial oxygen-sulfide countergradient builds up gradually. This cultivation system provides a particular advantage in that the bacteria of interest themselves precisely create the environmental conditions necessary for growth and resulted in the first isolation of pure cultures of *Beggiatoa alba* or magnetotactic cocci.

Iron is present in soils and sediments in a concentration range of several 10 mM. As a result, Fe(II) is an important electron-donating substrate in various habitats. Although most of these environments are circumneutral in pH, the majority of lithotrophic iron-oxidizing bacteria isolated to date are acidophilic. The underlying reason for this bias is the different stability of Fe(II) at different pH values. Below pH 4, Fe(II) is stable and consequently achieves high concentrations in oxic aqueous environments. Towards neutral pH, however, the chemical oxidation of Fe(II) becomes very rapid and competes with the biological process. This limits the growth of aerobic, neutrophilic iron-oxidizing bacteria to oxic-anoxic transition zones where Fe(II)-concentrations range between only tens and hundreds of micromolar and concentrations of dissolved oxygen reach tens of micromolar. Bacteria colonizing such gradients are adapted to the specific environmental conditions and often are microaerophiles. The low concentrations notwithstanding, bacteria in such gradients can form dense biomass accumulations due to the high fluxes of Fe(II) and O_2. In its simplest version, cultivation of neutrophilic Fe(II)-oxidizing bacteria is performed in gradient tubes which are set up by overlaying a mixture of ferrous ammonium sulfate $(Fe(NH_4)_2(SO_4)_2 \cdot 6H_2O)$ and FeS precipitates by an appropriate oxic liquid growth medium. In order to stabilize the chemical gradient and to maintain the position of the bacterial cells, the bottom layer is prepared in 1.5–3% agar or 1% high melt agarose and after solidification is overlaid with semi-solid mineral media, solidified by 0.8% agar or 0.15% low melt agarose.

Removal of Inhibitors and Avoiding the Formation of Toxic Compounds and Oxygen Radicals

Depending on the source, even drinking water may contain various compounds which inhibit the growth especially of freshwater or marine planktonic bacteria. While the purification by ion exchange or reverse osmosis may be sufficient for the cultivation of some prokaryotes, double quartz-distillation is the most reproducible way of maintaining a constantly high quality of water for the preparation of growth media. In addition, treatment of water with UV removes trace organic contaminants.

Another important element of cultivation is the proper cleaning of glassware. Sodium, silicate and borate are liberated from fresh glassware into the medium. Compared to other types, borosilicate glass releases only small quantities of free alkali, however, a special treatment for the removal of contaminants is still essential for the cultivation of fastidious bacteria. Detergents like Mucasol® aid in a proper cleaning, but if applied alone often do not remove all adsorbed impurities. Soaking in the appropriate detergent for 1 week and subsequent rinsing with double-distilled water, followed by 1 week of soaking in 0.5 N HCl, and then by repeated washings with double-distilled water often alleviates these problems. Afterwards, the glassware needs to be rinsed thoroughly with bidistilled water since even traces of detergents act inhibitory on various bacteria, like marine oligotrophic chemoorganotrophs or certain cyanobacteria.

The agents used for solidification of growth media include agar, Gelrite and silica gel. Agar is still employed most frequently, but commercial preparations contain variable amounts of impurities such as Ca, Mg, Fe and other minerals, as well as organics like long-chain fatty acids, phenolics, pigments. Such soluble nonpolymeric contaminants can be conveniently removed by repeated washing in distilled water (e.g., five times in a volume of double distilled water corresponding to 300% of the final volume). Alternatively, agar suspensions can be purified by subsequent extraction with 95% ethanol and analytical grade acetone. Inhibitory effects of agar can also be avoided by employing Gelrite (Gelan gum) as a solidifying agent. Gelrite is an anionic heteropolysaccharide consisting of glucose, glucuronic acid and rhamnose which is produced by *Sphingomonas* strains and starts to solidify upon addition of Ca^{2+}. In fact, it has recently been documented that bacteria which are inhibited by agar nevertheless grow on Gelrite media (Janssen et al. 2002). Other bacteria, like the reductively dechlorinating bacteria, though not growing on agar surfaces, do form colonies in semisolid medium prepared with 0.5–0.7% low-melting agarose. Finally, silica gel media have been developed for cases in which solid media free of any organic contaminations are needed, if a low pH of the medium is desired, or if agar-degrading microorganisms (some strains of *Cytophaga*, *Pseudomonas* or *Vibrio* spp.) are to be cultivated on solid media.

Trace elements typically are present as contaminations in the mineral salts used for media preparation or even as contaminants of glassware and water. For example, certain cyanobacteria like *Synechococcus* spp. or especially *Trichodesmium* spp. are highly sensitive to Cu at concentrations as little as ten nanomolar (!), such that cultivation has been found to be only successful if Cu is completely eliminated from the medium by the use of Chelex 100 columns. For the cultivation of such fastidious marine oligotrophic bacteria, special precautions therefore need to be taken to avoid organic and inorganic contaminants and can often be achieved only by employing non-glass cultivation systems. This includes the use of teflon vessels for the preparation of natural seawater which is sterilized by gentle Tyndallization (three times heating to 100°C for 30 min with intermittent cooling to room temperature) in a microwave and the cultivation of bacteria in polycarbonate flasks (e.g., Nalgene) (Waterbury 1991).

When autoclaved together, the carbonyl groups of reducing sugars react with free amino groups of primary amines which may result in the formation of toxic products like furfurals or furaldehyde. This so-called Maillard ("browning") reaction is accelerated by oxygen or oxidation products. Glucose and other sugars may form inhibitory sugar phosphates when autoclaved with phosphate salts. Sulfide and organic thiols such as cysteine will be oxidized by other medium ingredients during autoclaving and form toxic radicals. Accordingly, these reductants must be added separatedly afterwards.

Upon exposure to molecular oxygen, hydrogen peroxide (H_2O_2), superoxide (O_2^-), or hydroxyl radicals ($OH^.$) are produced by electron transport in the bacterial cells. In addition, these toxic oxygen intermediates are produced during autoclaving of growth media. Oxidative stress can be abolished by the addition of H_2O_2-degrading enzymes or compounds, like catalase, ascorbic acid, sodium pyruvate or 2-oxoglutaric acid. A lowering of the oxygen partial pressure to 1–2% O_2 vol/vol may exert similar effects. Catalase or pyruvate can be added as oxygen radical scavengers and can improve the growth of bacteria. Finally, activated charcoal has been added as a scavenger of toxic oxygen radicals and resulted in a significantly increased cultivation success for bacteria from deep terrestrial subsurface samples.

For the cultivation of obligate anaerobic bacteria, the Hungate technique still represents the fundamental and appropriate methodology (Hungate 1950, 1966). This technique has been further improved by the use of butyl rubber stoppers, and of serum bottles closed with crimp-closure aluminum seals holding butyl rubber stoppers, the syringe technique, and the use of pressurized tubes and vessels for culture of methanogens. Larger quantities (2–5 l) can be prepared anoxically in Widdel flasks. Traces of oxygen can be removed from the gas prior to use by leading the gas through hot copper wire. For the isolation in agar-solidified media, deep agar dilution series are frequently employed. Media are kept oxygen free by preparation under a N_2/CO_2 (80/20, v/v) atmosphere and by adding a reductant, most frequently sulfide or cysteine. In cases where free sulfide is toxic, amorphous ferrous sulfide can be employed as a reducing agent. A mixture of thioglycollate and ascorbate which has been used for the cultivation of *Clostridia*, yet may act inhibitory on other, more fastidious anaerobes. Sulfur-free reducing agents of especially low redox half-potential (−480 mV) are titanium (III) citrate or titanium nitriloacetic acid (NTA).

Removal or Selective Inhibition of Bacterial Competitors

Ultramicrobacteria reach a cell volume of < 0.1 μm^3 and prevail in oligotrophic waters. While several bacterial species produce ultramicrocells by reductive division upon nutrient deprivation, others like *Sphingomonas alaskensis*, some *Verrucomicrobiales* or *Pelagibacter ubique* constantly maintain a small cell volume of 0.03–0.06 μm^3. The small size of ultramicrobacteria can be exploited to selectively enrich them by separation from accompanying larger cells by filtration through 0.2, 0.45 or 0.8 μm pore size membrane filters (Hahn et al. 2004). Using the same principle, large filamentous bacteria can be mechanically enriched by passing cell suspensions through a copper grid (mesh size, 256 per mm^2) used for electron microscopy, which is fixed in a glass tube (Widdel 1983). This technique results in the removal of the often dominant smaller accompanying bacteria and can also readily be used for anaerobes.

In certain cases, accompanying but undesired bacteria can be selectively inhibited by supplementing the growth media with suitable inhibitors. Mixtures of rifampicin (final concentration, 2.5 mg·l^{-1}), kanamycin A (5 mg·l^{-1}), erythromycin (50 mg·l^{-1}), tetracyclin (100 μg·ml^{-1}), and penicillin G (50 mg·l^{-1}) or ampicillin (100 mg·l^{-1}) and fosfomycin (100 mg·l^{-1}) inhibit the growth of many *Bacteria*, while cycloheximide (100 mg·l^{-1}), tunicamycin (250 μg·l^{-1}), colchicin (20 μg·l^{-1}) and cordycepin (25 mg·l^{-1}) suppress growth of *Eucarya*. It has to be stressed, however, that resistant *Bacteria* or *Eukarya* occur even in pristine natural environments and usually overgrow target microorganisms like *Archaea* in such enrichments. Antibiotics interfering with cell wall biosynthesis inhibit cell division and growth but do not kill non-dividing cells. To ensure that sensitive populations are indeed eliminated, antibiotics need to be added to at least three consecutive culture transfers. Rifamycin and fosfomycin have been employed for selective enrichments of spirochetes.

Bile salts suppress the growth of many non-enteric bacteria. Compounds available for the selective inhibition of more specific physiological groups of bacteria comprise bromethane sulfonate (BES) and molybdate which can be employed for the selective inhibition of methanogenic Archaea and sulfate-reducing bacteria, respectively. Nitrapyrin (2-chloro-6-(trichloromethyl)-pyridine; N-serve) interacts with the active site of ammonia monooxygenase and thus inhibits nitrification at the level of NH$_4^+$-oxidation. In contrast, sodium chlorate (NaClO$_3$) inhibits the nitrate formation step of nitrification. Acetylene (C$_2$H$_2$) inhibits nitrification of chemo-lithoautotrophs at very low concentrations of 1–10 Pa. Ethylene can be employed against denitrifying bacteria and DCMU against oxygenic phototrophs.

Exploiting Positive Interactions Between Bacteria: Cocultivation and Dialysis Cultures

In their natural environment, bacterial cells are spaced only 1–112 μm apart, even if distributed homogenously in the environment (Overmann 2001). Over such small distances, metabolites like hydrogen or small organic carbon compounds diffuse rapidly within 0.0003 and 4.2 s, respectively. Due to the large diversity of bacteria in natural samples, the closest neighbour of a bacterial cell statistically would be expected to represent another species. These theoretical considerations predict that the exchange of metabolites and signal molecules between different bacteria is significant under natural conditions. Indeed, microcolonies, net-like structures, biofilms or aggregates consisting of up to 18 different prokaryotic genera are commonly observed in natural samples. In the pelagic environment, free-living bacteria seem to accumulate in 10–100 μm-large patches. In some cases, different bacteria are even capable of forming specific, highly structured and tight associations, so-called consortia (Overmann 2001). Cell-cell interactions therefore need to be taken into consideration for cultivation trials and may actually be decisive for a successful enrichment of some novel types of bacteria.

One of the most early recognized and best understood mutual interactions between prokaryotes is the interspecies hydrogen transfer (Schink 1991). Thermodynamically unfavorable substrates like propionate, butyrate, caproate, caprylate or benzoate are converted by fermenting bacteria to acetate, H_2 and CO_2 only in the presence of a hydrogen-scavenging partner. Growth of *Syntrophomonas wolfei* or *Syntrophobacter wolinii* is therefore only observed in coculture with a hydrogen-utilizing anaerobe, like a methanogen. Pure cultures of the fermenting bacteria can be obtained, however, using alternative substrates like crotonate. In sulfur syntrophy, cocultures of green sulfur bacteria with sulfur- or sulfate-reducing chemotrophic bacteria are capable of growing in the presence of an organic carbon source like acetate at limiting concentrations of sulfur compounds. Here, the sulfide-producing chemotroph relies on the sulfide-oxidizing phototroph for generation of its electron acceptor, while the activity of the green sulfur bacteria depends on the activity of the sulfide-producing organism (Pfennig 1980).

Cell-cell-signalling between cells of the same clone has been shown to be important for resuscitation of some bacteria. Autoinduction has been found in several actinobacteria including *Micrococcus luteus* (Mukamolova et al. 2006). In the latter, cells can be resuscitated by addition of a so-called resucitation promoting factor, a lytic transglycosylase excreted into the culture supernatant and active at picomolar concentrations.

Heterologous microbial interactions, even if unknown, can be exploited by embedding cells in agar, sandwiched between 0.03-μm pore-size membrane filters and incubation in the natural environment (Kaeberlein et al. 2002). Meanwhile, simple cultivation devices for the cocultivation of different bacteria have become available in which two different growth chambers are separated by a membrane which is permeable to metabolites (Fig. 3.1c).

Techniques for the Isolation of Individual Cells

Many approaches towards the molecular genetics and physiology of the bacterial cell still rely on the availability of a pure culture. A culture free of other microorganism (an axenic culture) represents a population derived from a single cell. Various methods have been established to physically separate individual bacterial cells, transfer it to a growth medium and incubate the culture under aseptic conditions. In soils and sediments, a significant fraction of bacterial cells are attached to particles. At least part of these cells can be detached by the addition of a mixture of the mild nontoxic detergent Tween 80 (final concentration 0.05% v/v) and sodium pyrophosphate (10 mM in 10 mM HEPES buffer) in order to complex bridging divalent cations which bind bacterial cells to mineral surfaces.

Classical isolation techniques rely on the sequential dilution of a sample containing bacterial cells in or on appropriate media. For cells capable of growing on solid media like agar plates, separation can be accomplished by streaking of a small sample using an inoculation loop or by evenly spreading a larger liquid volume (≤ 100 µl) with a Drigalski spatula. Liquid dilution methods comprise sequential liquid dilution series, agar pour plates, and deep agar dilution series (the latter for anaerobic bacteria). Among the latter three, the preparation of deep agar dilution series is the most demanding technique. Typically, test tubes containing 3 ml of 3% washed agar are autoclaved with cotton plugs on top. Afterwards, the plugs are replaced by sterile butyl rubber stoppers and the tubes kept at 60°C. Sulfide-reduced medium is heated to 40°C and 6 ml-aliquots are dispensed directly into the liquid agar. The tubes are kept at 43°C to avoid solidification. Some fastidious anaerobes have been reported to require highly reducing conditions for growth in these agar media. For this purpose, dithionite solution can be added to the medium at a final concentration of ≤ 200 µM (Overmann and Pfennig 1989). For dilution, the first tube is inoculated, mixed, and a tenth of its volume is transferred to the following tube. After dilution has been completed, the agar media are solidified in a 20°C water bath, rapidly gassed with N_2 or a mixture of N_2/CO_2 and sealed with the rubber stoppers. As soon as bacterial colonies appear in the agar, those colonies which are spaced apart from others can be isolated by means of sterile Pasteur pipets.

When liquid dilution series are to be used for the isolation of pure cultures, cell suspensions have to be diluted to the point where a culture tube receives just one single viable cell of the target bacterium. In these so called dilution-to-extinction approach, at least the final tube of each dilution series must not exhibit bacterial growth. Liquid dilutions are only suitable for isolation, if the target bacteria (1) outnumber all other bacteria growing in the growth medium and (2) are capable of dividing in the medium. A set of 3–10 parallel liquid dilution series can be used to estimate the most probable number (MPN) of dividing bacteria in a natural sample. The MPN is calculated from the pattern of tubes exhibiting positive growth based on probability values. If prepared in selective media, e.g. for benzoate-degrading or for sulfate-reducing bacteria, MPN series can be used to estimate the numbers of culturable cells of these particular physiological types.

With respect to the chances of isolating a pure bacterial culture, one drawback of the dilution-to-extinction method is that only few (3–10) potentially pure cultures are obtained from the highest positive dilutions. Recent attempts to enrich and isolate not-yet-cultivated bacteria indicated that such bacteria in fact can grow in carefully designed laboratory media, but that only a very low fraction of the target cells multiply and, on statistical grounds, would never end up in the last positive dilutions. Advanced high throughput techniques permit a separation of virtually each cell in a sample before cell division commences, thereby avoiding competition with faster growing accompanying bacteria. The inoculation can be performed in a highly efficient and automated manner employing the MicroDrop® AutoDrop microdispenser (Microdrop GmbH, Norderstedt, Germany) (Bruns et al. 2003a). With this device, microdroplets of 100–200 pl volume are generated by a piezoelectric actuator and positioned directly in individual wells of microtiter dishes (Fig. 3.1a). A large number of cultures is inoculated rapidly (100 inoculations in less than 1 min). Alternatively, microdroplets can also be positioned on the surface of suitable agar media. Provided that the average cultivation success for cells is known for a given sample, the number of pure cultures produced can be increased by orders of magnitude without increasing the total number of cultures inoculated. To this end, an appropriate amount of inoculum is chosen which statistically contains 0.5 viable cells, and many parallel tubes are inoculated. The number of positive wells or positions follows a Poisson distribution and MPN values can be calculated from the fraction p of positive tubes (the number of positive cultures divided by the total number of cultures n) according to

$$MPN = -\ln(1-p) \tag{3.5}$$

with the standard deviation of the MPN

$$S.D. = \sqrt{\frac{p}{[n\cdot(1-p)}} \tag{3.6}$$

In another approach, bacterial cells are encapsulated individually in agarose gel microdroplets, which are then packed into a column to be percolated with medium, e.g. sterilized seawater medium (Zengler et al. 2002). The rationale of this approach is to grow the cells of the same microbial community (1) in close proximity as they prevail under natural conditions, while (2) they are caged and thus can be separated afterwards, and (3) to provide low nutrient flux conditions. As few as five to seven doublings of the cells within the beads can be detected which permits a highly sensitive monitoring also of a slow growth response of many bacterial cells in high throughput. Microcapsules that contain colonies are subsequently sorted by flow cytometry into microtiter dishes containing organic-rich medium.

Even more tedious isolation procedures include the use of micromanipulator devices which are used to physically separate single bacterial cells from complex microbial communities using a microcapillary (diameter, 1–10 μm opening diameter) under an inverted light microscope and with computer-assisted electronic controls (Fröhlich and König 1999). The isolated cells are then transferred to sterile media. So-called optical tweezers represent a technically more costly technique for the separation of single cells. Individual bacteria suspended in a rectangular glass capillary are trapped in the strongly focussed beam of an infrared laser (Nd YAG, wavelength 1,064 nm, maximum power 1 W) and moved to one end of the capillary (Huber et al. 1995). Afterwards, the capillary is broken and the isolated cell transferred to appropriate growth medium. In contrast to the high throughput methods described above, these single cell-based techniques have been especially successful if morphologically conspicuous but less frequent bacteria are to be isolated from complex sample material.

References

Angle JS, McGrath SP, Chaney RL (1991) New culture medium containing ionic concentrations of nutrients similar to concentrations found in the soil solution. Appl Environ Microbiol 57:3674–3676

Balch WE, Wolfe RS (1976) New approach to the cultivation of methanogenic bacteria: 2-mercaptoethane sulfonic acid (HS-CoM)-dependent growth of *Methanobacterium ruminantium* in a pressurized atmosphere. Appl Environ Microbiol 32:781–791

Bartscht K, Cypionka H, Overmann J (1999) Evaluation of cell activity and of methods for the cultivation of bacteria from a natural lake community. FEMS Microbiol Ecol 28:249–259

Bast E (2001) Mikrobiologische methoden, 2nd edn. Gustav Fischer, Stuttgart, 429 pp

Bond DR, Holmes DE, Tender LM, Lovley DR (2002) Electrode-reducing microorganisms that harvest energy from marine sediments. Science 295:483–485

Bruns A, Hoffelner H, Overmann J (2003a) A novel approach for high throughput cultivation assays and the isolation of planktonic bacteria. FEMS Microbiol Ecol 45:161–171

Bruns A, Nübel U, Cypionka H, Overmann J (2003b) Effect of signal compounds and incubation conditions on the culturability of freshwater bacterioplankton. Appl Environ Microbiol 69:1980–1989

Button DK, Schut F, Quang P, Martin R, Robertson BR (1993) Viability and isolation of marine bacteria by dilution culture: theory, procedures, and initial results. Appl Environ Microbiol 59:881–891

Cho J-C, Giovannoni SJ (2004) Cultivation and growth characteristics of a diverse group of oligotrophic marine *Gammaproteobacteria*. Appl Environ Microbiol 70:432–440

Connon SA, Giovannoni SJ (2002) High-throughput methods for culturing microorganisms in very-low-nutrient media yield diverse new marine isolates. Appl Environ Microbiol 68:3878–3885

Coolen MJL, Overmann J (2000) Functional exoenzymes as indicators of metabolicallly active bacteria in 124,000-year-old sapropel layers of the eastern Mediterranean sea. Appl Environ Microbiol 66:2589–2598

Eilers H, Pernthaler J, Peplies J, Glöckner FO, Gerdts G, Amann R (2001) Isolation of novel pelagic bacteria from the German Bight and their seasonal contributions to surface picoplankton. Appl Environ Microbiol 67:5134–5142

Ferrari BC, Binnerup SJ, Gillings M (2005) Microcolony cultivation on a soil substrate membrane system selects for previously uncultured soil bacteria. Appl Environ Microbiol 71:8714–8720

Fröhlich J, König H (1999) Rapid isolation of single microbial cells from mixed natural and laboratory populations with the aid of a micromanipulator. System Appl Microbiol 22:249–257

Gans J, Wolinsky M, Dunbar J (2005) Computational improvements reveal great bacterial diversity and high metal toxicity in soil. Science 309:1387–1390

Hahn MW, Stadler P, Wu QL, Pöckl M (2004) The filtration-acclimatization method for isolation of an important fraction of the not readily cultivable bacteria. J Microbiol Meth 57:379–390

Huber R, Burggraf S, Mayer T, Barns SM, Rossnagel P, Stetter KO (1995) Isolation of a hyperthermophilic archaeum predicted by *in situ* RNA analysis. Nature 376:57–58

Hungate RE (1950) The anaerobic mesophilic cellulolytic bacteria. Bacteriol Rev 14:1–49

Hungate RE (1966) The rumen and its microbes. Academic, New York

Jannasch HW, Wirsen CO, Doherty KW (1996) A pressurized chemostat for the study of marine barophilic and oligotrophic bacteria. Appl Environ Microbiol 62:1593–1596

Janssen PH, Yates PS, Grinton BE, Taylor PM, Sait M (2002) Improved culturability of soil bacteria and isolation in pure culture of novel members of the divisions *Acidobacteria*, *Actinobacteria*, *Proteobacteria*, and *Verrucomicrobia*. Appl Environ Microbiol 68:2391–2396

Kaeberlein T, Lewis K, Epstein SS (2002) Isolating "uncultivable" microorganisms in pure culture in a simulated natural environment. Science 296:1127–1128

Leadbetter JR, Schmidt TM, Graber JR, Breznak JA (1999) Acetogenesis from H_2 plus CO_2 by spirochetes from termite guts. Science 283:686–689

Mukamolova GV, Murzin AG, Salina EG, Kell DB, Kaprelyants AS, Young M (2006) Muralytic activity of *Micrococcus luteus* Rpf and its relationship to physiological activity in promoting bacterial growth and resuscitation. Mol Microbiol 59:84–98

Musat N, Halm H, Winterholler B, Hoppe P, Peduzzi S, Hillion F, Horreard F, Amann R, Jørgensen BB, Kuypers MMM (2008) A single-cell view on the ecophyisology of anaerobic phototrophic bacteria. Proc Natl Acad Sci USA 105:17861–17866

Nedoma J, Vrba J (2006) Specific activity of cell-surface acid phosphatase in different bacterio-plankton morphotypes in an acidified mountain lake. Environ Microbiol 8:1271–1279

Ouverney CC, Fuhrman JA (1999) Combined microautoradiography – 16S rRNA probe technique for determination of radioisotope uptake by specific microbial cell types in situ. Appl Environ Microbiol 65:1746–1752

Overmann J (2001) Phototrophic consortia: a tight cooperation between non-related eubacteria. In: Seckbach J (ed) Symbiosis: mechanisms and model systems. Kluwer, Dordrecht, The Netherlands, pp 239–255

Overmann J (2002) Principles of enrichment, isolation, cultivation, and preservation of bacteria. In: Dworkin M et al (eds) The prokaryotes: an evolving electronic resource for the microbiological community, 3rd edn (latest update release 3.11, September 2002). Springer, New York, 2000 (http://ep.springer-ny.com:6336/contents/ or http://141.150.157.117:8080/prokPROD/index.htm)

Overmann J (2005) Chemotaxis and behavioural physiology of not-yet-cultivated microbes. Methods in enzymology, vol 397, Chapter II.8. Elsevier, San Diego, CA, pp 133–147

Overmann J, Pfennig N (1989) *Pelodictyon phaeoclathratiforme* sp. nov., a new brown-colored member of the Chlorobiaceae forming net-like colonies. Arch Microbiol 152:401–406

Pernthaler A, Pernthaler J, Schattenhofer M, Amann R (2002) Identification of DNA-synthesizing bacterial cells in coastal North Sea plankton. Appl Environ Microbiol 68:5728–5736

Pfennig N (1980) Syntrophic mixed cultures and symbiotic consortia with phototrophic bacteria: a review. In: Gottschalk G, Pfennig N, Werner H (eds) Anaerobes and anaerobic infections. Fischer, Stuttgart, pp 127–131

Radajewski S, Ineson P, Parekh NR, Murrell JC (2000) Stable-isotope probing as a tool in micro-bial ecology. Nature 403:646–649

Rappé MS, Connon SA, Vergin KL, Giovannoni SJ (2002) Cultivation of the ubiquitous SAR11 marine bacterioplankton clade. Nature 418:630–633

Schink B (1991) Syntrophism among prokaryotes. In: Balows A, Trüper HG, Dworkin M, Schleifer KH (eds) The prokaryotes, 2nd edn. Springer, Berlin/Heidelberg/New York, pp 276–299

Schink B, Friedrich M (2000) Phosphite oxidation by sulphate reduction. Nature 406:37

Schut F, Devries EJ, Gottschal JC, Robertson BR, Harder W, Prins RA, Button DK (1993) Isolation of typical marine bacteria by dilution culture: growth, maintenance, and characteris-tics of isolates under laboratory conditions. Appl Environ Microbiol 59:2150–2160

Schut F, Gottschal JC, Prins RA (1997) Isolation and characterization of the marine ultrami-crobacterium *Sphingomonas* sp. strain RB2256. FEMS Microbiol Rev 20:363–369

Staley JT, Konopka A (1985) Measurement of in situ activities of nonphotosynthetic microorgan-isms in aquatic and terrestrial habitats. Ann Rev Microbiol 39:321–346

Stevenson BS, Eichorst SA, Wertz JT, Schmidt TM, Breznak JA (2004) New strategies for cultiva-tion and detection of previously uncultured microbes. Appl Environ Microbiol 70:4748–4755

Thauer RK, Jungermann K, Decker K (1977) Energy conservation in chemotrophic anaerobic bacteria. Bacteriol Rev 41:100–180

Vogl K, Glaeser J, Pfannes KR, Wanner G, Overmann J (2006) *Chlorobium chlorochromatii* sp. nov., a symbiotic green sulfur bacterium isolated from the phototrophic consortium "*Chlorochromatium aggregatum*". Arch Microbiol 185:363–372

Waterbury JB (1991) The cyanobacteria – isolation, purification, and identification. P.2058 – 2078. In: Balows A, Trüper HG, Dworkin M, Schleifer KH (eds) The prokaryotes, 2nd edn. Springer, Berlin/Heidelberg/New York, pp 149–196

Whitman WB, Coleman DC, Wiebe WJ (1998) Prokaryotes: The unseen majority. Proc Natl Acad Sci USA 95:6578–6583

Widdel F (1983) Methods for the enrichment and pure culture isolation of filamentous gliding sulfate-reducing bacteria. Arch Microbiol 134:282–285

Widdel F, Kohring G-W, Mayer F (1983) Studies on dissimilatory sulfate-reducing bacteria that decompose fatty acids. III. Characterization of the filamentous gliding *Desulfonema limicola* gen. nov. sp. nov., and *Desulfonema magnum* sp. nov. Arch Microbiol 134:296–294

Zengler K, Toledo G, Rappe M, Elkins J, Mathur EJ, Short JM, Keller M (2002) Cultivating the uncultured. Proc Natl Acad Sci USA 99:15681–15686

Chapter 4
Environmental Proteomics: Studying Structure and Function of Microbial Communities

Thomas Schneider and Kathrin Riedel

Introduction

Open Questions in Microbial Ecology

Microbes such as bacteria, fungi and viruses are omnipresent. They play an essential role in biogeochemical cycles and can decompose virtually all natural compounds, thereby exerting a lasting effect on biosphere and climate. About 20 years ago microbial ecologists started to realize that activity and physiology of a certain environment is strongly dependent on the diversity of microbial communities and interactive processes such as nutrient competition, predation and cellular signalling (Brock 1987). However, the fact that more than 90% of the microorganisms in a given environment are un-cultivable (Amann et al. 1995) hampered investigations aiming towards a deeper insight into the structure and function of biological systems for a long time and individual contributions of different species to a certain environment remained largely unknown. The recent development of a multitude of novel molecular tools bypassing the need to isolate and culture individual microbial species has opened up a promising new field of research that might revolutionize our concepts of microbial diversity and physiology within complex consortia and up to entire ecosystems: (1) 16S rRNA sequencing approaches provide important

T. Schneider and K. Riedel (✉)
Department of Microbiology, Institute of Plant Biology, University of Zürich,
Winterthurerstrasse 190, CH-8057 Zurich, Switzerland
e-mail: kriedel@botinst.uzh.ch

L.L. Barton et al. (eds.), *Geomicrobiology: Molecular and Environmental Perspective*,
DOI 10.1007/978-90-481-9204-5_4, © Springer Science+Business Media B.V. 2010

information about species composition and evolution (reviewed in Schloss and Handelsman 2006), (2) novel shotgun sequencing techniques enable the mapping of whole metagenomes (reviewed in Vieites et al. 2009) as well as the study of transcriptional profiles of microbial consortia (see also Chapter 4), and (3) environmental proteomics techniques allow qualitative and quantitative assessment of the protein complement of an environment (reviewed in Keller and Hettich 2009; Wilmes and Bond 2009; Verberkmoes et al. 2009a; Schneider & Riedel 2009). The presented chapter starts with an overview of the historical development and potential of environmental proteomics, introduces the reader to state-of-the-art proteomics technologies, summarizes the most important publications in the field and closes with conclusions and future perspectives.

Historical Retrospective of "Omics" Technologies

Until 10 years ago global technologies to analyze microbial genomes, transcriptomes, proteomes, or even metabolomes were restricted to cultivable mono-species. Today, tremendous advances in "omics" technologies make it possible to study not only so far un-cultivable organisms, but complex microbial communities and even entire ecosystems. Figure 4.1 depicts the historical evolution and technical milestones of these global molecular approaches. At the turn of the millennium novel shotgun DNA sequencing technologies such as 454 pyrosequencing (Ahmadian et al. 2006) coupled with significant cost reduction have boosted culture-independent metagenomics approaches that suddenly opened up exciting perspectives on the diversity and distribution of indigenous microbial populations in natural environments (reviewed in Tringe et al. 2005). As yet metagenomics strategies can't elucidate the functionality of microorganisms present in the respective ecosystem. Moreover, an enormous number of newly identified ORFs with no homology to well-characterized genes still await functional assignment. These limitations have stimulated the development of environmental transcriptome analyses, although the short half-life of mRNA molecules, challenging extraction protocols due to interfering organic and inorganic compounds, and the often low correlation between transcript levels and actual protein expression still appear to be major drawbacks of metatranscriptome studies (Zhou and Thompson 2002). Finally, proteomics has emerged as a promising novel technique to characterize microbial activities at the molecular level. Proteomics, originally defined as "the large-scale study of proteins expressed by an organism" (Wilkins et al. 1995), started to develop in the 1970s when protein profiles of single organisms were analyzed by two-dimensional gel electrophoresis (O-Farrel 1975). At that time protein identification was, if at all possible, time consuming and cost-intensive due to a lack of genomic sequence information and advanced protein sequence analyses. Since the 1990s proteomics has become much more widespread, feasible and reliable thanks to three technical revolutions: (1) the enormous increase of genomic and metagenomic data provides a solid basis for protein identification, (2) tremendous progress in sensitivity and accuracy of mass spectrometers enables a correct, high-throughput protein identification, relative and absolute quantification of proteins, and the

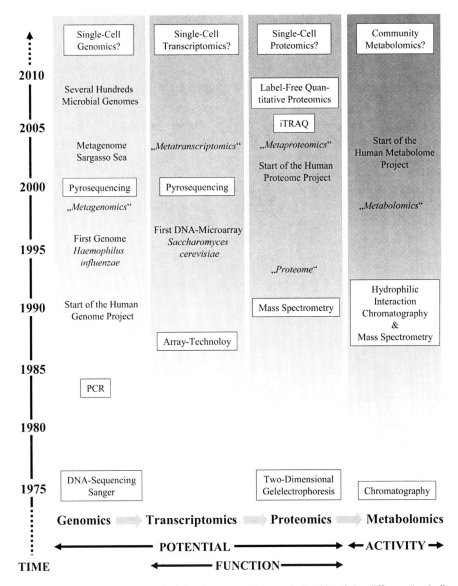

Fig. 4.1 Overview of the historical development and interrelationship of the different "omics" technologies including methodological innovations (*white boxes*), introduction of novel terms (*quotation marks*), and milestone publications or initiatives

determination of post-translational modifications, and (3) impressive improvements in computing power and bioinformatics allow processing and evaluation of substantial datasets. Global analyses of proteins involved in biotransformation, i.e. enzymes, finally allow a holistic characterization of microbial metabolic dynamics and shed light on the regulation of the metabolome, the complete set of metabolic intermediates, signalling molecules and secondary metabolites found within a biological sample

(Bochner 2009). State-of-the-art proteomics technologies and future perspectives will be discussed in the following paragraphs.

Environmental Proteomics – A Babylonian Confusion?

Less than 5 years ago Wilmes and Bond (2004) defined *metaproteomics* as "the large-scale characterization of the entire protein complement of environmental microbiota at a given point in time"; meanwhile rapid advances and multi-fold applications of high-throughput "omics" technologies have led to many novel denominations including *environmental proteomics, community proteomics,* or *community proteogenomics.* These terms are often used as synonyms; however, as rightly stated by Verberkmoes et al. (2009a), they stand in fact for slightly different experimental setups and outcomes. While *environmental proteomics* should be regarded as a generic term simply describing proteome analyses of environmental samples, *metaproteomics* comprises studies of highly complex biological systems which do not allow assigning large numbers of proteins to specific species within phylotypes. In contrast, the term *community proteomics* implies that most of the identified proteins can be related to specific members of the community; thus far such studies have been limited to low- or medium-complexity environments. The term *proteogenomics*, which was initially used to describe the application of proteomics for the enhancement of gene annotations, does nowadays also define the assessment of strain or species variations and the evolutionary development of the genomic makeup of certain environments (Verberkmoes et al. 2009a).

Potential Applications of Environmental Proteomics

In their natural habitat microorganisms are often facing expeditious and harsh changes of environmental parameters such as temperature, humidity, nutrient availability, and predators. Most microbes overcome these challenges rapidly by altering their protein expression profiles but not by genomic rearrangements. Consequently, the mere study of individual genes and their regulation is not sufficient to fully understand microbial adaptation strategies and post-genomic analyses including transcriptomics and proteomics are urgently needed to investigate the physiology of complex microbial consortia at a molecular level.

Even though still in its infancy, environmental proteomics already comprises a real "treasure chest" of technologies ranging from simple protein cataloging (e.g. by mapping the protein complement of an ecosystem at a certain time point), to comparative and quantitative proteomics (e.g. by evaluating how different environmental conditions affect protein expression), analyses of protein localizations, discovery of post-translational modifications which might affect protein functionality, investigation of protein-protein interactions, and even determination of amino-acid sequences and genotypes e.g. by strain-resolved proteogenomics.

Hence, potential applications of the above listed technologies in microbial ecology are numerous and include the description of novel functional genes, the identification of completely new catalytic enzymes or entire metabolic pathways and the description of functional bioindicators to monitor dynamics and sustainability of environment quality (reviewed in Maron et al. 2007). Further improvement and concerted usage of the complete set of "omics" technologies will allow us to revisit microbial ecology concepts by linking genetic and functional diversity in microbial communities and relating taxonomic and functional diversity to ecosystem stability.

State-of-the-Art Proteomics Technologies

A standard proteomics experiment typically comprises three basic steps (Fig. 4.2): (1) sample preparation including protein extraction, purification, and concentration, (2) protein or peptide separation and mass spectrometry (MS) analysis, and (3) protein identification based on the obtained MS and/or tandem mass spectrometry (MS/MS) data. A trustworthy proteome analysis includes furthermore the validation of the obtained results by complementary approaches e.g. transcriptome analyses, quantitative PCR, or (if applicable) phenotypical assays. Metaproteomics approaches employ basically similar experimental setups although they need to overcome additional challenges inherent in samples from natural environments e.g. high organism/protein complexity, over- or under-representation of certain organisms/ proteins, heterogeneity of organic and inorganic contaminants etc. The following paragraphs will give an overview of state-of-the-art proteomics techniques focussing on the requirements of environmental proteomics and will discuss weaknesses and strengths of different experimental strategies.

Sample Preparation

The first critical step in a metaproteome experiment is the comprehensive extraction of the entire protein complement of a given sample; thereby the respective protocol should be as efficient, non-biased and reproducible as possible. Moreover, it is crucial to avoid the addition of organic or inorganic compounds/solvents that might interfere with sequential protein separation and mass spectrometry. Dependent on sample type (e.g. activated sludge, ocean or ground water, soil) and complexity, different extraction strategies have to be employed. For activated sludge, an environment of moderate complexity, a multi-step protocol has been developed which included various washing buffers, French press lyses and precipitation procedures and resulted in a satisfying number of identified proteins (Wilmes and Bond 2004; Wilmes et al. 2008a, b). Protein extraction from highly complex samples such as soil, which is mainly hampered by the presence of perturbing matrix compounds

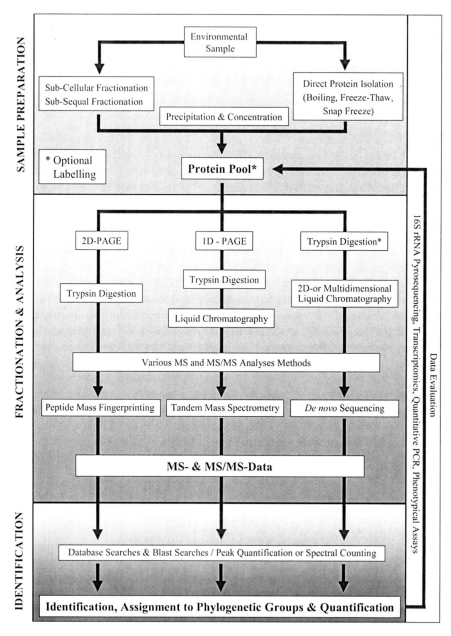

Fig. 4.2 Schematic flow-chart summarizing different environmental proteomics methodologies. Discrete steps i.e. sample preparation, protein fractionation and analysis, protein identification and optional quantification (indicated by *) and data evaluation are depicted in *boxes*. The various protein analysis strategies exhibit significant differences in throughput, sensitivity and reliability of identification, which are discussed in detail in the text. Abbreviations: 2D-PAGE, two-dimensional polyacrylamide gel electrophoresis; 1D-PAGE, one-dimensional polyacrylamide gel electrophoresis, MS, mass spectrometry; MS/MS tandem mass spectrometry

(e.g. humic acids), requires even harsher extraction procedures e.g. snap-freeze protein extraction (Singleton et al. 2003), hydrofluoric acid to dissolve soil minerals (Schulze et al. 2005), or NaOH extraction followed by phenol treatment (Benndorf et al. 2007, 2009). The relatively low number of proteins identified so far from soil-derived samples (see following paragraphs) demonstrates the need for improved protein extraction methods in order to obtain sufficiently concentrated and purified protein samples from complex environments for down-stream analyses.

Protein/Peptide Separation and Mass Spectrometry Analyses

Protein or peptide separation by gel-based or chromatographic techniques. Protein samples derived from natural environments do not lend themselves to direct MS analysis; rather, sample complexity has to be reduced first by gel-based or chromatographic techniques. This can either be accomplished on the protein level ("top-down approaches") or on the peptide level after proteolytic degradation of sample proteins ("bottom-up approach" or "shotgun" proteomics). For many years two-dimensional polyacrylamide gel electrophoresis (2D-PAGE) was regarded as the "gold standard" of proteomics research (O-Farrel 1975; Görg et al. 1988). With this method proteins are first separated along a pH gradient by isoelectric focussing, followed by a second separation according to mass on SDS-PAGE gels. In this way, over a thousand proteins can be resolved on a single gel as discrete spots. Staining the gels (e.g. with silver, Coomassie blue, or fluorescent dyes) allows the direct determination of protein expression rates based on protein spot size and intensity. Protein spots can be subsequently excised and digested in-gel (most commonly with trypsin) prior to mass spectrometric analysis. A significant improvement of this technology was introduced in the late 1990s, when it became possible to label different samples with fluorescent dyes and pool these samples before 2D-PAGE (Unlü et al. 1997), thereby reducing experimental variations. Nevertheless, 2D-PAGE, which has been employed in various environmental studies (see following paragraphs), suffers from numerous limitations: (1) separation of proteins with extreme molecular masses and isoelectric points as well as hydrophobic proteins (membrane protein) by 2D-PAGE is extremely difficult, (2) co-migration of proteins within the gel hampers accurate identification and quantification, and (3) the method is hardly automatable and consequently unsuited for high-throughput analyses. Recently, one- or multi-dimensional liquid chromatography (LC) coupled to mass spectrometry has emerged as a promising alterative to 2D-PAGE (reviewed in Lane 2005). An experimental strategy that has proven extremely useful for the analysis of membrane proteins or highly polluted samples (where contaminants might interfere with trypsin digestion) is the separation of proteins by 1D-PAGE, followed by in-gel digestion of excised protein bands and separation of the resulting peptides by reverse phase chromatography. Finally, the so-called shotgun proteomics employing two- or multidimensional chromatography (e.g. strong cation exchange chromatography followed by reverse-phase chromatography) to separate peptides generated by trypsin digestion of original protein samples (reviewed in Motoyama et al. 2008) gains more and more momentum for environmental proteomics studies. The major advantage of these approaches is their high level of automation, i.e. an online connection of chromatography and mass spectrometry enabling the generation of thousands of mass spectra per hour and thus greatly facilitating high-throughput analyses.

Mass spectrometry. After applying gel-based or chromatographic separation methods to reduce sample complexity sufficiently, peptide mixtures undergo mass spectrometry analyses. Generally a mass spectrometer consists of three components: an ion source, a mass-analyzer that measures the mass to charge (m/z) ratio of the ionized particles, and a detector that monitors the number of signals at each m/z value (reviewed in Aebersold and Mann 2003; Domon and Aebersold 2006) . Frequently used ionization techniques are (1) matrix-assisted laser desorption/ionization (MALDI), where the ionization of matrix-embedded peptides is triggered by a laser beam and (2) electrospray ionization (ESI), where the ionization is achieved by dispersing a peptide-containing liquid by electrospray; these ion sources can then be coupled to various mass analyzers, most commonly time-of-flight (Tof) or ion traps. Successful environmental proteomics studies require tandem mass spectrometers with high resolution, sensitivity and mass-accuracy; moreover, automation, e.g. coupling LC separation directly to ESI-MS, greatly facilitates the analyses of numerous, complex samples. State-of-the-art mass spectrometers such as hybride quadrupole time-of-flight (Q-Tof) analysers, Fourier-transformation ion cyclotron resonance (FT-ICR) mass spectrometers (Marshall et al. 1998), or LTQ-Orbitrap mass spectrometers (Hu et al. 2005) allow highly accurate mass determination in the low-ppm to sub-ppm range. Very recently, an ultra-high resolution Tof analyser (maXis) with thus far un-matched accuracy, resolution and sensitivity has been developed that will further boost the application of mass spectrometry for metaproteome analyses.

Data Analysis and Protein Identification

There are generally two main routes for protein identification: (1) peptide mass fingerprinting that matches peptide masses measured by MS with those calculated in silico for each protein entry in the database and (2) tandem mass spectrometry (MS/MS) that determines peptide masses and generates additional peptide sequence information.

Peptide mass fingerprinting (PMF). The identification of proteins based on PMF requires the presence of the respective protein sequence information in the reference database; such data are mainly generated by genome or metagenome sequencing projects. Consequently, PMF is generally not suited for organisms or microbial communities where little or no sequence information is available, which naturally limits its application for environmental proteomics studies.

Identification of proteins by MS and MS/MS data. MS/MS is emerging as the most reliable tool to identify proteins; it determines the masses of fragment ions that have been generated by the fragmentation of specific parent ions; this information can then be used to generate peptide sequence information if the reference database contains a protein/peptide with similar fragmentation characteristics (Hunt et al. 1986). Software packages such as Mascot (Perkins et al. 1999), SEQUEST (Eng et al. 1994), or X!tandem (Craig and Beavis 2004) allow high-throughput analyses of thousands of un-interpreted experimental MS and MS/MS spectra that can be generated by a single MS run. The newest generation of mass spectrometers provides MS/MS data with sufficient mass accuracy to deduce the exact amino acid sequence of peptides (Nesvizhskii et al. 2007). De-novo sequencing of peptides is especially useful for the identification/characterization of proteins for which no homologue in the database exists and thus highly suitable for metaproteome analyses of unexplored microbial communities.

Data Evaluation

Analogous to conventional proteome analyses, metaproteomics data should be evaluated by complementary approaches, e.g.16S rRNA-pyrosequencing or fluorescent in-situ hybridization (FISH) to assess community composition, and also transcriptome analyses and quantitative PCR approaches to confirm the expression of protein coding genes on the mRNA level. If applicable, phenotypical assays (e.g. enzymatic measurements) should be employed to confirm the expression and functionality of certain proteins. Naturally, the isolation of RNA and active proteins from environmental samples is hampered by sample complexity and interfering contaminants, thus a comprehensive validation of data obtained by metaproteome analyses remains extremely challenging.

Current Environmental Proteomics Studies – Where Are We So Far?

Even though microorganisms are of major importance for every biological system as they contribute to global nutrient cycling, organic matter decomposition, eutrophization, and many other processes, the application of community or metaproteome analyses to study structure and function of uncultivable microorganisms or microbial communities in their natural environment is still limited (Table 4.1). In the following paragraphs we will discuss milestone publications in detail, starting with low-complexity communities, e.g. mixed biofilms, followed by medium-complexity habitats, e.g. animal and human intestinal tract, up to highly complex habitats, e.g. aqueous or soil environments.

Community Proteomics of Marine Symbionts of Riftia pachyptila

The deep-sea tube worm R. pachyptila harbours a specialized organ, the trophosome, filled with sulphide-oxidizing endosymbiotic bacteria that provide the worm with carbon, nitrogen and other nutrients. An intracellular and membrane protein reference

Table 4.1 Overview of milestone proteomics approaches to study structure and function of microbial communities (Modified from Verberkmoes et al. 2009a). Studies are listed according to increasing habitat complexity. n.d. = not described

Environment	Estimated Nr. of species/phylotypes expressed proteins[a]	Methodology	Nr. of identified proteins	Reference
Trophosome of *Riftia pachyptila*	1 $3.0/10^3$	2D-PAGE & MALDI-ToF MS	220	Markert et al. 2007
Acid mine drainage biofilm	6	1D-PAGE, 2D- LC & Q-ToF MS/MS	2.033	Ram et al. 2005
Acid mine drainage biofilm	$1.8/10^4$	2D-LC & LTQ MS/MS	n.d.	Lo et al. 2007
Acid mine drainage biofilm		2D-LC & LTQ MS/MS	2.752	Denef et al. 2009
Waste water treatment reactor	17/268	2D-PAGE & MALDI-ToF MS/MS	109	Lacerda et al. 2007
Sludge EPS	$5.1/10^4/8.0/10^5$	1D-PAGE, LC & Qtrap MS/MS	10	Park et al. 2008
Sludge		2D-PAGE & MALDI-ToF MS, Q-ToF MS/MS	46	Wilmes et al. 2004
Sludge		2D-LC & LTQ MS/MS, Orbitrap, MS/MS	2.378	Wilmes et al. 2008a, b
Higher termite hindgut	~200 $6.0/10^5$	3D-LC & LCQ-MS/MS	n.d.	Warnecke et al. 2007
Sheep rumen	~20 dominant bacterial species $6.0/10^4$	1D-PAGE & MS/MS	4	Toyoda et al. 2009
Infant gastrointestinal tract	100/1,000	2D-PAGE & MALDI-Tof MS	1	Klaassens et al. 2007
Human distal gut	$3.0/10^5/3.0/10^6$	2D- LC & Orbitrap MS/MS	3.234	Verberkmoes et al. 2009b
Estuary	100/100,000	2D-PAGE + LC & Q-Tof MS/MS	3	Kan et al. 2005
Ocean	$3.0/10^5/3.0/10^7$	2D-LC & LTQ MS/MS	1.042	Sowell et al. 2009
Lake and soil	$1/10^6$	2D-LC & Q-Tof MS/MS	513	Schulze et al. 2005
Contaminated soil, groundwater	$3.0/10^9$	1D or 2D-PAGE + LC & MS/MS	59	Benndorf et al. 2007
Contaminated aquifer sediment		2D-PAGE & Nano-LC MS/MS + LTQ MS/MS	23	Benndorf et al. 2009

[a]Estimated numbers of species and proteins present in the respective environment are based on average environmental microbial genome size of 3 Mbp and 1 kbp of sequence coding for one gene (Modified from Wilmes and Bond 2006)

map based on metagenomic data of the endosymbionts (Robidart et al. 2008) was created by 2D-PAGE coupled to MALDI-Tof MS and 1D-PAGE combined with 2D-LC-MS/MS (Markert et al. 2007). It showed that the bacteria simultaneously express enzymes of the Calvin cycle and the reductive TCA cycle to fix CO_2. Moreover, the comparison of protein profiles derived from sulphide-rich and sulphide-depleted environments indicated that the *Riftia* endosymbionts repress the expression of energetically costly sulphide-oxidation-related enzymes and the key Calvin cycle enzyme RubisCo in favour of less ATP-consuming TCA cycle enzymes when H_2S is limited (Markert et al. 2007).

Whole-Community Proteomics of Richmond Acid Mine Drainage (AMD) Mixed Biofilms

An outstanding example for a comprehensive shotgun proteomics approach (2D-LC-MS/MS) is represented by the study of Ram et al. (2005) who investigated the protein complement of natural biofilms present in the acid drainage of the Richmond mine that exhibit a comparatively low complexity . The authors identified more than 2,000 proteins from the five most abundant species and obtained a remarkable 48% protein coverage for the dominant biofilm organism *Leptospirillum* group II. In further analyses of the same biofilms, Lo et al. (2007) were able to differentiate between peptides of discrete AMD populations and found strong evidence for interpopulation recombination – an approach strongly dependent on a database containing strain-specific genome information and referred to as "strain-resolving proteogenomics". The study was expanded by Denef et al. (2009), whose extensive analysis of 27 distinct AMD biofilm protein profiles revealed that specific environmental conditions select for particular recombinant types thus leading to a fine-scale tuning of microbial populations.

Proteome Analyses of Waste Water Treatment Plants and Activated Sludge

Lacerda et al. (2007) investigated the response of a natural community in a continuous-flow wastewater treatment bioreactor to an inhibitory level of cadmium by 2D-PAGE combined with MALDI-Tof/Tof MS and de-novo sequencing. The authors observed a significant shift of the community proteome after cadmium shock, as indicated by the differential expression of more than 100 proteins including ATPases, oxidoreductases and transport proteins. Park et al. (2008) analyzed the protein complement of extracellular polymeric substances of activated sludge flocs by 1D-PAGE combined with LC-MS/MS and identified a limited number of bacterial but also human polypeptides, among them proteins associated with bacterial defense, cell appendages, outer membrane proteins and a human elastase. In 2004, Wilmes and Bond (2004) studied the molecular mechanisms of enhanced biological phosphorus removal (EBPR) by a comparative metaproteome analysis of two laboratory wastewater sludge microbial communities with and without EBPR performance by 2D-PAGE combined to MALDI-Tof MS. Major differences in protein expression profiles between the two reactors were detected. A short time later, more than 2,300 proteins were identified by 2D-LC-MS/MS analyses of activated sludge (Wilmes et al. 2008a, b), aided by reference metagenomic data from studies of EBPR sludge (García Martín et al. 2006). The obtained data indicated that the uncultured polyphosphate-accumulating bacterium "Candidatus *Accumulibacter phosphatis*" is dominating the microbial community of the EBPR reactor and further enabled an extensive analysis of metabolic pathways, e.g. denitrification, fatty acid cycling and glyoxylate bypass, all central to EBPR.

Community Proteomics of Animal and Human Intestinal Tracts

Warnecke et al. (2007) employed a combined genomics and 3D-LC-MS/MS proteomics approach to investigate the microbial community present in the hindgut of higher wood-feeding termites. They reported the presence of a large set of bacterial enzymes involved in the degradation of cellulose and xylan and other important symbiotic functions such as H_2 metabolism, CO_2-reductive acetogenesis and N_2 fixation. In a more recent study Toyoda et al. (2009) used 1D-PAGE coupled to MS/MS to identify cellulose-binding proteins derived from sheep rumen; among these proteins were endoglucanase F of the cellulolytic bacterium *Fibrobacter succinogenes* and fungal exoglucanase Cel6A (*Piromyces equi*). Klaassens et al. (2007) studied the functionality of the uncultured microbiota of human infant stool samples by 2D-PAGE combined with MALDI-Tof MS. The authors observed time-dependent changes in the gut metaproteome, but were not able to identify more than one protein exhibiting high similarity to a bifidobacterial transaldolase due to at that time limited microbiome sequence information. Finally, Verberkmoes et al. (2009b) identified several thousand proteins present in human fecal samples by an extensive shotgun proteome analysis, among them bacterial proteins involved in well-known but also un-described microbial pathways and human antimicrobial peptides.

Metaproteome Analyses of Ocean Water

One of the very first metaproteome analyses was presented by Kan et al. (2005), who compared protein profiles from various sample origins of the Chesapeake Bay by 2D-PAGE and tried to identify protein spots excised from the gels by an LC-MS/MS approach; however, the obtained information was rather limited, as a substantial DNA sequence background was still lacking. Recently, Sowell et al. (2009) published a comprehensive study of the Sargasso Sea surface metaproteome. The authors employed 2D-LC coupled to MS/MS and identified over 1,000 proteins, among them an overwhelming number of SAR11 periplasmic substrate-binding proteins as well as *Prochlorococcus* and *Synechococcus* proteins involved in photosynthesis and carbon fixation. High abundance of SAR11 transporters suggests that cells endeavor to maximize nutrient uptake activity and thus gain a competitive advantage in nutrient-depleted environments.

Metaproteome Studies of Highly Complex Groundwater and Soil Environments

Schulze et al. (2005) presented an interesting functional insight into the complex micro-bial communities present in dissolved organic matter from lake water and seepage water adhering to soil micro-particles. Although the number of proteins identified by 2D-LC-MS/MS was comparatively low, the authors were able to assign functional proteins to broad taxonomic groups and observed rather unexpected seasonal variations of the protein complement. Notably, decomposing enzymes were only found among proteins extracted from soil particles, thereby indicating that the degradation of soil organic matter mainly takes place in biofilm-associated communities. More recently, Benndorf et al. (2007) published a metaproteome analysis of protein extracts from contaminated soil and groundwater employing either 1D- or 2-D-PAGE combined with LC and MS/MS. Proteome analyses of soils mainly suffer from numerous inorganic and organic contaminants, which hamper protein separation and identification; thus, only 59 proteins could be identified although the authors presented a multi-step purification protocol combining NaOH treatment and phenol extraction. A similar approach was employed to investigate the metaproteome of an anaerobic benzene degrading com-munity inhabiting aquifer sediments (Benndorf et al. 2009). Even though only a hand-ful of proteins were identified – among them an enoyl-CoA hydratase involved in the anoxic degradation of xenobiotics –, the authors demonstrated that their metaproteome extraction method is potentially valuable to investigate anoxic microbial communities.

Future Perspectives and Final Remarks

When viewed in relation to its enormous potential, the actual output of environmental proteomics appears so far to be disappointingly limited. Present studies have mainly focussed on microbial communities with a relatively low diversity or dominated by a particular phylogenetic group. The main obstacles towards a comprehensive metaproteome coverage seem to be (1) the irregular species distribution within environmental samples, (2) the wide range of protein expression levels within microbial cells and (3) the enormous genetic heterogeneity within microbial popu-lations (Wilmes and Bond 2009). It is encouraging to note, however, that constantly improving extraction methods alongside advances in down-stream mass spectrometry technology and a steadily growing pool of bioinformatics data might soon help to overcome the current challenges and limitations of metaproteomics research.

Improvements of Mass Spectrometer Sensitivity and Accuracy

A successful environmental proteomics experiment entails the reliable identification
not only of the predominant but also of low-abundance proteins, which intimately
depends on ultra-sensitive and highly accurate mass spectrometers. A significant
and foreseeable improvement in mass spectrometry performance will enable us to
(1) identify low-abundance but important gene products e.g. proteins involved in
transcriptional and translational regulation, (2) evaluate the spatial distribution of
proteins within complex habitats and (3) investigate proteins on such a fine scale
that we might even envision single cell proteome analyses.

Quantitative Metaproteomics – Dream or Reality?

Another important future line of research will take advantage of the increasing
power of metaproteomics approaches to quantitatively analyse/compare protein
expression rates in environmental samples. Despite its well-known drawbacks,
2D-PAGE has dominated quantitative protein expression studies until recently;
moreover, quantitative proteomics has been strictly limited to biological systems of
low complexity. Nowadays, 2D-gel free LC-MS-based technologies have emerged
as powerful tools for comparative/quantitative proteome studies and might be also
employed for in-depth, quantitative proteome profiling of complex environments.
A revolutionary development in the field of quantitative proteomics was the introduction of isotope- or isobar-tag based technologies e.g. ICAT (Gygi et al. 1999),
iTRAQ (Ross et al. 2004), and ANIBAL (Panchaud et al. 2008). However, so far
none of these methods was used to assess protein expression in complex environmental samples, which might be due to the fact that these techniques are relatively
costly, can only be applied to a limited number of samples and need considerable
post-processing of the original samples. These serious disadvantages might be
overcome by novel label-free techniques, which are either based on measuring
signal intensities of peptide precursor ions belonging to a particular protein (e.g.
Ono et al. 2006) or on counting fragment spectra of peptides used to identify a
certain protein (e.g. Gilchrist et al. 2006). The first approach relies on high-quality
mass-spectrometric data characterized by high mass accuracy to reduce interfering
signals and optimized peptide chromatographic reproducibility, which allows the
correlation of corresponding peptides in different samples. Promising tools to meet
these requirements are the newest generation electrospray ultra-high resolution
TOF mass spectrometers, which might even have the potential to generate reliable
quantitative data from complex environments. Another advantage of label-free
approaches is their comparable high dynamic range, which is of particular importance
when multifaceted and large protein changes within different samples have to
be anticipated (reviewed in Bantscheff et al. 2007). Even though promising, these
novel approaches still encounter certain limitations: On the one hand, they are less

accurate than 2D-PAGE or label-based quantitative technologies as samples have to be analyzed separately; to overcome this it is crucial that sample preparation and analysis become highly standardised and reproducible. On the other hand, the analysis of complex datasets places enormous demands on computational power. At present the development of commensurate hardware and, even more importantly, software tools lags behind the advances in mass spectrometry (Choi et al. 2008; Ono et al. 2006).

Final Remarks

Proteomics is one of today's fastest developing research areas and has contributed substantially to our understanding of individual organisms at the cellular level. Its attractiveness stems from being able to probe many protein functions and responses simultaneously, and seems also ideally suited to improve our knowledge of the complex interplay between the constitution of a habitat, diversity and architecture of microbial communities and ecosystem functioning. Recently, a limited number of studies describing large-scale proteome analyses of environmental samples have demonstrated the huge potential of metaproteomics to unveil the molecular mechanisms involved in function, interaction, physiology and evolution of microbial communities. Moreover, the rapidly growing number of genomic and metage-nomic sequences together with revolutionary advances in protein analysis and bioinformatics have opened up a completely new range of applications, e.g. studying the impact of environmental changes upon protein expression profiles of entire microbial communities ("quantitative metaproteomics") or measuring low-level protein expression differences in order to resolve the functional significance of spatial protein distribution within a given environment. In conclusion, the comprehensive knowledge gained by the concerted application of system-level approaches such as genomics, transcriptomics, proteomics and metabolomics will greatly advance our understanding of biogeochemical cycles and will facilitate the use of biotechnologies that rely on microbial communities or un-cultivable organisms.

Acknowledgement The authors would like to thank Alexander Grunau for critically reading the manuscript.

References

Aebersold R, Mann M (2003) Mass spectrometry-based proteomics. Nature 422:198–207
Ahmadian A, Ehn M, Hober S (2006) Pyrosequencing: history, biochemistry and future. Clin Chim Acta 363:83–94
Amann RI, Ludwig W, Schleifer KH (1995) Phylogenetic identification and *in situ* detection of individual microbial cells without cultivation. FEMS Microbiol Rev 59:143–169
Bantscheff M, Schirle M, Sweetman G, Rick J, Kuster B (2007) Quantitative mass spectrometry in proteomics: a critical review. Anal Bioanal Chem 389:1017–1031

Benndorf D, Balcke GU, Harms H, von Bergen M (2007) Functional metaproteome analysis of protein extracts from contaminated soil and groundwater. ISME J 1:224–234

Benndorf D, Vogt C, Jehmlich N, Schmidt Y, Thomas H, Woffendin G, Shevchenko A, Richnow HH, Von Bergen M (2009) Improving protein extraction and separation methods for investigating the metaproteome of anaerobic benzene communities within sediments. Biodegration 2009 Apr 21, Epub ahead of print

Bochner BR (2009) Global phenotypic characterization of bacteria. FEMS Microbiol Rev 33:191–205

Brock TD (1987) The study of microorganisms *in situ*: progress and problems. Symp Soc Gen Microbiol 41:1–17

Choi H, Fermin D, Nesvizhskii AI (2008) Significance analysis of spectral count data in label-free shotgun proteomics. Mol Cell Proteomics 7:2373–2385

Craig R, Beavis RC (2004) TANDEM: matching proteins with tandem mass spectra. Bioinformatics 20:1466–1467

Denef VJ, Verberkmoes NC, Shah MB, Abraham P, Lefsrud M, Hettich RL, Banfield JF (2009) Proteomics-inferred genome typing (PIGT) demonstrates inter-population recombination as a strategy for environmental adaptation. Environ Microbiol 11:313–325

Domon B, Aebersold R (2006) Mass spectrometry and protein analysis. Science 312:212–217

Eng JK, McCormack AL, Yates JR III (1994) An approach to correlate tandem mass spectral data of peptides with amino acid sequences in a protein database. J Am Soc Mass Spectrom 5:976–989

García Martín H, Ivanova N, Kunin V et al (2006) Metagenomic analysis of two enhanced biological phosphorus removal (EBPR) sludge communities. Nat Biotechnol 24:1263–1269

Gilchrist A, Au CE, Hiding J et al (2006) Quantitative proteomics analysis of the secretory pathway. Cell 127:1265–1281

Görg A, Postel W, Gunther S (1988) The current state of two-dimensional electrophoresis with immobilized pH gradients. Electrophoresis 9:531–546

Gygi SP, Rist B, Gerber SA, Turecek F, Gelb MH, Aebersold R (1999) Quantitative analysis of complex protein mixtures using isotope-coded affinity tags. Nat Biotechnol 17:994–999

Hu Q, Noll RJ, Li H, Makarov A, Hardman M, Graham Cooks R (2005) The Orbitrap: a new mass spectrometer. J Mass Spectrom 40:430–443

Hunt DF, Yates JR 3rd, Shabanowitz J, Winston S, Hauer CR (1986) Protein sequencing by tandem mass spectrometry. Proc Natl Acad Sci USA 83:6233–6237

Kan J, Hanson TE, Ginter JM, Wang K, Chen F (2005) Metaproteomic analysis of Chesapeake Bay microbial communities. Saline Syst 1:7–20

Keller M, Hettich R (2009) Environmental proteomics: a paradigm shift in characterizing microbial activities at the molecular level. Microbiol Mol Biol Rev 73:62–70

Klaassens ES, de Vos WM, Vaughan EE (2007) Metaproteomics approach to study the functionality of the microbiota in the human infant gastrointestinal tract. Appl Environ Microbiol 73:1388–1392

Lacerda CM, Choe LH, Reardon KF (2007) Metaproteomic analysis of a bacterial community response to cadmium exposure. J Proteome Res 6:1145–52

Lane CS (2005) Mass spectrometry-based proteomics in the life sciences. Cell Mol Life Sci 62:848–869

Lo I, Denef VJ, Verberkmoes NC et al (2007) Strain-resolved community proteomics reveals recombining genomes of acidophilic bacteria. Nature 446:537–541

Markert S, Arndt C, Felbeck H et al (2007) Physiological proteomics of the uncultured endosymbiont of *Riftia pachyptila*. Science 315:247–250

Maron PA, Ranjard L, Mougel C, Lemanceau P (2007) Metaproteomics: a new approach for studying functional microbial ecology. Microb Ecol 53:486–493

Marshall AG, Hendrickson CL, Jackson GS (1998) Fourier transform ion cyclotron resonance mass spectrometry: a primer. Mass Spectrom Rev 17:1–35

Motoyama A, John R, Yates JR III (2008) Multidimensional LC separations in shotgun proteomics. Anal Chem 80:7187–7193

Nesvizhskii AI, Vitek O, Aebersold R (2007) Analysis and validation of proteomic data generated by tandem mass spectrometry. Nat Methods 4:787–797

O-Farrel PH (1975) High resolution two-dimensional electrophoresis of proteins. J Biol Chem 250:4007–4021

Ono M, Shitashige M, Honda K, Isobe T, Kuwabara H, Matsuzuki H, Hirohashi S, Yamada T (2006) Label-free quantitative proteomics using large peptide data sets generated by nanoflow liquid chromatography and mass spectrometry. Mol Cell Proteomics 5:1338–1347

Panchaud A, Hansson J, Affolter M, Bel Rhlid R, Piu S, Moreillon P, Kussmann M (2008) ANIBAL, stable isotope-based quantitative proteomics by aniline and benzoic acid labeling of amino and carboxylic groups. Mol Cell Proteomics 7:800–812

Park C, Helm RF, Novak JT (2008) Investigating the fate of activated sludge extracellular proteins in sludge digestion using sodium dodecyl sulfate polyacrylamide gel electrophoresis. Water Environ Res 80:2219–2227

Perkins DN, Pappin DJ, Creasy DM, Cottrell JS (1999) Probability-based protein identification by searching sequence databases using mass spectrometry data. Electrophoresis 20:3551–67

Ram RJ, Verberkmoes NC, Thelen MP, Tyson GW, Baker BJ, Blake RC 2nd, Shah M, Hettich RL, Banfield JF (2005) Community proteomics of a natural microbial biofilm. Science 308:1915–1920

Robidart JC, Bench SR, Feldman RA, Novoradovsky A, Podell SB, Gaasterland T, Allen EE, Felbeck H (2008) Metabolic versatility of the Riftia pachyptila endosymbiont revealed through metagenomics. Environ Microbiol 10:727–737

Ross PL, Huang YN, Marchese JN et al (2004) Multiplexed protein quantitation in *Saccharomyces cerevisiae* using amine-reactive isobaric tagging reagents. Mol Cell Proteomics 3:1154–1169

Schloss PD, Handelsman J (2006) Toward a census of bacteria in soil. PLoS Comput Biol 2:e92

Schneider T, Riedel K (2009) Environmental proteomics: Analysis of structure and function of microbial communities. Proteomics 10:785–798

Schulze WX, Gleixner G, Kaiser K, Guggenberger G, Mann M, Schulze ED (2005) A proteomic fingerprint of dissolved organic carbon and of soil particles. Oecologia 142:335–343

Singleton I, Merrington G, Colvan S, Delahunty JS (2003) The potential of soil protein-based methods to indicate metal contamination. Appl Soil Ecol 23:25–32

Sowell SM, Wilhelm LJ, Norbeck AD, Lipton MS, Nicora CD, Barofsky DF, Carlson CA, Smith RD, Giovannoni SJ (2009) Transport functions dominate the SAR11 metaproteome at low-nutrient extremes in the Sargasso Sea. ISME J 3:93–105

Toyoda A, Iio W, Mitsumori M, Minato H (2009) Isolation and identification of cellulose-binding proteins from sheep rumen contents. Appl Environ Microbiol 75:1667–1673

Tringe SG, von Mering C, Kobayashi A et al (2005) Comparative metagenomics of microbial communities. Science 308:554–557

Unlü M, Morgan ME, Minden JS (1997) Difference gel electrophoresis: a single gel method for detecting changes in protein extracts. Electrophoresis 18:2071–2077

Verberkmoes NC, Denef VJ, Hettich RL, Banfield JF (2009a) Systems biology: functional analysis of natural microbial consortia using community proteomics. Nat Rev Microbiol 7:196–205

Verberkmoes NC, Russell AL, Shah M et al (2009b) Shotgun metaproteomics of the human distal gut microbiota. ISME J 3:179–189

Vieites JM, Guazzaroni ME, Beloqui A, Golyshin PN, Ferrer M (2009) Metagenomics approaches in systems microbiology. FEMS Microbiol Rev 33:236–255

Warnecke F, Luginbühl P, Ivanova N et al (2007) Metagenomic and functional analysis of hindgut microbiota of a wood-feeding higher termite. Nature 450:560–565

Wilkins MR, Sanchez JC, Gooley AA, Appel RD, Humphery-Smith I, Hochstrasser DF, Williams KL (1995) Progress with proteome projects: why all proteins expressed by a genome should be identified and how to do it. Biotechnol Genet Eng Rev 13:19–50

Wilmes P, Bond PL (2004) The application of two-dimensional polyacrylamide gel electrophoresis and downstream analyses to a mixed community of prokaryotic microorganisms. Environ Microbiol 6:911–920

Wilmes P, Bond PL (2006) Metaproteomics: studying functional gene expression in microbial ecosystems. Trends Microbiol 4:92–97

Wilmes P, Bond PL (2009) Microbial community proteomics: elucidating the catalysts and metabolic mechanisms that drive the Earth's biogeochemical cycles. Curr Opin Microbiol May 2, 2009. Epub ahead of print

Wilmes P, Andersson AF, Lefsrud MG, Wexler M, Shah M, Zhang B, Hettich RL, Bond PL, VerBerkmoes NC, Banfield JF (2008a) Community proteogenomics highlights microbial strain-variant protein expression within activated sludge performing enhanced biological phosphorus removal. ISME J 2:853–864

Wilmes P, Wexler M, Bond PL (2008b) Metaproteomics provides functional insight into activated sludge wastewater treatment. PLoS ONE 3:e1778

Zhou J, Thompson DK (2002) Challenges in applying microarrays to environmental studies. Curr Opin Biotechnol 13:204–207

Chapter 5
Analysis of Microbial Communities by Functional Gene Arrays

Joy D. Van Nostrand, Zhili He, and Jizhong Zhou

Introduction

A major hurdle to the study of microbial communities is that only about 1% of microorganisms are cultivated (Whitman et al. 1998). As such, culture-independent approaches are necessary in order to examine the vast majority of environmental microorganisms. Many molecular techniques are available for community analysis, and most of these techniques utilize phylogenetic markers such as the 16S rRNA or the DNA gyrase gene (*gyrB*) (Wilson et al. 1990; Yamamoto and Harayama 1995; Hugenholtz et al. 1998; Brodie et al. 2006). While the use of these genes provides information regarding phylogenic diversity and structure of a microbial community, they don't provide much, if any information relating to the functional potential and/or activity of the community. Functional genes have been used to examine both phylogenetic and functional diversities (e.g., McDonald et al. 1995; Braker et al. 1998). However, even if multiple functional genes are examined, conventional molecular techniques only provide information on a small fraction of the community. This is because conserved PCR primers cannot be designed for many functional genes of interest due to a lack of sequence homology or a lack of a sufficient number of sequences. Consequently, conventional PCR-based approaches cannot be used to detect and quantify many functional genes of interest. As such, a more comprehensive technique is required to provide a full picture of microbial community activity and dynamics in a rapid, parallel, and high-through-put manner.

Use of microarrays is one way to overcome these limitations. Microarrays provide a rapid and high-throughput method of examining thousands of functional

J.D. Van Nostrand, Z. He, and J. Zhou (✉)
Department of Botany and Microbiology, Institute for Environmental Genomics,
University of Oklahoma, 101 David L. Boren Blvd, Norman, OK 73019, USA
e-mail: jzhou@ou.edu

L.L. Barton et al. (eds.), *Geomicrobiology: Molecular and Environmental Perspective*,
DOI 10.1007/978-90-481-9204-5_5, © Springer Science+Business Media B.V. 2010

genes at one time without the need for conserved primers. Microarray technology was first developed for analyzing gene expression in the plant *Arabidopsis thaliana* (Schena et al. 1995). Such gene expression arrays are now commonly used to study individual microorganisms. However, the potential usefulness of microarray technology was greatly expanded when Guschin and colleagues (1997) proposed and tested the use of microarrays to study microbial communities. Now, several different types of arrays are available for the study of microbial communities in the environment (Zhou and Thompson 2002; Zhou 2003; Gentry et al. 2006). Phylogenetic oligonucleotide arrays (POA) are designed to determine community composition or phylogenetic relatedness using 16S rRNA or other phylogenetically informative genes (Small et al. 2001; Loy et al. 2001; Wilson et al. 2002). The most comprehensive POA is the PhyloChip, which contains almost 300,000 perfect-match and mismatch probes for 842 subfamilies (Brodie et al. 2006); although the 23S rRNA gene has also been used for POA construction (Lee et al. 2006). Community genome arrays (CGA) are used to examine the relatedness of microbial strains or to detect specific microorganisms in the environment using whole genomic DNA of individual species or strains as probes (Wu et al. 2004; Zhang et al. 2004; Wu et al. 2008). Metagenomic arrays (MGA) are made using clone libraries created from environmental DNA as probes (Sebat et al. 2003). This array was used as a high-throughput screening method for clone libraries. Whole-genome open reading frame (ORF) arrays (WGA) are comprised of probes for all ORFs in one or more genomes (Wilson et al. 1999). This type of array has been used to examine the diversity and relatedness of several metal-reducing *Shewanella* strains (Murray et al. 2001). A WGA of 353 virulence factors was used to evaluate over 100 strains of *Pseudomonas syringae* to determine those genes associated with host specificity and several genes were identified that were statistically associated with specific hosts (Sarkar et al. 2006). Functional gene arrays (FGA) are composed of probes for key genes involved in microbial functional processes of interest (Wu et al. 2001; Gentry et al. 2006; He et al. 2007). FGAs allow for the simultaneous examination of many functional gene groups (Wu et al. 2001; Zhou and Thompson 2002; Gentry et al. 2006; Wu et al. 2006; He et al. 2007; Wagner et al. 2007; Zhou et al. 2008; Wang et al. 2009) unlike PCR-based techniques, which limit the number of genes that can be examined at once. The focus of this chapter will be the development and application of FGAs.

Functional Gene Array Development

The first reported FGA contained ~100 PCR-amplicon probes targeting N-cycling genes (*nirS*, *nirK*, *amoA*, and *pmoA*) (Wu et al. 2001). However, the use of PCR-based probes limits the comprehensiveness of an array since a very large number of diverse bacterial strains and environmental clones would be required. Another issue with the use of PCR probes is the need to develop conserved primers for each gene or gene group, which would pose a problem as no conserved PCR primers can be designed for many functional genes of interest. In addition, although fairly conserved primers can be designed for some functional genes of interest, it is sometimes difficult to amplify these genes from environmental samples. To overcome these technical challenges, oligonucleotide probes have been used instead of PCR amplicons. Oligonucleotide probes have higher specificity but lower sensitivity than PCR-based probes (Zhou 2003), can be easily customized allowing more targeted probe design (Denef et al. 2003; Zhou 2003; Gentry et al. 2006), and are relatively inexpensive. As such, FGAs are often constructed using oligonucleotide probes.

In the decade since the first FGA was reported, several different FGAs have been developed (Cho and Tiedje 2002; Bodrossy et al. 2003; Rhee et al. 2004; Gentry et al. 2006; Zhang et al. 2006). Some have targeted specific functional groups or genes, such as antibiotic resistance (Call et al. 2003), organic contaminant degradation and metal resistance (Rhee et al. 2004), N-cycling (Taroncher-Oldenburg et al. 2003; Steward et al. 2004; Tiquia et al. 2004), methanotrophs (Stralis-Pavese et al. 2004), virulence factors and pathogen-specific markers (Miller et al. 2008; Palka-Santini et al. 2009), *nodC* variants (Bontemps et al. 2005), or specific locations like acid mine drainage sites (Yin et al. 2007). To date, the most comprehensive FGA reported is the GeoChip 2.0 (He et al. 2007), a high density FGA, with 24,243 50-mer oligo-nucleotide probes, targeting ~10,000 functional genes from 150 gene families involved in the geochemical cycling of C, N, and P cycling, sulfate reduction, metal reduction and resistance, and organic contaminant degradation. The GeoChip was designed to provide sufficient oligonucleotide probe specificity for genes that have high homologies and to provide a truly comprehensive FGA probe set, both of which were lacking in previous FGAs (He et al. 2007). A newer version, GeoChip 3.0, which covers ~47,000 sequences from 292 gene families, covering twice as many functional gene groups as GeoChip 2.0, including the phylogenetic marker *gyrB*, has been developed (He et al. 2010a).

Comparison of FGA to Other High-Throughput Genomic Technologies

Several high-throughput methods are available for microbial community studies in addition to GeoChip, including the PhyloChip and barcode-based high-throughput sequencing. In contrast to 16S rRNA gene-based microarrays (e.g., PhyloChip) and sequencing technologies (e.g., 454), functional gene arrays (e.g., GeoChips) have several advantages: (a) Detecting functions. While PhyloChip is a powerful tool for examining microbial communities, this array only has probes for the 16S rRNA gene to detect the presence of a strain but not its functional activity. Sequencing allows for obtaining hundreds of thousands of sequences, but requires an initial PCR step. As mentioned above, the use of PCR primer based methods are limited since primers are only available for a limited number of functional genes. In addition, the sheer number of reactions would be prohibitive. Therefore, an examination of a wide variety of functional genes of interest is nearly impossible with this method. FGAs, however, use functional gene markers and thus provide information on the potential metabolic functions of a community (Wu et al. 2001; He et al. 2007). As such, GeoChip, which contains probes for thousands of functional genes, is particularly useful in linking microbial community structure to community function (He et al. 2007; 2010a). (b) Higher resolution. GeoChip can provide resolution at the species-strain level (Tiquia et al. 2004); whereas the resolution of PhyloChip is at the family-subfamily level (DeSantis et al. 2007) and 16S rRNA gene-based sequencing can generally provide resolution at the genus-species level (He Z et al., unpublished data). (c) Quantitation. Many ecological studies require quantitative information regarding microbial abundance. Since PhyloChip and sequencing require PCR amplification steps, these techniques may not provide quantitative results because amplification bias is a well-known phenomenon in PCR (Warnecke et al. 1997; Lueders and Friedrich 2003; Suzuki and Giovannoni 1996). In contrast, GeoChip does not rely on PCR amplification for detection. Previous studies with FGAs have shown that hybridization of template DNA with or without random amplification are quite quantitative (Wu et al. 2001; Tiquia et al. 2004; Rhee et al. 2004; Wu et al. 2006; Gao et al. 2007). In addition, whole community genome amplification (WCGA), which can be used to increase the amount of DNA available for hybridization has been shown to produce minimal bias (Wu et al. 2006).

In addition to the process differences discussed above, these technologies also have advantages and disadvantages specific to the study of microbial community structure and dynamics. These include (a) Random sampling errors. Although metagenomic technologies are able to provide a lot of data, only a small portion of the microbial community is actually sampled in most studies. Mckenna et al. (2008) obtained only ~1,400 rRNA gene sequences per sample from the gut microbiome of Rhesus macaques using 454 sequencing, an underestimate of the gut community based on Chao1 estimates. With an estimated density of 10^{11}–10^{12} bacterial cells per mL (Whitman et al. 1998) or 10^9–10^{10} g^{-1} of stool (Palmer et al. 2007) in the human gut, results of 454 pyrosequencing would be expected to greatly underestimate the microbial community. If the sampling process is completely random, theoretically, the probability of sampling the same portion of a community over multiple sampling events would be small (Zhou et al. 2008). Although dominant populations would, in all probability, be sampled multiple times, it is still not possible to ensure that the same populations of a microbial community are measured across different sampling events. As such, the estimated species richness would be subjected to random sampling errors. In contrast, microarray-based approaches compare all samples against the same set of probes (i.e., those contained on the array), ensuring that the same population are sampled for comparison across all samples in a study. As a result, the artifact due to the nature of random sampling can be minimized if not eliminated. (b) Relative abundance. Unlike pure cultures, the abundance of different species in a microbial community varies greatly. Sequencing-based approaches will be very sensitive to the distribution of species abundance. For instance, if the abundance of species A is 5% of the abundance of species B; theoretically, only one of the 20 molecules sequenced will be from species A while the remainder would be from species B. To have a 1x sequence coverage for species A, 20-fold more sequencing effort would be required. Thus, detecting all of the rare species within a community would not be cost effective or feasible even with next-generation sequencing technologies.

Microarray-based detection approaches are not affected by the relative abundance of species as long as their abundance is above the detection limit. (c) Detecting new sequences. Sequenced-based technologies can find new or novel sequences since there are no limitations on what sequences are sequenced. However, because microarrays are limited to detecting only sequences covered by the probe sequences on an array, detecting new sequences is impossible. (d) Community comparison. Randomly sequencing a small portion of the microbial communities from different environments or conditions may not be informative or meaningful for comparative purposes due to random sampling errors (Zhou et al. 2008), as discussed above, unless sufficient sequencing coverage is achieved. In contrast, since microarrays interrogate communities with the same set of probes across samples, comparisons between samples and environments can easily be made. (e) Cost. In addition, although sequencing technology has developed rapidly and the cost per base pair has decreased considerably, sequencing capacity and cost are still limiting factors when entire microbial communities and/or multiple communities are considered. In contrast, after the initial output for printing and imaging equipment, microarray analysis is much less expensive than 454 pyroequencing, even for multiple samples with a barcode approach.

Design and Development of Geochip

Probe Design

A flowchart of the basic design protocol is shown in Fig. 5.1. First, specific functional genes for key functional processes of interest are selected. Genes should be

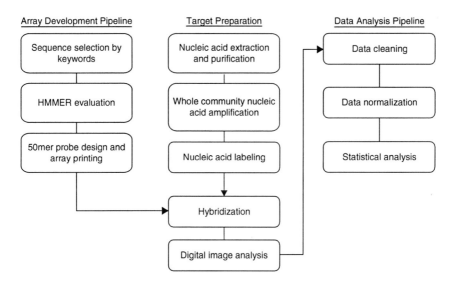

Fig. 5.1 Major steps for GeoChip design and use. See text for full explanation of all steps

chosen for key enzymes or proteins that are vital to pathways or functions of interest. Public databases (e.g., GenBank) are searched automatically using selected keywords and resulting sequences are downloaded. Care should be taken in selecting keywords as genes may be annotated differently in different microorganisms or may have a more general or specific description so using very broad key words is often best. Second, the downloaded sequences are evaluated using HMMER alignment (http://hmmer.wustl.edu/) with seed sequences, which have had protein identity and function and experimentally confirmed. The selection of seed sequences is a critical step in probe design and care should be taken in choosing appropriate sequences. The seed sequences are stored in a database for later array updates. Sequences passing HMMER alignment are deposited to a local sequence database. Third, gene-specific or group-specific 50-mer oligonucleotide probes are designed with CommOligo (Li et al. 2005) using experimentally determined criteria based on sequence homology ($\leq$90% identity for gene-specific; $\geq$96% for group-specific), continuous stretch length ($\leq$20 bases for gene-specific; $\geq$35 for group-specific), and free energy ($\geq$ −35 kJ mol^{-1} for gene-specific; $\leq$ −60 kJ mol^{-1} for group-specific (He et al. 2005b; Leibich et al. 2006). In addition, to ensure specificity, all designed probes are screened against the GenBank database. Finally, the resultant probes are then commercially synthesized and used for array construction. Probes can be spotted onto glass slides (Taroncher-Oldenburg et al. 2003; Tiquia et al. 2004; Rhee et al. 2004) or nylon membranes (Steward et al. 2004). Glass slides are generally used since they produce less background fluorescence (Schena et al. 1995, 1996) and allow higher probe density (Ehrenreich 2006).

Target Preparation

An important factor in obtaining reliable microarray data is to use high quality DNA or RNA. The key steps in target preparation are shown in Fig. 5.1. Microbial community DNA from environmental samples is generally extracted and purified using a well-established freeze-grind method since it results in large fragments of genomic DNA (Zhou et al. 1996; Hurt et al. 2001) which are important if the DNA needs to be amplified. The purified DNA should have A_{260}:A_{280} > 1.8 and A_{260}:A_{230} > 1.7. We have had success with the use of agarose gel purification followed by a phenol-chloroform-butanol extraction (Liang et al. 2009b). Impurities in the DNA can inhibit subsequent amplification, labeling and hybridization processes. If < 2.0 µg of DNA is obtained, WCGA can be used to increase the amount of DNA available for hybridization with small quantities of DNA (1–100 ng) (Wu et al. 2006). WCGA provides a sensitive (10 fg detection limit) and representative amplification (<0.5% of amplified genes showed >twofold different from unamplified) (Wu et al. 2006).

While the use of DNA provides information on the community structure and the functional potential of the microbial community, it does not provide information on the activity of the community. To examine activity of microbial communities, mRNA can also be used with FGAs. However, two major challenges in using RNA are the relatively low abundance of mRNA in environmental samples, and short turnover rates of mRNA. Several methods are available for extraction of community RNA from environmental samples, including simultaneous extraction of DNA and RNA (Hurt et al. 2001), RNA extraction via a bead-beating method (Burgmann et al. 2003), or use of gel electrophoresis to isolate mRNA from total RNA (McGrath et al. 2008). Purified RNA should have $A_{260}{:}A_{280} > 1.90$ and $A_{260}{:}A_{230} > 1.70$. Since only a small portion of the total RNA is mRNA, a large quantity of RNA (10–20 μg) is required for hybridization. However, environmental samples often do not provide a sufficient quantity of RNA, so whole community RNA amplification (WCRA) (Gao et al. 2007) may be required. WCRA employs a fusion primer comprised of a short (6–9) set of random nucleotides and a T7 promoter. Amplification of 50–100 ng of total RNA resulted in a representative amplification that maintained the original relationship of mRNA (Gao et al. 2007). Another option is the use of stable isotope probing of active community members (Leigh et al. 2007).

The DNA or RNA is then labeled with fluorescent dyes (e.g., Cy3, Cy5). For DNA, random priming with the Klenow fragment of DNA polymerase is used (Wu et al. 2006). RNA is labeled using Superscript™ II/III RNase H-reverse transcriptase (He et al. 2005b). The labeled nucleic acids are then purified and dried for hybridization.

Hybridization

Labeled DNA or RNA is suspended in hybridization buffer for hybridization. GeoChips can be hybridized at 42–50°C and 50% formamide (He et al. 2007; Mason et al. 2009; Liang et al. 2009a, b; Van Nostrand et al. 2009; Waldron et al. 2009). The hybridization temperature and formamide concentration can be adjusted to increase or decrease stringency in order to detect more or less diverse sequences. The effective hybridization temperature can be increased by the use of formamide (0.6°C for every 1%).

Hybridizations using glass arrays can be carried out manually or using automated or semi-automated hybridization stations. Manual hybridizations are performed using a water bath or hybridization oven and specially designed hybridization chambers that help maintain humidity levels within the chamber. Several hybridization stations provide incubation at controlled temperatures and mixing (e.g., Mai Tai® from SciGene, SlideBooster from Advalytix, Maui from BioMicro Systems). Washing after hybridization can be accomplished manually or using an automated wash station (e.g., Maui Wash Station, BioMicro Systems). Other systems are completely automated from pre-hybridization through post-hybridization washes (e.g., Tecan HS4800Pro, TECAN US).

Image Analysis

After hybridization, the array is imaged using a microarray scanner with a resolution of 10 μm or better. The image is then digitally analyzed by quantifying the pixel density (intensity) of each spot using microarray-analysis software. The analysis software can also be used to evaluate spot quality using predetermined criteria and flag poor or low quality spots for later removal. Distinguishing a positive spot from background noise is generally based on signal-to-noise ratio [SNR; SNR = (signal mean − background mean)/background standard deviation]. However, other calculations can be used instead of SNR. He and Zhou (2008) developed a signal-to-both-standard-deviations ratio [SSDR; SSDR = (signal mean − background mean)/(signal standard deviation − background standard deviation)] which resulted in fewer false positives and negatives than the SNR calculation.

This raw data is then uploaded to the GeoChip data analysis pipeline (http://ieg. ou.edu/) and evaluated. The quality of individual spots, evenness of control spot hybridization signals across the slide surface, and background levels are assessed. Poor and low quality spots are removed along with outliers. Outliers are determined based on the signal intensities of replicate arrays and are defined as those positive spots with (signal − mean signal intensity of all replicate spots) is greater than three times the replicate spots' signal standard deviation (He and Zhou 2008). The signal intensities are then normalized and the data is stored in an experiment database for further statistical analysis using the data analysis pipeline.

Data Analysis

The most difficult task with FGAs, especially GeoChip, is data analysis due to the seemingly overwhelming amount of data obtained. A few data analysis methods have been used frequently and include relative abundance of gene groups based on gene number or total signal intensity, richness and diversity indices based on gene number, percent of gene overlap between samples, and response ratios. Methods commonly used for statistical analysis of microarray data include principal component analysis (PCA), cluster analysis (CA), and neural network analysis (NNA) (He et al. 2008). PCA is a multivariate statistical method which reduces the dimensionality of variables to maximize the visible variability of the data. The major advantage of PCA is that it identifies outliers (e.g., genes) in the data set that behave differently from most of the genes across a set of experiments. CA is used to identify groups with similar gene profiles, and it can help establish functionally related groups of genes to gain insights into structure and function of a given microbial community. NNA is a relatively new analysis technique for FGA data but can be used to examine gene relationships. Response ratios, which compare community response (e.g., gene levels or signal intensity) between conditions (e.g., control versus treatment, contaminated versus uncontaminated) (Luo et al. 2006), has been used to compare the community response to varying levels of oil contamination (Liang et al., 2009a). In addition, if environmental variables are available, canonical correspondence analysis (CCA) (ter Braak 1986), variation partitioning analysis (VPA) (Økland and Eilertsen 1994; Ramette and Tiedje 2007), and other correlation analyses (e.g., Mantel test) can be used to correlate environmental conditions with the community structure for further understanding of the relationship between the microbial community and ecosystem functioning. CCA has been used in several GeoChip-based studies to better understand how environmental factors are affecting community structure (Yergeau et al. 2007; Wu et al. 2008; Zhou et al. 2008; Waldron et al. 2009; Van Nostrand et al. 2009). VPA is used to determine the relative influence of environmental parameters on the microbial community structure and is based on results of the CCA. The Mantel test has been used to correlate environmental factors with functional genes detected with GeoChip (He et al. 2007; Wu et al. 2008; Van Nostrand et al. 2009; Waldron et al. 2009).

Important Issues for Microarray Application

A great deal of progress has been made over the past decade with regards to the development of microarray technology for studying environmental communities (Wu et al. 2001, 2004, 2006; Adey et al. 2002; Rhee et al. 2004; Leibich et al. 2006; Gao et al. 2007; He et al. 2007; He and Zhou 2008). However, several challenges and key issues remain.

Nucleic acid quality. One of the most important steps for successful FGA analysis is obtaining high-quality DNA or/and RNA from environmental samples. Our lab has successfully used an established freeze-grind extraction method (Zhou et al. 1996; Hurt et al. 2001) followed by agarose gel purification (Liang et al. 2009b). However, some samples are still difficult to purify to the necessary level. In addition, gel purification only works for fresh DNA and for samples which yield a relatively large amount of DNA, since some portion of DNA may be lost with this method. Use of mRNA presents an even greater challenge. Isolation of mRNA from environmental samples is difficult due to the low abundance and instability of the mRNA. Very few studies have used FGAs for environmental mRNA analysis (Hurt et al. 2001; Dennis et al. 2003; Gao et al. 2007).

Probe coverage. One of the drawbacks of earlier FGA versions is that they lacked a comprehensive probe set and focused on only a few genes or gene groups. GeoChip 2.0 provided the most comprehensive FGA currently available and covered >150 gene groups (He et al. 2007). GeoChip 3.0 provides coverage of 292 gene groups and about four times as many genes as GeoChip 2.0 (He et al. 2010a). However, regardless of how comprehensive the GeoChip is now, sequences are constantly being added to public databases, leading to an exponential increase in the number of functional genes as well as the number of sequences for each particular functional gene. As such, continual updates of GeoChip are necessary. The probe design system used in our lab has an automatic update feature using predetermined keywords and seed sequences. However, even with the advances in probe design software, this process is still time consuming due to the large number of sequences and probes that must be designed and tested.

Specificity. Specificity is an important attribute of gene probes, especially for those designed to analyze environmental samples. However, a difficulty in designing specific probes is that so many environmental sequences are unknown, and that many homologous genes are highly similar. Careful design of oligonucleotide probes can provide highly specific hybridizations. Criteria based on similarity, stretch and free energy have been used to design specific probes (He et al. 2005a, 2007; Leibich et al. 2006). Our evaluation of GeoChip 2.0 probes designed using these criteria revealed that a minimal number of false positives (0.002–0.004%) were observed (He et al. 2007).

In addition to probe design criteria, specificity can be adjusted by changing the hybridization conditions to increase or decrease stringency. Hybridization stringency is generally controlled by temperature or formamide concentration. At 65°C, hybridization occurred for sequences with similarities >87% (Wu et al. 2001). At 45°C, hybridization occurred for sequences with similarities as low as 70–75% (Wu et al. 2001). Hybridizations using 50-mer oligonucleotide FGAs at 50°C and 50% formamide (effective temperature, 80°C) were able to discriminate sequences with <88 to 94% similarities (Rhee et al. 2004; Leibich et al. 2006; Deng et al. 2008).

Signal intensity can be affected by sequence divergence as well as by sequence abundance; therefore, strategies need to be developed to determine which condition is occurring. One option to determine a true signal is the use of mismatched probes. Deng et al. (2008) found that probes with mismatches (3–5) distributed evenly across the probe were better able to distinguish perfect-matched targets versus mismatched targets than randomly distributed mismatched nucleotides. In addition, using relative comparisons across samples (i.e., comparing signal intensities from test samples to a control or background sample) rather than absolute comparisons will minimize or eliminate the effects of potential cross-hybridization (He et al. 2007). Assuming that test and control samples have similar community composition, using the ratios of test to control samples will cancel out any cross-hybridization (He et al. 2007).

Sensitivity. Another important aspect of microarrays is sensitivity, especially for environmental samples which often have complex communities with many strains in low abundance. Based on current FGA technology, the detection limit is 5% of the microbial community (Bodrossy et al. 2003), which provides coverage for only the dominant community members. PCR-based probes had sensitivity of 1 ng of pure genomic DNA or 25 ng of community DNA (Wu et al. 2001). Similar detection limits were observed for 50 mer oligonucleotide probes (Rhee et al. 2004; Tiquia et al. 2004). Several strategies could be used to increase sensitivity although these also decrease specificity. For example, increasing the length of probes, increases sensitivity (Denef et al. 2003; He et al. 2005a), but at the cost of specificity (Relógio et al. 2002). Another strategy that has been suggested is to increase the amount of probe per spot (Cho and Tiedje 2002; Relógio et al. 2002; Zhou and Thompson 2002) since membrane based arrays are generally more sensitive, due to the higher probe concentrations on the array surface (>1 µg/spot for membranes; <20 pg/spot for glass slides) (Cho and Tiedje 2002). Although increasing the probe concentration may result in a lower signal intensity (Denef et al. 2003), which would effectively counteract any gain in sensitivity.

In addition to probe or array design strategies, sensitivity can also be increased by utilizing several sample preparation and hybridization strategies. Whole community genome amplification (WCGA) can increase the concentration of all community

DNA; including low abundance sequences (Wu et al. 2006). WCGA is able to representatively amplify 1 to 250 ng of community DNA (Wu et al. 2006). This amplification method showed a detection limit of 10 pg of the original DNA, although with a much higher amplification bias than observed with 1 ng of DNA. Amplification via multiplex PCR, using primers for all genes contained on an array, has been used to increase the amount of array-specific DNA for a pathogen array (Palka-Santini et al. 2009). While this strategy may work for some FGAs, especially those focused on relatively small numbers of genes, this would not work for GeoChip because of the difficulty in designing primers for all genes on the array. Another option is the use of more sensitive labeling techniques. Using cyanine dye-doped nanoparticles (Zhou and Zhou 2004) or tyramide signal amplification labeling (Denef et al. 2003) can increase sensitivity up to tenfold. A final strategy is to develop more sensitive signal detection systems (Cho and Tiedje 2002; Zhou and Thompson 2002).

Quantitative applications. A major goal for microarray analysis is to be able to provide quantitative information. Some studies have shown a correlation between signal intensity and DNA concentration. PCR probes showed a correlation ($r = 0.94$) between signal intensity and DNA quantity over the concentration range of 0.5–100 ng (Wu et al. 2001). Oligonucleotide probes (50-mer) provided a linear relationship ($r = 0.98$–0.99) over a concentration range of 8–1,000 ng (Tiquia et al. 2004). Hybridizations with RNA have also been shown to be linear over the range of 50–100 ng (Gao et al. 2007).

Activity. Using DNA with FGAs provides information on population changes which can be used to infer microbial activity; but does not provide absolute evidence of that activity. The use of community mRNA would provide information on which community members or functional processes are active, similar to transcriptional arrays for pure cultures (Dennis et al. 2003; Bodrossy et al. 2006; Gao et al. 2007). However, our current difficulties with community mRNA are the low amount and the stability of environmental mRNA. WCRA overcomes the problem of low abundance (Gao et al. 2007) although the current amplification protocol is complex and time consuming. Therefore, improved methods for RNA extraction from environmental samples and mRNA amplification are needed. Another option for determining microbial activity is SIP analysis (Leigh et al. 2007).

Application of GeoChip for Microbial Community Analysis

The GeoChip has been used in numerous studies to examine the functional community structure and dynamics of microbial communities. Most of these studies have utilized community DNA to measure gene abundance although RNA or stable isotope probing (SIP) can also be used to examine gene expression. These studies have shown the power of GeoChip to link microbial community functional structure to biogeochemical, ecological, and environmental processes.

U(VI) contaminated environments. GeoChip has been used in several studies to examine U-contaminated groundwater at the US DOE's Field Research Center

(FRC) in Oak Ridge, TN. GeoChip 1.0, a prototype array containing 2006 (50-mer) oligonucleotide probes (Rhee et al. 2004; Tiquia et al. 2004), was used to examine communities within the FRC, which is contaminated with nitrate, uranium, and organic compounds. This study examined samples from contaminated and uncontaminated areas and observed higher gene numbers in uncontaminated sites compared to the contaminated sites, indicating the deleterious effect of the contaminants on the microbial communities (Wu et al. 2006). Using GeoChip 1.0, another study examined samples with a range of contamination levels and found that more genes were detected from the uncontaminated control site compared to the numbers detected from contaminated wells (Waldron et al. 2009). Microbial communities were examined within a pilot-scale test system established for the biostimulation of U(VI) reduction in the subsurface by injection of ethanol. A significant correlation ($r = 0.73$, $p < 0.05$) was observed between the U(VI) concentration and the amount of cytochrome genes detected, indicating the importance of cytochrome containing microorganisms in U(VI) reduction at this site (He et al. 2007). In the same bioremediation system, the effects of dissolved oxygen (DO) and ethanol amendment on the microbial community were examined, and the results showed that ethanol was a much stronger driver in controlling community structure than U(VI) or DO (Van Nostrand et al. 2009).

Hydrocarbon contaminated sites. GeoChip 2.0 was used to examine the microbial community of a bioremediation system designed for the remediation of diesel fuel in Vega Baja, Puerto Rico (Rodríguez-Martínez et al. 2006). Genes involved in the degradation of diesel and organic contaminants (phthalate, biphenyl, cyclohexanol, benzoate, and naphthalene degradation genes) were detected. The amount of anaerobic degradation genes increased over time, suggesting that, consistent with other evidence, the system shifted to an anaerobic process. Liang et al. (2009b) used GeoChip 2.0 to examine the microbial community of contaminated oil fields before and after bioremediation treatment. Ozonation treatment resulted in a decrease in most functional gene categories, including carbon, nitrogen, and sulfur cycling and organic contaminant degradation genes, but all gene categories recovered after treatment. Another study examined microbial communities from a contaminated oil field in China and found that higher levels of oil contamination resulted in lower diversity and a decreased number of functional genes detected (Liang et al., 2009a). In addition, genes associated with oil degradation, including catechol, protocatechuate, biphenyl degradation, increased under a moderate oil contamination level.

Soil communities. GeoChip 1.0 was used to examine functional changes of microbial communities under different land use strategies and found that diversity and functional gene numbers increased as soil organic carbon increased (Zhang et al. 2007). In another study, Yergeau et al. (2007) examined Antarctic sediments and found that cellulose degradation and denitrification genes were positively correlated with soil temperature. Additionally, Zhou et al. (2008) used GeoChip 2.0 to assess the gene-area relationship of microbial communities of forest soils, and the results suggest that the forest soil microbial community demonstrated a relatively flat gene-area relationship with less turnover than observed for plants and animals (Zhou et al. 2008).

Marine environments. Sediments from the Gulf of Mexico were examined using GeoChip 1.0 (Wu et al. 2008). Genes for carbon degradation, nitrification, denitrification, nitrogen fixation, sulfur reduction, phosphorus utilization, contaminant degradation, and metal resistance were detected and communities become more distinctive as depth increased. The environmental parameters, depth, porosity, and concentrations of ammonium, phosphate, Mn(II), and silicic acid appeared to be important drivers in determining the structure of the microbial communities in this environment. Another study characterized microbial communities from deep sea hydrothermal vents, including a mature chimney and the inner and outer portions of a 5-day-old chimney with GeoChip 2.0, and the results showed communities from the inner chimney were less diverse than those from the outer portion of the 5-day-old chimney or the mature chimney (Wang et al. 2009). GeoChip 2.0 has also been used to examine microbial communities from deep sea basalt and genes involved in carbon fixation, methane oxidation, methanogenesis, and nitrogen fixation, processes not previously associated with this environment, were detected (Mason et al. 2009).

Climate change. The latest version of GeoChip (GeoChip 3.0) has been used to study the effects of elevated CO_2 on microbial communities at a multifactor grassland experiment site, BioCON (Biodiversity, CO_2, and Nitrogen deposition) at the Cedar Creek Ecosystem Science Reserve in Minnesota, USA. This study suggests that elevated CO_2 significantly affects the soil microbial community (He et al., 2010b). In addition, the soil microbial community appears to influence global carbon and nitrogen cycling and may moderate the effects of climate change. These studies show the power and utility of GeoChip for analyzing microbial community functional structure from a variety of environments.

Other. In addition to these community DNA-based studies, several other studies have used GeoChip 2.0 to examine pure cultures and microbial activity within communities. Van Nostrand et al. (2007) used GeoChip 2.0 to probe four metal-resistant (Ni, Co, Cd, Zn) actinomycetes for metal resistance genes. Multiple metal resistance genes were detected including some implicated in Ni, Co, Cd, and Zn resistance. Leigh et al. (2007) used stable isotope probing (biphenyl) in conjunction with GeoChip to detect active PCB-degrading microbial populations within a hydrocarbon-contaminated aquifer. Gao et al. (2007) used amplified community mRNA to examine the activity of microbial communities from a denitrifying fluidized bed reactor at a uranium contaminated site. Genes expected to be active at this site were detected including nitrate reduction genes and several organic contaminant degradation genes.

Summary

Over the past decade great advances have been made in microarray technology and in FGA development. The GeoChip 2.0 has garnered a great deal of attention, and numerous studies over the last couple of years have demonstrated its applications in the study of microbial ecology and linking microbial communities and geochemistry. These studies have demonstrated GeoChip's ability to provide sensitive, specific, and potentially quantitative information concerning microbial communities from a wide range of environments. This high-throughput, cutting edge technology is expected to revolutionize the field of microbial ecology and the study of microbial community functional structure and dynamics.

However, there are still technical, experimental and analysis challenges that need to be overcome. These include increasing sensitivity either through new technologies and methods in array printing or novel and/or improved target labeling methods to better detect functional genes at a low abundance. Strategies must be developed to improve the quantitative accuracy of FGA hybridizations. Bioinformatic tools and techniques are needed to assist in analysis, evaluation, and interpretation of the vast amounts of data resulting from FGA analysis. New tools are also needed for sequence retrieval, evaluation, and probe design. Novel analytical techniques are needed to fully utilize the FGA data. In addition, strategies and techniques must be developed in order to be able to compare data sets across samples, experiments, and labs.

Acknowledgements The effort for preparing this review was supported by the Virtual Institute for Microbial Stress and Survival (http://VIMSS.lbl.gov) supported by the U. S. Department of Energy, Office of Science, Office of Biological and Environmental Research, Genomics Program:GTL through contract DE-AC02-05CH11231 between Lawrence Berkeley National Laboratory and the U. S. Department of Energy, Environmental Remediation Science Program (ERSP), Office of Biological and Environmental Research, Office of Science, and Oklahoma Applied Research Support (OARS), Oklahoma Center for the Advancement of Science and Technology (OCAST), the State of Oklahoma through the Project AR062-034.

References

Adey NB, Lei M, Howard MT et al (2002) Gains in sensitivity with a device that mixes microarray hybridization solution in a 25-μm-think chamber. Anal Chem 74:6413–6417

Bodrossy L, Stralis-Pavese N, Murrell JC, Radajewski S, Weilharter A, Sessitsch A (2003) Development and validation of a diagnostic microbial microarray for methanotrophs. Environ Microbiol 5:566–582

Bodrossy L, Stralis-Pavese N, Konrad-Köszler M, Weilharter A, Reichenauer TG, Schöfer SA (2006) mRNA-based parallel detection of active methanotroph populations by use of a diagnostic microarray. Appl Environ Microbiol 72:1672–1676

Bontemps C, Goldier G, Gris-Liebe C, Carere S, Talini L, Boivin-Masson C (2005) Microarray-based detection and typing of the rhizobium nodulation gene *nodC*: potential of DNA arrays to diagnose biological functions of interest. Appl Environ Microbiol 71:8042–8048

Braker G, Fesefeldt A, Witzel KP (1998) Development of PCR primer systems for amplification of nitrite reductase genes (*nirK* and *nirS*) to detect denitrifying bacteria in environmental samples. Appl Environ Microbiol 64:3769–3775

Brodie EL, DeSantis TZ, Joyner DC et al (2006) Application of a high-density oligonucleotide microarray approach to study bacterial population dynamics during uranium reduction and reoxidation. Appl Environ Microbiol 72:6288–6298

Burgmann H, Widmer F, Sigler WV, Zeyer J (2003) mRNA extraction and reverse transcription-PCR protocol for detection of *nifH* gene expression by *Azotobacter vinelandii* in soil. Appl Environ Microbiol 69:1928–1935

Call DR, Bakko MK, Krug MJ, Roberts MC (2003) Identifying antimicrobial resistance genes with DNA microarrays. Antimicrob Agents Chemother 47:3290–3295

Cho JC, Tiedje JM (2002) Quantitative detection of microbial genes by using DNA microarrays. Appl Environ Microbiol 58:1425–1430

Denef VJ, Park J, Rodrigues JLM, Tsoi TV, Hashsham SA, Tiedje JM (2003) Validation of a more sensitive method for using spotted oligonucleotide DNA microarrays for functional genomics studies on bacterial communities. Environ Micrbiol 5:933–943

Deng Y, He Z, Van Nostrand JD, Zhou J (2008) Design and analysis of mismatch probes for long oligonucleotide microarrays. BMC Genomics 9:491–503

Dennis P, Edwards EA, Liss SN, Fulthorpe R (2003) Monitoring gene expression in mixed microbial communities by using DNA microarrays. Appl Environ Microbiol 69:769–778

DeSantis TZ, Brodie EL, Moberg JP, Zubieta IX, Piceno YM, Andersen GL (2007) High-density universal 16S rRNA microarray analysis reveals broader diversity than typical clone library when sampling the environment. Microb Ecol 53:371–383

Ehrenreich A (2006) DNA microarray technology for the microbiologist: an overview. Appl Microbiol Biotechnol 73:255–273

Gao H, Yang ZK, Gentry TJ, Wu L, Schadt CW, Zhou J (2007) Microarray-based analysis of microbial community RNAs by whole-community RNA amplification. Appl Environ Microbiol 73:563–571

Gentry TJ, Wickham GS, Schadt CW, He Z, Zhou J (2006) Microarray application in microbial ecology research. Microb Ecol 52:159–175

Guschin DY, Mobarry BK, Proudnikov D, Stahl DA, Rittmann BE, Mirzabekov AD (1997) Oligonucleotide microchips as genosensors for determinative and environmental studies in microbiology. Appl Environ Microbiol 63:2397–2402

He Z, Zhou J (2008) Empirical evaluation of a new method for calculating signal to noise ratio (SNR) for microarray data analysis. Appl Environ Microbiol 74:2957–2966

He Z, Wu L, Fields MW, Zhou J (2005a) Use of microarrays with different probe sizes for monitoring gene expression. Appl Environ Microbiol 71:5154–5162

He Z, Wu LY, Li XY, Fields MW, Zhou JZ (2005b) Empirical establishment of oligonucleotide probe design criteria. Appl Environ Microbiol 71:3753–3760

He Z, Gentry TJ, Schadt CW et al (2007) GeoChip: a comprehensive microarray for investigating biogeochemical, ecological and environmental processes. ISME J 1:67–77

He Z, Van Nostrand JD, Wu L, Zhou J (2008) Development and application of functional gene arrays for microbial community analysis. T Nonferr Metal Soc 18:1319–1327

He Z, Deng Y, Van Nostrand JD, Tu Q, Xu M, Hemme CL, Li X, Wu L, Gentry TJ, Yin Y, Leibich J, Hazen TC, Zhou J (2010a) GeoChip 3.0 as a high-throughput tool for analyzing microbial community composition, structure, and functional activity. ISME J doi:10.1038/ismej.2010.46

He Z, Xu M, Deng Y, Kang S, Kellogg L, Wu L, Van Nostrand JD, Hobbie SE, Reich PB, Zhou J (2010b) Metagenomic analysis reveals a marked divergence in the functional structure of belowground microbial communities at elevated CO2. Ecol Lett 13:564–575

Hugenholtz P, Goebel BM, Pace NR (1998) Impact of culture-independent studies on the emerging phylogenetic view of bacterial diversity. J Bacteriol 180:4765–4774

Hurt RA, Qiu X, Wu L, Roh Y, Palumbo AV, Tiedje JM, Zhou Z (2001) Simultaneous recovery of RNA and DNA from soils and sediments. Appl Environ Microbiol 67:4495–4503

Lee DY, Shannon K, Beaudette LA (2006) Detection of bacterial pathogens in municipal wastewater using an oligonucleotide microarray and real-time quantitative PCR. J Microbiol Meth 65:453–467

Leigh MB, Pellizari VH, Uhlík O, Sutka R, Rodrigues J, Ostrom NE, Zhou J, Tiedje JM (2007) Biphenyl-utilizing bacteria and their functional genes in a pine root zone contaminated with polychlorinated biphenyls (PCBs). ISME J 1:134–148

Li X, He Z, Zhou J (2005) Selection of optimal oligonucleotide probes for microarrays using multiple criteria, global alignment and parameter estimation. Nucleic Acids Res 33:6114–6123

Liang Y, Li G, Van Nostrand JD, He Z, Wu L, Deng Y, Zhang X, Zhou J (2009a) Microarray-based analysis of microbial functional diversity along an oil contamination gradient in oil field. FEMS Microbiol Ecol 70:324–333

Liang Y, Wang J, Van Nostrand JD, Zhou J, Zhang X, Li G (2009b) Microarray-based functional gene analysis of soil microbial communities in ozonation and biodegradation of crude oil. Chemosphere 75:193–199

Leibich J, Schadt CW, Chong SC, He Z, Rhee SK, Zhou J (2006) Improvement of oligonucleotide probe design criteria for functional gene microarrays in environmental applications. Appl Environ Microbiol 72:1688–1691

Loy A, Lehner A, Lee N, Adamczyk J, Meier H, Ernst J, Schleifer KH, Wagner M (2001) Oligonucleotide microarray for 16S rRNA gene-based detection of all recognized lineages of sulfate-reducing prokaryotes in the environment. Appl Environ Microbiol 68:5064–5081

Lueders T, Friedrich MW (2003) Evaluation of PCR amplification bias by terminal restriction fragment length polymorphism Analysis of small-subunit rRNA and mcrA genes by using defined template mixtures of methanogenic pure cultures and soil DNA extracts. Appl Envrion Microbiol 69:320–326

Luo Y, Hui D, Zhang D (2006) Elevated CO_2 stimulates net accumulations of carbon and nitrogen in land ecosystems: a meta-analysis. Ecology 87:53–63

Mason OU, DiMeo-Savoie CA, Van Nostrand JD, Zhou J, Fisk MR, Giovannoni SJ (2009) Prokaryotic diversity, distribution, and preliminary insights into their role in biogeochemical cycling in marine basalts. ISME J 3:231–242

McDonald IR, Kenna EM, Murrell JC (1995) Detection of methanotrophic bacteria in environmental samples with the PCR. Appl Environ Microbiol 61:116–121

McGrath KC, Thomas-Hall SR, Cheng CT, Leo L, Alexa A, Schmidt S, Schenk PM (2008) Isolation and analysis of mRNA from environmental microbial communities. J Microbiol Meth 75:172–176

McKenna P, Hoffmann C, Minkah N, Aye PP, Lackner A, Liu Z, Lozupone CA, Hamady M, Knight R, Bushman FD (2008) The Macaque gut microbiome in health, lentiviral infection, and chronic enterocolitis. PLoS Pathog 4:e20

Miller SM, Tourlousse DM, Stedtfeld RD, Baushke SW, Herzog AB, Wick LM, Rouillard JM, Gulari E, Tiedje JM, Hashsham SA (2008) In situ-synthesized virulence and marker gene biochip for detection of bacterial pathogens in water. Appl Environ Microbiol 74: 2200–2209

Murray AE, Lies D, Li G, Nealson K, Zhou J, Tiedje JM (2001) DNA–DNA hybridization to microarrays reveals gene-specific differences between closely related microbial genomes. Proc Natl Acad Sci USA 98:9853–9858

Økland RH, Eilertsen O (1994) Canonical correspondence analysis with variation partitioning: some comments and an application. J Veg Sci 5:117–126

Palka-Santini M, Cleven BE, Eichinger L, Krönke M, Krut O (2009) Large scale multiplex PCR improves pathogen detection by DNA microarrays. BMC Microbiol 9:1

Palmer C, Bik EM, Digiulio DB, Relman DA, Brown PO (2007) Development of the human infant intestinal microbiota. PLoS Biol 5:e177

Ramette A, Tiedje JM (2007) Multiscale responses of microbial life in spatial distance and environmental heterogeneity in a patchy ecosystem. Proc Natl Acad Sci USA 104:2761–2766

Relógio A, Schwager C, Richter A, Ansorge W, Valcárcel J (2002) Optimization of oligonucleotide-based DNA microarrays. Nucleic Acids Res 30:e51

Rhee SK, Liu X, Wu L, Chong SC, Wan X, Zhou J (2004) Detection of genes involved in biodegradation and biotransformation in microbial communities by using 50-mer oligonucleotide microarrays. Appl Environ Microbiol 70:4303–4317

Rodríguez-Martínez EM, Pérez EX, Schadt CW, Zhou J, Massol-Deyá AA (2006) Microbial diversity and bioremediation of a hydrocarbon-contaminated aquifer (Vega Baja, Puerto Rico). Int J Environ Res Public Health 3:292–300

Sarkar SF, Gordon JS, Martin GB, Guttman DS (2006) Comparative genomics of host-specific virulence in Pseudomonas syringae. Genetics 147:1041–1056

Schena M, Shalon D, Davis RW, Brown PO (1995) Quantitative monitoring of gene expression patterns with a complementary DNA microarray. Science 270:467–470

Schena M, Shalon D, Heller R, Chai A, Brown PO (1996) Parallel human genome analysis: microarray-based expression monitoring of 1000 genes. Proc Natl Acad Sci USA 93:10614–10619

Sebat JL, Colwell FS, Crawford RL (2003) Metagenomic profiling: microarray analysis of an environmental genomic library. Appl Environ Microbiol 69:4927–4934

Small J, Call DR, Brockman FJ, Straub TM, Chandler DP (2001) Direct detection of 16S rRNA in soil extracts by using oligonucleotide microarrays. Appl Environ Microbiol 67:4708–4716

Steward GF, Jenkins BD, Ward BB, Zehr JP (2004) Development and testing of a DNA macroarray to assess nitrogenase (nifH) gene diversity. Appl Environ Microbiol 70:1455–1465

Stralis-Pavese N, Sessitsch A, Weilharter A, Reichenauer T, Riesing J, Csontos J, Murrell JC, Bodrossy L (2004) Optimization of diagnostic microarray for application in analysing landfill methanotroph communities under different plant covers. Environ Microbiol 6:347–363

Suzuki MT, Giovannoni SJ (1996) Bias caused by template annealing in the amplification of mixtures of 16S rRNA genes by PCR. Appl Envrion Microbiol 62:625–630

Taroncher-Oldenburg G, Griner EM, Francis CA, Ward BB (2003) Oligonucleotide microarray for the study of functional gene diversity in the nitrogen cycle in the environment. Appl Environ Microbiol 69:1159–1171

ter Braak CJF (1986) Canonical correspondence analysis: a new eigenvector technique for multivariate direct gradient analysis. Ecology 67:1167–1179

Tiquia SM, Wu L, Chong SC, Passovets S, Xu D, Ying Xu, Zhou J (2004) Evaluation of 50-mer oligonucleotide arrays for detecting microbial populations in environmental samples. Biotechniques 36:1–8

Van Nostrand JD, Khijniak TV, Gentry TJ, Novak MT, Sowder AG, Zhou JZ, Bertsch PM, Morris PJ (2007) Isolation and characterization of four gram-positive nickel-tolerant microorganisms from contaminated sediments. Microb Ecol 53:670–682

Van Nostrand JD, Wu WM, Wu L et al (2009) GeoChip-based analysis of functional microbial communities during the reoxidation of a bioreduced uranium-contaminated aquifer. Environ Microbiol 11:2611–2626

Wagner M, Smidt H, Loy A, Zhou Z (2007) Unravelling microbial communities with DNA-microarrays: challenges and future directions. Microb Ecol 53:498–506

Waldron PJ, Wu L, Van Nostrand JD, Schadt C, Watson D, Jardine P, Palumbo T, Hazen TC, Zhou J (2009) Functional gene array-based analysis of microbial community structure in groundwaters with a gradient of contaminant levels. Environ Sci Technol 43:3529–3534

Wang F, Zhou H, Meng J, Peng X, Jiang L, Sun P, Zhang C, Van Nostrand JD, Deng Y, He Z, Wu L, Zhou J, Xiao X (2009) GeoChip-based analysis of metabolic diversity of microbial communities at the Juan de Fuca Ridge hydrothermal vent. Proc Natl Acad Sci USA 106:4840–4845

Warnecke PM, Stirzaker C, Melki JR, Millar DS, Paul CL, Clark SJ (1997) Detection and measurement of PCR bias in quantitative methylation analysis of bisulphite-treated DNA. Nucleic Acids Res 25:4422–4426

Whitman WB, Coleman DC, Wiebe WJ (1998) Prokaryotes: the unseen majority. Proc Natl Acad Sci USA 95:6578–6583

Wilson KH, Blitchington RB, Greene RC (1990) Amplification of bacterial 16S ribosomal DNA with polymerase chain reaction. J Clin Microbiol 28:1942–1946

Wilson M, DeRisi J, Kristensen HH, Imboden P, Rane S, Brown PO, Schoolnik GK (1999) Exploring drug-induced alterations in gene expression in mycobacterium tuberculosis by microarray hybridization. Proc Natl Acad Sci USA 96:12833–12838

Wilson WJ, Strout CL, DeSantis TZ, Stilwell JL, Carrano AV, Andersen GL (2002) Sequence-specific identification of 18 pathogenic microorganisms using microarray technology. Mol Cell Probes 16:119–127

Wu L, Thompson DK, Li G, Hurt RA, Tiedje JM, Zhou J (2001) Development and evaluation of functional gene arrays for detection of selected genes in the environment. Appl Environ Microbiol 67:5780–5790

Wu L, Thompson DK, Liu X, Fields MW, Bagwell CE, Tiedje JM, Zhou J (2004) Development and evaluation of microarray-based whole genome hybridization for detection of microorganisms within the context of environmental applications. Environ Sci Technol 38:6775–6782

Wu L, Liu X, Schadt CW, Zhou J (2006) Microarray-based analysis of submicrogram quantities of microbial community DNAs by using whole-community genome amplification. Appl Environ Microbiol 72:4931–4941

Wu L, Kellogg L, Devol AH, Tiedje JM, Zhou J (2008) Microarray-based characterization of microbial community functional structure and heterogeneity in marine sediments from the Gulf of Mexico. Appl Environ Microbiol 74:4516–4529

Yamamoto S, Harayama S (1995) PCR amplification and direct sequencing of gyrB genes with universal primers and their application to the detection and taxonomic analysis of Pseudomonas putida strains. Appl Environ Microbiol 61:1104–1109

Yergeau E, Kang S, He Z, Zhou J, Kowalchuk GA (2007) Functional microarray analysis of nitrogen and carbon cycling genes across an Antarctic latitude transect. ISME J 1:1–17

Yin H, Cao L, Qiu G, Wang D, Kellogg L, Zhou J, Dai Z, Liu X (2007) Development and evaluation of 50-mer oligonucleotide arrays for detecting microbial populations in Acid Mine Drainages and bioleaching systems. J Microbiol Meth 70:165–178

Zhang L, Srinivasan U, Marrs CF, Ghosh D, Gilsdorf JR, Foxman B (2004) Library on a slide for bacterial comparative genomics. BMC Microbiol 4:12–18

Zhang K, Martiny AC, Reppas NB, Barry KW, Malek J, Chisholm SW, Church GM (2006) Sequencing genomes from single cells by polymerase cloning. Nat Biotechnol 24:680–686

Zhang Y, Zhang X, Liu X, Xiao Y, Qu L, Wu L, Zhou J (2007) Microarray-based analysis of changes in diversity of microbial genes involved in organic carbon decomposition following land use/cover changes. FEMS Lett 266:144–151

Zhou J (2003) Microarrays for bacterial detection and microbial community analysis. Curr Opin Microbiol 6:288–294

Zhou J, Thompson DK (2002) Challenges in applying microarrays to environmental studies. Curr Opin Biotech 13:204–207

Zhou X, Zhou J (2004) Improving the signal sensitivity and photostability of DNA hybridizations on microarrays by using dye-doped core-shell silica nanoparticles. Anal Chem 76:5302–5312

Zhou J, Bruns MA, Tiedje JM (1996) DNA recovery from soils of diverse composition. Appl Environ Microbiol 62:316–322

Zhou J, Kang S, Schadt CW, Garten CT Jr (2008) Spatial scaling of functional gene diversity across various microbial taxa. Proc Natl Acad Sci USA 105:7768–7773

Chapter 6
Probing Identity and Physiology of Uncultured Microorganisms with Isotope Labeling Techniques

Alexander Loy and Michael Pester

Introduction

The global nutrient cycles are mainly driven by microorganisms, which are the caretakers of our past, present, and future biosphere. Microorganisms also interact intimately with the geosphere, thereby mediating fluxes of elements from the abiotic to the biotic world. While the functional importance of microorganisms for sustaining life on Earth is generally acknowledged, we are just beginning to understand how physiological activities of specific microbial guilds and taxa contribute to large-scale processes on the ecosystem and the global levels. In the last 2 decades, molecular methods targeting ribosomal RNA (rRNA) or rRNA genes for microbial identification and quantification in environmental samples have been instrumental in deciphering the vast natural diversity of microorganisms. More than 85 new bacterial phyla were discovered with this approach, but for most not a single cultivated representative is available and virtually nothing is known about the metabolic capabilities of their constituents (Achtman and Wagner 2008). It is expected that this gap in knowledge will further increase because the discovery of novel microbial diversity has not yet come to an end (Sogin et al. 2006), fueled by new high-throughput sequencing technologies that generate an unprecedented amount of data such as the GS FLX Titanium sequencer from 454 Life Sciences Corporation. A single run of the GS FLX Titanium sequencer produces about 1,250,000 sequences of 400–450 bp from a complex mixture of bacterial 16S rRNA genes (e.g., recovered by PCR or reverse transcriptase PCR from an environmental DNA or RNA extract, respectively), which is more than have been generated during 20 years of application of the 16S rRNA approach and traditional Sanger sequencing. In addition, metagenomics and genome sequencing of recently isolated microorganisms continually reveal the presence of variants of

A. Loy (✉) and M. Pester
Department for Microbial Ecology, Faculty of Life Sciences, University of Vienna, Althanstrasse 14, A-1090 Wien, Austria
e-mail: loy@microbial-ecology.net

L.L. Barton et al. (eds.), *Geomicrobiology: Molecular and Environmental Perspective*,
DOI 10.1007/978-90-481-9204-5_6, © Springer Science+Business Media B.V. 2010

phylogenetic marker genes which are not covered by the "universal" primer sets widely used in phylogenetic studies. In terms of the natural 16S rRNA sequence diversity, one can thus expect a continuous filling of the missing branches in the tree of life. While a comprehensive analysis of microbial richness in environments having high microbial diversity (such as soils, sediments, and the gastrointestinal tracts of humans and animals) now seems within reach, a major conceptual and technical challenge confronting the field of microbiology is to uncover the physiological capabilities and ecological functions of the many microorganisms for whom only the 16S rRNA sequence is presently known. Of great importance for detecting the physiological properties and activity of uncultivated microorganisms are holistic molecular approaches targeting whole natural communities rather than individual community members, such as metagenomics, metatranscriptomics, and environmental proteomics. There are, however, limitations with the aforementioned methods because neither the mere presence of genes coding for specific metabolic functions nor their expression (transcription and translation) in an environmental sample definitively proves the existence of a specific physiological processes (Wagner 2009). Furthermore, the function of many genes and proteins recovered by these methods is still unknown, and therefore the absence of known functional genes involved in a specific metabolic pathway does not prove that a particular metabolic capability is absent, but rather that there may exist another, not yet elucidated, pathway.

Substrate-mediated isotope labeling techniques combined with molecular methods for identification of microorganisms are powerful approaches to fill this gap. By linking physiological activity to organismal phylogeny, isotopic labeling methods are used both as discovery tools as well as to confirm hypotheses generated by the above-mentioned "meta-omics" technologies. Some isotope tracer techniques also enable quantitative analysis of the contributions of specific microbial groups to certain ecosystem processes.

This chapter provides a concise overview of isotopic labeling methods and provides guidance for integrated application of these techniques in light of recent advances in state-of-the-art instrumentation such as the NanoSIMS and confocal Raman-microspectroscope. For further details, the reader is referred to several excellent reviews (e.g. Boschker and Middelburg 2002; Dumont and Murrell 2005; Friedrich 2006; Madsen 2006; Neufeld et al. 2007a, c; Orphan, 2009; Wagner et al. 2006; Wagner 2009; Whiteley et al. 2006).

The Principle of Substrate-Mediated Isotope Labeling Techniques

"You (= microorganism) are what you eat (= isotope-labeled substrate) under certain conditions" is the simplified concept behind isotope labeling techniques. In an experiment, a natural microbial community is exposed to an isotope-labeled compound under conditions designed to stimulate microbes that are capable of carrying out a process of interest. In addition to measuring uptake of the isotope label and its incorporation into the cellular biomass, the identities of the labeled microorganisms can also be determined by molecular methods.

Depending on the downstream analysis, substrates are usually fully labeled (>99.5% of the respective atoms) with either a radioactive (mostly ^{14}C, ^{33}P or ^{3}H) or a stable (mostly ^{13}C, ^{15}N or ^{18}O) isotope. The labeling experiments are often performed in micro- or mesocosms in which environmental samples are incubated under (nearly) in situ conditions. Ideally the incubations reproduce in situ conditions, realized through the addition of the labeled compound at low, environmentally-relevant concentrations and/or by simulating the inherently dynamic conditions in natural habitats (e.g., gradients or fluctuating concentrations of nutrients and other environmental parameters). Stable isotope compounds offer greater flexibility in experimental design of tracer studies because radioactive isotopes of many biologically important elements, such as nitrogen and oxygen, have an impractically short half-life (<1 day). An added benefit of stable isotope compounds is that they can be directly applied in the environment (Jeon et al. 2003; Liou et al. 2008; Pumphrey and Madsen 2008) or administered to animals or humans without safety concerns.

No matter which type of isotope is used, if identification of primary consumers of the isotope-labeled substrate is desired it is advisable to minimize the incubation time as much as possible. Time series experiments can be used to optimize the incubation time such that the extent of cell labeling just exceeds detection limits (Neufeld et al. 2007a). Prolonged incubation times can lead to cross-feeding of labeled degradation products by secondary and tertiary consumers, creating a blurring effect that could result in the miss-identification of microorganisms involved in the metabolic process of interest. While cross-feeding was initially considered a problem, it is now also recognized as an opportunity to trace the flow of the isotopic label through different trophic levels. Time course incubation experiments, during which samples for isotopic and molecular analyses are taken at appropriate time intervals, need to be carefully planned and performed in order to obtain insights into nutritional interdependencies of microorganisms. While such analyses of the microbial food web or of dietary interactions among symbiotic organisms are very compelling (Lu and Conrad 2005; Lueders et al. 2004b), the effort associated with the analyses of many samples has so far restricted widespread application of this approach. Rapid analysis of multiple samples is now, however, becoming feasible with the application of the highly parallel microarray format for simultaneous microbial identification and tracking of isotope incorporation into rRNA (see isotope array below).

Substrate-mediated isotope labeling studies can be broadly classified as either exploratory community screening approaches or directed approaches employing phylogenetic probe hybridization for identification of (groups of) organisms under surveillance (Fig. 6.1, Table 6.1). The former approaches are also known as stable isotope probing (SIP) and generally rely on the analysis of cellular biomarkers in which the stable isotope label has been incorporated. Nucleic acids-based SIP approaches are often used as discovery tools to search for new organisms that mediate a certain environmental process. In contrast, the focus of directed approaches is the ecophysiology of defined groups of microorganisms for which specific rRNA-targeted probes are either available (Loy et al. 2007) or can be developed. The two approaches

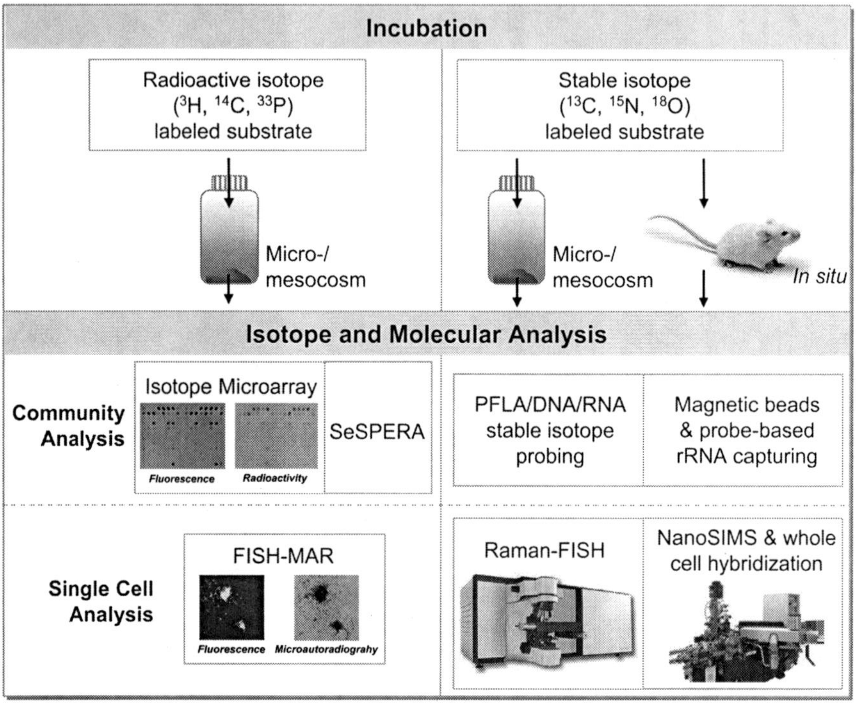

Fig. 6.1 Schematic overview of radioactive and stable isotope-labeling techniques

are complementary and generally produce the most reliable and interesting results when used in combination (Ginige et al. 2004; Huang et al. 2009a).

Community-Wide Screening Approaches

Stable Isotope Probing of Phospholipid-Derived Fatty Acids

The first biomarkers exploited for SIP were membrane lipids, i.e. phospholipid-derived fatty acids (PLFA), extracted from environmental samples after incubation with labeled substrates. Stable isotope ratios of PLFA are analyzed via gas chromatography combustion isotope ratio mass spectrometry (GC/C/IRMS). This analytical setup is very sensitive; detecting even very small changes in stable isotope composition (e.g., changes in relative ^{13}C content of ca. 0.001%), enabling the amendment of labeled compounds to SIP incubations at levels that mirror or even constitute only a fraction of the natural substrate pool. PLFA-SIP can provide three pieces of information: (a) identity of the microorganisms (assessed from the composition of the environmental PLFA profile), (b) amount of biomass (PLFA concentration),

Table 6.1 Selected features of isotope-labeling techniques

Method	Features
Community-Wide Screening	
PLFA-SIP	• Rapid label incorporation, short incubation times
	• Very sensitive (detection limit 1–2‰ $\delta^{13}C$ enrichment)
	• Estimates on growth rates and yields of functional populations
	• Limited phylogenetic resolution
	• PLFA composition of uncultivated microorganisms is unknown
DNA-SIP	• Prolonged incubation times compared to PLFA- and RNA-SIP
	• Lower sensitivity (>10–30 atom% ^{13}C, >40 atom% ^{15}N)
	• Discovery tool for organisms performing specific metabolic functions
	• High sequence/primer-dependent phylogenetic resolution
	• Combination with microarrays and metagenomics possible
	• Work intensive
	• Metabolic rate measurements for defined target organisms are not possible
RNA-SIP	• Short incubation times
	• Lower sensitivity (>10–30 atom% ^{13}C, >40 atom% ^{15}N)
	• Discovery tool for organisms performing specific metabolic functions
	• High sequence/primer-dependent phylogenetic resolution
	• Combination with microarrays and metatranscriptomics possible
	• Work intensive
	• Metabolic rate measurements for defined target organisms are not possible
Magnetic beads-probe-based rRNA capturing	• Capture probe-dependent phylogenetic resolution; probe specificity depends on hybridization conditions
	• High isotope detection sensitivity (using micro-elemental analyzer-IRMS)
	• Dependent on larger rRNA quantities; limited to more abundant taxa
SeSPERA	• Primer-dependent phylogenetic resolution; primer specificity depends on hybridization conditions in solution
	• Low multiplexing (i.e., parallel analysis of a few primer-dependent target populations)
	• Discovery tool, if primers are designed based on environmental rRNA gene libraries
	• Limited to radioactive isotopes and their application field
Isotope microarray	• Short incubation time
	• Sensitive due to radioactive isotopes
	• High multiplexing, high-throughput screening (i.e., parallel analysis of multiple probe-dependent target populations)
	• Probe-dependent phylogenetic resolution; probe specificity depends on hybridization conditions

(continued)

Table 6.1 (continued)

Method	Features
	• Discovery tool, if microarray probes are designed based on environmental rRNA gene libraries • Limited number of probes (up to hundreds) due to large probe spots (Ø 500–700 µm) • Limited to radioactive isotopes and their application field • Substrate incorporation rate measurements possible
Single cell analyses	• **Spatial (co-)localization within the natural environment (requires appropriate sample preservation and preparation)** • **Probe-dependent phylogenetic resolution; probe specificity depends on hybridization conditions** • **Discovery tool, if probes are designed based on environmental rRNA gene libraries**
MAR-FISH	• Short incubation time • Sensitive due to radioactive isotopes • Lateral resolution 0.5–2 µm; limited single cell resolution in microbial aggregates • Limited to radioactive isotopes and their application field • Substrate uptake measurements (affinity) possible
Raman-FISH	• Lower sensitivity (>10 atom% ^{13}C) • Lateral resolution 1 µm • Analysis of label incorporation into different compound classes (e.g., nucleic acids vs. proteins vs. carbohydrates) • Non-destructive method • Combination with downstream single cell methods possible, e.g., single cell genomics • Cost intensive instrument (approximately € 200,000) • Low throughput (typically one-cell-after-another analysis)
NanoSIMS-FISH	• Very sensitive (approximately 1,000 times more sensitive than MAR) • Superb lateral resolution (up to ~ 50 nm) • Imaging of multiple isotopes in parallel • Very cost intensive instrument (approximately €3,000,000) • Very time-consuming sample preparation and screening (low throughput)

and (c) level of activity (isotope ratios, i.e., relative amount of isotope incorporated into PLFA) (Boschker and Middelburg 2002).

The first application of PLFA-SIP revealed that bacteria related to the Gram-positive *Desulfotomaculum acetoxidans* were the main acetate consuming sulfate reducers in estuarine and brackish sediment samples (Boschker et al. 1998). In the same study, type I methanotrophs were identified as the primary methane oxidizers in a freshwater sediment sample. While these applications nicely showed the potential of PLFA-SIP, the main constraint with this method is that the PLFA composition of uncultivated microorganisms is entirely unknown. Environmental PLFA profiles, typically consisting of a few tens of dominant peaks, can thus only be interpreted in view of the PLFA composition of cultivated microorganisms, which, along with the limited phylogenetic resolution of PLFA, has thus far restricted wide application of PLFA-SIP in complex environmental samples.

Stable Isotope Probing of Nucleic Acids (DNA/RNA-SIP)

Nucleic acid biomarkers provide a higher phylogenetic resolution than PLFA, allowing for a more refined assignment in SIP experiments of functional properties to certain phylogenetic groups of microorganisms. No sophisticated and expensive instrumentation is usually needed for DNA- and RNA-SIP, making these techniques very popular for metabolic analyses of complex microbial communities. After sample incubation with a labeled substrate (most studies use ^{13}C-compounds), the extracted mixture of heavy and light nucleic acids is simply separated by isopycnic ultracentrifugation in cesium chloride (DNA) or cesium trifluoroacetate (RNA) gradients. When the DNA-SIP technique was initially developed, separated DNA was visualized by ethidium bromide staining/UV illumination, which facilitated the selective retrieval of heavy and light DNA fractions with a needle and syringe for subsequent molecular analysis. This is not, however, an adequate option to retrieve partially labeled DNA that results from experiments designed with short incubation times because of the overlap of unlabeled high-GC content DNA and partially labeled low/intermediate-GC content DNA. In addition, RNA, which can be loaded only to a maximum amount of 500 ng, can hardly be visualized by such an approach. The preferred alternative is thus the complete fractionation of density gradients and subsequent analysis of changes in the (phylo)genetic composition over all retrieved DNA or RNA fractions (Lueders et al. 2004a).

Important to the success of DNA/RNA-SIP are the rate and degree of isotopic enrichment of nucleic acids, which must be great enough to enable sufficient separation during density gradient centrifugation. Isotope incorporation in DNA requires replication and growth of the respective organisms, which is potentially problematic for the application of DNA-SIP because this could result in community shift biases during prolonged incubations (i.e., enrichment through selective overgrowth of certain microorganisms). In contrast, RNA is a more sensitive biomarker for SIP because isotopic enrichment of RNA depends on RNA turnover rather than replication, and thus proceeds significantly faster than isotopic enrichment of DNA (Manefield et al. 2002). Incubation times can be further shortened via approaches that increase the detection sensitivity of heavy target nucleic acids. Increased sensitivity can be achieved by improving the recovery efficiency of heavy target nucleic acids through the addition of ^{13}C-labeled carrier nucleic acids (synthetic or from non-target organisms) during gradient centrifugation (Gallagher et al. 2005) or the use of glycogen as a carrier during nucleic acid precipitation from gradient fractions (Neufeld et al. 2007b). These simple tricks can shorten incubation times from days to hours, thereby reducing possible community shift biases.

The great majority of DNA/RNA-SIP studies rely on the administration of ^{13}C-labeled tracers, but the use of other isotopes such as ^{15}N and ^{18}O greatly expand the suite of possible SIP applications. ^{15}N$_2$-DNA-SIP has been used in combination with 16S rRNA gene and *nifH* (a marker gene for nitrogen fixing microorganisms) to identify free-living diazotrophs in soil, which provided the first evidence of nitrogen fixation by uncultivated *Actinobacteria* outside of the order *Actinomycetales* (Buckley et al. 2007a). $H_2^{18}O$, which is not used as an energy, nitrogen, or carbon source by microorganisms, represents a general activity marker for the identification of all growing cells in DNA-SIP experiments, without the need to add additional substrates (Schwartz 2007). $H_2^{18}O$-based SIP is also an excellent option for studying the impact of recurring wetting and drying events on microbial activities, e.g., in natural wetland systems or floodplains. The major issue with use of isotopes other than ^{13}C is reduced sensitivity; for example labeled pure culture DNA must contain at least ~40 atom% ^{15}N for sufficient separation in the density gradient (Cadisch et al. 2005).

An additional complication in separation of nucleic acids is that the buoyant density of nucleic acids is not only dependent on the level of isotopic enrichment, but also on GC content. Thus, unlabeled high-GC content nucleic acids can display buoyant densities similar to labeled low-GC content nucleic acids (Lueders et al. 2004a). This problem is of even greater relevance for ^{15}N studies because of the lower number of N atoms in nucleic acids. Optimization of ultracentrifugation conditions is an important first step to minimize this problem, but additional strategies have also been employed for the detection and amelioration of cross-contamination. A useful indicator for cross-contamination of 'heavy' nucleic acids fractions with 'light' nucleic acids is the addition of unlabeled, high-GC content nucleic acids from an organism that does not occur in the sample (Singleton et al. 2007). The confounding effect of unlabeled, high-GC content DNA can be alleviated by application of a sophisticated, two-step ultracentrifugation protocol. Addition of the DNA-binding ligand bis-benzimide to the second centrifugation step ensures enhanced separation of labeled and unlabeled DNA (Buckley et al. 2007b). Another important control is parallel incubation under exactly the same conditions but with unlabeled (e.g., ^{12}C) substrate. Comparative analysis of the nucleic acids SIP fractions obtained from the labeled and unlabeled substrate incubations helps to assess potential cross-contamination of 'heavy' fractions with unlabeled GC-rich nucleic acids.

There are many options for molecular analysis following fractionation and recovery of nucleic acids from SIP gradients. Differences in the phylogenetic composition of the many fractions (as assessed from rRNA or the respective genes) can be analyzed by fingerprinting techniques (such as denaturing gradient gel electrophoresis and terminal restriction fragment length polymorphism), DNA microarrays, and quantitative (reverse transcriptase) PCR. Phylogenetic analyses of sequenced genes have often only been performed for selected fractions because of the effort associated with the preparation and analysis of clone libraries, but traditional phylogenetic analyses will likely increasingly be complemented by deep pyrosequencing of PCR products, which does not require the tedious preparation of clone libraries and offers higher sampling depth and a superior phylogenetic resolution compared to fingerprinting techniques.

An advantage of DNA-SIP is that additional genetic information can be obtained from metabolically active microorganisms, either by PCR amplification of selected genes that encode characteristic metabolic functions or by metagenomic analysis

(Dumont et al. 2006). Preparation of large-insert metagenomic clone libraries requires significant amounts of labeled DNA, which can often only be obtained after long incubation with artificially high substrate concentrations, and so microorganisms that are less relevant in situ might be enriched in the SIP incubations, leading to biased community compositions. A promising alternative is whole genome amplification of low amounts of labeled DNA based on phi29-DNA polymerase. However, like other enzymatic DNA amplifications, phi29-based multiple displacement amplification is subject to selective amplification and chimera formation (Lasken and Stockwell 2007), though impact of the latter can be minimized by subsequent enzymatic treatment (Zhang et al. 2006). This approach was successfully applied to retrieve the near complete genome of the novel methylotroph *Methylotenera mobilis* from a lake sediment (Kalyuzhnaya et al. 2008) and to snapshot the genomic repertoires of uncultivated *Methylocystis* species in acidic peatlands (Chen et al. 2008).

Although there would be great value in assigning transcribed genes to microorganisms that are active under certain circumstances, molecular analysis of RNA-SIP fractions has so far been restricted to rRNA, with one exception. In a ground-breaking study, expression of a certain napthalene dioxygenase gene type was detected in heavy RNA-SIP fractions by RT-PCR after exposing groundwater samples aerobically to *in situ* concentrations of [$^{13}C_{10}$]napthalene (Huang et al. 2009a). The expressed napthalene dioxygenase gene could subsequently be linked to an uncultivated *Acidovorax* species by combining rRNA/mRNA-SIP with Raman microspectroscopy-fluorescence in situ hybridization (FISH) of single cells (see below). Low recovery of mRNA from environmental samples is generally considered one of the major constraints of mRNA-SIP (Neufeld et al. 2007a). In a total RNA extract, mRNA represents only a minor fraction among the bulk of rRNA. Selective amplification of mRNA (Frias-Lopez et al. 2008) or depletion of rRNA (Urich et al. 2008) from heavy RNA fractions coupled with high-throughput pyrosequencing are potential strategies for (a) increased sensitivity of future mRNA-SIP experiments and (b) selective metatranscriptomic analysis of microbial populations with specific functional properties.

Directed Phylogenetic Oligonucleotide Probe-Based Approaches: From Communities to Single Cells

What to Keep in Mind When Using rRNA-Targeted Oligonucleotide Probes/Primers

A suite of published rRNA-targeted probes is available that target microorganisms at different phylogenetic levels (Loy et al. 2007). New probes can also easily be designed if necessary using appropriate computer programs (Ludwig et al. 2004). Owing to the ever-increasing number of rRNA sequences, periodic checks of the *in silico* coverage and specificity of existing and newly designed probes

(Loy et al. 2008) is essential for pre-selection of adequate probes. However, current *in silico* predictions of probe behavior fall short of addressing the physically and chemically variable conditions of different hybridization-based techniques such as microarrays or FISH. Optimal hybridization conditions must be determined empirically for each hybridization format using adequate target and non-targets to ensure high specificity and sensitivity of a probe (set). To increase the reliability of identification, it is recommended that hierarchically nested multiple probes (i.e., a target group is covered by more than one probe) are used when the hybridization format supports simultaneous use of multiple probes. Rigorous computational and empirical evaluation of a probe (set) for a particular hybridization application is necessary before the probe can be used for identification of target microorganisms that have incorporated an isotope-labeled substrate in one of the approaches described below.

The Magnetic Beads-Probe-Based rRNA Capture Approach

A conceptually elegant method for focusing the analysis of isotopic labeled rRNA to a defined phylogenetic group of microorganisms is to employ an oligonucleotide probe as bait in a molecular 'fishing' approach (MacGregor et al. 2002). After incubation of a sample with a stable isotope-labeled substrate, total RNA is extracted and hybridized in solution with a biotin-labeled probe. Probe-target hybrids are subsequently bound to streptavidin-coated paramagnetic beads and separated from non-target rRNA with the help of a magnet. The yield of captured rRNA can be estimated by polyacrylamide gel electrophoresis. Elemental analyzer-IRMS is used to determine the relative isotopic composition of captured rRNA. A first application of this approach used domain-specific rRNA-targeted probes and demonstrated that *Bacteria* and not *Eukarya* were the main consumers of ^{13}C-labeled acetate, propionate, amino acids or glucose in North Sea sediments. Use of a microelemental analyzer-IRMS increases the sensitivity of the technique approximately 10-fold and has allowed insights into ^{13}C-substrate utilization by *Desulfobacteraceae* in marine sediments (Miyatake et al. 2009). While ^{13}C-labeling in the range of natural abundance delta[13]C ratios can be studied, it remains to be shown whether low amounts of labeled rRNA from rare taxa can also be effectively captured and analyzed with this approach.

Sequence Specific Primer Extension RNA Analysis (SeSPERA)

The SeSPERA method allows simultaneous investigation of a small number of different phylogenetic groups for their ability to assimilate a radioactively-labeled substrate (Franchini et al. 2009). A key feature of this multiplex method is the use of taxon-specific primers that bind to different positions on target rRNA. Primer elongation with a reverse transcriptase lacking RNase H activity and subsequent digestion of single stranded rRNA overhangs with RNase T1 yields taxon-specific rRNA-DNA hybrids of different length. Size-dependent separation and identification of the fragments is performed by gel electrophoresis. Radioactive rRNA-DNA hybrids from active microbes are determined by electrotransfer from the gel onto a membrane, followed by membrane autoradiography or other radioactive imaging techniques. This relatively simple method does not require expensive instrumentation. The chief disadvantage, however, is that multiplex analysis is limited by the number of primers that can hybridize specifically to target rRNA under monostringent conditions and by the number of rRNA-DNA bands that can be effectively separated via gel electrophoresis.

Isotope Microarrays

Phylogenetic DNA microarrays (phylochips), utilizing up to thousands of different rRNA-targeted probes immobilized on a miniaturized, solid support like a glass microscope slide, are used to simultaneously analyze many microbial community members. The same principle is also modified in the isotope microarray not solely for microbial identification, but also to uncover the active metabolism of microorganisms through the isotopic labeling of their rRNA (Wagner et al. 2007). This isotope microarray approach begins with the exposure of an environmental sample to a radioactive (^{14}C) labeled compound. In the next step, total RNA is extracted from the sample, tagged with fluorescence dye molecules, fragmented chemically, and hybridized with a phylochip. Sequential scanning for fluorescence and radioactivity and subsequent digital image analysis distinguishes microorganisms that incorporated ^{14}C in their rRNA from community members that did not. An additional, attractive feature of the isotope microarray is that the efficiency of isotope incorporation by a probe-defined target group can be inferred from changes in the ratio between radioactivity and fluorescence intensity of a probe recorded over different incubation times (Adamczyk et al. 2003). Isotope microarrays contain probe spots of larger sizes than most other DNA microarray applications (>500 µm in diameter) (Loy and Bodrossy 2006) due to the lower resolution of the beta-imager used for radioactivity scanning. This currently restricts the number of spots on an isotope microarray and thus the number of target groups that can be analyzed in a single hybridization to a few hundred.

The isotope microarray and other complementary methods were recently used to reveal the substrate utilization profiles of diverse *Rhodocyclales* community members in activated sludge from a wastewater treatment plant (Hesselsoe et al. 2009). Three different substrates were separately added to activated sludge samples and incubated under oxic or anoxic, nitrate-reducing conditions. Each incubation contained an inhibitor of autotrophic ammonia oxidation together with ^{14}C-bicarbonate as a general activity marker for heterotrophic carbon assimilation (Hesselsoe et al. 2005). Comparisons with control incubations (performed under identical conditions, but without addition of substrates) revealed that most detected *Rhodocyclales* groups were actively involved in nitrate reduction, but varied in their consumption of propionate, butyrate or toluene. Furthermore, radioactive signals were recorded for probes targeting *Rhodocyclales* groups that were present in the activated sludge at a low relative abundance of approximately 1% of the total bacterial community (as analyzed by quantitative FISH). This study demonstrates that the high-throughput capacity of the isotope array is of unique value for analysis of many samples and target organisms. This is particularly relevant for exploring potential substrate competition and niche partitioning of different community members, which necessitates the study of many different incubation regimes.

Fluorescence In Situ Hybridization Combined with Microautoradiography

Approximately 20 years ago, hybridization with fluorescently-labeled rRNA-targeted oligonucleotides was used for the first time for the identification of phylogenetically distinct groups of chemically preserved microorganisms under the microscope (DeLong et al. 1989). The invention of FISH was the start of a success story that revolutionized the field of microbiology, yielding unparalleled information about microbial abundances, the stereoscopic organization of microscopic architectures such as microbial biofilms and flocs, and spatial interdependence of trophically co-acting or competing microbes (Daims et al. 2006; Maixner et al. 2006). Microautoradiography (MAR), the microscopic visualization of silver grain formation in a photographic emulsion above individual cells that have become radioactive after assimilation or uptake of a radioactive isotope-labeled substrate, has an even longer history (Brock and Brock 1966). However, it took about a decade of co-existence of FISH and MAR before the techniques were merged (Lee et al. 1999; Ouverney and Fuhrman 1999) into what is today, besides DNA/RNA-SIP, the most widely applied "bread and butter" technique for physiological studies of uncultivated microbes (Wagner et al. 2006; Wagner 2009). With a lateral resolution of 0.5–2 μm (depending on the radioactive isotope and equipment used) FISH-MAR was the first single-cell technique that allowed linking the phylogenetic identity of individual microbial cells to their physiological properties. In comparison to DNA/RNA-SIP, FISH-MAR is more sensitive because the total cellular radioactivity and not only the isotope-labeled DNA or RNA is measured. The cross-feeding problem is thus less of a concern with FISH-MAR, because an environmental sample can be incubated for only a few hours with a low dose of a radioactive compound. During the FISH-MAR procedure, samples are first fixed with paraformaldehyde or ethanol and then overlaid with the autoradiographic emulsion. Samples with multidimensional architectures of aggregating microbes such as biofilms additionally require cryosectioning or alternative cell dispersal strategies to obtain a thin layer of cells on the slide. This ensures that silver grain formation is correctly attributed to individual radioactive cells or cell clusters whose corresponding identities are determined in the microscope by FISH. Under certain circumstances, the number of silver grains per cell is even a measure for the amount of substrate that was assimilated by the cell (Nielsen et al. 2003). The quantification of substrate-assimilation rates unquestionably adds to our understanding of the ecophysiology of the microorganisms of interest. However, quantitative FISH-MAR is laborious and requires the establishment of a standard curve from known microbes exposed to different doses of a radioactive compound, which in many cases is not feasible (Nielsen et al. 2003; Wagner et al. 2006).

Another limitation of FISH-MAR is that many, especially large, organic molecules with a uniform isotope-labeling are either not available or available only at exceptionally high prices, limiting the range of metabolic features that can be analyzed in isotope-labeling experiments. Heterotrophic FISH-MAR circumvents this limitation by using inexpensive $^{14}CO_2$ for labeling and comparison of MAR-active cells after incubation with and without unlabeled substrates (Hesselsoe et al. 2005). Most heterotrophic microorganisms incorporate the ^{14}C from CO_2 into biomass via diverse carboxylation reactions during growth. Undesired labeling of autotrophs can be excluded by the addition of appropriate inhibitors.

Fluorescence In situ Hybridization Combined with Raman-Microspectroscopy

Raman-microspectroscopy is based on inelastic scattering of the incident monochromatic laser light by the specific chemical bonds of the molecules in a small sample volume (approx. 1 μm in diameter, depending on the width of the incident laser beam). Application of this technique to a single microbial cell yields a Raman spectrum that represents the overlaying individual Raman spectra of the different cellular molecules and can thus be regarded as a proxy for the biochemical composition of the cell (Wagner 2009). While inherently complex, several major compound classes of a microbial cell can be distinguished in a Raman spectrum (such as nucleic acids, proteins, lipids, and carbohydrates) (Huang et al. 2004). In addition, it is possible to delineate specific molecules that have the capacity for pronounced Raman scattering {such as phenylalanine, polyhydroxybutyrate (De Gelder et al. 2008), cytochrome C, and elemental sulfur (Pasteris et al. 2001)}. Due to these characteristics, Raman spectra of individual cells were initially used for biochemical fingerprinting and identification of cultivated microorganisms. However, even more interesting for microbiologists is that assimilation of heavy isotope substrates produces significant shifts in certain bands of the Raman spectra of individual microorganisms (so far shown for ^{13}C and ^{15}N), compared to microorganisms fed with the corresponding light isotope substrates (Huang et al. 2004). Using a customized confocal Raman-microspectroscope that is additionally equipped with an epifluorescence microscope unit, the identity of an uncultivated microorganism can be determined through hybridization with a specific FISH probe, whereas the physiology of this microbe can additionally be inferred from its Raman spectrum. In such studies, the pronounced phenylalanine peak is used as diagnostic region, because a significant and isotope tracer abundance-dependent shift in the position and height of the peak is visible in the Raman spectrum of ^{13}C-labeled cells. The ratio of the light (unlabeled) and heavy (labeled) phenylalanine peaks at 1,003 and 967 cm^{-1}, respectively, shows a nearly linear relationship with the ^{13}C-content of living and fluorescent probe-stained pure culture cells (Huang et al. 2007). Such reference curves of pure cultures can thus be applied to quantify the ^{13}C-content from phenylalanine Raman peak shift ratios of corresponding probe-defined cells in complex environmental samples. The detection limit for the phenylalanine peak shift is a labeling level of approximately 10 atom% ^{13}C, which is equal to or better than the detection limit of RNA- or DNA-SIP.

In an initial application of the Raman-FISH approach, members of the genus *Pseudomonas* were demonstrated to be involved in aerobic naphthalene degradation in groundwater microcosms (Huang et al. 2007). From differences in the ^{13}C-content of individual *Pseudomonas* cells, it was further evident that metabolic activities varied within the probe-defined population in situ. The subsequent study of the same system by Whiteley and colleagues provides an excellent example how different isotope-labeling techniques can be integrated under the same experimental umbrella (Huang et al. 2009a). Combining rRNA- and mRNA-SIP with Raman-FISH, it was shown that also members of an uncultivated *Acidovorax* species were performing naphthalene degradation, but occupied a different niche in the groundwater system as they were selectively active under a much lower naphthalene concentration. In another pioneering study, Raman-microspectroscopy was combined with laser tweezer sorting to separate individual living ^{13}C-labeled cells for subsequent cultivation or whole genome amplification (Huang et al. 2009b); opening the door for genomics of single cells with defined physiological properties.

Whole Cell Hybridization and Secondary Ion Mass Spectrometry

Another very promising option for detecting differences in the isotopic composition of microorganisms is secondary ion mass spectrometry (SIMS). Combination of FISH and SIMS was first introduced by DeLong and colleagues, but because of limitations in resolution (~5 μm) was initially restricted to the analysis of larger cell aggregates such as those formed by syntrophic archaea and sulfate-reducing bacteria performing anaerobic oxidation of methane (Orphan et al. 2001). However, the latest breakthrough in developing an instrument with nano-scale resolution (up to ~50 nm) for imaging multiple isotopes simultaneously (NanoSIMS) not only allows SIMS analysis of single microbial cells, but also promises a level of sensitivity unmatched by any other isotopic-labeling detection technique (for an overview of biological NanoSIMS applications see Lechene et al. 2006; Wagner 2009). Isotope ratios of different ion masses obtained in parallel from the same biological sample area provide quantitative data and, if ratios are above the natural level, directly confirm incorporation of the isotope tracer. The first studies utilizing this approach investigated (a) the in situ fixation of ^{15}N$_2$ by the bacterial symbiont *Teredinibacter turnerae* thriving in shipworm gills, which provided evidence for transfer of bacterial N-compounds to the host (Lechene et al. 2007), (b) $^{13}CO_2$ and $^{15}N_2$ fixation and metabolite exchange in and between individual cells of filamenteous *Anabaena oscillarioides* (Popa et al. 2007), and (c) subcellular and intraday segregation of $^{13}CO_2$ and $^{15}N_2$ fixation in the cyanobacterium *Trichodesmium* (Finzi-Hart et al. 2009). Combination of whole cell hybridization for microbial identification with NanoSIMS is achieved by using a halogen (such as iodine, bromine, and fluorine with low natural background levels) or stable isotope label for visualizing cells that hybridize to phylogenetic probes. The halogen label is introduced into the cell either directly with the halogenated probe (Li et al. 2008) or by using the catalyzed reporter deposition system. In the latter approach, detection sensitivity is significantly improved by enhancing halogen deposition in the cell via a combination of horseradish peroxidase-tagged probes with signal amplification using tyramides labeled with halogen-containing fluorescent dyes (Behrens et al. 2008; Musat et al. 2008). This technique (a) proved that ^{13}C- and ^{15}N-metabolites that are initially assimilated by an *Anabaena* cyanobacterium are shuttled to a co-cultivated and physically interacting *Rhizobium* strain, and (b) identified that members of the phylum *Bacteroidetes* incorporate ^{13}C-labeled amino acids in an oral biofilm from a healthy human individual (Behrens et al. 2008). Another intriguing NanoSIMS in situ hybridization study explored the role of anaerobic, phototrophic bacteria in an oligotrophic, meromictic lake and identified a potential keystone species, which is present at low abundance, but contributes significantly to ecosystem function (Musat et al. 2008). *Chromatium okenii* was the least abundant (~0.3% of total cell number) of three targeted species, but was responsible for more than 40% of the total uptake of ^{15}N-ammonium and for 70% of the total uptake of $H^{13}CO_3^-$ in the system. Furthermore, metabolic rates varied greatly between individual cells of the same species, potentially resulting from considerable genomic and/or physiological heterogeneity between different populations.

Conclusions

There is an increasing flood of information about the (phylo)genetic composition of natural microbial communities, obtained by using current- and next-generation sequencing technologies for 16S rRNA gene and metagenomic surveys, which is crucial for informing hypotheses on the ecological function of microorganisms in their natural environment. It is time to complement these large-scale efforts in inventorying the structure and dynamics of the microbial metagenome in time and space with targeted follow-up studies aimed at addressing the actual physiological and metabolic status of microbial community members. The tool box of isotope-labeling techniques offers many opportunities that can be customized and combined to address various physiological aspects of different phylogenetic groups of microbes at the level of all group members or of individual cells. Such cultivation-independent insights into organismal physiology are important for elucidating interrelationships between different groups or cells of microorganisms and their natural environment. Low sample throughput remains the main bottleneck of isotope-labeling techniques, restricting time- and space-dependent analyses that require analysis of many samples. The parallelized analysis of many community members via the isotope array approach is a promising solution and thus justifies continued efforts in further developing this technology.

Acknowledgements The authors would like to thank Michael Wagner, Holger Daims, Matthias Horn, Markus Schmid, and all other colleagues from the Department of Microbial Ecology for the excellent cooperation and support over the last years. Per Halkjær Nielsen, Andrew Whiteley, Marcel Kuypers, and their group members are acknowledged for past and present collaboration on the development of isotopic labeling methods. Special thanks go to David Berry for proof reading. Work of the authors is financed by the Austrian Science Fund (projects P18836-B17 and P20185-B17), the Austrian Federal Ministry of Science and Research (GEN-AU III, InflammoBiota), the German Federal Ministry of Science and Education (BIOLOG/BIOTA project 01LC0621D), and the Alexander von Humboldt-foundation.

References

Achtman M, Wagner M (2008) Microbial diversity and the genetic nature of microbial species. Nat Rev Microbiol 6:431–440

Adamczyk J, Hesselsoe M, Iversen N, Horn M, Lehner A, Nielsen PH, Schloter M, Roslev P, Wagner M (2003) The isotope array, a new tool that employs substrate-mediated labeling of rRNA for determination of microbial community structure and function. Appl Environ Microbiol 69:6875–6887

Behrens S, Losekann T, Pett-Ridge J, Weber PK, Ng WO, Stevenson BS, Hutcheon ID, Relman DA, Spormann AM (2008) Linking microbial phylogeny to metabolic activity at the single-cell level by using enhanced element labeling-catalyzed reporter deposition fluorescence in situ hybridization (EL-FISH) and NanoSIMS. Appl Environ Microbiol 74:3143–3150

Boschker HTS, Middelburg JJ (2002) Stable isotopes and biomarkers in microbial ecology. FEMS Microbiol Ecol 40:85–95

Boschker HTS, Nold SC, Wellsbury P, Bos D, de Graaf W, Pel R, Parkes RJ, Cappenberg TE (1998) Direct linking of microbial populations to specific biogeochemical processes by ^{13}C-labelling of biomarkers. Nature 392:801–805

Brock TD, Brock ML (1966) Autoradiography as a tool in microbial ecology. Nature 209:734–736

Buckley DH, Huangyutitham V, Hsu SF, Nelson TA (2007a) Stable isotope probing with $^{15}N_2$ reveals novel noncultivated diazotrophs in soil. Appl Environ Microbiol 73:3196–3204

Buckley DH, Huangyutitham V, Hsu SF, Nelson TA (2007b) Stable isotope probing with ^{15}N achieved by disentangling the effects of genome G+C content and isotope enrichment on DNA density. Appl Environ Microbiol 73:3189–3195

Cadisch G, Espana M, Causey R, Richter M, Shaw E, Morgan JA, Rahn C, Bending GD (2005) Technical considerations for the use of ^{15}N-DNA stable-isotope probing for functional microbial activity in soils. Rapid Commun Mass Spectrom 19:1424–1428

Chen Y, Dumont MG, Neufeld JD, Bodrossy L, Stralis-Pavese N, McNamara NP, Ostle N, Briones MJ, Murrell JC (2008) Revealing the uncultivated majority: combining DNA stable-isotope probing, multiple displacement amplification and metagenomic analyses of uncultivated Methylocystis in acidic peatlands. Environ Microbiol 10:2609–2622

Daims H, Lücker S, Wagner M (2006) Daime, a novel image analysis program for microbial ecology and biofilm research. Environ Microbiol 8:200–213

De Gelder J, Willemse-Erix D, Scholtes MJ, Sanchez JI, Maquelin K, Vandenabeele P, Boever PD, Puppels GJ, Moens L, Vos PD (2008) Monitoring poly(3-hydroxybutyrate) production in cupriavidus necator DSM 428 (H16) with raman spectroscopy. Anal Chem 80:2155–2160

DeLong EF, Wickham GS, Pace NR (1989) Phylogenetic stains: ribosomal RNA-based probes for the identification of single cells. Science 243:1360–1363

Dumont MG, Murrell JC (2005) Stable isotope probing – linking microbial identity to function. Nat Rev Microbiol 3:499–504

Dumont MG, Radajewski SM, Miguez CB, McDonald IR, Murrell JC (2006) Identification of a complete methane monooxygenase operon from soil by combining stable isotope probing and metagenomic analysis. Environ Microbiol 8:1240–1250

Finzi-Hart JA, Pett-Ridge J, Weber PK, Popa R, Fallon SJ, Gunderson T, Hutcheon ID, Nealson KH, Capone DG (2009) Fixation and fate of C and N in the cyanobacterium *Trichodesmium* using nanometer-scale secondary ion mass spectrometry. Proc Natl Acad Sci USA 106:6345–6350

Franchini AG, Nikolausz M, Kästner M (2009) Sequence Specific Primer Extension RNA Analysis (SeSPERA) for the investigation of substrate utilization of microbial communities. J Microbiol Methods 79:111–113

Frias-Lopez J, Shi Y, Tyson GW, Coleman ML, Schuster SC, Chisholm SW, Delong EF (2008) Microbial community gene expression in ocean surface waters. Proc Natl Acad Sci USA 105:3805–3810

Friedrich MW (2006) Stable-isotope probing of DNA: insights into the function of uncultivated microorganisms from isotopically labeled metagenomes. Curr Opin Biotechnol 17:59–66

Gallagher E, McGuinness L, Phelps C, Young LY, Kerkhof LJ (2005) ^{13}C-carrier DNA shortens the incubation time needed to detect benzoate-utilizing denitrifying bacteria by Stable-Isotope Probing. Appl Environ Microbiol 71:5192–5196

Ginige MP, Hugenholtz P, Daims H, Wagner M, Keller J, Blackall LL (2004) Use of stable-isotope probing, full-cycle rRNA analysis, and fluorescence in situ hybridization-microautoradiography to study a methanol-fed denitrifying microbial community. Appl Environ Microbiol 70:588–596

Hesselsoe M, Nielsen JL, Roslev P, Nielsen PH (2005) Isotope labeling and microautoradiography of active heterotrophic bacteria on the basis of assimilation of [14]CO_2. Appl Environ Microbiol 71:646–655

Hesselsoe M, Fureder S, Schloter M, Bodrossy L, Iversen N, Roslev P, Nielsen PH, Wagner M, Loy A (2009) Isotope array analysis of *Rhodocyclales* uncovers functional redundancy and versatility in an activated sludge. ISME J 3:1349–1364

Huang WE, Griffiths RI, Thompson IP, Bailey MJ, Whiteley AS (2004) Raman microscopic analysis of single microbial cells. Anal Chem 76:4452–4458

Huang WE, Stoecker K, Griffiths R, Newbold L, Daims H, Whiteley AS, Wagner M (2007) Raman-FISH: combining stable-isotope Raman spectroscopy and fluorescence in situ hybridization for the single cell analysis of identity and function. Environ Microbiol 9:1878–1889

Huang WE, Ferguson A, Singer AC, Lawson K, Thompson IP, Kalin RM, Larkin MJ, Bailey MJ, Whiteley AS (2009a) Resolving genetic functions within microbial populations: in situ analyses using rRNA and mRNA stable isotope probing coupled with single-cell raman-fluorescence in situ hybridization. Appl Environ Microbiol 75:234–241

Huang WE, Ward AD, Whiteley AS (2009b) Raman tweezers sorting of single microbial cells. Environ Microbiol Rep 1:44–49

Jeon CO, Park W, Padmanabhan P, DeRito C, Snape JR, Madsen EL (2003) Discovery of a bacterium, with distinctive dioxygenase, that is responsible for in situ biodegradation in contaminated sediment. Proc Natl Acad Sci USA 100:13591–13596

Kalyuzhnaya MG, Lapidus A, Ivanova N et al (2008) High-resolution metagenomics targets specific functional types in complex microbial communities. Nat Biotechnol 26:1029–1034

Lasken RS, Stockwell TB (2007) Mechanism of chimera formation during the multiple displacement amplification reaction. BMC Biotechnol 7:19

Lechene C, Hillion F, McMahon G et al (2006) High-resolution quantitative imaging of mammalian and bacterial cells using stable isotope mass spectrometry. J Biol 5:20

Lechene CP, Luyten Y, McMahon G, Distel DL (2007) Quantitative imaging of nitrogen fixation by individual bacteria within animal cells. Science 317:1563–1566

Lee N, Nielsen PH, Andreasen KH, Juretschko S, Nielsen JL, Schleifer K-H, Wagner M (1999) Combination of fluorescent in situ hybridization and microautoradiography – a new tool for structure-function analyses in microbial ecology. Appl Environ Microbiol 65:1289–1297

Li T, Wu TD, Mazeas L, Toffin L, Guerquin-Kern JL, Leblon G, Bouchez T (2008) Simultaneous analysis of microbial identity and function using NanoSIMS. Environ Microbiol 10:580–588

Liou JS, Derito CM, Madsen EL (2008) Field-based and laboratory stable isotope probing surveys of the identities of both aerobic and anaerobic benzene-metabolizing microorganisms in freshwater sediment. Environ Microbiol 10:1964–1977

Loy A, Bodrossy L (2006) Highly parallel microbial diagnostics using oligonucleotide microarrays. Clin Chim Acta 363:106–119

Loy A, Maixner F, Wagner M, Horn M (2007) probeBase–an online resource for rRNA-targeted oligonucleotide probes: new features 2007. Nucleic Acids Res 35:D800–D804

Loy A, Arnold R, Tischler P, Rattei T, Wagner M, Horn M (2008) probeCheck – a central resource for evaluating oligonucleotide probe coverage and specificity. Environ Microbiol 10:2894–2896

Lu Y, Conrad R (2005) In situ stable isotope probing of methanogenic archaea in the rice rhizosphere. Science 309:1088–1090

Ludwig W, Strunk O, Westram R et al (2004) ARB: a software environment for sequence data. Nucleic Acids Res 32:1363–1371

Lueders T, Manefield M, Friedrich MW (2004a) Enhanced sensitivity of DNA- and rRNA-based stable isotope probing by fractionation and quantitative analysis of isopycnic centrifugation gradients. Environ Microbiol 6:73–78

Lueders T, Wagner B, Claus P, Friedrich MW (2004b) Stable isotope probing of rRNA and DNA reveals a dynamic methylotroph community and trophic interactions with fungi and protozoa in oxic rice field soil. Environ Microbiol 6:60–72

MacGregor BJ, Bruchert V, Fleischer S, Amann R (2002) Isolation of small-subunit rRNA for stable isotopic characterization. Environ Microbiol 4:451–464

Madsen EL (2006) The use of stable isotope probing techniques in bioreactor and field studies on bioremediation. Curr Opin Biotechnol 17:92–97

Maixner F, Noguera DR, Anneser B, Stoecker K, Wegl G, Wagner M, Daims H (2006) Nitrite concentration influences the population structure of *Nitrospira*-like bacteria. Environ Microbiol 8:1487–1495

Manefield M, Whiteley AS, Griffiths RI, Bailey MJ (2002) RNA stable isotope probing, a novel means of linking microbial community function to phylogeny. Appl Environ Microbiol 68:5367–5373

Miyatake T, Macgregor BJ, Boschker HT (2009) Stable isotope probing combined with magnetic bead capture hybridization of 16S rRNA: linking microbial community function to phylogeny of sulfate-reducing Deltaproteobacteria in marine sediments. Appl Environ Microbiol 75:4927–4935

Musat N, Halm H, Winterholler B, Hoppe P, Peduzzi S, Hillion F, Horreard F, Amann R, Jorgensen BB, Kuypers MM (2008) A single-cell view on the ecophysiology of anaerobic phototrophic bacteria. Proc Natl Acad Sci USA 105:17861–17866

Neufeld JD, Dumont MG, Vohra J, Murrell JC (2007a) Methodological considerations for the use of stable isotope probing in microbial ecology. Microb Ecol 53:435–442

Neufeld JD, Vohra J, Dumont MG, Lueders T, Manefield M, Friedrich MW, Murrell JC (2007b) DNA stable-isotope probing. Nat Protoc 2:860–866

Neufeld JD, Wagner M, Murrell JC (2007c) Who eats what, where and when? Isotope-labelling experiments are coming of age. ISME J 1:103–110

Nielsen JL, Christensen D, Kloppenborg M, Nielsen PH (2003) Quantification of cell-specific substrate uptake by probe-defined bacteria under in situ conditions by microautoradiography and fluorescence in situ hybridization. Environ Microbiol 5:202–211

Orphan VJ (2009) Methods for unveiling cryptic microbial partnerships in nature. Curr Opin Microbiol 12:231–237

Orphan VJ, House CH, Hinrichs KU, McKeegan KD, DeLong EF (2001) Methane-consuming archaea revealed by directly coupled isotopic and phylogenetic analysis. Science 293:484–487

Ouverney CC, Fuhrman JA (1999) Combined microautoradiography-16S rRNA probe technique for determination of radioisotope uptake by specific microbial cell types *in situ*. Appl Environ Microbiol 65:1746–1752

Pasteris JD, Freeman JJ, Goffredi S, Buck KR (2001) Raman spectroscopic and laser scanning confocal microscopic analysis of sulfur in living sulfur-precipitating marine bacteria. Chem Geol 180:3–18

Popa R, Weber PK, Pett-Ridge J, Finzi JA, Fallon SJ, Hutcheon ID, Nealson KH, Capone DG (2007) Carbon and nitrogen fixation and metabolite exchange in and between individual cells of *Anabaena oscillarioides*. ISME J 1:354–360

Pumphrey GM, Madsen EL (2008) Field-based stable isotope probing reveals the identities of benzoic acid-metabolizing microorganisms and their in situ growth in agricultural soil. Appl Environ Microbiol 74:4111–4118

Schwartz E (2007) Characterization of growing microorganisms in soil by stable isotope probing with H$_2$^{18}O. Appl Environ Microbiol 73:2541–2546

Singleton DR, Hunt M, Powell SN, Frontera-Suau R, Aitken MD (2007) Stable-isotope probing with multiple growth substrates to determine substrate specificity of uncultivated bacteria. J Microbiol Methods 69:180–187

Sogin ML, Morrison HG, Huber JA, Welch DM, Huse SM, Neal PR, Arrieta JM, Herndl GJ (2006) Microbial diversity in the deep sea and the underexplored "rare biosphere". Proc Natl Acad Sci USA 103:12115–12120

Urich T, Lanzen A, Qi J, Huson DH, Schleper C, Schuster SC (2008) Simultaneous assessment of soil microbial community structure and function through analysis of the meta-transcriptome. PLoS ONE 3:e2527

Wagner M (2009) Single-cell ecophysiology of microbes as revealed by Raman microspectros-copy or secondary ion mass spectrometry imaging. Annu Rev Microbiol 63:411–429

Wagner M, Nielsen PH, Loy A, Nielsen JL, Daims H (2006) Linking microbial community structure with function: fluorescence in situ hybridization-microautoradiography and isotope arrays. Curr Opin Biotechnol 17:1–9

Wagner M, Smidt H, Loy A, Zhou J (2007) Unravelling microbial communities with DNA-microarrays: challenges and future directions. Microb Ecol 53:498–506

Whiteley AS, Manefield M, Lueders T (2006) Unlocking the 'microbial black box' using RNA-based stable isotope probing technologies. Curr Opin Biotechnol 17:67–71

Zhang K, Martiny AC, Reppas NB, Barry KW, Malek J, Chisholm SW, Church GM (2006) Sequencing genomes from single cells by polymerase cloning. Nat Biotechnol 24:680–686

Chapter 7
The Geomicrobiology of Arsenic

Rhesa N. Ledbetter and Timothy S. Magnuson

Introduction

Since the beginning of life on Earth, microorganisms have had the remarkable ability to evolve the necessary molecular machinery to cope with and even benefit from high concentrations of toxic metals in the environment. Many metals (in trace amount) play an integral role in biological processes; however, many of the same metals, as well as those not required in biological systems, can be quite harmful. This is most commonly a consequence of the metal concentration; however, speciation and physicochemical form of the element are added factors. Some metals that microorganisms depend on in low concentrations include arsenic, calcium, cobalt, chromium, copper, iron, potassium, magnesium, manganese, sodium, nickel, and zinc; whereas, aluminum, cadmium, gold, lead, mercury, and silver are not known to be part of cellular structures or processes (Bruins et al. 2000; Stolz et al. 2002). The essential metals have been found to be crucial components in redox processes, gene expression, biomolecule activity, cellular osmotic balance, and protein and bacterial cell wall structures (Hughes and Poole 1989; Ji and Silver 1995; Poole and Gadd 1989). Yet, if any of these metals exceed certain concentrations, microorganisms must use resistance mechanisms to survive the 'metal stress'. Because many environments inhabited by microorganisms have continuously contained poisonous elements, resistance mechanisms most likely developed shortly after the evolution of prokaryotic life.

Over one third of the elements on the periodic table are transformed by microbes (Stolz et al. 2002). One such element is arsenic, and until recently its role and importance in biogeochemical cycling had not been fully understood (Oremland et al. 2004, 2005; Oremland and Stolz 2003; Rhine et al. 2005). Arsenic is not considered an abundant element; however, in certain environments, such as hydrothermal, sulfidic, evaporitic, and iron hydroxide rich systems, it can reach significant

R.N. Ledbetter and T.S. Magnuson (✉)
Department of Biological Sciences, Idaho State University, P.O. Box 8007,
Pocatello, ID 83209, USA
e-mail: magntimo@isu.edu

L.L. Barton et al. (eds.), *Geomicrobiology: Molecular and Environmental Perspective*,   147
DOI 10.1007/978-90-481-9204-5_7, © Springer Science+Business Media B.V. 2010

concentrations (Stolz et al. 2002). Microorganisms living in arsenic-rich environments possess unique metabolic pathways for surviving in and even benefiting from the presence of this toxic metal. Furthermore, arsenic-transforming microbes play a notable role in the cycling of arsenic and overall, contribute to the balance of the ecosystem.

Arsenic in the Environment

Arsenic is relatively abundant in soils and natural waters, and although it is primarily a consequence of anthropogenic sources, it also occurs naturally in the Earth's crust, minerals, and in groundwaters in contact with geologic formations containing high levels of arsenic. In nature, there are four oxidation states of arsenic. They include arsenide [As(-III)], elemental arsenic [As(0)], arsenite [As(III)], and arsenate [As(V)] (Table 7.1). Arsines are primarily found in fungal cultures and strongly reducing environments, and elemental arsenic rarely occurs (Bentley and Chasteen 2002; Stolz et al. 2006). Arsenite and arsenate are the two primary forms of inorganic arsenic in the environment. Arsenate is most often bound with iron oxides on mineral surfaces (Matera et al. 2003), while arsenite is more toxic and soluble. Arsenic-transforming microbes and/or the physicochemical conditions of the environment can alter the valence state and mobility of arsenic (Oremland and Stolz 2005; Stolz et al. 2006). For example, arsenate trapped in sediment and minerals can be readily reduced and liberated in the form of arsenite (Ahmann et al. 1997; Jones et al. 2000; Zobrist et al. 2000). Abiotic and biotic reduction of arsenate in

Table 7.1 Chemical speciation and formulas of arsenic (Adapted from Paez-Espino et al. 2009)

Arsenic species	Oxidation state	Chemical formula
Inorganic		
Elemental Arsenic	As(0)	As
Arsine	As(-III)	AsH_3
Arsenite	As(III)	$As(OH)_3$
Arsenate	As(V)	AsO_4H_3
Organoarsenicals		
Methylarsine	MMA(III)	CH_3AsH_2
Dimethylarsine	DMA(III)	$(CH_3)_2AsH$
Trimethylarsine	TMA(III)	$(CH_3)_3As$
Methylarsonic acid	MMA(V)	$CH_3AsO(OH)_2$
Dimethylarsonic acid	DMA(V)	$(CH_3)_2AsO(OH)$
Trimethylarsine oxide	TMAO	$(CH_3)_3AsO$
Tetramethylarsonium ion	TMA^+	$(CH_3)_4As^+$
Arsenocholine		$(CH_3)_3As^+CH_2CH_2OH$
Arsenobetaine		$(CH_3)_3As^+CH_2COOH$

the presence of sulfide can also occur, resulting in the formation of arsenic-sulfide minerals (Eary 1992; Ledbetter et al. 2007; Newman et al. 1997a). Furthermore, reduction of minerals (i.e. iron oxides) can release bound arsenate, making it more available for reduction throughout the system (Lovley 1993).

Research on inorganic arsenic in bodies of water has shown that there are distinct zones of arsenic speciation. Data from several lakes have revealed a distinct transition zone between arsenate and arsenite throughout the oxycline (Hollibaugh et al. 2005; Peterson and Carpenter 1983; Seyler and Martin 1989). Studies in arsenic rich thermal springs showed a change in the distribution of arsenite and arsenate down an outflow channel (Fig. 7.1) with microbial communities all along the flow path harboring arsenic transforming genera (Fig. 7.2) (Connon et al. 2008).

Although inorganic species are the most prevalent forms of arsenic, there are also several organoarsenicals found in the environment (Table 7.1). Organoarsenicals are typically less toxic than their inorganic counterparts and are frequently associated with marine environments. These arsenic compounds have also been widely used in medicine, herbicides, and animal feed additives (Mukhopadhyay et al. 2002). For example, organoarsenicals are added to chicken feedstock to enhance the growth of chicken muscle by promoting vascularization (a good example of arsenic being a beneficial toxin). Considering how much chicken is produced and consumed around the world, the amount of arsenic released into the environment in chicken production and processing is substantial (Jackson and Bertsch 2001).

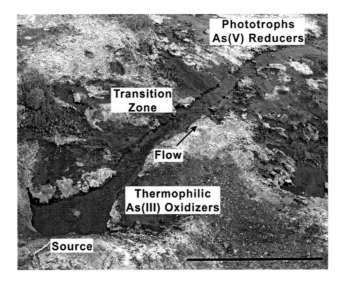

Fig. 7.1 Outflow channel of Alvord Hot Spring A03C. Thermal waters containing primarily As(III) emerge from the source, and microbial biofilms are established all along the flow channel. Mineral deposits in the transition zone (*dark* and *light* materials) include iron oxides, which can bind to arsenic species and contain abundant arsenic-reducing bacteria. Scale bar equals 1 m

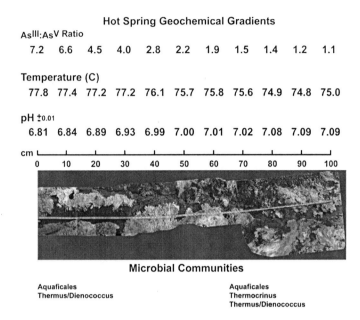

Hot Spring Geochemical Gradients

As^{III}:As^V Ratio

| 7.2 | 6.6 | 4.5 | 4.0 | 2.8 | 2.2 | 1.9 | 1.5 | 1.4 | 1.2 | 1.1 |

Temperature (C)

| 77.8 | 77.4 | 77.2 | 77.2 | 76.1 | 75.7 | 75.8 | 75.6 | 74.9 | 74.8 | 75.0 |

pH $\pm$0.01

| 6.81 | 6.84 | 6.89 | 6.93 | 6.99 | 7.00 | 7.01 | 7.02 | 7.08 | 7.09 | 7.09 |

cm

| 0 | 10 | 20 | 30 | 40 | 50 | 60 | 70 | 80 | 90 | 100 |

Microbial Communities

Aquaficales
Thermus/Dienococcus

Aquaficales
Thermocrinus
Thermus/Dienococcus

Fig. 7.2 Field data from Alvord Hot Spring A03C. There is a thermal and pH gradient from source to end, as well as an As(III) gradient which decreases along the flow path. Microbial metabolism coupled to physicochemical factors drives the biogeochemistry found along the flow channel. Prominent phylogenic groups found in each area are shown

Arsenic in Biological Systems

Living cells are most commonly exposed to and affected by arsenite and arsenate (Silver and Phung 2005b). Although human cells cannot tolerate high concentrations of arsenic, many microbial cells have developed the necessary machinery to manage the toxicity due to their continuous association with the metal. Arsenite has a high affinity for thiol groups and subsequently affects essential metabolic reactions (NRC 1999), whereas arsenate is an analog of phosphate and manifests its poisonous affects by uncoupling oxidative phosphorylation (Hindmarsh 2000). Microorganisms can control the toxicity of arsenic by either actively exporting it out of the cell or by chemically altering it to a less toxic form (Ellis et al. 2001). Some microorganisms can use arsenic to their advantage and employ it in energy generating reactions (Gihring and Banfield 2001; Macy et al. 1996, 2000; Newman et al. 1997b; Santini et al. 2000; Silver and Phung 2005b; Stolz and Oremland 1999). Specific processes used to transform arsenic include oxidation, reduction, methylation, and demethylation. To accomplish these processes, energy generating or not, microbes must possess the proper uptake systems and enzymes.

Uptake of Arsenic from the Environment

Even though arsenic is considered an 'essential' toxin, it is not required for nutrition or metabolism intracellularly (Paez-Espino et al. 2009). Arsenic is essential in that may be used as an osmolite in some marine environments (Stolz et al. 2002). It has also been found in polysaccharide materials from algae (Stolz and Oremland 1999). The fundamental role of arsenic as an 'essential' toxin is still quite unclear to date, and benefits of this toxic metal at low concentrations are still being discovered.

Because arsenic plays no role in the cytoplasm, existing transport systems on the cell surface are used to uptake arsenic from the environment (Rosen and Liu 2009). Arsenite is brought into the cells by aqua-glyceroporins, which are part of the aquaporin superfamily of transporters (Meng et al. 2004; Mukhopadhyay et al. 2002). At physiological pH, arsenite resembles glycerol (Mukhopadhyay et al. 2002). The aqua-glyceroporin in *E. coli*, known as GlpF, channels glycerol into the cell but can also utilize arsenite as a substrate' doesn't seem to fit correctly. Homologues of this transporter have been found in other organisms (Gourbal et al. 2004) and are also thought to take up reduced arsenic. Arsenate, being an analog of phosphate, is taken up using phosphate transport membrane systems (Harold and Baarda 1966; Rosen and Liu 2009), such as the Pit (phosphate transporter) and Pst (specific phosphate transporter) systems of *Escherichia coli* (Rosenberg et al. 1977; Willsky and Malamy 1980). These mechanisms of arsenic uptake illustrate how substrate specificity may be somewhat non-specific, accommodating several different substrates. This is reasonable, as phosphate and glycerol molecules have a distinct structural similarity when examined side by side with arsenate and arsenite respectively.

Arsenic Tranformation Mechanisms

Arsenic Oxidation

Arsenite oxidase catalyzes the oxidation of highly toxic As(III) to less toxic As(V) and is most commonly used as an electron donor in the electron transport chain for energy generation; however, it can also function in detoxification processes with no net energy yield. The arsenite oxidase enzyme was first purified and characterized from the bacterium *Alcaligenes faecalis* (Anderson et al. 1992), and its crystal structure was determined (Ellis et al. 2001; Silver and Phung 2005b). The data demonstrated that the arsenite oxidase belongs to the dimethylsulfoxide (DMSO) class of enzymes and is a periplasmic or membrane associated heterodimer consisting of both a large (88 kDa) and a small subunit (14 kDa) (Fig. 7.3) (Silver and Phung 2005b; Stolz et al. 2002). The large 'A' subunit contains a molybdopterin cofactor and a [3Fe-4S] cluster, whereas the small 'B' subunit encompasses a

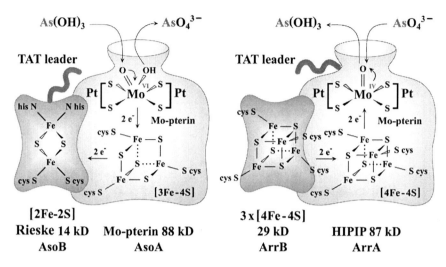

Fig. 7.3 Models of arsenic oxidase (*left*) and arsenic reductase (*right*) enzyme complexes, showing catalytic centers and substrate binding 'funnel' (Modified from Silver and Phung 2005b)

Rieske type [2Fe-2S] cluster that appears to serve as an electron shuttle (Ellis et al. 2001). *aso*A and *aso*B genes encode the large and small subunit respectively and are found on an "arsenic gene island" comprised of more than 20 additional genes that are also considered to be a part of arsenic metabolism (Lloyd and Oremland 2006; Silver and Phung 2005b). A Twin Arginine Transporter (TAT) leader sequence is found on AsoB and serves to transport the folded protein across the cytoplasmic membrane (Silver and Phung 2005a, b). Once the protein has reached the periplasmic space, the leader sequence is cleaved, and the substrate As(III) can come into contact with the enzyme. As(III) enters through a shallow, 'funnel' shaped cavity found on the large subunit, upon which it binds to the molybdenum center (Silver and Phung 2005a, b). Upon contact, the molybdenum subsequently becomes reduced, arsenate is released from the cavity, and a succession of two electron transfer events occurs (Hoke et al. 2004; Silver and Phung 2005a, b). They include the transfer of electrons to the [3Fe-4S] in the large subunit, then onto the [2Fe-2S] center of the Rieske type subunit, and finally into the respiratory chain (Fig. 7.3). It is important to note that the nomenclature for these enzymes varies depending on specific process, with Aox referring to the oxidase used by chemoheterotrophs, and Aro referring to that used by chemolithoautotrophs. Regardless, the protein designations Aso, Aox, and Aro all refer to arsenite oxidase.

Recently, an arsenite oxidase was purified and characterized from an *Arthrobacter* sp. that had similar subunit composition to that of *Alcaligenes*, with the exception of an unusual FeS protein in the small subunit (Prasad et al. 2009). Upon sequencing of the small subunit gene, little homology was found when compared to other known *aso*B sequences, raising the possibility that the biochemical behavior of arsenite oxidases could be modulated through sequence changes in the small subunit. Despite the difference in structure and sequence homology, the function of the enzyme remains the same.

Genetic analysis of two arsenic oxidizing bacteria (*Achromobacter* sp. SY8 and *Pseudomonas* sp. TS44) revealed distinct arsenic oxidation and resistance operons, both of which were inducible by arsenite and arsenate respectively (Cai et al. 2009). These result suggesting that this type of organization and regulation might be present in other arsenic oxidizing microbes.

An arsenic oxidizing, nitrate-reducing chemolithotroph, *Alkalilimnicola ehrlichii*, was described in physiologic and genomic detail (Richey et al. 2009), and while the organism lacked expected arsenic oxidase genes, it did possess two respiratory arsenate reductase (*arr*) operons, one of which appears to be evolutionarily adapted for arsenic oxidation. Catalytically, the enzyme resulting from expression of this operon is biochemically reversible and can catalyze both arsenic oxidation and reduction reactions. These findings represent an interesting evolutionary possibility, in that oxidation capability arose from reductase enzymes.

Phylogenetically and physiologically, arsenic oxidizing bacteria are very diverse and include representatives from the Alpha-, Beta-, and Gammaproteobacteria, Deinocci (i.e. *Thermus*), and Crenarchaeota (Stolz et al. 2002). The fact that this type of metabolism is so widespread supports the idea that microbial arsenic metabolism is an ancient process that has spread to many kinds of microbes.

Arsenic Reduction

There are two distinct types of arsenate reductases, namely, respiratory (or dissimilatory) and cytoplasmic detoxification enzymes. The respiratory arsenate reductase is used as a terminal electron acceptor in the electron transport chain, and this metabolism is found in a number of microorganisms (referred to as *d*issimilatory *a*rsenic *r*educing *p*rokaryotes, or DARPs) distributed throughout the prokaryotic domain (Stolz et al. 2002). As with arsenite oxidase, it falls into the DMSO class of enzymes and is a periplasmic or membrane associated heterodimer consisting of both a large (87 kDa) and a small (29 kDa) subunit (Fig. 7.3) (Silver and Phung 2005b; Stolz et al. 2002). The large subunit contains a molybdopterin cofactor and a [4Fe-4S] cluster, while the small subunit contains up to four [4Fe-4S] clusters (Silver and Phung 2005b). In addition, the small subunit is twice the size (Afkar et al. 2003; Krafft and Macy 1998; Saltikov and Newman 2003) and non- homologous to the Rieske-type small subunit of the arsenite oxidase (Silver and Phung 2005b), providing more evidence that oxidation and reduction can be altered by the structure of the small subunit. The genes encoding for the respiratory arsenate reductase subunits are *arr*A for the large subunit and *arr*B for the small subunit. A TAT sequence is located on ArrA and provides the same function as described for arsenite oxidase enzyme (Silver and Phung 2005a, b). The series of events for the respiratory arsenate reductase occur opposite in comparison to the oxidation reaction (Silver and Phung 2005a). Because arsenate is used as a terminal electron acceptor, the electrons are carried from the end of the chain, to the enzyme, onto the molybdopterin cofactor, and finally to the substrate As(V), which in turn reduces the As(V) to As(III) (Fig. 7.3) (Silver and Phung 2005a).

The ArsC cytoplasmic arsenate reductase, used for detoxification of arsenic, is a small monomeric, cytoplasmic protein that requires either thioredoxin or glutaredoxin linked reactions for activity (Mukhopadhyay and Rosen 2002; Silver and Phung 2005b). The glutathione-glutaredoxin coupled reaction or thioredoxin provides the reducing power to convert As(V) to As(III) and in the end results in the removal of As(III) by an ATPase (Fig. 7.4) (Mukhopadhyay and Rosen 2002). ArsC arsenate reductases have been found in phylogenetically diverse groups of microbes and in nearly all bacteria with genomes larger than two megabases (Silver and Phung 2005b). Additionally, these genes are often found on plasmids, making them available for lateral gene transfer and distribution within arsenic bearing environments (Paez-Espino et al. 2009). It has even been noted that this activity appears to be more predominant in prokaryotes than the activity for tryptophan biosynthesis (Silver and Phung 2005b).

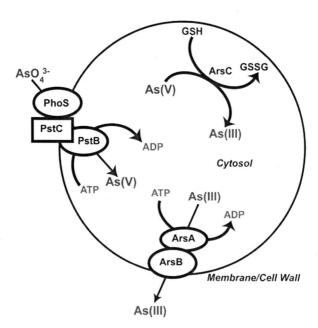

Fig. 7.4 Arsenic detoxification systems in *E. coli*. Transporters bring arsenate (AsO_4^{3-}) into the cell, where it is then reduced by ArsC, with glutathione as electron donor. The resultant As(III) is then exported via ATP-dependent ArsAB

Methylation

Although detoxification and energy generating processes are the most common mechanisms employed in the transformation of arsenic, both prokaryotic and eukaryotic organisms can methylate arsenic (Paez-Espino et al. 2009; Stolz et al. 2006). Methylation is a process by which organisms synthesize organic arsenic compounds (C–As) and produce mono-, di-, and trimethylated arsenicals (Thomas et al. 2007). This mechanism has been well studied in eukaryotes (Bentley and Chasteen 2002; Gadd 1993); however, much less is known about bacterial methylation (Stolz et al. 2006). Challenger was the first to describe methylation in fungi (Challenger 1945). His work with *Scopulariopsis brevicaulis* demonstrated that the organism reduced As(V) to As(III), then oxidatively added a methyl group to the reduced arsenic (Challenger 1951; Dombrowski et al. 2005). This process created various methylated arsenic compounds including methyl arsenite, dimethyl arsenate, dimethyl arsenite, and trimethyl arsine oxide (Fig. 7.5) (Paez-Espino et al. 2009; Stolz et al. 2006). This mechanism proposed by Challenger is predicted to be very similar in bacterial species. *S*-adenosylmethionine (SAM) typically provides the methyl groups, but in some cases, anaerobic bacteria may use methylcobalamin as the donor instead (Gadd 1993; Krautler 1990; Stupperich 1993). Additionally, glutathione and other thiol-containing compounds supply reducing power during various methylation steps (Paez-Espino et al. 2009). It is important to note that different organisms will yield various methylated products. For example, bacteria tend to produce more arsines (Bentley and Chasteen 2002).

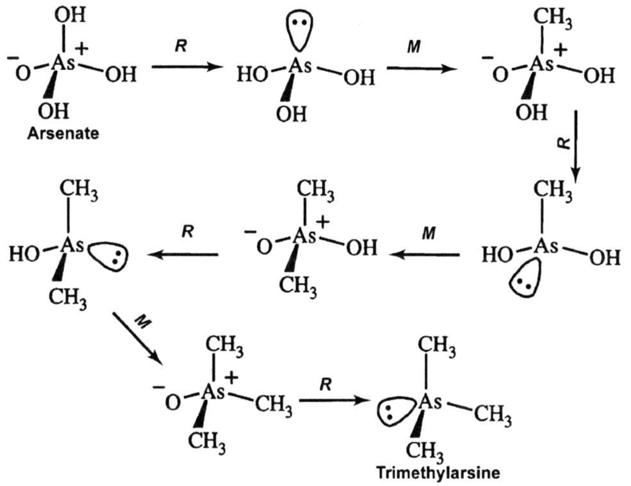

Fig. 7.5 Microbial methylation of arsenic. The initial reduction of arsenate supplies arsenite to feed the pathway. The arsines are typically volatile. R: Reduction step; M: Methylation step

Most methylated arsenicals are less toxic than inorganic arsenic species, but trivalent arsenicals are more noxious (Qin et al. 2006); thus, it is debated as to whether or not this is a detoxification mechanism. In spite of this debate, these trivalent organic species are volatile, which means they are immediately released into the atmosphere upon assimilation, then quickly re-oxidized into water-soluble arsenic species that are less toxic (Paez-Espino et al. 2009).

The enzyme Cyt19 catalyzes As(III)-S-adenosylmethyltransferase activity in rats and humans (Thomas et al. 2004; Waters et al. 2004a, b). More recently though a homolog of that enzyme, ArsM (30 kDa) (Qin et al. 2006), has been identified in over 120 bacteria and 16 archaea (Stolz et al. 2006). The *ars*M genes are found downstream of an *ars*R gene, encoding the archetypical arsenic-responsive transcriptional repressor that controls expression of *ars* operons (Wu and Rosen 1993). The location of the *ars*M genes strongly suggests that ArsM is directly related to arsenic resistance in bacteria (Qin et al. 2006).

Demethylation

Some organisms can utilize methylated arsenic compounds as carbon and energy sources through cleavage of the C–As bond (Stolz et al. 2006) in a process known as demethylation. Challenger, who discovered methylation, also first described demethylation in fungi (Challenger 1945). Bacteria from the genera *Alcaligenes, Pseudomonas,* and *Mycobacterium* have been found to demethylate mono- and dimethyl arsenic compounds (Bentley and Chasteen 2002). Additionally, Quinn and McMullan isolated a Gram-negative bacterium (strain ASV2) and found that it was able to use arsonoacetate and arsonochloroacetate as its sole carbon and energy sources and upon degradation release arsenate into the environment (Quinn and McMullan 1995). Although several organisms have been implicated in demethylation, very little is known concerning the specific process(es) and enzymes involved (Stolz et al. 2006), and more research in this area is warranted.

Applying Molecular and Microbiological Tools to the Study of Microbial Arsenic Transformation

Despite continuous efforts at developing new and innovative techniques for uncovering and examining microbes that can mediate metal transformations, current methodologies are still demonstrating a lack of specificity or deficiencies in sensitivity. Further research is necessary for the development of methods that will complement and extend present approaches. Currently, there are four major areas that are employed to study microbe metal transformations. They include cultivation, functional gene analysis, (meta)proteomics, and (meta)genomics.

Cultivation Methods

It has been estimated that less than 1% of the microorganisms from the environment can be cultured (Amann et al. 1990). This certainly limits the knowledge and understanding of microbial physiology, genetics, and ecology (Schloss and Handelsman 2005) and in fact, most likely presents a skewed view of the phylogenetic and physiologic groups that are most dominant in specific environments. Scientific researchers have been actively attempting to develop new methods that increase the culturability of microbes, and one approach in particular that seems to have some effect on the level of cultivation success is the unique design of a growth medium that selects for the physiologic group(s) of interest. Having the ability to accurately simulate the organisms' environment would be ideal; however, it is a very difficult task due to the gap of knowledge not only in the understanding of natural habitats but also in the lack of tools for measuring each component, significant or not, contributing to the system. Nonetheless, the more the laboratory environment mimics the natural environment, the higher the probability of culturing the microorganisms that are sought after (Kaeberlein et al. 2002). In the case of arsenic transforming microorganisms, Connon and coworkers developed a synthetic hot spring medium containing a solidifying agent and successfully isolated several arsenic oxidizing and reducing *Thermus* strains from thermal areas in Oregon within a time of 2 weeks (Connon et al. 2008). Successful application of unique cultivation approaches has also resulted in the discovery of new modes of energy production. The discovery of organisms capable of coupling the oxidation of arsenite (electron donor) to the reduction of nitrate (electron acceptor) was achieved in spite of a low energy yielding respiration and extreme haloalkaliphilic conditions (Hoeft et al. 2007). Physiologic description of '*Halarsenatibacter silvermanii*' SLAS-1(T) (Blum et al. 2009) revealed a novel mode of motility and an *arr* operon conferring arsenate respiration ability, demonstrating the scientific benefits of cultivating new members of arsenic transforming organisms for study.

Additional strategies that have been implemented and contributed to some improvement in cultivation include increased incubation times (Davis et al. 2005; Joseph et al. 2003; Stevenson et al. 2004), the addition of signaling compounds and inhibitors (Bruns et al. 2002, 2003; Stevenson et al. 2004), and the use of solid-phase minerals as a growth substratum (Reardon et al. 2004). Even with advances in strategic culturing approaches, it is evident that there are major challenges to overcome and improvements to be made before this classical method can be used exclusively to look at microbes that are playing a role in arsenic transformation.

Culture Independent Methods

Because of the limitations with cultivation studies, the development of improved techniques is vital to the understanding of what phylogenetic and physiologic groups are present and actively participating in metabolic and biogeochemical processes. Continued development of methods that do not rely on cultivation but rather on community analysis provide better insight into microbial structure and function in the natural environment.

Functional Gene Detection

PCR based functional gene analysis is utilized to amplify genes that encode specific proteins. However, the genes encoding the enzyme of interest are widely distributed among microbes and typically possess great sequence diversity. The lack of highly conserved sequences prevents the development of gene specific universal primers for the detection of these genes in new isolates (Silver and Phung 2005b). For example, Fig. 7.6 represents the genes for the arsenite oxidase enzyme in four different isolates and illustrates the variability in homology when compared to *A. faecalis*. The highest percentages of amino acid sequence similarity for the large and small subunit are 73% and 62%, respectively. The lowest percentage for the large subunit is 45% and 43% for the small subunit. Despite the recognized variability in sequence homology, new primers that claim enhanced detection of arsenite oxidase genes have been developed (Inskeep et al. 2007; Rhine et al. 2007). Cloning and analysis of *Hydrogenobaculum aox*AB genes led to a means of examining *aox* gene expression in situ along physicochemical gradients in an acidic thermal stream in Yellowstone National Park (Clingenpeel et al. 2009). Recent examples of successful detection and amplification of respiratory arsenic reduction genes (*arr*A) are also in the literature (Malasarn et al. 2004; Song et al. 2009). With the increasing variety of functional gene primers becoming available, detection and characterization of these genes in a wide variety of environments is much more dependable. However, it is important to realize the potential problems. The primers may be too degenerate or too specific for microbes with novel enzymes that have yet to be discovered. By using this method in conjunction with other approaches, added reliability and accuracy can be obtained.

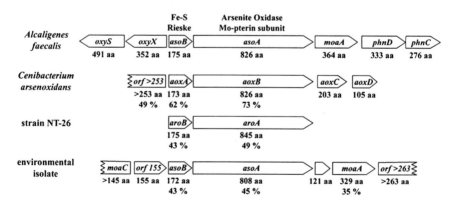

Fig. 7.6 Arsenite oxidase operons from four different isolates. Gene designations and protein size and are given, and percentages represent homology to the *A. faecalis* sequences (Adapted from Silver and Phung 2005b)

(Meta)proteomics

One area of promise in the research on arsenic geomicrobiology is the use of proteomics tools to examine complex samples from environments such as microbial mat biofilms. This method is carried out by extracting a total protein fraction from an environmental sample such as soil, biofilm, or water, digesting it with trypsin, and then using shotgun mass spectrometry to sequence the fragmented peptides (Ram et al. 2005). The approach is relatively simple, in that one can either analyze environmental samples directly (environmental proteomics) or analyze a subset of proteins obtained from protein electrophoresis gels. A study using a combination of these approaches was done using a thermophilic pure culture and biofilms from arsenic rich hydrothermal springs in Oregon (Connon et al. 2008; Ledbetter et al. 2007). Arsenite oxidase and arsenate reductase activities were initially detected using zymography (Fig. 7.7) and then confirmed using proteomic analysis as described above. Peptide hits obtained from proteomic analysis corresponded to

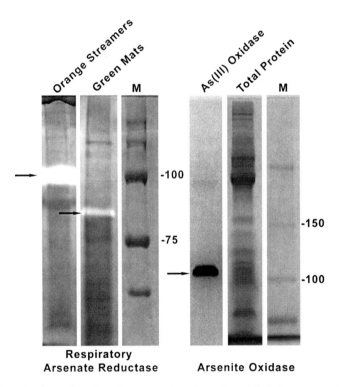

Fig. 7.7 Protein electrophoresis and zymography of arsenic metabolizing enzymes from pure cultures and microbial biofilms. *Left Panel*: Protein extracts from thermal biofilms resolved by SDS-PAGE. The gel was stained for activity, followed by total protein staining with Coomassie Blue. Note that two different sizes of arsenate reductase are detected in different biofilm samples. *Right panel*: Protein extract from *Thermus* A03C resolved by SDS-PAGE. The gel was stained for As(III) oxidase and then for total protein. Note that in both cases, the bands detected are of the expected size for these enzymes

arsenic transforming enzymes, verifying that the enzyme activities seen on the gel were indeed arsenite oxidase and arsenate reductase activities (Table 7.2). When applied to pure cultures, proteomics can reveal much about how arsenic affects cellular metabolism and gene expression. A study on an arsenic hypertolerant Pseudomonad revealed the expected *ars* operons, but proteomics showed that a number of 'hypothetical' proteins were also expressed with arsenic, as were cellular arsenic binding proteins not previously known to be associated with arsenic transformation (Patel et al. 2007). Proteomic analysis is a powerful complementary tool to current existing approaches used to characterize complex microbial communities. When used in combination with metagenomics, one can not only assess the phylogenetic and functional gene diversity of an environment but also the metabolism taking place with both spatial and temporal resolution comparable to nucleic acid based methods.

(Meta)genomics

Metagenomics is a culture independent method based either on sequencing or expression and is used to analyze genetic information extracted directly from the environment (Riesenfeld et al. 2004; Schloss and Handelsman 2003). This tool has been applied in a variety of metal-impacted habitats including the Sargasso Sea (Venter et al. 2004), an acid mine drainage system (Tyson et al. 2004), soil (Tringe et al. 2005), and on sunken whale skeletons (Tringe et al. 2005). Although there are challenges associated with these metagenome studies due to the number of sequence assemblies and in the distinguishing of heterogeneities between organisms, it appears to be a logical method for examining the genome sequences of microbial communities (Schloss and Handelsman 2005). The overall foundation of this method is quite powerful and not only provides insight into taxonomy but can also be applied to the study of functional genes using expression libraries (Handelsman 2004). Functional genes and their associated protein can be determined through the transcription and translation of the metagenome into peptide sequences. The acquired amino acid sequences can be entered into a database to search for homologous proteins; however, if the protein is novel, no putative identity can be assigned. Another possibility is that the new protein will not be expressed in the host bacterium that was selected for cloning (Handelsman 2004).

A combined 'proteogenomic' analysis was applied to the study of an acid mine drainage microbial community (Tyson et al. 2004). The spectrometry data results were compared to the metagenome sequence in order to identify proteins that were part of the community structure. This study determined that approximately 49% of the open reading frames from the five major genomes found in their metagenome data were linked to one or more of the peptide sequences. Even with the originality of this work, it still poses problems with the discovery of novel proteins since these cannot always be identified by database matching (Banfield et al. 2005). With additional optimization, these techniques hold great promise when looking for undiscovered enzymes along with characterized enzymes that catalyze metal transformations.

Table 7.2 Proteomic results obtained from activity staining and bulk analysis of arsenic transforming pure cultures and microbial communities from thermal springs. Closest BLAST hits are shown, along with the number of peptides corresponding to the protein

Significant hits	Number of peptides
Thermus arsenite oxidase gel band	
Transporter periplasmic component *Thermus thermophilus*	6
Polynucleotide phosphorylase *Thermus thermophilus*	5
Arsenite oxidase large subunit *Thermus thermophilus*	4
NAD dependent glutamate dehydrogenase *Thermus thermophilus*	3
60 kDa chaperonin *Thermus thermophilus*	2
Delta aminolevulinic acid dehydratase *Thermus thermophilus*	1
Acetyl CoA acetyltransferase *Thermus thermophilus*	1
Probable transaldolase *Thermus thermophilus*	1
Thermus arsenate reductase gel band	
Molybdopterin oxidoreductase *Chloroflexus aurantiacus* J 10 fl	2
Molybdopterin oxidoreductase *Chloroflexus aurantiacus* J 10 fl	2
Environmental sample	
Arsenite oxidase Mo-pterin subunit (Fragment), Environmental sequence	3
Arsenite oxidase Mo-pterin subunit (Fragment), environmental sequence	2

By combining tools commonly used in the study of microbial ecology and by taking advantage of the most contemporary (meta)genomic and (meta)proteomic techniques, great strides can be made in the achieving a greater understanding of arsenic-transforming microbes and their ecophysiologic roles. Bioinformatics and *in silico* tools also have an important place in the genomic study of arsenic transforming microbes. Since there are now hundreds of sequenced microbial genomes, one can query the data repositories for any number of genes related to arsenic metabolism. One such useful tool is the Joint Genome Institute (US Department of Energy) Integrated Microbial Genomes interface (http://www.jgi.doe.gov), where both pure culture genomes and metagenomes from a variety of environments can be queried.

Contributions from the Study of Microbial Arsenic Transformation

Microorganisms can facilitate the transformation of a considerable number of elements (Stolz et al. 2002) and through these chemical alterations play a central role in the geochemistry of Earth environments. Several positive roles metal-transforming bacteria play have just recently been recognized and the appreciation for these organisms and processes has proliferated among scientists. As local, national, and world news sources have popularized on global warming, clean up of radioactive waste, and life on Mars, even the general public is beginning to connect the importance of microorganisms with world changing processes and exobiology.

Biogeochemical Cycling

Biogeochemical cycling is the movement (cycling) of elements through both biological and geological systems. A variety of elements are cycled, many of which are essential components of life. The significance of microorganisms in the carbon, oxygen, nitrogen, and phosphorus cycles has long been established; however, biological roles in the cycling of some trace elements including arsenic have just recently been elucidated (Oremland et al. 2004; Oremland and Stolz 2005; Stolz et al. 2002; Stolz and Oremland 1999). Figure 7.8 demonstrates the coupled biological cycling of carbon and arsenic in Mono Lake, CA, USA (Oremland et al. 2004). This is not only a prime example of a microbial arsenic cycle in nature but also an illustration of how arsenic metabolism contributes to the cycling of organic and inorganic carbon as well.

Bioremediation

The cleanup of contaminants using biological agents is largely regarded as an attractive treatment method since it can be carried out with limited disturbance to the contaminated area (Head 1998). Microorganisms have the metabolic capabilities to convert toxic compounds into relatively non-toxic products, hence eliminating contaminants (Head 1998). Physical mechanisms of clean up do not destroy contaminants but rather consolidate them into certain locations (Head 1998). Enzyme mediated microbial metal transformations, for example reductive immobilization, play a vital

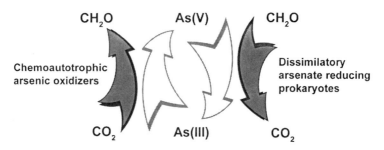

Fig. 7.8 Coupled carbon-arsenic biogeochemical cycle. Key ecophysiologic reactions are shown. Note how two distinct groups of arsenic transforming organisms (oxidizers and reducers) can interact and exchange carbon as well as arsenic

role in the bioremediation of pollutants (Lovley and Coates 1997). Because arsenic holds a number of varying valence states, which differ in toxicity and mobility, remediation strategies are complicated, due to the continuous transformation of arsenic, biologically or chemically (pH, redox potential, etc.). This means a deep understanding of microbial populations and how they behave and interact in the natural environment is fundamental to the development of efficient and effective mechanisms of clean up in arsenic contaminated environments.

Exobiology

The notion that life may be or was present on other worlds has always been considered a viable concept, and research in this area has given some evidence to support this possibility. However, due to the absence of unidentifiable remains, mineralogical 'biosignatures' are now being used as a means of searching for signs of life (Banfield et al. 2001). Many minerals are associated with biological activity and can often form indirectly as byproducts of microbial metal metabolism. Examples of biogenic minerals include arsenic sulfides, iron sulfides, iron oxides, manganese oxides, carbonates, phosphates, halides, oxalates, sulfates, and silica (Lowenstam 1981; Stolz and Oremland 1999). *Desulfotomaculm auripigmentum* forms the arsenic sulfide mineral, orpiment (As_2S_3) by reducing arsenate and sulfate during growth in culture (Stolz and Oremland 1999). This mineral, along with many other minerals, have abiotic counterparts (Banfield et al. 2001; Lowenstam 1981; Stolz and Oremland 1999), thus their occurrence in the environment is not always indicative of life. ß-realgar (ß-As_4S_4) produced by the arsenate reducing thermophile strain YeAs is unique in structure, and the first description of biologically produced ß-realgar (Fig. 7.9) (Ledbetter et al. 2007). Additionally, the abiotic formation of this mineral is thermodynamically unfavorable and has only been produced abiotically at temperatures above 250°C. In many instances, there are certain attributes (chemical or physical) that distinguish between the abiotic and biotic forms, as evidenced with β-realgar. Although it is imperative that non-biological activity be positively identified when looking for evidence of life in these extreme environments, previously classified biogenic minerals along with newly unearthed signatures could undoubtedly result in the discovery of life on Mars and other worlds.

Conclusions

Regardless of the exact method of arsenic metabolism, the notable role microorganisms play in the biogeochemical cycling of arsenic and numerous other elements cannot go unnoticed. The microbial 'eating and breathing' (Nealson et al. 2002) of metals has considerably impacted our world and certainly contributes to the operation and stability of ecosystems. Arsenic oxidation and reduction appear to

Fig. 7.9 Biogenic mineral production in a thermophilic anaerobe. *Left Panel*: Colonies of arsenate reducing bacteria isolated from Murky Pot hot spring, OR. The colonies contain conspicuous arsenic sulfide minerals in association with the cells. *Right Panel*: A culture of strain YeAs grown anaerobically with arsenate. Note the abundant arsenic sulfide mineral precipitate. This material was determined to be ß-realgar

have evolved in several directions, with distinct biochemical mechanisms required depending on whether the organism is gaining energy or detoxifying the metal. Arsenic methylation is a ubiquitous process stemming from humans to Archaea, whereas demethylation is strictly found in microorganisms. Whether building or breaking arsenic compounds, both methylation and demethylation contribute to the coupled cycling of arsenic and carbon, and represent additional forms of arsenic required by some microbes. Genomics and proteomics have opened new vistas onto microbial arsenic metabolism and have revealed unexpected mechanisms and adaptations for survival with arsenic. Microorganisms capable of transforming the elements are making extensive contributions on a global scale and even though their vast metabolic capabilities are difficult to completely comprehend, the more understanding that is gained only enhances the admiration and appreciation for these intricate and multifaceted life forms.

Acknowledgements The authors express their gratitude to the National Science Foundation, the Department of Energy, and the National Aeronautics and Space Administration for funding various aspects of our research on the ecophysiology of arsenic.

References

Afkar E, Lisak J, Saltikov C, Basu P, Oremland RS, Stolz JF (2003) The respiratory arsenate reductase from *Bacillus selenitireducens* strain MLS10. FEMS Microbiol Lett 226:107–112

Ahmann D, Krumholz LR, Hemond H, Lovley DR, Morel FMM (1997) Microbial mobilization of arsenic from sediments of the Aberjona Watershed. Environ Sci Technol 31:2923–2930

Amann RI, Binder BJ, Olson RJ, Chisholm SW, Devereux R, Stahl DA (1990) Combination of 16S rRNA-targeted oligonucleotide probes with flow cytometry for analyzing mixed microbial populations. Appl Environ Microbiol 56:1919–1925

Anderson GL, Williams J, Hille R (1992) The purification and characterization of arsenite oxidase from *Alcaligenes faecalis*, a molybdenum-containing hydroxylase. J Biol Chem 267: 23674–23682

Banfield JF, Moreau JW, Chan CS, Welch SA, Little B (2001) Mineralogical biosignatures and the search for life on Mars. Astrobiology 1:447–465

Banfield JF, Verberkmoes NC, Hettich RL, Thelen MP (2005) Proteogenomic approaches for the molecular characterization of natural microbial communities. Omics 9:301–333

Bentley R, Chasteen TG (2002) Microbial methylation of metalloids: arsenic, antimony, and bismuth. Microbiol Mol Biol Rev 66:250–271

Blum JS, Han S, Lanoil B, Saltikov C, Witte B, Tabita FR, Langley S, Beveridge TJ, Jahnke L, Oremland RS (2009) Ecophysiology of "*Halarsenatibacter silvermanii*" Strain SLAS-1(T), gen. nov., sp nov., a facultative chemoautotrophic arsenate respirer from salt-saturated Searles lake, California. Appl Environ Microbiol 75:1950–1960

Bruins M, Kapil S, Oehme FW (2000) Microbial resistance to metals in the environment. Ecotoxicol Environ Saf 45:198–207

Bruns A, Cypionka H, Overmann J (2002) Cyclic AMP and acyl homoserine lactones increase the cultivation efficiency of heterotrophic bacteria from the central Baltic sea. Appl Environ Microbiol 68:3978–3987

Bruns A, Nubel U, Cypionka H, Overmann J (2003) Effect of signal compounds and incubation conditions on the culturability of freshwater bacterioplankton. Appl Environ Microbiol 69:1980–1989

Cai L, Rensing C, Li X, Wang G (2009) Novel gene clusters involved in arsenite oxidation and resistance in two arsenite oxidizers: *Achromobacter sp.* SY8 and *Pseudomonas sp.* TS44. Appl Microbiol Biotechnol 83:715–725

Challenger F (1945) Biological methylation. Chem Rev 36:315–361

Challenger F (1951) Biological methylation. Adv Enzymol Relat Subj Biochem 12:429–491

Clingenpeel SR, D'Imperio S, Oduro H, Druschel GK, McDermott TR (2009) Cloning and in situ expression studies of the *Hydrogenobaculum* arsenite oxidase genes. Appl Environ Microbiol 75:3362–3365

Connon SA, Koski AK, Neal AL, Wood SA, Magnuson TS (2008) Ecophysiology and geochemistry of microbial arsenic oxidation within a high arsenic, circumneutral hot spring system of the Alvord desert. FEMS Microbiol Ecol 64:117–128

Davis KE, Joseph SJ, Janssen PH (2005) Effects of growth medium, inoculum size, and incubation time on culturability and isolation of soil bacteria. Appl Environ Microbiol 71:826–834

Dombrowski PM, Long W, Farley KJ, Mahony JD, Capitani JF, Di Toro DM (2005) Thermodynamic analysis of arsenic methylation. Environ Sci Technol 39:2169–2176

Eary LE (1992) The solubility of amorphous As_2S_3 from 25 to 90°C. Geochim Cosmochim Acta 56:2267–2278

Ellis PJ, Conrads T, Hille R, Kuhn P (2001) Crystal structure of the 100 kDa arsenite oxidase from *Alcaligenes faecalis* in two crystal forms at 1.64 A and 2.03 A. Structure 9:125–132

Gadd G (1993) Microbial formation and transformation of organometallic and organometalloid compounds. FEMS Microbiol Rev 11:297–316

Gihring TM, Banfield JF (2001) Arsenite oxidation and arsenate respiration by a new *Thermus* isolate. FEMS Microbiol Lett 204:335–340

Gourbal B, Sonuc N, Bhattacharjee H, Legare D, Sundar S, Ouellette M, Rosen BP, Mukhopadhyay R (2004) Drug uptake and modulation of drug resistance in *Leishmania* by an aquaglyceroporin. J Biol Chem 279:31010–31017

Handelsman J (2004) Metagenomics: application of genomics to uncultured microorganisms. Microbiol Mol Biol Rev 68:669–685

Harold FM, Baarda JR (1966) Interaction of arsenate with phosphate-transport systems in wild-type and mutant *Streptococcus faecalis*. J Bacteriol 91:2257–2262

Head M (1998) Bioremediation: towards a credible technology. Microbiology 144:599–608

Hindmarsh JT (2000) Arsenic, its clinical and environmental significance. J Trace Elem Exp Med 13:165–172

Hoeft SE, Blum JS, Stolz JF, Tabita FR, Witte B, King GM, Santini JM, Oremland RS (2007) *Alkalilimnicola ehrlichii* sp nov., a novel, arsenite-oxidizing haloalkaliphilic gammaproteobacterium capable of chemoautotrophic or heterotrophic growth with nitrate or oxygen as the electron acceptor. Int J Syst Evol Microbiol 57:504–512

Hoke KR, Cobb N, Armstrong FA, Hille R (2004) Electrochemical studies of arsenite oxidase: an unusual example of a highly cooperative two-electron molybdenum center. Biochemistry 43:1667–1674

Hollibaugh JT, Carini S, Gurleyuk H, Jellison R, Joye SB, LeCleir G, Meile C, Vasquez L, Wallschlager D (2005) Arsenic speciation in Mono lake, California: response to seasonal stratification and anoxia. Geochim Cosmochim Acta 69:1925–1937

Hughes MN, Poole RK (1989) Metals and micro-organism. Chapman & Hall, London, pp 280–285

Inskeep WP, Mancur RE, Hamamura N, Warelow TP, Ward SA, Santini JM (2007) Detection, diversity and expression of aerobic bacterial arsenite oxidase genes. Environ Microbiol 9:934–943

Jackson BP, Bertsch PM (2001) Determination of arsenic speciation in poultry wastes by IC-ICP-MS. Environ Sci Technol 35:4868–4873

Ji G, Silver S (1995) Bacterial resistance mechanisms for heavy metals of environmental concern. J Ind Microbiol 14:61–75

Jones CA, Langner HW, Anderson K, McDermott TR, Inskeep WP (2000) Rates of microbially mediated arsenate reduction and solubilization. Soil Sci Soc Am J 64:600–608

Joseph SJ, Hugenholtz P, Sangwan P, Osborne CA, Janssen PH (2003) Laboratory cultivation of widespread and previously uncultured soil bacteria. Appl Environ Microbiol 69:7210–7215

Kaeberlein T, Lewis K, Epstein SS (2002) Isolating "uncultivable" microorganisms in pure culture in a simulated natural environment. Science 296:1127–1129

Krafft T, Macy JM (1998) Purification and characterization of the respiratory arsenate reductase of *Chrysiogenes arsenatis*. Eur J Biochem 255:647–653

Krautler B (1990) Chemistry of methylcorrinoids related to their roles in bacterial C1 metabolism. FEMS Microbiol Rev 7:349–354

Ledbetter RN, Connon SA, Neal AL, Dohnalkova A, Magnuson TS (2007) Biogenic mineral production by a novel arsenic-metabolizing thermophilic bacterium from the Alvord basin, Oregon. Appl Environ Microbiol 73:5928–5936

Lloyd JR, Oremland RS (2006) Microbial transformations of arsenic in the environment: from soda lakes to aquifers. Elements 2:85–90

Lovley DR (1993) Dissimilatory metal reduction. Annu Rev Microbiol 47:263–290

Lovley DR, Coates JD (1997) Bioremediation of metal contamination. Curr Opin Biotechnol 8:285–289

Lowenstam HA (1981) Minerals formed by organisms. Science 211:1126–1131

Macy JM, Nunan K, Hagen KD, Dixon DR, Harbour PJ, Cahill M, Sly LI (1996) *Chrysiogenes arsenatis* gen. nov., sp. nov., a new arsenate-respiring bacterium isolated from gold mine wastewater. Int J Syst Bacteriol 46:1153–1157

Macy JM, Santini JM, Pauling BV, O'Neill AH, Sly LI (2000) Two new arsenate/sulfate-reducing bacteria: mechanisms of arsenate reduction. Arch Microbiol 173:49–57

Malasarn D, Saltikov CW, Campbell KM, Santini JM, Hering JG, Newman DK (2004) *arr*A is a reliable marker for As(V) respiration. Science 306:455

Matera V, Le Hecho I, Laboudigue A, Thomas P, Tellier S, Astruc M (2003) A methodological approach for the identification of arsenic bearing phases in polluted soils. Environ Pollut 126:51–64

Meng YL, Liu Z, Rosen BP (2004) As(III) and Sb(III) uptake by GlpF and efflux by ArsB in *Escherichia coli*. J Biol Chem 279:18334–18341

Mukhopadhyay R, Rosen BP (2002) Arsenate reductases in prokaryotes and eukaryotes. Environ Health Perspect 110(Suppl 5):745–748

Mukhopadhyay R, Rosen BP, Phung LT, Silver S (2002) Microbial arsenic: from geocycles to genes and enzymes. FEMS Microbiol Rev 26:311–325

Nealson KH, Belz A, McKee B (2002) Breathing metals as a way of life: geobiology in action. Antonie Leeuwenhoek 81:215–222

Newman DK, Beveridge TJ, Morel F (1997a) Precipitation of arsenic trisulfide by *Desulfotomaculum auripigmentum*. Appl Environ Microbiol 63:2022–2028

Newman DK, Kennedy EK, Coates JD, Ahmann D, Ellis DJ, Lovley DR, Morel FM (1997b) Dissimilatory arsenate and sulfate reduction in *Desulfotomaculum auripigmentum* sp. nov. Arch Microbiol 168:380–388

NRC (1999) National Reseach Council Report: arsenic in drinking water. National Academy Press, Washington, DC

Oremland RS, Stolz JF (2003) The ecology of arsenic. Science 300:939–944

Oremland RS, Stolz JF (2005) Arsenic, microbes and contaminated aquifers. Trends Microbiol 13:45–49

Oremland RS, Stolz JF, Hollibaugh JT (2004) The microbial arsenic cycle in Mono lake, CA. FEMS Microbiol Ecol 48:15–27

Oremland RS, Kulp TR, Blum JS, Hoeft SE, Baesman S, Miller LG, Stolz JF (2005) A microbial arsenic cycle in a salt-saturated, extreme environment. Science 308:1305–1308

Paez-Espino D, Tamames J, de Lorenzo V, Canovas D (2009) Microbial responses to environmental arsenic. Biometals 22:117–130

Patel PC, Goulhen F, Boothman C, Gault AG, Charnock JM, Kalia K, Lloyd JR (2007) Arsenate detoxification in a Pseudomonad hypertolerant to arsenic. Arch Microbiol 187:171–183

Peterson ML, Carpenter R (1983) Biogeochemical processes affecting total arsenic and arsenic species distributions in an intermittently stratified fjord. Mar Chem 12:295–321

Poole RK, Gadd GM (1989) Metals: microbe interactions. IRL Press, Oxford, pp 1–37

Prasad KS, Subramanian V, Paul J (2009) Purification and characterization of arsenite oxidase from *Arthrobacter* sp. Biometals 22(5):711–721

Qin J, Rosen BP, Zhang Y, Wang G, Franke S, Rensing C (2006) Arsenic detoxification and evolution of trimethylarsine gas by a microbial arsenite S-adenosylmethionine methyltransferase. Proc Natl Acad Sci USA 103:2075–2080

Quinn JP, McMullan G (1995) Carbon-arsenic bond cleavage by a newly isolated gram-negative bacterium, strain ASV2. Microbiology 141(Pt 3):721–725

Ram RJ, Verberkmoes NC, Thelen MP, Tyson GW, Baker BJ, Blake RC 2nd, Shah M, Hettich RL, Banfield JF (2005) Community proteomics of a natural microbial biofilm. Science 308:1915–1920

Reardon CL, Cummings DE, Petzke LM, Kinsall BL, Watson DB, Peyton BM, Geesey GG (2004) Composition and diversity of microbial communities recovered from surrogate minerals incubated in an acidic uranium-contaminated aquifer. Appl Environ Microbiol 70:6037–6046

Rhine ED, Garcia-Dominguez E, Phelps CD, Young LY (2005) Environmental microbes can speciate and cycle arsenic. Environ Sci Technol 39:9569–9573

Rhine ED, Ni Chadhain SM, Zylstra GJ, Young LY (2007) The arsenite oxidase genes (*aro*AB) in novel chemoautotrophic arsenite oxidizers. Biochem Biophys Res Commun 354:662–667

Richey C, Chovanec P, Hoeft SE, Oremland RS, Basu P, Stolz JF (2009) Respiratory arsenate reductase as a bidirectional enzyme. Biochem Biophys Res Commun 382:298–302

Riesenfeld CS, Schloss PD, Handelsman J (2004) Metagenomics: genomic analysis of microbial communities. Annu Rev Genet 38:525–552

Rosen BP, Liu Z (2009) Transport pathways for arsenic and selenium: a minireview. Environ Int 35:512–515

Rosenberg H, Gerdes RG, Chegwidden K (1977) Two systems for the uptake of phosphate in *Escherichia coli*. J Bacteriol 131:505–511

Saltikov CW, Newman DK (2003) Genetic identification of a respiratory arsenate reductase. Proc Natl Acad Sci USA 100:10983–10988

Santini JM, Sly LI, Schnagl RD, Macy JM (2000) A new chemolithoautotrophic arsenite-oxidizing bacterium isolated from a gold mine: phylogenetic, physiological, and preliminary biochemical studies. Appl Environ Microbiol 66:92–97

Schloss PD, Handelsman J (2003) Biotechnological prospects from metagenomics. Curr Opin Biotechnol 14:303–310

Schloss PD, Handelsman J (2005) Metagenomics for studying unculturable microorganisms: cutting the Gordian knot. Genome Biol 6:229

Seyler P, Martin JM (1989) Biogeochemical processes affecting total arsenic and arsenic species distribution in a permanently stratified lake. Envriron Sci Technol 23:1258–1263

Silver S, Phung LT (2005a) A bacterial view of the periodic table: genes and proteins for toxic inorganic ions. J Ind Microbiol Biotechnol 32:587–605

Silver S, Phung LT (2005b) Genes and enzymes involved in bacterial oxidation and reduction of inorganic arsenic. Appl Environ Microbiol 71:599–608

Song B, Chyun E, Jaffe PR, Ward BB (2009) Molecular methods to detect and monitor dissimilatory arsenate-respiring bacteria (DARB) in sediments. FEMS Microbiol Ecol 68:108–117

Stevenson BS, Eichorst SA, Wertz JT, Schmidt TM, Breznak JA (2004) New strategies for cultivation and detection of previously uncultured microbes. Appl Environ Microbiol 70:4748–4755

Stolz JF, Oremland RS (1999) Bacterial respiration of arsenic and selenium. FEMS Microbiol Rev 23:615–627

Stolz JF, Basu P, Oremland RS (2002) Microbial transformation of elements: the case of arsenic and selenium. Int Microbiol 5:201–207

Stolz JF, Basu P, Santini JM, Oremland RS (2006) Arsenic and selenium in microbial metabolism. Annu Rev Microbiol 60:107–130

Stupperich E (1993) Recent advances in elucidation of biological corrinoid functions. FEMS Microbiol Rev 12:349–365

Thomas DJ, Waters SB, Styblo M (2004) Elucidating the pathway for arsenic methylation. Toxicol Appl Pharmacol 198:319–326

Thomas DJ, Li J, Waters SB, Xing W, Adair BM, Drobna Z, Devesa V, Styblo M (2007) Arsenic (+3 oxidation state) methyltransferase and the methylation of arsenicals. Exp Biol Med (Maywood) 232:3–13

Tringe SG, von Mering C, Kobayashi A et al (2005) Comparative metagenomics of microbial communities. Science 308:554–557

Tyson GW, Chapman J, Hugenholtz P et al (2004) Community structure and metabolism through reconstruction of microbial genomes from the environment. Nature 428:37–43

Venter JC, Remington K, Heidelberg JF et al (2004) Environmental genome shotgun sequencing of the Sargasso sea. Science 304:66–74

Waters SB, Devesa V, Del Razo LM, Styblo M, Thomas DJ (2004a) Endogenous reductants support the catalytic function of recombinant rat cyt19, an arsenic methyltransferase. Chem Res Toxicol 17:404–409

Waters SB, Devesa V, Fricke MW, Creed JT, Styblo M, Thomas DJ (2004b) Glutathione modulates recombinant rat arsenic (+3 oxidation state) methyltransferase-catalyzed formation of trimethylarsine oxide and trimethylarsine. Chem Res Toxicol 17:1621–1629

Willsky GR, Malamy MH (1980) Characterization of two genetically separable inorganic phosphate transport systems in *Escherichia coli*. J Bacteriol 144:356–365

Wu J, Rosen BP (1993) Metalloregulated expression of the ars operon. J Biol Chem 268:52–58

Zobrist J, Dowdle PR, Davis JA, Oremland RS (2000) Mobilization of arsenite by dissimilatory reduction of adsorbed arsenate. Environ Sci Technol 34:4747–4775

Chapter 8
Bioinformatics and Genomics of Iron- and Sulfur-Oxidizing Acidophiles

Violaine Bonnefoy

Important protagonists in geomicrobiology are the "biomining" microorganisms which are used to recover valuable metals from mineral ores and concentrates. These microorganisms either convert insoluble metal sulfides to soluble metal sulfates, a process referred to as bioleaching, or weaken the ore by removing iron and/or sulfur making the valuable metal accessible to subsequent chemical treatment, a process known as biooxidation (Rawlings 2005; Rawlings and Johnson 2007). The drawback of this industrial biotechnology is the formation of acid mine drainage (AMD) from uncontrolled abandoned mines, mine dumps or tailing dams, and acid rock drainage when sulfide-rich ores are exposed to air and weathering.

This process is linked indirectly to the capacity of acidophilic prokaryotes to oxidize ferrous iron and reduced inorganic sulfur compounds (RISC) where sulfur has different oxidation states. The products of these oxidations, that is ferric iron and sulfuric acid respectively, then chemically attack the mineral (Rohwerder and Sand 2007) releasing the metal, Fe(II) and RISC (Fig. 8.1a).

The identification of the microorganisms involved in metal solubilization and the understanding of the molecular mechanisms involved in Fe(II) and RISC oxidation are therefore of outstanding importance. A number of reviews have been published about the microbial ecology of bioleaching operations (Hallberg and Johnson 2001; Rawlings 2002, 2005; Norris 2007; Schippers 2007; Rawlings and Johnson 2007 and Chapter 19 of this book). The physiological characteristics of Fe(II) and/or RISC oxidizing acidophilic microorganisms which genome sequence has been analysed are summarized in Table 8.1.

Despite multidisciplinary approaches, knowledge of Fe(II) and RISC oxidation in bioleaching microorganisms remains rudimentary. A bottleneck in advancing it, is the lack of well established genetic systems that permit the construction of mutants

V. Bonnefoy (✉)

Institut de Microbiologie de la Méditerranée, Laboratoire de Chimie Bactérienne, C.N.R.S., Université de la Méditerranée, 31 chemin Joseph Aiguier, 13402 Marseille Cedex 20, France
e-mail: bonnefoy@ifr88.cnrs-mrs.fr

L.L. Barton et al. (eds.), *Geomicrobiology: Molecular and Environmental Perspective*,
DOI 10.1007/978-90-481-9204-5_8, © Springer Science+Business Media B.V. 2010

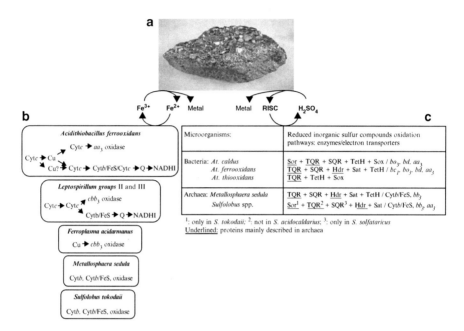

Fig. 8.1 Schematic representation of the metal sulfide bioleaching (**a**), the iron (**b**) and RISC (**c**) oxidation pathways. Cu: copper protein. For the other abbreviations, see the text

to test existing hypotheses. Furthermore, marker exchange mutagenesis can only be applied for non essential genes already characterized. Till recently, the analysis of a metabolic pathway was therefore limited to few specific genes/proteins. However, these last decades have seen the development of "global" analysis owed to rapid advances in technologies such as genomics, transcriptomics and proteomics. How these new tools were applied to mining biotechnologies and how these new concepts helped deciphering, not only the molecular mechanisms involved in Fe(II) or RISC oxidation of known bioleaching microorganisms, but also of the microbial communities during the process are the main topics addressed in this chapter.

"Omics", What Does that Mean?

The suffix "-*om*-" comes from "genome" and "chromosome" and designs different large-scale quantitative biology fields, such as "transcriptome" and "proteome". These approaches are global rather than targeted and aim to explain the highly complex regulatory and metabolic networks inside the living cell. Since few years, such technologies have been implemented for Fe(II) and/or RISC oxidizing acidophiles and have been reviewed by Holmes and Bonnefoy (2007), Quatrini et al. (2007c) and Jerez (2007).

Table 8.1 Characteristics of the microorganisms discussed in this chapter

Eubacteria	Phylum	pH minimum-(optimum)-maximum	Temperature-minimum-(optimum)-maximum (°C)	Oxidation Fe(II)	Oxidation S⁰	Carbon source[a]	Genome[b]	Transcriptome	(Meta) Proteome
Acidithiobacillus caldus	γ-Proteobacteria	1.0-(2.0–2.5)-3.5	32-(45)-52	–	+	I, O	I	–	–
Acidithiobacillus ferrooxidans	γ-Proteobacteria	1.3-(2.5)-4.5	10-(30–35)-37	+	+	I	C	+	+
Acidithiobacillus thiooxidans	γ-Proteobacteria	0.5-(2.0–3.0)-5.5	10-(28–30)-37	–	+	I	I	–	–
Leptospirillum ferrooxidans (group I)	Nitrospira	1.3-(1.5–3.0)-4.0	nd-(28–30)-nd	+	–	I	I	+	–
Leptospirillum ferrooxidans (group II)	Nitrospira	nd-(1.3–1.8)-nd	nd-(30–37)-45	+	–	I	I	–	(+)
"Leptospirillum. rubarum" (group II)	Nitrospira	nd	nd	nd	nd	nd	I	–	(+)
"Leptospirillum ferrodiazotrophum" (group III)	Nitrospira	nd	nd	+	nd	nd	I	–	(+)
Archaebacteria									
"Ferroplasma acidarmanus"	Euryarchaeota	<0-(1.2)-1.5	23-(42)-46	+	–	I, O	I	–	+
Ferroplasma type I	Euryarchaeota	nd	nd	nd	nd	nd	I	–	(+)
Ferroplasma type II	Euryarchaeota	nd	nd	nd	nd	nd	I	–	(+)
Metallosphaera sedula	Crenarchaeota	1-(2–3)-4.5	50-(75)-80	+	+	I, O	C	+	–
Sulfolobus acidocaldarius	Crenarchaeota	1-(2-3)-6	55-(70–75)-85	–	–[c]	O	C	+	–
Sulfolobus metallicus	Crenarchaeota	1-(3)-4.5	50-(65)-75	+	+	I	I	–	–
Sulfolobus solfataricus	Crenarchaeota	2-(3–4.5)-5.5	50-(85)-87	–	–[c]	O	C	+	+
Sulfolobus tokodaii	Crenarchaeota	2-(2.5–3)-5	70-(80)-85	+	+	O	C	–	–

nd: not determined

[a] "I" stands for inorganic and "O" for organic

[b] "C" stands for complete and "I" for incomplete

[c] Previously described as S⁰ oxidizers (see text)

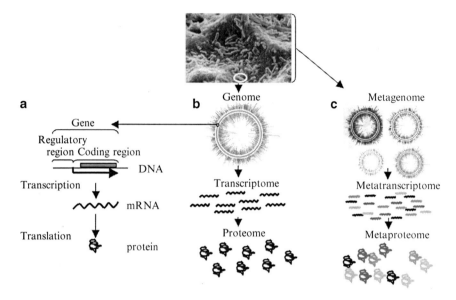

Fig. 8.2 Schematic overview of the different "omics" approaches

Few basic knowledges are required to understand these techniques. A gene is a portion of deoxyribonucleic acid (DNA) that contains both the "coding" sequence that determines a transcript, and the "regulatory" sequence that determines when the gene is expressed (Fig. 8.2a). When a gene is switched on, the coding sequence is transcribed into ribonucleic acid (RNA) (Fig. 8.2a). Two sorts of RNA exist: the "coding" RNA (messenger mRNA (mRNA)), and the non-coding RNA which play roles in the regulation of DNA expression and in the structure of cell components (such as the ribosomal and transfer RNA in the ribosomes). The ribosomes "read" the mRNA and synthesize the proteins (translation) (Fig. 8.2a).

Genomics

Genomics is the analysis of the entire DNA sequence of an organism (genome) with informatic software programs (bioinformatics) (Fig. 8.2b). It allows when possible: (a) the identification of genes, (b) the assignment of their biological function, (c) the reconstruction in silico of metabolic pathways, (d) the prediction of regulatory circuits in which different processes are connected depending on the environmental conditions. The comparison of the genome and of the metabolism between different microorganisms (comparative genomics) could help understanding the general mechanism of evolution but also deciphering metabolic pathways. Metagenomics is the culture-independent genomic analysis of microbial communities directly from their natural environment (Fig. 8.2c), allowing to bypass the isolation and cultivation of individual organisms and to study the unculturable ones which could represent up to more than 99% of the micro-organisms in some environments. It provides clues about how the different organisms interact and benefit each other living in the same biotope.

Transcriptomics

Transcriptome is the totality of the mRNA present in one cell population at a given time under defined conditions (Fig. 8.2b). It reflects the active expression of a sub-group of genes and varies according to the growth conditions. Higher the expression of a gene in one condition, the more likely that gene is involved in the corresponding metabolism. Consequently, transcriptomics can help deciphering the role of genes that do not have an assigned function.

Transcriptomics is generally based on DNA microarray. In this high throughput technology, single-stranded DNA corresponding to each gene of a genome are spotted on a solid substrate (nylon membrane or glass slide). These probes are used to capture (hybridize) selectively their complementary labeled transcripts (target), acting as "molecular sensors" for quantitative measurements (Dharmadi and Gonzalez 2004). First the RNA are extracted from the bacteria grown under the two conditions to be compared (Fig. 8.3a). They are retro-transcribed into labeled cDNA, with radioactivity or fluorochromes (Fig. 8.3a). The labeled single stranded cDNA are then hybridized to the DNA microarrays. They bind specifically to their complementary sequence with high affinity. Probe-target hybridization is detected and

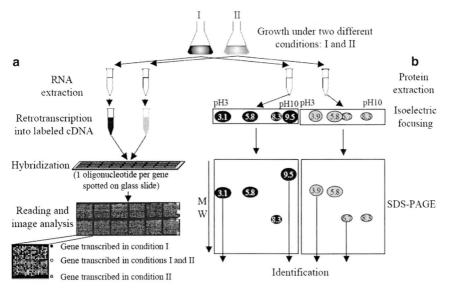

Fig. 8.3 Schematic representation of the transcriptomics (**a**) and proteomics (**b**). MW: molecular weight

quantified to determine relative abundance of nucleic acid sequences in the target (Fig. 8.3a). The signal intensity of each spot on the microarray gives an estimate of the abundance of the corresponding target (cDNA) in the given environmental condition and therefore of the expression level of each gene.

Metatranscriptomics is the study of the metatranscriptome of a microbial consortia (Fig. 8.2c). It provides information about the response of the community to varying environmental conditions, in particular which genes are more highly expressed, and therefore required for adaptation.

Proteomics

The proteome is the set of synthesized proteins at a given time under defined conditions (Blackstock and Weir 1999) (Fig. 8.2b). Proteomics allows to detect specific qualitative and quantitative protein changes, i.e. to determine the presence and the relative abundance of proteins in certain growth conditions and to detect post-translational modifications such as phosphorylation and methylation. This approach gives the conditions in which a protein is synthesized, thus giving clues to its function.

In proteomics, the proteins are separated by two dimensional polyacrylamide gel electrophoresis (2D-PAGE). Complex mixtures of up to 1,800 proteins can be resolved in a single gel. Separation of proteins by 2D-PAGE relies on their charge and their molecular weight (MW). In the first dimension, the proteins are separated according to their isoelectric point (pI) through a pH gradient (isoelectric focusing). A protein stops migrating when its net electrical charge is neutral, i.e. when it reaches a pH value corresponding to its pI (Fig. 8.3b). Proteins are then further resolved according to their MW by SDS-PAGE (Fig. 8.3b). Separated proteins are visualized by gel staining. The proteins that changed their level of synthesis during growth on the conditions analyzed are identified (N-terminal amino acid sequencing, mass spectrometry, peptide mass fingerprinting, tandem mass spectrometry, etc.).

Metaproteomics is the study of all the proteins synthesized in one microbial ecosystem with cultivation-independent molecular genomic approaches (Wilmes and Bond 2004, 2006) (Fig. 8.2c). It has been used to monitor the functional products of the microbiota.

Acidithiobacillus ferrooxidans

Genomics

The sequence of *At. ferrooxidans* ATCC23270 (type strain) genome has been determined by the J. Craig Venter Institute (JCVI) and annotated with the Center for Bioinformatics and Genome Biology (CBGB) (Valdés et al. 2008a). Two other

At. ferrooxidans genomes have been sequenced: ATCC53993, previously characterized as *Leptospirillum ferrooxidans* (DOE JGI), and DSM16786 owned by
BioSigma, SA. They will not be discussed here, the first one being not published
and the second one being not publicly available (Levicán et al. 2008). The circular
chromosome of ATCC23270 strain carries 2,982,397 bp (58.77% G+C content) and
encodes 3,217 putative proteins, of which 1,147 are hypothetical or of unknown
function (Valdés et al. 2008a). No plasmids were detected. Lateral gene transfers
have been predicted and notably, the *petI* operon involved in Fe(II) oxidation
(Valdés and Holmes 2009). It has to be pointed out also that the genes and operons
involved in Fe(II) (*rus* and *petI*) and RISC (*tetH, dox* and *sqr*) oxidation are located
near by the origin of replication suggesting that they are highly transcribed early in
the cell division cycle (Valdés and Holmes 2009).

From this genome sequence, analysis have been performed to search for genes
encoding cytochromes *c* (Yarzábal et al. 2002a), sulfurtransferases (Acosta et al.
2005), acyl-homoserine lactone (AHL) synthases (Farah et al. 2005; Rivas et al. 2007),
and small regulatory RNAs (Shmaryahu and Holmes 2007; Shmaryahu et al. 2009).
Some metabolic pathways have been confirmed or extended such as carbon metabolism (Appia-Ayme et al. 2006; Esparza et al. 2009 and Holmes, personal communication), amino acid metabolism (Selkov et al. 2000; Barreto et al. 2003), sulfur uptake
and assimilation (Valdés et al. 2003; Quatrini et al. 2007c), hydrogen metabolism
(Barreto et al. 2003) and toxic compound fluxes (Barreto et al. 2003), while others
were reconstructed *in silico* including phosphate starvation response (Vera et al. 2003),
iron homeostasis (Quatrini et al. 2005, 2007a, b; Osorio et al. 2008a, b), biofilm formation (Barreto et al. 2005a, b; Quatrini et al. 2007c), quorum sensing (Farah et al. 2005,
Valenzuela et al. 2007) and c-di-GMP pathway (Ruiz et al. 2007; Castro et al. 2009;
Ruiz et al., in preparation). Models were presented for anaerobic metabolism (Osorio
et al. 2009), nitrogen fixation, stress responses, DNA repair and pH tolerance mechanisms (Valdés et al. 2008a). Concerning the topic of this review, Fe(II) and RISC
oxidation, most of the components of the electron transport chains which have been
characterized in different *At. ferrooxidans* strains (Appia-Ayme et al. 1999; Levicán
et al. 2002; Brasseur et al. 2002, 2004; Yarzábal et al. 2004; Bruscella et al. 2007), have
been confirmed for the type strain (Quatrini et al. 2006, 2009).

Proteomics

Proteomics in *At. ferrooxidans* has been performed long before the genome
sequence was available thanks to the pioneering work of Jerez's group. The effects
of the external pH (Amaro et al. 1991), heat shock (Varela and Jerez 1992), phosphate starvation (Seeger and Jerez 1993a, b; Jerez et al. 1995; Vera et al. 2003; He
et al. 2005a) and attachment to solid surfaces (Jerez et al. 1995; Valenzuela et al.
2006) on *At. ferrooxidans* ATCC19859 and/or R2 strain proteome were reported.
Response to copper or other heavy metals of private strains was also analyzed
(Novo et al. 2000, 2003; Paulino et al. 2002; Felício et al. 2003).

By a high throughput proteomic analysis, 131 proteins located in the periplasm of the type strain grown in thiosulfate have been identified (Chi et al. 2007). Most of them were very basic (pI > 7) in agreement with the strict acidophily of *At. ferrooxidans*. Among the proteins identified, six were involved in energy metabolism. Two were encoded by the *petII* operon proposed to be involved in sulfur (S^0) oxidation pathway (Bruscella et al. 2007): the cytochrome c_4 CycA2, and the HiPIP protein Hip (Bruscella et al. 2005, 2007). While reported as an outer membrane protein, the tetrathionate hydrolase (Buonfiglio et al. 1999; Kanao et al. 2007) was detected in the periplasm, suggesting that it is loosely bound to the outer membrane. Finally, the cytochromes c_4 CycA1 and Cyc1, the outer membrane cytochrome Cyc2 and the rusticyanin were identified.

The differential protein expression of several strains of *At. ferrooxidans* (ATCC23270, ATCC19859, CCM4253 and a private strain) grown on different energy sources, including, Fe(II), S^0 or metal sulfides, has been analyzed by following proteomics (Osorio et al. 1993; Ramirez et al. 2002, 2004; Jerez et al. 1995; He et al. 2005b; Valenzuela et al. 2005, 2006; Bouchal et al. 2006; Jerez 2007), paving the way for the understanding of oxic Fe(II) and RISC respiratory pathways. Among the redox proteins which are more expressed in Fe(II) than S^0, were the outer membrane cytochrome *c* Cyc2, the cytochrome c_4 Cyc1, the rusticyanin, the aa_3 cytochrome oxidase subunit II, the Rieske subunit of the bc_1 complex I (PetA1), supporting the models in which the proteins encoded by *rus* and *petI* operons are involved in Fe(II) oxidation. On the other side, a periplasmic rhodanese-like protein (P21), a periplasmic putative thiosulfate sulfur transferase protein, a thiosulfate/sulfate/molybdate binding protein and sulfide quinone reductase were more expressed in S^0 than in Fe(II).

Transcriptomics

From the annotated and curated *At. ferrooxidans* ATCC23270 genome sequence (Valdés et al. 2008a), an internal 50-mer oligonucleotide was designed for each of the predicted ORF and spotted on glass slides (Quatrini et al. 2006), opening the way to comparison of gene expression profiles of *At. ferrooxidans* cells grown on different conditions.

Preliminary genomic expression profiles of planktonic and sessile cells of *At. ferrooxidans* grown on pyrite showed different gene expression patterns (Vera et al. 2009). Most notably, there was clearly a parallel between planktonic state and Fe(II) oxidation on one hand, and sessile state and RISC oxidation on the other hand, indicating that the planktonic cells oxidized the soluble Fe(II) released by pyrite bioleaching and the sessile cells oxidized insoluble RISC attached to pyrite.

The expression profiles of the *At. ferrooxidans* cells grown on Fe(II) with those grown on S^0 were compared, focusing on carbon metabolism (Appia-Ayme et al. 2006) and on the energy metabolism (Quatrini et al. 2006, 2009). Interestingly, genes involved in CO_2 fixation, carboxysome formation, 2P-glycolate detoxification, glycogen biosynthesis and glycogen branching/debranching system are upregulated in S^0 medium, whereas those proposed to be required for glycogen utilization are upregulated in Fe(II) medium. S^0 being more energetic than Fe(II), it was proposed that *At. ferrooxidans* fixes CO_2 and stores some of the fixed carbon as glycogen in these energetically favorable conditions while it uses glycogen as carbon and energy sources in leaner conditions such as when Fe(II) is the only energy source available.

This transcriptomic analysis allows to extend our knowledge on the Fe(II) and RISC oxidation pathways within *At. ferrooxidans* (Quatrini et al. 2006, 2009). As shown in Figs. 8.1 and 8.4, these pathways are quite complex being both branched and redundant providing *At. ferrooxidans* a flexible respiratory system allowing it to adapt efficiently to environmental changes. The electrons from the oxidation of Fe(II) are transferred to oxygen (downhill pathway) by the proteins encoded by the *rus* operon and to NAD(P) (uphill pathway) through the electron carriers encoded by the *petI* operon (Fig. 8.4a). The proteins encoded by the *rus* operon have been proposed to constitute an "electron wire" spanning both the outer and the inner membranes to conduct electrons from metal sulfides, the natural substrate of *At. ferrooxidans*, to oxygen (Appia-Ayme et al. 1999; Yarzábal et al. 2002b, 2004). The outer membrane cytochrome *c*, Cyc2, transfers electrons from Fe(II) to the periplasmic blue copper rusticyanin, which passes them to the membrane-bound cytochrome c_4 Cyc1 and from there to the cytochrome oxidase CoxBACD where oxygen is reduced to water (Fig. 8.4a). Cup, a putative copper protein, was proposed to be involved in electron transfer, perhaps between Cyc2 and Cyc1 bypassing rusticyanin and providing an alternative route for electron flow during Fe(II) oxidation and an additional point for its regulation (Quatrini et al. 2009). The *petI* operon encodes the proteins involved in the uphill pathway in which electrons from the cytochrome c_4 CycA1 are transferred via the bc_1 complex to a NADH complex (Fig. 8.4a) driven energetically by the proton motive force (Levicán et al. 2002;

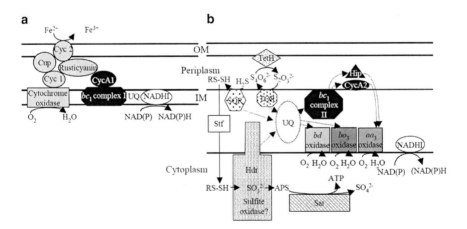

Fig. 8.4 Model for the iron (**a**) and RISC (**b**) oxidation pathways in *Acidithiobacillus ferrooxidans*. The proteins encoded in the same transcriptional unit are represented with the same motif. In (**b**), the redox proteins which give the electrons to the quinone pool are in a stippled motif while the ones which receive the electrons from the quinone pool are in a solid motif. In (**b**) electron transfer is indicated as *dotted arrows*. Notice that in (**b**) the inner membrane (IM) is larger than in (**a**) for convenience. OM: outer membrane. See the text for the abbreviations

Quatrini et al. 2006, 2009; Bruscella et al. 2007). The bifurcation in the electron flow from Fe(II) to NAD(P) (uphill) or O_2 (downhill) has been proposed to occur at the level of rusticyanin (Fig. 8.4a) (Quatrini et al. 2006, 2009; Bruscella et al. 2007). By adjusting the electron flow at this branch point, At. ferrooxidans could balance its requirements for NAD(P)H and ATP.

RISC oxidation pathways are predicted to involve various enzymes and electron carriers because, in contrast to iron that occurs in only two oxidation states, +2 and +3, sulfur exists in multiple states from −2 to +6. In addition, abiotic oxidation of some sulfur compounds can occur complicating identification of relevant enzymes and products. Microarray transcript profiling results predict that RISC oxidation pathways involve proteins in all the cellular compartments: in the outer membrane facing the periplasm (tetrathionate reductase, TetH), in the periplasm (high potential iron-sulfur protein, Hip), attached to the cytoplasmic membrane on the periplasmic side (cytochrome c, CycA2), in the cytoplasmic membrane (sulfide quinone reductase [SQR], thiosulfate quinone reductase [TQR], bc_1 complex, NADH complex I, aa_3, bd and bo_3 terminal oxidases) and in the cytoplasm (heterodisulfide reductase [Hdr], and ATP sulfurylase [Sat]) (Fig. 8.4b) (Quatrini et al. 2009). S^0 is converted to sulfane sulfate (GSSH) which is then transferred to the cytoplasmic heterodisulfide reductase (Hdr) through a cascade of sulfur transferases (Stf). Electrons coming from RISC enter the respiratory chain at the level of the quinol pool (QH_2) through SQR, TQR or Hdr and are transferred either (1) directly to terminal oxidases bd or bo_3, or indirectly throught a bc_1 complex and cytochrome c CycA2/high potential iron–sulfur protein Hip probably to the aa_3 oxidase, where O_2 reduction takes place or (2) to NADH complex I to generate reducing power (Fig. 8.4b) (Quatrini et al. 2009).

Possible regulators allowing coordinated modulation of Fe(II) and RISC oxidation gene expression according to the growth conditions have been proposed: (a) CtaR predicted to belong to the iron responsive regulator Rrf2 family and (b) a two component sensor-regulator of the RegB-RegA family that may respond to the redox state of the quinone pool and could play a role in switching between Fe(II) and S^0 oxidation, between aerobic and anaerobic oxidation or could help in making regulatory changes to balance electron equivalents between uphill and downhill electron flow (Quatrini et al. 2009). CtaR and regBA expression is induced by Fe(II) and not repressed by S^0, like rus and petI operons (Amouric et al. 2009). Furthermore, RegA binds to the regulatory region of the rus and cta operons, strongly suggesting that this protein regulates their expression (Amouric et al. 2009).

Acidithiobacillus thiooxidans and Acidithiobacillus caldus

Acidithiobacillus caldus

A draft genome sequence of the At. caldus type strain (ATCC51756) has been annotated and analysed (Valdés et al. 2009). It carries 2,982,397 bp (61.4% G+C content).

Genes necessary for amino acid, nucleotide and prosthetic group synthesis have been predicted as well as the genes involved in carbon fixation and ammonia uptake. Contrarily to *At. ferrooxidans*, the genes for flagella formation and chemotaxis have been identified but not those necessary for nitrogen fixation (Valdés et al. 2009).

Two *cyoBACD*, six *cydAB* and one *quoxBACD* operons encoding bo_3, *bd* and aa_3 quinol oxidases respectively, have been detected suggesting that *At. caldus* is able to adapt to a large range of oxygen tensions. As expected, the *At. ferrooxidans rus* and *petI* operons involved in Fe(II) oxidation pathways are not present. Concerning RISC oxidation, some genes already identified in *At. ferrooxidans* are present, i.e. *sqr* and *tetH* encoding sulfur quinone reductase and tetrathionate hydrolase, as well as *doxD* encoding the thiosulfate quinone oxidoreductase previously characterized in this strain (Rzhepishevska et al. 2007) (Fig. 8.1c). Very interestingly, the *sox* genes involved in thiosulfate oxidation in most neutrophilic and mesophilic sulfur oxidizers and the *sor* gene encoding sulfur oxygenase:reductase identified so far only in thermophiles are present (Fig. 8.1c). The presence of *sor* gene in different *At. caldus* strains has recently been reported (Janosch et al. 2009). All together, these data indicate that the RISC oxidation pathways in *At. caldus* and *At. ferrooxidans* are different.

Comparative Genomics Between the Acidithiobacilli

Draft genome sequence of the *At. thiooxidans* type strain (ATCC19377) has been annotated by Holmes' group (unpublished data). The DSM17318 strain has also been sequenced but will not be discussed here since it is owned by Biosigma, SA (Levicán et al. 2008). Comparison between the *At. thiooxidans*, *At. caldus* and *At. ferrooxidans* genome sequences has been performed to predict genes involved in respiratory pathways, carbon and nitrogen fixation, quorum sensing, flagella formation, chemotaxis, and iron assimilation (Valdés et al. 2008b) as well as in c-di-GMP pathway (Castro et al. 2009). In the genome of the three bacteria, a large number of genes encoding iron transporters, in particular TonB-dependent Fe(III)-siderophore transporters (OMR) typical of iron scavengers, are present. Interestingly, the isoelectric point of these OMR varies in broad range, likely allowing the Acidithiobacilli to grow in environments with different pH and iron availability (Osorio et al. 2008a, b). *At. ferrooxidans* is particularly more fit out Fe(III) siderophore transporters which could explain its greater sensitivity to Fe(III). Surprisingly, no genes for classical siderophore production have been detected in the three species. Instead, Fe(III) uptake by phosphate chelation or dicitrate uptake system has been proposed (Osorio et al. 2008a, b). Main differences between the three Acidithiobacilli lie in (a) the absence of nitrogen fixation genes in *At. caldus* and *At. thiooxidans*, (b) the presence of flagella formation and chemotaxis signaling genes in *At. caldus* and *At. thiooxidans* but not *At. ferrooxidans*, (c) the absence of AHL-based quorum sensing genes in *At. caldus* and *At. thiooxidans* (d) the absence in these two Acidithiobacilli of the *At. ferrooxidans rus* and *petI* operons necessary for Fe(II) oxidation, and (e) the presence in

At. caldus and *At. thiooxidans*, but not in *At. ferrooxidans*, of the *sox* genes involved in thiosulfate oxidation (Fig. 8.1c) suggesting some differences in the RISC oxidation mechanisms between the three Acidithiobacilli, even if all the three have the genes encoding tetrathionate hydrolase (*tetH*) and thiosulfate quinone oxidoreductase (*doxD*) (Fig. 8.1c) (Valdés et al. 2008a, b, 2009; Valdés and Holmes, personal communication).

Leptospirillum spp.

Before *Leptospirilla* genome sequences were accessible, Parro and Moreno-Paz (2003; 2004; 2006) have studied the global nitrogen fixation pathway in *L. ferrooxidans*. Gene expression analysis by shotgun DNA microarray (i.e., a random genomic library printed on a slide) between *L. ferrooxidans* grown with and without ammonium allowed them to identify most of the genes known to be involved in nitrogen fixation.

A *L. ferriphilum* strain (*Leptospirillum* sp. group II) was sequenced, but this strain and its genome sequence are owned by Biosigma, SA and will not be discussed here (Levicán et al. 2008). Nearly complete genome sequences of two *Leptospirillum* group II (one related to *L. ferriphilum* [2.72 Mbp] and *"L. rubarum"* [2.64 Mbp]) and of *"L. ferrodiazotrophum"* from group III (2.66 Mbp) were reconstructed directly from the environment (Tyson et al. 2004; Lo et al. 2007; Simmons et al. 2008; Goltsman et al. 2009). Surprisingly, *Leptospirillum* group II has no genes involved in nitrogen fixation while they have been detected in *Leptospirillum* group III. In addition, more proteins involved in biofilm formation (glycosyltransferases, polysaccharide export proteins, cellulose synthase) from group III than from group II have been detected by metaproteomic analysis (Ram et al. 2005; Lo et al. 2007; Goltsman et al. 2009) suggesting that *Leptospirillum* group III plays a key role in the microbial community. Both groups possess all the informations for a chemotaxis system and flagella formation. Leptospirilla encode a number of metal resistances. Contrary to the Acidithiobacilli, Leptospirilla have no known Fe(II) transporters and it was proposed that this was a strategy to evade iron stress (Quatrini et al. 2007b, c; Osorio et al. 2008a, b). On the other hand, Leptospirilla have less Fe(III) siderophore receptors and these receptors are in a much restricted pIs range than the Acidithiobacilli, likely because Leptospirilla grow in a more acidic and narrow pH range.

Concerning energetic metabolism, *Leptospirillum* groups II and III have genes encoding a high number of cytochromes *c*, a *bc*$_1$ complex, NADH:ubiquinone dehydrogenase, two cytochrome *cbb*$_3$-type heme-copper oxidases and a cytochrome *bd*-type quinol oxidase (Tyson et al. 2004; Goltsman et al. 2009). Among the cytochromes, two were proposed to play a role in Fe(II) oxidation: Cyc$_{572}$ and Cyc$_{579}$ (see Fig. 8.1b and below in the "From the microbial communities to protein analysis" section).

Surprisingly, the genes encoding the *bd* oxidase are clustered with a gene encoding a sulfide-quinone reductase and the genes encoding formate-hydrogen lyase have been detected suggesting that Leptospirilla can oxidize hydrogen sulfide and grow in anaerobic conditions (Goltsman et al. 2009), while they have been considered as "specialist" oxic iron oxidizers using Fe(II) as their sole electron donor and oxygen as their electron acceptor.

Ferroplasma spp.

Three genome sequences derived from the Iron Mountain AMD are available for *Ferroplasma* spp.: a draft from the isolate *Fp. acidarmanus* fer1 (1.94 Mbp) (Allen et al. 2007) and two partial from *Ferroplasma* type I (1.48 Mbp) and II (1.82 Mbp) (Tyson et al. 2004).

From the analysis of these sequences, a number of pathways have been proposed (Tyson et al. 2004), supported in some cases by proteomics (Baker-Austin et al. 2005; Dopson et al. 2005, 2007; Baker-Austin et al. 2007) or metaproteomics (Ram et al. 2005). Nitrogen fixation and flagella formation genes have not been detected in *Ferroplasma* spp. (Tyson et al. 2004). Surprisingly, the genes encoding the key enzymes of the three known archaea CO_2 fixation pathways (the 3-hydroxypropionate, the reductive acetyl coenzyme A and the reverse TCA cycles) are not present. Metabolic reconstruction strategies predict a novel and chimaeric CO/CO_2 fixation pathway with steps from the reductive Acetyl-CoA and serine pathways (Cardenas et al. 2009). However, no up-regulation of proteins typically associated with CO_2 fixation was evident from proteomic analysis between cells grown with or without yeast extract (Dopson et al. 2005). Since a plethora of ABC-type sugar and amino acid transporters encoded in the *Ferroplasma* type I and II genomes have been found, these organisms may prefer the heterotrophic to autotrophic lifestyle (Tyson et al. 2004). Several metal resistance genes have been identified (Gihring et al. 2003; Tyson et al. 2004; Baker-Austin et al. 2005). Proteomic analysis have shown DNA repair and stress proteins upregulation in response to high levels of copper (Baker-Austin et al. 2005) or arsenic (Baker-Austin et al. 2007). It has to be pointed out that *Ferroplasma* spp., while extremely resistant to arsenic, lack *arsC* encoding arsenate reductase suggesting that they have an alternative arsenate resistance mechanism.

The cellular machinery of *Fp. acidiphilum* (which genome has not been sequenced) is iron-protein-dominated (Ferrer et al. 2007) as shown by proteomics, the iron atom acting as "iron rivets" organizing and stabilizing the three-dimensional structure of the proteins.

No cytochrome *c* genes have been detected while heme-copper terminal oxidases, cytochrome *b*, Rieske iron-sulfur proteins, and sulfocyanin (a blue copper protein) genes are present suggesting that they may form a terminal oxidase super complex (Tyson et al. 2004). Based on biochemical, genomic and proteomic analysis, a model for oxic Fe(II) oxidation in which the sulfocyanin receives the electron from Fe(II) and transfers it to the cbb_3 oxidase (Fig. 8.1b) (Dopson et al. 2005) and a model for the anoxic Fe(III) reduction mediated by the reductive TCA cycle and an electron transport involving cytochromes b and c (Dopson et al. 2007) have been proposed.

Metallosphaera sedula

The circular chromosome of the extremely thermoacidophilic archaeon *Metallosphaera sedula* type strain (DSM535) carries 2.2 Mbp (46% G+C content) predicted to encode 2,258 proteins, 35% of which are annotated as either "hypothetical protein" or "protein of unknown function" (Auernik et al. 2008). Adhesion and metal tolerance genes have been detected as well as the genes involved in carbon fixation via a modified 3-hydroxypropionate cycle (Auernik et al. 2008).

From genomic combined with transcriptomic analysis (Auernik et al. 2008; Auernik and Kelly 2008), components involved in Fe(II) and RISC oxidation electron transfer chains have been identified. Obviously, the *M. sedula* Fe(II) iron oxidation pathway is different from that described in *At. ferrooxidans*. Four genes encoding blue copper proteins, two of which containing the recognized protein signatures for rusticyanin, have been detected (Auernik et al. 2008). However, none of them responded to Fe(II) (Auernik and Kelly 2008). Instead, SoxNL-CbsAB (putative Rieske protein and cytochrome b of the bc_1 complex-cytochromes b_{558} and b_{562}) and Fox cluster (*soxHB*-like and *cbsAB*-like genes encoding cytochrome oxidase subunits I and II and cytochromes b_{558} and b_{562}, respectively) are important for Fe(II) oxidation (Fig. 8.1b). The *fox* cluster was proposed to be involved in the reverse (uphill) direction to NAD(P).

The genes encoding a thiosulfate:quinone reductase (DoxDA), involved in the oxidation of thiosulfate to tetrathionate, are present in the *M. sedula* genome sequence (Auernik et al. 2008), but, surprisingly, they are not differentially expressed in RISC compared to Fe(II) conditions (Auernik and Kelly 2008). On the contrary, and therefore likely important for the RISC oxidation (Auernik and Kelly 2008), are (a) the *soxLDD'-ABC* gene cluster encoding the Rieske protein and the cytochrome b of a putative bc_1 complex (SoxLC), the cytochrome oxidase subunits I and II (SoxAB) and two membrane proteins (SoxDD'), (b) the *tetH* gene encoding a putative tetrathionate hydrolase, (c) a locus encoding a novel polysulfide/sulfur/dimethyl sulfoxide reductase-like complex, and (d) a large locus encoding sulfurtransferases and a heterodisulfide reductase complex which has been proposed to transfer the sulfur as sulfane sulfur to the heterodisulfide reductase which could work in reverse and oxidize disulfide to sulfite and/or adenosine-5'phosphosulfate (APS) (see the *Acidithiobacillus ferrooxidans* section above and Quatrini et al. 2009) (Fig. 8.1c).

Sulfolobus spp.

Sulfolobus acidocaldarius

The type strain circular chromosome of 2,225,959 bp (36.7% G+C content) encodes 2,292 proteins of which 50% were either exclusive to *S. acidocaldarius* or to *Sulfolobus* (Chen et al. 2005). The stable genome organization of *S. acidocaldarius* compared to *S. solfataricus* and *S. tokodaii* is probably due to the absence of active mobile elements. The genes involved in the production of sulfuric acid from hydrogen sulfide (i.e. sulfite reductase, phosphoadenosine phosphosulfate reductase, and sulfate adenylyltransferase) have been detected (Chen et al. 2005), but not those encoding sulfur oxygenase:reductase, thiosulfate and sulfide quinone reductases (Fig. 8.1c). A *hdr* cluster encoding a novel heterodisulfide reductase-like complex also present in *At. ferrooxidans* (Quatrini et al. 2009) has been identified (Fig. 8.1c).

Cell cycle (Lundgren and Bernander 2007) and UV irradiation response (Götz et al. 2007), but not S^0 metabolism, of this archaeon were analysed by transcriptomics.

Sulfolobus solfataricus

The type strain single chromosome is 2,992,245 bp long (35% G+C content) encoding 2,997 proteins of which one third have no homologs (She et al. 2001). Putative mobile elements represent 11% of the genome, conferring it a high plasticity.

The genes encoding the enzymes involved in RISC oxidation to sulfate have been identified. S^0 and thiosulfate are converted to sulfate via the formation of adenylylsulfate. The electrons from sulfide and thiosulfate oxidation reduce the caldariella-quinone pool via the sulfide and thiosulfate quinone reductases, respectively (Fig. 8.1c). The sulfur oxygenase:reductase gene was not detected. The caldariella-quinol then transfer electrons to two archaeal specific cytochrome complexes, SoxABCD (a heme aa_3-Cu oxidase) and Sox EFGHIM (heme bb_3-Cu oxidase) (Fig. 8.1c). Recently a cluster of genes encoding sulfur transferases and a heterodisulfide reductase complex, proposed to be involved in S^0 oxidation to sulfite or APS, has been detected (see *At. ferrooxidans* section above and Quatrini et al. 2009) (Fig. 8.1c).

The *S. solfataricus* proteome was analysed for n-propanol assimilation (Chong et al. 2007a), ethanol metabolism (Chong et al. 2007b), stress to nickel (Salzano et al. 2007). Response to UV irradiation was studied by a whole-genome microarray approach (Götz et al. 2007). By combining genomic, proteomic, transcriptomic and biochemical data, Snijders et al. (2006) have reconstructed the central carbon metabolism (glycolysis, gluconeogenesis and tricarboxylic acid cycle). Unfortunately, RISC metabolism has not be studied.

Sulfolobus tokodaii

The size of the type strain circular chromosome is 2,694,756 bp size (G+C content of 32.8%) (Kawarabayasi et al. 2001). Duplication of genomic regions, rearrangement of genomic structure and plasmid integration suggest a high level of plasticity. The genome encodes 2,826 potential proteins, 32.2% with assigned function. Among them, are the genes involved in hydrogen sulfide oxidation to sulfate: flavocytochrome *c* sulfide dehydrogenase, sulfite oxidase, thiosulfate quinone reductase, thiosulfate sulfurtransferase, sulfite reductase, phosphoadenosine phosphosulfate reductase, and sulfate adenyltransferase (Fig. 8.1c). As with the other *Sulfolobus* spp. which genome has been sequenced, the locus encoding a heterodisulfide reductase complex and sulfurtransferases has been detected (Quatrini et al. 2009) (Fig. 8.1c). Noteworthy, it is the only *Sulfolobus* spp. encoding a sulfur oxygenase:reductase (Chen et al. 2005) (Fig. 8.1c).

The *fox* locus, up-regulated in Fe(II) conditions in *S. metallicus* (Bathe and Norris 2007) and encoding SoxB/M-like (quinol oxidase subunit I), SoxH-like (quinol oxidase subunit II), CbsA-like (cytochrome *b*) and iron-sulfur proteins, was detected in *S. tokodaii* (Fig. 8.1c) and, indeed, *S. tokodaii* was shown to grow by oxidizing Fe(II) (Bathe and Norris 2007).

From the Microbial Communities to Protein Analysis

This section illustrates how metagenomics and metaproteomics can help (a) to study a microbial community directly in its environment and to show who is doing what, (b) to find the growth conditions to isolate an as yet unculturable microorganism (c) to improve cell yield by predicting nutritional requirement and (d) to purify directly from natural microbial consortia proteins which play a key role in their environment and to characterize them, bypassing the microorganism isolation step.

Richmond Mine at Iron Mountain (California, USA) represents a self-contained biogeochemical system characterized by tight coupling between microbial iron oxidation and acidification due to pyrite dissolution with extremely acid water (pH 0.5–1.0) and high levels of metals. Banfield and coworkers focused their studies on a pink biofilm grown hundred of feets underground on the surface of sulfuric acid-rich solutions. Because of these harsh conditions, the microbial communities have a low diversity and contain *Leptospirillum* group II (75%), *Leptospirillum* group III (10%), archaea (1%), including *Ferroplasma* spp., and eucaryotic species (4%) (Tyson et al. 2004). From this natural acidophilic biofilm, genome sequences from two *Leptospirillum* group II (one related to *L. ferriphilum* (Tyson et al. 2004) and *L. rubarum* (Lo et al. 2007; Goltsman et al. 2009)), one from group III (*L. ferrodiazotrophum* (Goltsman et al. 2009), from *F. acidarmanus* fer1 (Allen et al. 2007), and *Ferroplasma* types I and II (Tyson et al. 2004; Allen et al. 2007) were reconstructed. These data combined with metaproteomic analysis (Ram et al. 2005; Lo et al. 2007; Goltsman et al. 2009) provide insights in the community metabolic network. Noteworthy, *L. ferrodiaz-otrophum* (group III) carbohydrate metabolism was shown by proteomics (Ram et al. 2005) to be considerably greater than in *Leptospirillum* group II and in *Ferroplasma* type II where genes for extracellular polymer substances have not been clearly identified. In addition, this bacterium, which at first sight was considered as a minor member, is the only one which could fix nitrogen since the *nif* genes were detected only in its genome sequence (Tyson et al. 2004; Lo et al. 2007; Allen et al. 2007; Goltsman et al. 2009). Therefore, while in low abundance in the microbial community, *L. ferrodiazotrophum* (group III) played a central role in biofilm formation and nitrogen fixation.

Based on this last prediction, one representative was isolated in nitrogen-free liquid medium from an AMD biofim in which *Leptospirillum* group III was abundant (Tyson et al. 2005). This Fe(II) oxidizing, free-living diazotroph has been tentatively named *"L. ferrodiazotrophum"*.

Because a number of genes encoding metabolic transporters for Ni^{2+}, sugars, and amino acids have been predicted from *Fp. acidarmanus* reconstructed genome, it was hypothesized that growth could be improved by increasing the yeast extract concentration and by adding nickel to the medium. Indeed, yield increased more than 100-fold compared to the usual laboratory medium (Baumler et al. 2005).

Finally, two unusual cytochromes *c*, Cyt_{572} and Cyt_{579}, which have been shown to be abundant in the biofilm by proteomic analysis (Ram et al. 2005) have been isolated directly from natural samples (Jeans et al. 2008; Singer et al. 2008). Both are encoded by groups II and III Leptospirilla (Goltsman et al. 2009). Insignificant sequence similarity was found for both cytochromes in protein data banks. Cyt_{572} is located in the outer membrane and Cyt_{579} likely in the periplasm. While Fe(II) oxidation was not favored thermodynamically at a pH < 3 by Cyt_{579}, Cyt_{572} oxidized it at these physiologically relevant values. All together, these results indicate that Cyt_{572} is likely the primary electron acceptor from Fe(II) and transfers them to the periplasmic Cyt_{579} which in turn shuttles them to the cbb_3 terminal oxidase complex embedded into the cytoplasmic membrane according to the following pathway (Fig. 8.1b):

$$Fe(II) \rightarrow Cyc_{572} \rightarrow Cyc_{579} \rightarrow cbb_3\text{-type oxidase} \rightarrow O_2$$

Conclusion

Most of the acidophilic Fe(II) and/or RISC oxidizers are difficult to handle experimentally (slow growth, poor cell yield, etc.). In addition, genetic manipulation in these microorganisms is a real challenge and even when developed, is fastidious. As shown in this review, not only the "omics" approaches can help to overcome these difficulties but also give a global answer and therefore more informations. A number of metabolic pathways have been reconstructed *in silico*, sometimes supported by proteomics and/or transcriptomics. One of the most obvious conclusion that can be drawn from the data obtained so far is that different Fe(II) and RISC oxidation pathways exist (Fig. 8.1b, c).

The electrons coming from Fe(II) are transferred to the terminal acceptor through different redox proteins depending on the microorganism. In some cases, cytochromes c are involved (*At. ferrooxidans*, *Leptospirillum* groups II and III), in others copper protein (*At. ferrooxidans*, *Fp. acidarmanus*) or a complete/incomplete bc_1 complex (*At. ferrooxidans*, *Leptospirillum* groups II and III, *M. sedula*, *S. tokodaii*) (Fig. 8.1b). The terminal oxidase where oxygen is reduced can be aa_3 (*At. ferrooxidans*) or cbb_3-type (*Leptospirillum* group II and III, *Fp. acidarmanus*) (Fig. 8.1b).

The most characterized RISC oxidation pathway is the Sox system (Friedrich et al. 2005) which is present in the mesophilic neutrophiles but also in *At. caldus*, and *At. thiooxidans* (Fig. 8.1c). Noteworthy, it is absent in *At. ferrooxidans*. In a number of thermophiles, S^0 is oxidized by a sulfur oxygenase:reductase encoded by the *sor* gene (Urich et al. 2004). This gene is also present in *At. caldus*, which is a moderate thermophile, and in the extreme thermophile *S. tokodaii* (Fig. 8.1c). In *At. ferrooxidans*, none of these characterized systems are present (Fig. 8.1c). However, the *dorDA*, encoding thiosulfate quinone reductase, which has been described till now only in Archaea (Muller et al. 2004) has been detected in the three Acidithiobacilli (Fig. 8.1c). As far as we can say for the moment, a bc_1 and a (caldariella) quinol oxidase complexes are involved in the respiratory chain from RISC to oxygen (*At. ferrooxidans*, *Sulfolobus* spp. and *M. sedula*) (Fig. 8.1c).

This would suggest that biological Fe(II) oxidation has evolved separately many times while RISC oxidation have evolved only few times and lateral gene transfer occurred giving a "mosaic" pathway (like for example in Acidithiobacilli in which some genes mainly described in archaea were detected [*sor, doxDA* and *hdr*]). The answer will likely being given from the genome sequence analysis of more acidophilic Fe(II) and/or RISC oxidizers.

Acknowledgements The author acknowledges DS Holmes (Universidad Andres Bello, Santiago, Chile) and N Guiliani (Universidad de Chile, Santiago, Chile) to give her access to information before publication.

References

Acosta M, Beard S, Ponce J, Vera M, Mobarec JC, Jerez CA (2005) Identification of putative sulfurtransferase genes in the extremophilic *Acidithiobacillus ferrooxidans* ATCC23270 genome: structural and functional characterization of the proteins. OMICS 9:13–29
Allen EE, Tyson GW, Whitaker RJ, Detter JC, Richardson PM, Banfield JF (2007) Genome dynamics in a natural archaeal population. Proc Natl Acad Sci USA 104:1883–1888
Amaro AM, Chamorro D, Seeger M, Arredondo R, Peirano I, Jerez CA (1991) Effect of external pH perturbations on *in vivo* protein synthesis by the acidophilic bacterium *Thiobacillus ferrooxidans*. J Bacteriol 173:910–915

Amouric A, Appia-Ayme C, Yarzabal A, Bonnefoy V (2009) Regulation of the iron and sulfur oxidation pathways in the acidophilic *Acidithiobacillus ferrooxidans*. Adv Mat Res 71–73:163–166

Appia-Ayme C, Guiliani N, Ratouchniak J, Bonnefoy V (1999) Characterization of an operon encoding two *c*-type cytochromes, an *aa*(3)-type cytochrome oxidase, and rusticyanin in *Thiobacillus ferrooxidans* ATCC 33020. Appl Environ Microbiol 65:4781–4787

Appia-Ayme C, Quatrini R, Denis Y et al (2006) Microarray and bioinformatic analyses suggest models for carbon metabolism in the autotroph *Acidithiobacillus ferrooxidans*. Hydrometallurgy 83:273–280

Auernik KS, Kelly RM (2008) Identification of components of electron transport chains in the extremely thermoacidophilic crenarchaeon *Metallosphaera sedula* through iron and sulfur compound oxidation transcriptomes. Appl Environ Microbiol 74:7723–7732

Auernik KS, Maezato Y, Blum PH, Kelly RM (2008) The genome sequence of the metal-mobilizing, extremely thermoacidophilic archaeon *Metallosphaera sedula* provides insights into bioleaching-associated metabolism. Appl Environ Microbiol 74:682–692

Baker-Austin C, Dopson M, Wexler M, Sawers RG, Bond PL (2005) Molecular insight into extreme copper resistance in the extremophilic archaeon '*Ferroplasma acidarmanus*' Fer1. Microbiology 151:2637–2646

Baker-Austin C, Dopson M, Wexler M, Sawers RG, Stemmler A, Rosen BP, Bond PL (2007) Extreme arsenic resistance by the acidophilic archaeon '*Ferroplasma acidarmanus*' Fer1. Extremophiles 11:425–434

Barreto M, Quatrini R, Bueno S, Arriagada C, Valdes J, Silver S, Jedlicki E, Holmes DS (2003) Aspects of the predicted physiology of *Acidithiobacillus ferrooxidans* deduced from an analysis of its partial genome sequence. Hydrometallurgy 71:97–105

Barreto M, Gehrke T, Harneit K, Sand W, Jedlicki E, Holmes DS (2005a) Unexpected insights into biofilm formation by *Acidithiobacillus ferrooxidans* revealed by genome analysis and experimental approaches. In: Harrison STL, Rawlings DE, Petersen J (eds) Proceedings of the 16th international biohydrometallurgy symposium. Compress, Cape Town, South Africa, pp 817–825

Barreto M, Jedlicki E, Holmes DS (2005b) Identification of a gene cluster for the formation of extracellular polysaccharide precursors in the chemolithoautotroph *Acidithiobacillus ferrooxidans*. Appl Environ Microbiol 71:2902–2909

Bathe S, Norris PR (2007) Ferrous iron- and sulfur-induced genes in *Sulfolobus metallicus*. Appl Environ Microbiol 73:2491–2497

Baumler DJ, Jeong KC, Fox BG, Banfield JF, Kaspar CW (2005) Sulfate requirement for heterotrophic growth of "*Ferroplasma acidarmanus*" strain fer1. Res Microbiol 156:492–498

Blackstock WP, Weir MP (1999) Proteomics: quantitative and physical mapping of cellular proteins. Trends Biotechnol 17:121–127

Bouchal P, Zdrahal Z, Helanova S, Janiczek O, Hallberg KB, Mandl M (2006) Proteomic and bioinformatic analysis of iron- and sulfur-oxidizing *Acidithiobacillus ferrooxidans* using immobilized pH gradients and mass spectrometry. Proteomics 15:4278–4285

Brasseur G, Bruscella P, Bonnefoy V, Lemesle-Meunier D (2002) The *bc*(1) complex of the iron-grown acidophilic chemolithotrophic bacterium *Acidithiobacillus ferrooxidans* functions in the reverse but not in the forward direction. Is there a second *bc*(1) complex? Biochim Biophys Acta 1555:37–43

Brasseur G, Levican G, Bonnefoy V, Holmes D, Jedlicki E, Lemesle-Meunier D (2004) Apparent redundancy of electron transfer pathways via *bc*(1) complexes and terminal oxidases in the extremophilic chemolithoautotrophic *Acidithiobacillus ferrooxidans*. Biochim Biophys Acta 1656:114–126

Bruscella P, Cassagnaud L, Ratouchniak J, Brasseur G, Lojou E, Amils R, Bonnefoy V (2005) The HiPIP from the acidophilic *Acidithiobacillus ferrooxidans* is correctly processed and translocated in *Escherichia coli*, in spite of the periplasm pH difference between these two microorganisms. Microbiology 151:1421–1431

Bruscella P, Appia-Ayme C, Levican G, Ratouchniak J, Jedlicki E, Holmes DS, Bonnefoy V (2007) Differential expression of two bc_1 complexes in the strict acidophilic chemolithoautotrophic bacterium *Acidithiobacillus ferrooxidans* suggests a model for their respective roles in iron or sulfur oxidation. Microbiology 153:102–110

Buonfiglio V, Polidoro M, Soyer F, Valenti P, Shively J (1999) A novel gene encoding a sulfur-regulated outer membrane protein in *Thiobacillus ferrooxidans*. J Biotechnol 72:85–93

Cardenas JP, Martinez V, Covarrubias PC, Holmes DS, Quatrini R (2009) Predicted CO/CO_2 fixation in *Ferroplasma* spp. via a novel chimaeric pathway. Adv Mat Res 71–73:219–222

Castro M, Ruiz LM, Díaz M, Mamani S, Jerez CA, Holmes DS, Guiliani N (2009) c-di-GMP pathway in biomining bacteria. Adv Mat Res 71–73:223–226

Chen L, Brugger K, Skovgaard M et al (2005) The genome of *Sulfolobus acidocaldarius*, a model organism of the Crenarchaeota. J Bacteriol 187:4992–4999

Chi A, Valenzuela L, Beard S, Mackey AJ, Shabanowitz J, Hunt DF, Jerez CA (2007) Periplasmic proteins of the extremophile *Acidithiobacillus ferrooxidans*: a high throughput proteomics analysis. Mol Cell Proteomics 6:2239–2251

Chong PK, Burja AM, Radianingtyas H, Fazeli A, Wright PC (2007a) Proteome analysis of *Sulfolobus solfataricus* P2 propanol metabolism. J Proteome Res 6:1430–1439

Chong PK, Burja AM, Radianingtyas H, Fazeli A, Wright PC (2007b) Proteome and transcriptional analysis of ethanol-grown *Sulfolobus solfataricus* P2 reveals ADH2, a potential alcohol dehydrogenase. J Proteome Res 6:3985–3994

Dharmadi Y, Gonzalez R (2004) DNA microarrays: experimental issues, data analysis, and application to bacterial systems. Biotechnol Prog 20:1309–1324

Dopson M, Baker-Austin C, Bond PL (2005) Analysis of differential protein expression during growth states of *Ferroplasma* strains and insights into electron transport for iron oxidation. Microbiology 151:4127–4137

Dopson M, Baker-Austin C, Bond P (2007) Towards determining details of anaerobic growth coupled to ferric iron reduction by the acidophilic archaeon '*Ferroplasma acidarmanus*' Fer1. Extremophiles 11:159–168

Esparza M, Bowien B, Holmes DS, Jedlicki E (2009) Gene organization and CO_2-responsive expression of four *cbb* operons in the biomining bacterium *Acidithiobacillus ferrooxidans*. Adv Mat Res 71–73:207–210

Farah C, Vera M, Morin D, Haras D, Jerez CA, Guiliani N (2005) Evidence for a functional quorum-sensing type AI-1 system in the extremophilic bacterium *Acidithiobacillus ferrooxidans*. Appl Environ Microbiol 71:7033–7040

Felício AP, Garcia Junior O, Bertolini MC, Ottoboni LMM, Novo MTM (2003) The effects of copper ions on the synthesis of periplasmic and membrane proteins in *Acidithiobacillus ferrooxidans* as analyzed by SDS-PAGE and 2D-PAGE. Hydrometallurgy 71:165–171

Ferrer M, Golyshina OV, Beloqui A, Golyshin PN, Timmis KN (2007) The cellular machinery of *Ferroplasma acidiphilum* is iron-protein-dominated. Nature 445:91–94

Friedrich CG, Bardischewsky F, Rother D, Quentmeier A, Fischer J (2005) Prokaryotic sulfur oxidation. Curr Opin Microbiol 8:253–259

Gihring TM, Bond PL, Peters SC, Banfield JF (2003) Arsenic resistance in the archaeon "*Ferroplasma acidarmanus*": new insights into the structure and evolution of the *ars* genes. Extremophiles 7:123–130

Goltsman DS, Denef VJ, Singer SW et al (2009) Community genomic and proteomic analyses of chemoautotrophic iron-oxidizing "*Leptospirillum rubarum*" (Group II) and "*Leptospirillum ferrodiazotrophum*" (Group III) bacteria in acid mine drainage biofilms. Appl Environ Microbiol 75:4599–4615

Götz D, Paytubi S, Munro S, Lundgren M, Bernander R, White MF (2007) Responses of hyperthermophilic crenarchaea to UV irradiation. Genome Biol 8:R220

Hallberg KB, Johnson DB (2001) Biodiversity of acidophilic prokaryotes. Adv Appl Microbiol 49:37–84

He Z, Zhong H, Hu Y, Xiao S, Liu J, Xu J, Li G (2005a) Analysis of differential-expressed proteins of *Acidithiobacillus ferrooxidans* grown under phosphate starvation. J Biochem Mol Biol 38:545–549

He ZG, Hu YH, Zhong H, Hu WX, Xu J (2005b) Preliminary proteomic analysis of *Thiobacillus ferrooxidans* growing on elemental sulphur and Fe^{2+} separately. J Biochem Mol Biol 38:307–313

Holmes DS, Bonnefoy V (2007) Genetic and bioinformatic insights into iron and sulfur oxidation mechanisms of bioleaching organisms. In: Rawlings DE, Johnson DB (eds) Biomining. Springer, Berlin/Heidelberg, pp 281–307

Janosch C, Thyssen C, Vera M, Bonnefoy V, Rohwerder T, Sand W (2009) Sulfur oxygenase reductase in different *Acidithiobacillus caldus*-like strains. Adv Mat Res 71–73:239–242

Jeans C, Singer SW, Chan CS, Verberkmoes NC, Shah M, Hettich RL, Banfield JF, Thelen MP (2008) Cytochrome 572 is a conspicuous membrane protein with iron oxidation activity purified directly from a natural acidophilic microbial community. ISME J 2:542–550

Jerez CA (2007) Proteomics and metaproteomics applied to biomining microorganisms. In: Donati ER, Sand W (eds) Microbial processing of metal sulfides. Springer, Dordrecht, The Netherlands, pp 241–251

Jerez CA, Varela P, Osorio G, Seeger M, Amaro AM, Toledo H (1995) Differential gene expression of *Thiobacillus ferrooxidans* under different environmental conditions. In: Holmes DS, Smith RW (eds) Mineral bioprocessing (vol 2). The Minerals, Metals and Materials Society, Warrendale, PA, pp 111–121

Kanao T, Kamimura K, Sugio T (2007) Identification of a gene encoding a tetrathionate hydrolase in *Acidithiobacillus ferrooxidans*. J Biotechnol 132:16–22

Kawarabayasi Y, Hino Y, Horikawa H et al (2001) Complete genome sequence of an aerobic thermoacidophilic crenarchaeon, *Sulfolobus tokodaii* strain7. DNA Res 8:123–140

Levicán G, Bruscella P, Guacunano M, Inostroza C, Bonnefoy V, Holmes DS, Jedlicki E (2002) Characterization of the *petI* and *res* operons of *Acidithiobacillus ferrooxidans*. J Bacteriol 184:1498–1501

Levicán G, Ugalde JA, Ehrenfeld N, Maass A, Parada P (2008) Comparative genomic analysis of carbon and nitrogen assimilation mechanisms in three indigenous bioleaching bacteria: predictions and validations. BMC Genomics 9:581

Lo I, Denef VJ, Verberkmoes NC et al (2007) Strain-resolved community proteomics reveals recombining genomes of acidophilic bacteria. Nature 446:537–541

Lundgren M, Bernander R (2007) Genome-wide transcription map of an archaeal cell cycle. Proc Natl Acad Sci USA 104:2939–2944

Moreno-Paz M, Parro V (2006) Amplification of low quantity bacterial RNA for microarray studies: time-course analysis of *Leptospirillum ferrooxidans* under nitrogen-fixing conditions. Environ Microbiol 8:1064–1073

Muller FH, Bandeiras TM, Urich T, Teixeira M, Gomes CM, Kletzin A (2004) Coupling of the pathway of sulphur oxidation to dioxygen reduction: characterization of a novel membrane-bound thiosulphate:quinone oxidoreductase. Mol Microbiol 53:1147–1160

Norris PR (2007) Acidophile diversity in mineral sulfide oxidation. In: Rawlings DE, Johnson DB (eds) Biomining. Springer, Berlin/Heidelberg, pp 199–216

Novo MT, da Silva AC, Moreto R, Cabral PC, Costacurta A, Garcia O Jr, Ottoboni LM (2000) *Thiobacillus ferrooxidans* response to copper and other heavy metals: growth, protein synthesis and protein phosphorylation. Antonie Leeuwenhoek 77:187–195

Novo MT, Garcia Junior O, Ottoboni LM (2003) Protein profile of *Acidithiobacillus ferrooxidans* strains exhibiting different levels of tolerance to metal sulfates. Curr Microbiol 47:492–496

Osorio G, Varela P, Arredondo R, Seeger M, Amaro AM, Jerez CA (1993) Changes in global gene expression of *Thiobacillus ferrooxidans* when grown on elemental sulfur. In: Torma AE, Apel ML, Brierley CL (eds) Biohydrometallurgical technologies (vol 2). The Minerals, Metals and Material Society, Warrendale, PA, pp 565–575

Osorio H, Martinez V, Nieto PA, Holmes DS, Quatrini R (2008a) Microbial iron management mechanisms in extremely acidic environments: comparative genomics evidence for diversity and versatility. BMC Microbiol 8:203

Osorio H, Martinez V, Veloso FA, Pedroso I, Valdes J, Jedlicki E, Holmes DS, Quatrini R (2008b) Iron homeostasis strategies in acidophilic iron oxidizers: studies in *Acidithiobacillus* and *Leptospirillum*. Hydrometallurgy 94:175–179

Osorio H, Cárdenas JP, Valdés J, Holmes DS (2009) Prediction of Fnr regulated genes and meta-
    bolic pathways potentially involved in anaerobic growth of *Acidithiobacillus ferrooxidans*.
    Adv Mat Res 71–73:195–198
Parro V, Moreno-Paz M (2003) Gene function analysis in environmental isolates: the *nif* regulon
    of the strict iron oxidizing bacterium *Leptospirillum ferrooxidans*. Proc Natl Acad Sci USA
    100:7883–7888
Parro V, Moreno-Paz M (2004) Nitrogen fixation in acidophile iron-oxidizing bacteria: the *nif*
    regulon of *Leptospirillum ferrooxidans*. Res Microbiol 155:703–709
Paulino LC, de Mello MP, Ottoboni LM (2002) Differential gene expression in response to copper
    in *Acidithiobacillus ferrooxidans* analyzed by RNA arbitrarily primed polymerase chain reac-
    tion. Electrophoresis 23:520–527
Quatrini R, Jedlicki E, Holmes DS (2005) Genomic insights into the iron uptake mechanisms of
    the biomining microorganism *Acidithiobacillus ferrooxidans*. J Ind Microbiol Biotechnol
    32:606–614
Quatrini R, Appia-Ayme C, Denis Y et al (2006) Insights into the iron and sulfur energetic
    metabolism of *Acidithiobacillus ferrooxidans* by microarray transcriptome profiling.
    Hydrometallurgy 83:263–272
Quatrini R, Lefimil C, Veloso FA, Pedroso I, Holmes DS, Jedlicki E (2007a) Bioinformatic predic-
    tion and experimental verification of Fur-regulated genes in the extreme acidophile
    Acidithiobacillus ferrooxidans. Nucleic Acids Res 35:2153–2166
Quatrini R, Martinez V, Osorio H, Veloso FA, Pedroso I, Valdés J, Jedlicki E, Holmes DS (2007b)
    Iron homeostasis strategies in acidophilic iron oxidizers: comparative genomic analyses. Adv
    Mat Res 20–21:531–534
Quatrini R, Valdés J, Jedlicki E, Holmes DS (2007c) The use of bioinformatics and genomic biology
    to advance our understanding of bioleaching microorganisms. In: Donati ER, Sand W (eds)
    Microbial processing of metal sulfides. Springer, Dordrecht, The Netherlands, pp 221–239
Quatrini R, Appia-Ayme C, Denis Y, Jedlicki E, Holmes DS, Bonnefoy V (2009) Extending the
    models for iron and sulfur oxidation in the extreme acidophile *Acidithiobacillus ferrooxidans*.
    BMC Genomics 10:394
Ram RJ, Verberkmoes NC, Thelen MP, Tyson GW, Baker BJ, Blake RC 2nd, Shah M, Hettich RL,
    Banfield JF (2005) Community proteomics of a natural microbial biofilm. Science
    308:1915–1920
Ramirez P, Toledo H, Guiliani N, Jerez CA (2002) An exported rhodanese-like protein is induced
    during growth of *Acidithiobacillus ferrooxidans* in metal sulfides and different sulfur com-
    pounds. Appl Environ Microbiol 68:1837–1845
Ramirez P, Guiliani N, Valenzuela L, Beard S, Jerez CA (2004) Differential protein expression
    during growth of *Acidithiobacillus ferrooxidans* on ferrous iron, sulfur compounds, or metal
    sulfides. Appl Environ Microbiol 70:4491–4498
Rawlings DE (2002) Heavy metal mining using microbes. Annu Rev Microbiol 56:65–91
Rawlings DE (2005) Characteristics and adaptability of iron- and sulfur-oxidizing microorganisms
    used for the recovery of metals from minerals and their concentrates. Microb Cell Fact 4:13
Rawlings DE, Johnson DB (2007) The microbiology of biomining: development and optimization
    of mineral-oxidizing microbial consortia. Microbiology 153:315–324
Rivas M, Seeger M, Jedlicki E, Holmes DS (2007) Second acyl homoserine lactone production
    system in the extreme acidophile *Acidithiobacillus ferrooxidans*. Appl Environ Microbiol
    73:3225–3231
Rohwerder T, Sand W (2007) Mechanisms and biochemical fundamentals of bacterial metal sul-
    fide oxidation. In: Donati ER, Sand W (eds) Microbial processing of metal sulfides. Springer,
    Dordrecht, The Netherlands, pp 35–58
Ruiz LM, Sand W, Jerez CA, Guiliani N (2007) c-di-GMP pathway in *Acidithiobacillus ferrooxi-
    dans*: analysis of putative diguanylate cyclases (DGCs) and phosphodiesterases (PDEs)
    bifunctional proteins. Adv Mat Res 20–21:551–555
Rzhepishevska OI, Valdés J, Marcinkeviciene L, Gallardo CA, Meskys R, Bonnefoy V, Holmes
    DS, Dopson M (2007) Regulation of a novel *Acidithiobacillus caldus* gene cluster involved

in metabolism of reduced inorganic sulfur compounds. Appl Environ Microbiol 73:7367–7372

Salzano AM, Febbraio F, Farias T, Cetrangolo GP, Nucci R, Scaloni A, Manco G (2007) Redox stress proteins are involved in adaptation response of the hyperthermoacidophilic archaeon *Sulfolobus solfataricus* to nickel challenge. Microb Cell Fact 6:25

Schippers A (2007) Microorganisms involved in bioleaching and nucleic acid-based molecular methods for their identification and quantification. In: Donati ER, Sand W (eds) Microbial processing of metal sulfides. Springer, Dordrecht, The Netherlands, pp 3–33

Seeger M, Jerez CA (1993a) Phosphate-starvation induced changes in *Thiobacillus ferrooxidans*. FEMS Microbiol Lett 108:35–41

Seeger M, Jerez CA (1993b) Response of *Thiobacillus ferrooxidans* to phosphate limitation. FEMS Microbiol Rev 11:37–42

Selkov E, Overbeek R, Kogan Y, Chu L, Vonstein V, Holmes D, Silver S, Haselkorn R, Fonstein M (2000) Functional analysis of gapped microbial genomes: amino acid metabolism of *Thiobacillus ferrooxidans*. Proc Natl Acad Sci USA 97:3509–3514

She Q, Singh RK, Confalonieri F et al (2001) The complete genome of the crenarchaeon *Sulfolobus solfataricus* P2. Proc Natl Acad Sci USA 98:7835–7840

Shmaryahu A, Holmes DS (2007) Discovery of small regulatory RNAs in the extremophile *Acidithiobacillus* genus suggests novel genetic regulation. Adv Mat Res 20–21:535–538

Shmaryahu A, Lefimil C, Jedlicki E, Holmes DS (2009) Small regulatory RNA genes in *Acidithiobacillus ferrooxidans*: case studies of 6S RNA and Frr. Adv Mat Res 71–73:223–226

Simmons SL, Dibartolo G, Denef VJ, Goltsman DS, Thelen MP, Banfield JF (2008) Population genomic analysis of strain variation in *Leptospirillum* group II bacteria involved in acid mine drainage formation. PLoS Biol 6:e177

Singer SW, Chan CS, Zemla A, VerBerkmoes NC, Hwang M, Hettich RL, Banfield JF, Thelen MP (2008) Characterization of cytochrome 579, an unusual cytochrome isolated from an iron-oxidizing microbial community. Appl Environ Microbiol 74:4454–4462

Snijders AP, Walther J, Peter S, Kinnman I, de Vos MG, van de Werken HJ, Brouns SJ, van der Oost J, Wright PC (2006) Reconstruction of central carbon metabolism in *Sulfolobus solfataricus* using a two-dimensional gel electrophoresis map, stable isotope labelling and DNA microarray analysis. Proteomics 6:1518–1529

Tyson GW, Chapman J, Hugenholtz P et al (2004) Community structure and metabolism through reconstruction of microbial genomes from the environment. Nature 428:37–43

Tyson GW, Lo I, Baker BJ, Allen EE, Hugenholtz P, Banfield JF (2005) Genome-directed isolation of the key nitrogen fixer *Leptospirillum ferrodiazotrophum* sp. nov. from an acidophilic microbial community. Appl Environ Microbiol 71:6319–6324

Urich T, Bandeiras TM, Leal SS et al (2004) The sulphur oxygenase reductase from *Acidianus ambivalens* is a multimeric protein containing a low-potential mononuclear non-haem iron centre. Biochem J 381:137–146

Valdés JH, Holmes DS (2009) Genomic lessons from biomining organisms: case study of the *Acidithiobacillus* genus. Adv Mat Res 71–73:215–218

Valdés J, Veloso F, Jedlicki E, Holmes D (2003) Metabolic reconstruction of sulfur assimilation in the extremophile *Acidithiobacillus ferrooxidans* based on genome analysis. BMC Genomics 4:51

Valdés J, Pedroso I, Quatrini R, Dodson RJ, Tettelin H, Blake R 2nd, Eisen JA, Holmes DS (2008a) *Acidithiobacillus ferrooxidans* metabolism: from genome sequence to industrial applications. BMC Genomics 9:597

Valdés J, Pedroso I, Quatrini R, Holmes DS (2008b) Comparative genome analysis of *Acidithiobacillus ferrooxidans*, *A. thiooxidans* and *A. caldus*: insights into their metabolism and ecophysiology. Hydrometallurgy 94:180–184

Valdés J, Quatrini R, Hallberg K, Mangold S, Dopson M, Valenzuela PD, Holmes DS (2009) Draft genome sequence of the extremely acidophilic bacterium *Acidithiobacillus caldus* ATCC 51756 reveals metabolic versatility in the *Acidithiobacillus* genus. J Bacteriol 191:614–622

Valenzuela L, Beard S, Guiliani N, Jerez CA (2005) Differencial expression proteomics of *Acidithiobacillus ferrooxidans* grown in different oxidizable substrates: study of the sulfate/

thiosulfate/molybdate binding proteins. In: Harrison STL, Rawlings DE, Petersen J (eds) 16th International biohydrometallurgy symposium. Compress Cape Town, South Africa, Cape Town, South Africa, pp 773–780

Valenzuela L, Chi A, Beard S, Orell A, Guiliani N, Shabanowitz J, Hunt DF, Jerez CA (2006) Genomics, metagenomics and proteomics in biomining microorganisms. Biotechnol Adv 24:197–211

Valenzuela S, Banderas A, Jerez CA, Guiliani N (2007) Cell-cell communication in bacteria. In: Donati ER, Sand W (eds) Microbial processing of metal sulfides. Springer, Dordrecht, The Netherlands, pp 253–264

Varela P, Jerez CA (1992) Identification and characterization of GroEL and DnaK homologues in *Thiobacillus ferrooxidans*. FEMS Microbiol Lett 77:149–153

Vera M, Guiliani N, Jerez CA (2003) Proteomic and genomic analysis of the phosphate starvation response of *Acidithiobacillus ferrooxidans*. Hydrometallurgy 71:125–132

Vera M, Rohwerder T, Bellenberg S, Sand W, Denis Y, Bonnefoy V (2009) Characterization of biofilm formation by the bioleaching acidophilic bacterium *Acidithiobacillus ferrooxidans* by a microarray transcriptome analysis. Adv Mat Res 71–73:175–178

Wilmes P, Bond PL (2004) The application of two-dimensional polyacrylamide gel electrophoresis and downstream analyses to a mixed community of prokaryotic microorganisms. Environ Microbiol 6:911–920

Wilmes P, Bond PL (2006) Metaproteomics: studying functional gene expression in microbial ecosystems. Trends Microbiol 14:92–97

Yarzábal A, Brasseur G, Bonnefoy V (2002a) Cytochromes *c* of *Acidithiobacillus ferrooxidans*. FEMS Microbiol Lett 209:189–195

Yarzábal A, Brasseur G, Ratouchniak J, Lund K, Lemesle-Meunier D, DeMoss JA, Bonnefoy V (2002b) The high-molecular-weight cytochrome *c* Cyc2 of *Acidithiobacillus ferrooxidans* is an outer membrane protein. J Bacteriol 184:313–317

Yarzábal A, Appia-Ayme C, Ratouchniak J, Bonnefoy V (2004) Regulation of the expression of the *Acidithiobacillus ferrooxidans rus* operon encoding two cytochromes *c*, a cytochrome oxidase and rusticyanin. Microbiology 150:2113–2123

Chapter 9
The Geomicrobiology of Catastrophe: A Comparison of Microbial Colonization in Post-volcanic and Impact Environments

Charles Seaton Cockell

During microbial life's tenure on the Earth it has been subject to catastrophic disturbances both at the local and global scale. The number of mechanisms for these disturbances is very large and they include: storms, fires, earthquakes, ocean turnover and disease. However, two mechanisms of change have had a particularly profound influence exerted through geological changes wrought from within and outside the Earth – volcanism and asteroid and comet impact events, respectively. Both of these mechanisms of geological change have been linked to past mass extinctions (Alvarez et al. 1980; Wignall 2001). Although there is often a focus on the negative consequences of these changes and unravelling their effects on the global scale is necessary to understand their influence on the course of biological evolution, an equally pertinent line of enquiry is to understand the opportunities created in post-volcanic and impact environments and thus the way in which devastation caused by these events might provide new possibilities for life's persistence on the Earth through time.

Volcanism is a continuous process and at any given time there are active volcanoes in some location on the Earth (Schminke 2004). By contrast, asteroid and comet impact events, at least on the present-day Earth, are sporadic. An impact event that caused the 1 km-diameter Barringer crater in Arizona, USA, for example, is thought to occur about once every 1,000 years (Toon et al. 1997). Furthermore, although the global scale effects of volcanic activity and impact events attract justified attention, it is worthwhile to remember that most volcanic activity and impact events alter environmental conditions on a local scale (Toon et al. 1997; Kring 1997, 2003).

Understanding the way in which life can take advantage of post-volcanic and impact environments has relevance for assessing the possibility of life on other planets. Many of the perturbations that occur to ecosystems on the Earth, such as fire and storm damage, cannot convincingly be described as universal phenomena since they depend on planetary conditions. Fire and storm damage, for instance,

C.S. Cockell (✉)
CEPSAR, Open University, Milton Keynes, MK7 6AA, UK
e-mail: c.s.cockell@open.ac.uk

L.L. Barton et al. (eds.), *Geomicrobiology: Molecular and Environmental Perspective*,
DOI 10.1007/978-90-481-9204-5_9, © Springer Science+Business Media B.V. 2010

depend upon combustible matter and sufficient oxygen in the atmosphere to support
fire in the former case (Agee 1993; del Pino et al. 2007) and atmospheric conditions
suitable for sufficiently sized storms to alter or threaten surface ecosystems in the
latter (Tester et al. 2003). Similar planet-specific arguments can be advanced for
disease and ocean turnover, for instance. Insofar as no solar system-forming process is known that does not leave behind debris, impact events can be considered to
be a universal phenomenon. Similarly on any planet that has not completely cooled
down, volcanism would be expected to occur. As plate tectonics, which is one
mechanism for generating volcanic activity, might be required to create conditions
suitable for life (van Thienen et al. 2007), then it may be the case that volcanism is
inextricably linked to the phenomenon of life. Thus, understanding the geomicrobiology of volcanic and impact environments has particular astrobiological
significance.

This chapter will focus on the geomicrobiology of rocky environments created
in volcanic and impact environments. The chapter will focus on phototrophs. The
arrival of phototrophs in disturbed environments has for a long time been recognised to be one of the first events in the re-establishment of the carbon cycle and
subsequently higher trophic levels (Carson and Brown 1978). Hydrothermal systems in impact craters and volcanic environments, created by thermal perturbation,
also provide new opportunities for microorganisms (Atkinson et al. 2000; Osinski
et al. 2001; 2005a; Donachie et al. 2002; Koeberl and Reimold 2004; Hode et al.
2008). The ecosystems sustained in these systems can be remarkably diverse, such
as those encountered in Yellowstone National Park (Barns et al. 1994). However,
this review will focus on mesophilic biota associated with terrestrial rocky substrates generated in devastated volcanic and impact environments, which are present long after hydrothermal systems have cooled.

The Geological Context

Igneous petrology is a vast area of research, but from a geomicrobiological perspective, several points can be made about volcanism that are essential to the ensuing
discussion. Rocks can be broadly split into different groups depending upon their
silica and alkali content. This convention, established by the International Union of
Geological Sciences (IUGS) is shown in Fig. 9.1 (Le Bas et al. 1992). Igneous
rocks are split into 15 fields. These categories are established, among other factors,
by the melting temperature of the magma. Rocks of high silica content are formed
at lower temperatures (typically about 700°C) than basaltic and ultramafic rocks,
which are formed at higher melting temperatures (between about 1,000–1,500°C).
High silica rocks (also referred to as 'acidic' rocks) generally have a higher content
of orthoclase (feldspar) and quartz and a lower content of pyroxenes, olivines and
plagioclase compared to mafic and ultramafic rocks (referred to as 'basic'). Not all
igneous rocks can be classified on the TAS (Total Alkali-Silica) diagram and there
are variants on this diagram, for example the QAPF (Quartz, Feldspar, Plagioclase,

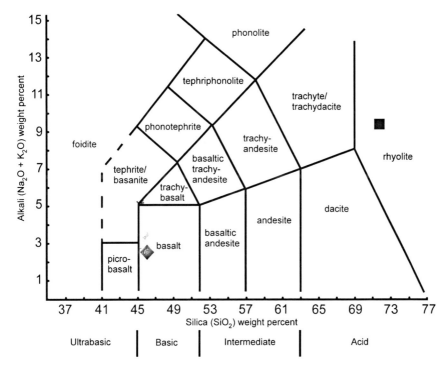

Fig. 9.1 Total Alkali-Silica (TAS) diagram of volcanic rocks. The composition of the basaltic glass described in this chapter is shown with a *diamond*, the composition of the obsidian with a *square*

Feldspathoid[1]) double triangle, but the TAS diagram is useful to categorise most igneous rocks. Although many of these rock types could potentially be studied, one approach to understanding the geomicrobiology of volcanic terrains is to examine the end member mineralogies (high and relatively low silica rocks).

Another factor that is not expressed by the TAS diagram is the glassiness of the material. Volcanic lava that comes into contact with ice or water is rapidly quenched before crystals can form, resulting in a generally homogeneous glass, in contrast to other rocks that are formed subaerially and cool slowly, forming crystalline materials. Biologically, this may be an important difference from two perspectives. Firstly, glasses will present to a biota a material with a more homogenous and mixed composition, whereas in crystalline materials bioessential elements will be localised to particular crystals, meaning that less of the total number of bioessential elements required can be found in one location at the microscale. Secondly, basaltic glass, in contrast to crystalline basalt, has the distinctive feature of weathering to palagonite, a clay-like substance (Thorseth et al. 1991; Stroncik and Schminke 2002)

---

[1] The **feldspathoids** are a group of Silicate minerals which resemble feldspars but have a different structure and much lower silica content.

which tends to form within vesicles and fractures within the material and may have implications for the enclosed biota, as will be discussed later.

In contrast to volcanism, impact events are an exogenous perturbation (French 2004). The first stage of a meteorite impact is a contact and compression stage during which the impactor makes contact with the ground or water, followed by an excavation phase during which a bowl-shaped "transient" crater cavity is formed. For diameters greater than two to four kilometres on Earth subsequent modification can occur, resulting in a central peak and/or peak ring, depending on the magnitude of the event (Melosh 1989). The kinetic energy of such an event is large. The mean impact velocity with the Earth is about 21 km/s; Stuart & Binzel 2004). The energy will be released in large part as heat and the longevity of this thermal excursion will depend, *inter alia*, upon the target lithology, the availability of water and local climate (Naumov 2005; Osinski et al. 2005a; Versh et al. 2005). The formation of a crater cavity during the excavation and modification stages will influence the hydrological cycle, particularly through the formation of long-lived water bodies. Many impact craters known on Earth today with surface expressions host some type of intra-crater water body. Lakes are eventually drained by breach of the crater rim or infilling of the crater (they are, of course, not relevant for impacts in the marine environment), making them, in most cases, more short-lived than changes in the target geology. It is during the contact and compression phase that the target material experiences temperatures that can exceed a 1,000°C and pressures on the order of tens of gigapascals (Melosh 1989). Bulk properties of rocks are drastically changed depending upon the target lithology.

From the point of view of geomicrobiology, the important distinction between volcanism and impacts is that volcanism always generates a well-defined mineral suite that is linked to repeatable mixing and cooling processes in magma chambers, leading to the universally applicable TAS diagram. In contrast, impacts generate diverse mineral suites and melts that will be linked to quite specific lithologies at the site of impact, the way in which they are mixed (particular if glasses are formed) and the specific conditions of temperature, pressure, presence of water, etc., in any given location within an impact structure itself. Impact events can occur into any type of target lithology: sedimentary, metamorphic or igneous. Despite these statements, specific conditions of heat and pressure do lead broadly to well-defined types of shocked minerals in given target materials that can be classified into particular groups, just one example being the shocked gneisses of the Haughton impact structure, Canada, which have been classified (Metzler et al. 1988).

Phototrophs and Endolithic Habitats in the Post-volcanic Environment

Phototrophs are already known to be some of the first colonists of post-volcanic environments, colonizing the surface of new volcanic substrates and soils and initiating rock weathering (Adamo and Violante 1991; Adamo et al. 1993). However, volcanic rocks are generally opaque and considered poor substrates for the establishment of phototroph populations within the substrate itself. Phototrophic colonization of hydrothermal deposits laid down in geothermal regions (Gross et al. 1998; Gaylarde et al. 2006; Walker et al. 2005) has previously been reported and in these cases the geothermal substrates formed around hot springs, which are often iron-poor and silica rich, probably provide favourable light penetration for phototrophs.

To understand better the factors that control colonization of volcanic rocks by microorganisms, and particularly phototrophs, we examined two end member mineralogies: a silica-rich glass (obsidian) and basaltic glass. Figure 9.1 shows the TAS (Total Alkaline-Silica) graph and the mineralogical position of the samples examined. Silica-rich volcanic materials weather more slowly than basaltic materials (comparatively low silica content) (Wolff-Boenisch et al. 2004, 2006) and have very different cation concentrations. Basaltic glass weathers to the clay-like material, palagonite. Although obsidian can also weather to a hydrous material called perlite, this is not the case in many exposed obsidians and recent outcrops in volcanic environments. We hypothesized that the differences between these two end member minerals would allow us to understand the factors that influence endolithic colonization in the post-volcanic environment. This review focuses on the glass because we have a greater amount of data with respect to the phototrophs and the crystalline materials have less contrasting colonization patterns and so are less useful for making comparisons and drawing lessons with respect to plausible mechanisms of colonization.

Samples of obsidian were collected from outcrops in Iceland (64°2.01'N, 19°7.75'W; Fig. 9.2), formed during the ~A.D. 150–300 Dómadalshraun lava flow-forming eruptions. Samples of weathered basalt glass (hyaloclastite) were collected from Valafell, near Hekla volcano, at location 64°4.83'N, 19°32.53'W in the south of Iceland (Fig. 9.2). The material is a mixture of basaltic glass and its weathering product, palagonite (Cockell et al. 2009a). The material is of Pleistocene age (<0.8 Ma old) and forms exposed outcrops throughout the study area (Fig. 9.2) (Jacobsson and Gudmundsson 2008). Samples of between 1 and 10 cm diameter were collected. Rocks were collected into sterile plastic bags without handling. Electron microprobe analyses of both sample types are shown in Table 9.1.

Culture-Independent Analysis

Culture independent methods were used to investigate the microbial inhabitants of the material. 16S rDNA clone libraries were constructed from DNA extracted from both types of material (Herrera and Cockell 2007; Herrera et al. 2008; Cockell et al. 2009a, b). The DNA was extracted from pooled samples and included the surface of rocks and material to a depth of ~1 cm into each rock type.

The 16S rDNA libraries (Fig. 9.3) were used to determine whether sequences corresponding to phototrophs could be detected, but also to understand the diversity of the total bacterial community. In the obsidian, 47 bacterial sequences were

a

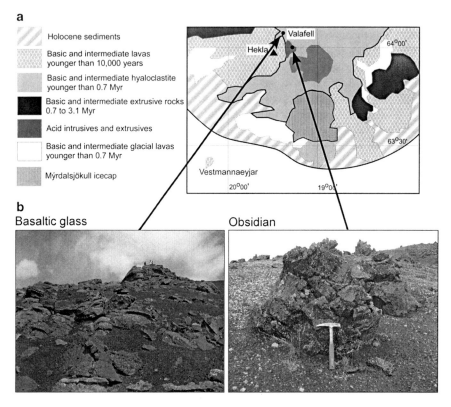

Holocene sediments

Basic and intermediate lavas
younger than 10,000 years

Basic and intermediate hyaloclastite
younger than 0.7 Myr

Basic and intermediate extrusive rocks
0.7 to 3.1 Myr

Acid intrusives and extrusives

Basic and intermediate glacial lavas
younger than 0.7 Myr

Mýrdalsjökull icecap

Valafell
Hekla
64°00'
63°30'
Vestmannaeyjar
20°00'
19°00'

b

Basaltic glass   Obsidian

Fig. 9.2 (**a**) Location of the two samples sites where basaltic glass and obsidian was collected in southern Iceland in relation to major geological outcrops. (**b**) Images of rock outcrops from which samples ere obtained

affiliated to six phyla. The most dominant phylum was the Actinobacteria (19%). The second most abundant, accounting for 14% of the total sequences, was the Acidobacteria. Ten percent of the total sequences were related to the Verrucomicrobia. The Proteobacteria sequences, representing 9% of the total sequences, fell into the α-subdivision. Six sequences were related to phototrophs.

Many of these sequences were related to organisms or sequences previously obtained in rock environments. For example, two Actinobacteria showed closest similarity with an uncultured actinobacterium 16S rDNA clone isolated from an endolithic community in the Rocky Mountains (Walker et al. 2005). Four other 16S rDNA sequences also showed closest similarity to an uncultured alpha proteobacterium 16S rDNA clone isolated from cryptoendolithic communities present in the Dry Valleys of Antarctica (de la Torre et al. 2003), another alpha proteobacterium sequence was found to be similar to an uncultured bacterium identified from a deglaciated soil (Nemergut et al. 2007). One sequence was most similar to *Geodermatophilus* sp. isolated from rocks, monument surfaces and dried soils (Eppard et al. 1996).

Table 9.1 Data showing mean composition of basaltic glass and obsidian from Iceland described in this chapter (data obtained with electron microprobe; see Herrera et al. 2008; Cockell et al. 2009a), and mean composition of unshocked and impact-shocked gneiss from the Haughton impact structure, Canada (Data from Fike et al. 2003)

	Na_2O	K_2O	MgO	CaO	MnO	FeO	Al_2O_3	SiO_2	TiO_2	Total
Volcanic rocks										
Basaltic glass										
Mean (n = 10)	2.57	0.39	7.09	10.71	0.20	13.20	15.31	46.10	2.63	98.23
SD	0.03	0.01	0.13	00.4	0.01	0.06	0.04	0.22	0.03	0.21
Obsidian										
Mean (n = 10)	4.88	4.71	0.17	0.76	0.07	2.44	14.77	72.66	0.24	100.73
SD	0.06	0.04	0.01	0.03	0.02	0.06	0.14	0.45	0.02	0.62
Impact rocks										
Unshocked gneiss										
Mean (n = 6)	2.37	6.05	1.56	2.07	0.04	3.56	14.40	67.97	0.43	98.61
SD	0.60	2.77	0.72	0.44	0.01	1.74	0.99	2.78	0.27	2.02
Shocked gneiss										
Mean (n = 10)	1.44	4.58	0.24	2.18	0	0.70	10.37	77.49	0.13	97.15
SD	0.64	1.38	0.09	0.86	0	0.23	1.40	4.18	0.04	0.81

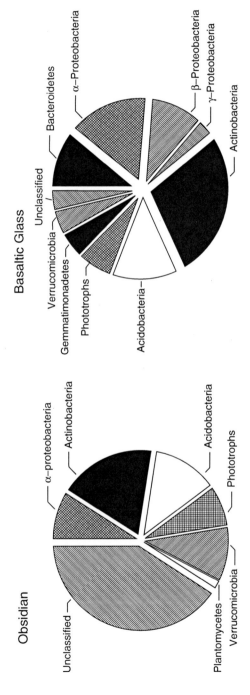

Fig. 9.3 Pie charts showing composition at phylum level of 16S rDNA libraries constructed with genomic DNA from basaltic glass and obsidian

The other 33 sequences were identified as "unclassified" by RDP II database analysis. These sequences showed similarity with only a few sequences from unidentified bacterial clones or *Chloroflexi* related clones and they showed less than 85% similarity with known 16S rDNA sequences from isolated bacteria in GenBank.

Molecular analyses of the obsidian revealed the presence of phototrophs. 16S rDNA cyanobacteria-related and plastid sequences (Table 9.2) were evident. Three sequences (Genbank accession numbers AM773380, AM773352, AM773416) showed a closest similarity (99%) with the cyanobacterium *Chamaesiphon* sp. 16S rDNA (AY170472) (Turner 1997) and 98% similarity with an uncultured cyanobacterium 16S rDNA clone (DQ514063) isolated from a deglaciated soil (Nemergut et al. 2007). Three other sequences (AM773393, AM773369 and AM773349) showed closest similarity (99%) with an uncultured chlorophyte 16S rDNA clone (EF522360) isolated from an endolithic community inhabiting the Rocky Mountains (Walker et al. 2005). Their closest similarity (90%) to a cultured 5organism was *Koliella* sp. (AF278747) (Katana et al. 2001).

In the basaltic glass one hundred and ten clones were examined and the phylotype comparison to the obsidian is shown in Figs. 9.3 and 9.4. The community within the material was dominated by phylotypes belonging to Actinobacteria (30%), Proteobacteria (26%), which fell into the α, β and γ subdivision, Bacteroidetes (11%). Seven of the sequences (6%) were related to phototrophs.

Thirty-one of the clones had a closest sequence match to organisms previously described in cold soil and/or endolithic environments (12 from Antarctic soils [represented by three Bacteroidetes, one Acidobacteria, five Actinobacteria, one α- and one β-proteobacteria, and one unclassified], seven from Antarctic endoliths

Table 9.2 Phototrophs obtained in basaltic glass and obsidian by culture-independent methods (16S rDNA analysis; which picks up cyanobacteria and plastids)

Accession number	Nearest cultured sequences	Percent similarity	Reference
Obsidian			
AM773380		99%	(Turner 1997)
AM773352	*Chamaesiphon* sp. (shrub-like colonies)		
AM773416			
AM773393		90%	(Katana et al. 2001)
AM773369	*Koliella* sp. (filamentous)		
AM773349			
Basaltic glass			
EU621975		99%	(Turmel et al. 2002)
FJ360644	*Klebsormidium* sp. (filamentous)		
EU621980	*Stichococcus* sp. (rod)	97%	(Katana et al. 2001)
EU621967	*Closteriopsis* sp. (filamentous)	86%	(Ustinova et al. 2001)
FJ360671		95–95%	(Jezberova 2006)
FJ360653	*Acaryochloris* sp. (coccoid)		
FJ360664			

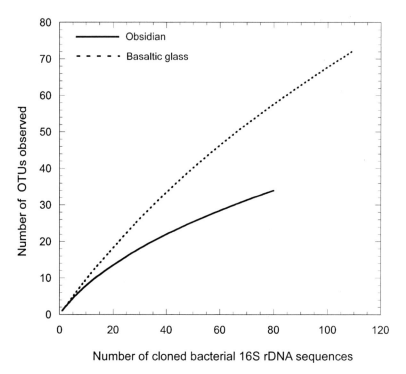

Fig. 9.4 Rarefaction curves (97% cut off) of basaltic glass and obsidian clone libraries

[all α-proteobacteria], five from glaciers and deglaciated soils [three Bacteroidetes, one α- and one β-proteobacteria], three from the Atacama desert [all Actinobacteria] and four from other endolithic environments [two actinobacteria and one α- and one β-proteobacteria] Cockell et al. 2009a).

Molecular analyses of the basalt glass revealed the presence of phototrophs. 16S rDNA cyanobacteria-related and plastid sequences (Table 9.2) were identified. Two sequences (EU621975, FJ360644) showed a closest similarity (99%) to *Klebsormidium* sp. (AF393600) (Turmel et al. 2002). One sequence (EU621980) had a closest similarity (97%) to *Stichococcus* sp. (AF278751) (Katana et al. 2001). One sequence (EU621967) had a closest similarity (96%) to an uncultured eubacterium (AJ292689) found in soil (Nogales et al. 2001). Its closest similarity (86%) to a cultured organism was *Closteriopsis* sp. (Y17632) (Ustinova et al. 2001). Three sequences (FJ360671, FJ360653, FJ360664) showed a closest similarity (96–97%) to an uncultured cyanobacterium found under quartz substrates in the Atacama Desert, Chile (FJ890990). Their closest similarity (95–96%) to a cultured organism was *Acaryochloris* sp. (AM710387) (Jezberova 2006).

Among the non-phototrophic clones, Actinobacteria and Proteobacteria were found in both rock types and have previously been shown to be abundant in endolithic habitats (de la Torre et al. 2003; Walker et al. 2005). They are important groups in soils. Bacteroidetes (formally Cytophaga-Flexibacter-Bacteroidetes) were abundant in the basalt glass and have previously been associated with soil crusts and soils (Shivaji et al. 2004; Nagy et al. 2005; Gundlapally and Garcia-Pichel 2006), although they are also found in freshwater (Wu et al. 2007), marine (O'Sullivan et al. 2004; Murray and Grzymski 2007) and microbial mat (Abed et al. 2007) environments and they may be associated with the degradation of organic material.

Some of the DNA sequences that we observed may belong to inactive organisms that have become incorporated in the vesicular rocky material. For example, hyperthermophiles can be cultured from within the rock (unpublished data) whose provenance may be hot springs in Iceland. However, both Bacteroidetes and Actinobacteria isolates can be cultured from basaltic glass using crushed basalt glass (and its indigenous carbon sources) as the growth substrate (Cockell et al. 2009b), showing that there exists a core population of these organisms within the rocks that represent the metabolically active components. These organisms exhibit resistant to desiccation, temperature fluctuations and some to heavy metals, attributes that would be expected from volcanic rock-dwelling organisms. The addition of a small amount of carbon in the form of yeast extract yielded many other actinobacteria and *Bacillus* spp. (Cockell et al. 2009b).

All of the sequences obtained were associated with heterotrophic clades. None clearly matched previously recognised chemolithotrophic clades. These results show that the majority of the microbial population that is active within both rock types must depend on carbon input, whether in precipitation, in the form of other dead and decaying organisms, or carbon produced by phototrophs. These results show the important role for phototrophs in initiating a carbon cycle within volcanic environments.

Culture-Dependent Observations on the Phototroph Population in Volcanic Glasses

Both rock types exhibit sequences corresponding to phototrophs using culture-independent methods (16S rDNA libraries). However, in the obsidian, the phototrophs, which comprise both algal and cyanobacterial components, can be directly observed to form layers of growth within the glass, which can be visualised by SEM (Fig. 9.5). They can also be observed using FISH (Fluorescent In-situ Hybridisation) in association with presumptive heterotrophic members of the population (Herrera et al. 2008). The endolithic phototrophs are found under glassy layers on the obsidian which can be removed from the rock surface with a sharp blade. The endolithic growth of the phototroph population bears some similarities to phototrophic endolithic communities in sedimentary rocks such as sandstones and limestones (Friedmann 1980, 1982; Saiz-Jimenez et al. 1990; Büdel and Wessels 1991; Bell 1993; Weber et al. 1996; Büdel et al. 2004; Omelon et al. 2006), however the obsidian phototrophic endoliths are not widespread and do not form

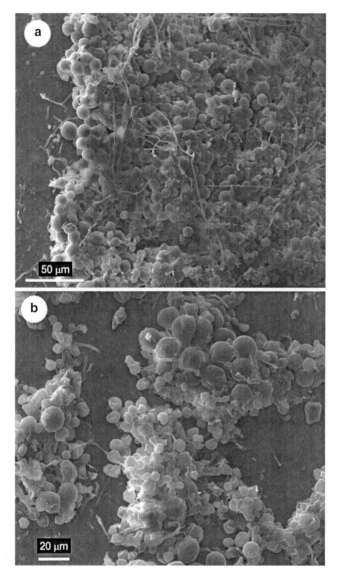

Fig. 9.5 Secondary Scanning Electron Microscopy (SEM) (**a** and **b**) of phototrophs colonizing obsidian and forming well-defined layered colonies within glass

coherent layers throughout the rock substrate. Instead they form localised layers within the rock where vesicles and exfoliation of the glass allow them to penetrate the substrate and grow under sheet-like layers of glass.

These observations are in contrast to the basaltic glass where phototrophs are observed in the 16S rDNA library, but they are not observed to form obvious layers on or within the material. The presence of viable phototrophs can be confirmed and tested for both rock types by simple culturing experiments. Pieces (1 cm^3) of the obsidian and basaltic glass were incubated at room temperature (21°C) in 5 mL of BG-11 medium to culture phototrophs. After incubation for 1 month, the phototrophic microorganisms growing in each sample were characterized by light microscopy using a Leica DMRP microscope. After 1 month incubation microscopic observations revealed the presence of filamentous and coccoid phototrophs, whose morphology resembled *Anabaena* and *Chlorella* (Herrera et al. 2009).

A question is how long these organisms take to colonize the rocks. We investigated this by laying out four microscope slides (2.54 × 7.62 cm) in each of the obsidian and basalt glass locations in July 2007 and collecting them in July 2008. The slides were placed on exposed rock surfaces. After collection, the slides were transferred to 50-mL tubes with 30 mL BG11 in the field and left to incubate under a natural light/dark cycle at 21°C. After 1 month all of the tubes showed visible growth of phototrophs. All cultured organisms from the slides were coccoid. They were exclusively found in two morphotypes (Fig. 9.6). One morphotype was a coccoid alga (Fig. 9.6a) and corresponded in size to large cells observed in the endolithic habitat by SEM (Fig. 9.5). The second morphotype was smaller, lighter green organisms (Fig. 9.6b) which resemble coccoid cyanobacteria and also corresponded in size to the small cells observed in the endolithic habitat by SEM. Intriguingly, we did not observe filamentous forms, which dominate the phototroph sequences in the clone library (Table 9.2) and occur in cultures from the rocks. These data suggest that coccoid organisms are the dominant early aerial colonists of the volcanic rocks, which might reflect more efficient aerial transport than filamentous organisms.

These data show that: (1) viable populations of phototrophs exist within the obsidian and on the basaltic glass, but that in the obsidian they are capable of forming well-defined endolithic colonies inside the glass itself, and (2) phototrophic pioneers are being constantly deposited on volcanic substrates in Iceland and new substrates will be rapidly colonized.

Effects of Position in TAS Diagram on Endolithic Colonization

These observations can be better understood when the chemical weathering characteristics of these rocks is considered. Although weathering of the obsidian occurs, it is much slower than the weathering of basaltic glass (Wolff-Boenisch et al. 2004, 2006). The higher silica concentration impedes the degradation of the material since Si–O bonds are much more difficult to break than other bonds, and the silica will tend to retard leaching of cations from the rock. Thus, the habitat is more long-lived and the photosynthetic zone in which the organisms live preserved over longer time periods compared to basaltic glasses. In contrast the basaltic glass is comprised of vesiculated glass which weathers to the clay-like material, palagonite, which forms rinds on the glass (Cockell et al. 2009a), which may be less stable as

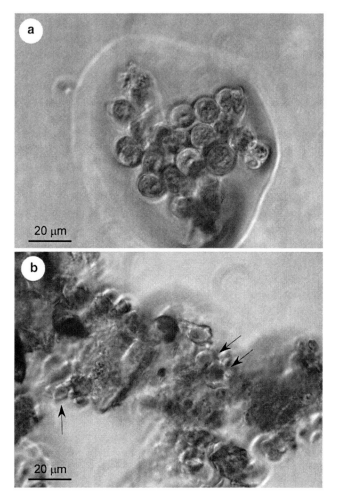

Fig. 9.6 Bright field micrographs of phototrophs cultured from microscope slides left at the sample location for 1 year. These examples are from slides left at the basaltic glass location, but identical morphotypes were observed at the obsidian site. (**a**) Presumptive algae based on size and pigmentation. (**b**) Presumptive cyanobacteria

a substrate for phototrophs inside the rock and, as discussed later, probably reduces light penetration into the material.

These differences may go someway to explaining the different rarefaction data for the bacterial community as a whole. The nutrients in the basaltic glass clay-like palagonite are likely to be more accessible to a biota than the solid obsidian rock matrix, creating a relatively copiotrophic environment compared with obsidian. The palagonite is nearly 20% water (Cockell et al. 2009a). Although the basaltic glass has lower concentrations of sodium and potassium than the obsidian, it has higher concentrations of iron, magnesium and calcium and if the nutrients in palagonite are more accessible than in solid obsidian rock, overall a larger and more diverse community would be predicted to be sustained. This would be consistent with observations of microbial abundance. Cockell et al. (2009a) observed high cell numbers in basalt glass samples with abundances of ~10^7 cells g^{-1}. By contrast, Herrera et al. (2008) found that cells were highly localised to altered regions in the obsidian glass and subsequent attempts to gather cell number data (unpublished) find total cell numbers less than 10^4 cells/gram. Because the cells are so localised within the obsidian, whole rock cell numbers were unreliable, but nevertheless the results show that cell numbers are generally less in the obsidian than the basaltic glass samples.

Phototrophs and Endolithic Habitats in the Post-impact Environment

We now turn to post-impact environments to consider how rocks in these environments are colonised. The most obvious influence of impacts on the local geology is to fracture rocks, increasing their permeability. The advantages to a biota to be gained from the impact-induced fracturing of rock are particularly evident in surface environments where, if conditions are hostile, the fractures inside rocks provide spaces for microorganisms to grow that are protected from the external conditions. In the 39 million year old Haughton impact structure located in Nunavut, Canada, abundant phototrophic colonization of impact-fractured rocks can be observed (Cockell et al. 2005). Although these 'chasmoendolithic' habitats evidence an increase in space for microbial growth, they are not unique to impacts since freeze-thaw, and tectonic activity over the longer-term, for instance, can fracture rocks. Impact events will merely have the effect of increasing the abundance of fractures for microorganisms.

An effect more specific to impacts is the heating and pressurization of rocks over very short time scales, leading to gross physical changes in the bulk characteristics of the rock that are quite distinct from other geological processes. In the Haughton structure this is manifested most clearly in the cryptoendolithic colonization of shocked gneisses, which are colonized by cyanobacterial genera that inhabit the pore spaces within shocked material exposed on the carbonate-rich melt rock hills (Cockell et al. 2002, 2003a, 2005). One of the highest abundances of cryptoendolithic habitats is on Anomaly Hill (Fig. 9.7), a localised area with a high abundance of heavily shocked clasts (Osinski et al. 2005b). In contrast to the impact-induced chasmoendolithic habitats, the shocked gneisses have a highly porous glassy structure (Fig. 9.8) (Metzler et al. 1988) with sufficient permeability to allow the organisms not merely to invade the rock along any fractures connected to the surface, but also to grow through the spaces within the rock, forming coherent bands. The increase in porosity provides space for microorganisms to colonise the rock. The density and porosity of shocked rocks (shocked to >20 GPa) is on the order of 1×10^3 kg m^{-3} and 18.3%, respectively. The low-shocked or unshocked rocks have a more typical density and porosity of >2.5×10^3 kg m^3 and 9.3%, respectively.

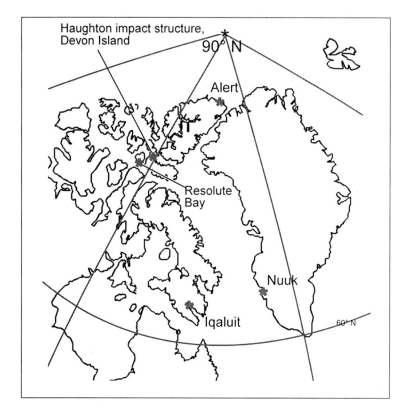

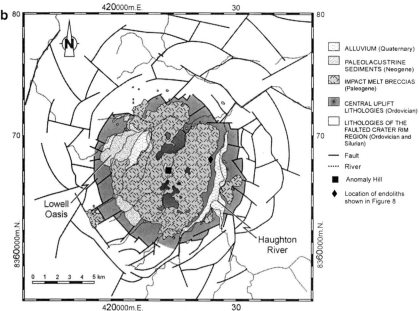

Fig. 9.7 Location of the Haughton impact structure (**a**) in the Canadian High Arctic and geology of the structure (**b**) showing location of Anomaly Hill (marked with a *square*), which has the highest abundance of shocked gneisses colonized by phototrophs. However, colonized gneisses can found throughout the crater on the meltrock hills. The examples shown in Fig. 9.8 were from the location marked with a *star*

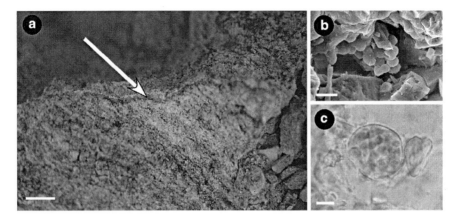

Fig. 9.8 Cryptoendolithic colonization of impact shocked gneiss. (**a**) Band of cryptoendolithic colonization of shocked gneiss (scale bar 1 cm). (**b**) Secondary scanning electron image of cyanobacterial colony within cryptoendolithic zone (scale bar 10 μm). (**c**) Bright field micrograph of presumptive *Chroococcidiopsis* sp. colony in cryptoendolithic zone (scale bar 10 μm)

Colonization by cyanobacteria is of specific interest in the context of impact shock because their ability to colonise the rock will depend upon the availability of light. The effects of element volatilization and the increase in permeability increased the translucence of the shocked rock. Transmission at 680 nm, the chlorophyll *a* absorption maximum, was increased by approximately an order of magnitude in the shocked rocks compared to the unshocked material (Cockell et al. 2002). In the shocked rocks the lower limit of colonization, set by the minimum light level required for photosynthesis, was calculated to be 3.6 mm, although this will clearly vary within rocks (caused by heterogeneous fracturing and porosity) and with shock level. In unshocked rocks the transmission of light is insufficient to allow interior colonization of the rock by cyanobacteria, even if the porosity was sufficient for the cells to grow within it. The cyanobacteria that inhabit the rocks are not unique organisms to the crater, but are found on the surface and underside of other rocks in the Arctic (Cockell and Stokes 2004, 2006). Within the endolithic habitat species are dominated by *Chroococcidiopsis* and *Gloeocapsa* spp. morphotypes.

Impacts also alter the chemical characteristics of target rock. One effect would be predicted to be the impact-volatilisation of elements from target material, and thus their depletion, at the high temperatures and pressures associated with impact, rendering the target area impoverished. Evidence for depletion of bioessential cations can be found in the shocked gneiss compared to the unshocked material in the Haughton impact structure (Table 9.1; data adapted from Fike et al. 2003). The gneiss is comprised of diverse minerals which yield variations in the bulk composition determined by Inductively Coupled Plasma-Atomic Emission Spectroscopy (ICP-AES). The shocked gneiss (samples shock to above 20 GPa) is depleted in all major bioessential cations and phosphorus and resembles a silicate-rich glass.

The chemical evidence from the impact-shocked gneiss must be reconciled with the observations of enhanced phototrophic endolithic colonization of the shocked material in the field. The result can be explained by the following observations: (1) if the unshocked material lacks sufficiently sized or interconnected pore spaces for colonization then clearly the presence of bioessential cations and anions within the material is irrelevant, except for weathering populations of microorganisms on the surface of the material, (2) the shocked material, despite its nutrient depleted conditions, probably receives low concentrations of nutrient input from rainwater, snowmelt and leaching of more nutrient-rich rocks in the melt sheet. Thus, although the shocked material is nutrient poor compared to the unshocked material, the organisms receive enough nutrients to colonize the material. These data show that even oligotrophic impoverished impact environments can ultimately be colonised by microorganisms that rely on allochthonous nutrient sources.

Although the gneissic habitats resemble endolithic habitats found in sedimentary rocks a significant factor is that they represent the formation of an endolithic habitat in crystalline rocks – substrates which are usually unsuitable for endolithic colonization. This observation is important because it shows that the impact event has not merely increased the abundance of a geomicrobiological habitat that is already to be found in similar rocks elsewhere (as is the case for chasmoendolithic habitats), but that it has generated an entirely new type of endolithic habitat.

Some of the energy from impact must be consumed in physical disruption of the target material, as well as that released as heat, sound and light. For low porosity rocks, such as the gneisses just described, this energy will result in rock fracturing, vesicularisation, etc., as discussed above. Although this would be expected to be the case for other types of rocks, in the case of rocks that are initially very porous, such as certain sedimentary rocks, some of the energy will be taken up in pore collapse (Kieffer et al. 1976). Thus, a prediction would be that rocks that are initially very favourable habitats could become less so.

Field observations of sandstones, again from the Haughton structure (Osinski 2007), provide a direct comparison to the gneissic habitats (Cockell and Osinski 2007). A coccoid cyanobacterium (*Chroococcidiopsis*) and the Gram-positive bacterium, *Bacillus subtilis* were used to colonise samples of sandstones collected from the Haughton structure and shocked to different levels. Low shock pressures up to just above ~5 GPa cause pore collapse as would be expected, rendering the rocks less suitable for endolithic colonization. However, higher shock pressures caused vesicularisation of the rocks, generating interconnected spaces that allow for the natural colonization of the rocks. Class 3 sandstones shocked to greater than 10 GPa, but less than 20–25 GPa show natural endolithic colonization by cyanobacteria such as *Gloeocapsa* and *Chroococcidiopsis* (Cockell and Osinski 2007). At shock pressures higher than about 30 GPa, solid glass is formed, which is impenetrable to microorganisms. The effects are further complicated by the fact that many of these rocks are emplaced in the meltsheets and would have been subjected to heating during the cooling of the crater. The long term heating and annealing of the rocks changes the characteristics of the pore spaces and their interconnectedness.

Comparison of Endolithic Colonization in the Post-volcanic and Impact Environments

Despite the obvious differences just elaborated between volcanic and impact environments there are some similarities from the point of view of geomicrobiology. An obvious common theme that links these two environments is that the interior of rocks provide space and protection from environmental extremes for endolithic inhabitants (including phototrophs, but also their associated heterotrophic populations) during and subsequent to initial colonization of devastated post-volcanic and impact environments.

One of these extremes is protection from UV radiation. Measurements of light penetration through the obsidian and basaltic glass (Herrera et al. 2008; Cockell et al. 2009a) showed that UV radiation was hardly measurable with standard spectrophotometric equipment. The high iron content of volcanic materials (Table 9.1) probably accounts for the high UV absorbance of the material as iron is known to be an efficient absorber of short wavelength UV radiation (Pierson et al. 1993). In addition the vesiculated rock substrate scatters the UV radiation.

We evaluated exposure to UV radiation within the impact shocked gneiss in more in-depth experiments with a modification of the DLR (Deutsches Zentrum für Luft- und Raumfahrt)-Biofilm (Cockell et al. 2003b). The DLR-biofilm dosimeter uses the inactivation of *Bacillus subtilis* (HA101) spores to quantify UV exposure (Quintern et al. 1992, 1994). DLR-Biofilms were placed under sealed rock sections. After 24 h, the (uncovered) control dosimeter spores were all killed as they were for longer time periods. After 10 days only the thinnest section of impact-shocked gneiss (0.08 ± 0.02 mm) had measurable spore inactivation. If one assumes that the dosimeters exposed for 24 h are only just fully exposed (a conservative estimate), then the data show that a 1 mm thick rock section is providing at least a two order of magnitude reduction in UV exposure. Over the summer season in the Arctic, organisms under even 1 mm of impact-shocked gneiss would receive a total UV radiation dose not much greater than that received in 1 day of exposure to unattenuated UV radiation.

Taken together, these data show that both volcanic and impact shocked rocks provide efficient screening of UV radiation. In the case of volcanic rocks this is partly contributed to by the presence of iron in the material. However, despite the low concentrations of iron in the shocked gneiss it is effective at UV screening, showing that the vesiculated and porous structures of materials also account for substantial UV screening caused primarily by light scattering.

Absorption and scattering of UV radiation also occurs concomitantly with absorption and scattering of light in the photosynthetically active region (PAR). This will establish the lowest depth at which phototrophs can be sustained. In the shocked gneisses this depth is highly variable depending on fracturing and shock levels, but in the most colonized gneisses it is typically a value of about 3–5 mm. This depth is similar to that previously reported for sedimentary endolithic habitats (Nienow et al. 1988; Matthes et al. 2001). The depth is much less in the volcanic substrates where the dark iron rich material, and in the basaltic glass, the presence of palagonite, effectively attenuates the light. In both of these substrates the depth of minimum light is less than 250 μm (Cockell et al. 2009a; Herrera et al. 2009). These measurements were made with solid volcanic rock and the presence of vesicles and fractures is likely to extend the depth at which phototrophs can persist. However, as previously discussed, coherent layers of endolithic communities of phototrophs have only been observed in the obsidian (Herrera et al. 2009).

Another advantage to be gained by living within rocks is an altered temperature regimen. Observations of temperature changes were made in the previously described obsidian outcrop in Iceland from June 11 to 16, 2008. Temperatures were made at approximately 10 cm depth within a natural crack in the rock, under a thin obsidian rock resting on soil, which was approximately 2 cm thick, and on the surface of the soils near the rock outcrop. The temperatures 10 cm deep within the rock outcrop only exceeded 20°C on one of the measurement days. The mean temperature over the measurement period was 12.1°C. Under 2 cm of rock temperatures exceeded 30°C for the 3 days. Temperatures on the surface of the soils exceeded 40°C for 3 of the 5 days (Herrera et al. 2008). The high temperatures recorded on the soil surface can be accounted for by the dark colored volcanic soil which effectively absorbs solar radiation. This will also be the case for rock surfaces. These data show that the habitat deep within volcanic rock can provide some protection against heating experienced on the soil surface. However, for phototrophs, which must be near the surface of the rock to be active, the data show that dark volcanic rocks and soils, even in the relatively cool environment of Iceland, might achieve temperatures within the range of thermophily and potentially exert stress on mesophilic phototrophs. Nevertheless, temperatures within the rock are generally higher than the corresponding air temperatures during summer (Herrera et al. 2008) and so in some cases, when air temperatures are very cold, such as during spring and autumn, it is likely that microorganisms within the rocks would gain advantage from the warming of the dark volcanic materials.

Temperatures within impact shocked gneiss in the Haughton structure were measured from July 18 to 26, 2002. During the 9 days of measurements the mean air temperature was 4.51°C, the mean rock surface temperature was 5.49°C and the mean temperature in the endolithic habitat (at a depth of ~2 mm in the rock) was 5.93°C. The highest temperature recorded in the endolithic habitat was 21.78°C at 13:17 on 20 July. The corresponding air temperature was 10.57°C and the rock surface temperature was 16.11°C. On 18 July, when the air temperature was below freezing, the endolithic temperature was 1.25°C and it remained above freezing, showing how the endolithic habitat can protect against freeze-thaw.

The impact-shocked rocks are not as dark as volcanic material and so extreme heating was not observed. Thus, the impact shocked rocks share the property of providing thermally improved conditions on cold days for their inhabitants, but are less susceptible to high temperatures caused by solar absorption of dark volcanic material during sunny days.

Yet another advantage to be gained by living inside a rock is access to trapped water, particularly in environments where precipitation may be infrequent. We have not carried out experiments on volcanic rocks, but experiments on the impact-shocked rocks illustrate the general principle. Two pieces of shocked gneiss with different masses (207.2 and 68.9 g) were immersed in a pan of water for 4 h to simulate immersion in snowmelt. The rocks could retain water for days after saturation. After artificial immersion, the larger of the specimens had drawn up 8.4% of its dry mass in water and the smaller specimen 6.1%. Following exposure to field conditions without rain, the larger specimen had retained 3.3% its mass in water after 48 h and 1.4% after 5 days. The smaller specimen (with a smaller surface area, but a larger surface area to volume ratio) had retained 0.9% of its mass in water after 48 h and 0.4% after 5 days.

Post-volcanic and Impact Succession

The data yield insights into the way in which post-catastrophe rocky environments can be colonized and a carbon cycle established by phototrophs. In the case of silica-rich volcanic materials, the long weathering times allow for phototroph populations to become established within the material in analogy to previously described endolithic communities in sedimentary rocks. FISH data from the obsidian (Herrera et al. 2008) show presumptive heterotrophs in association with the phototrophs, showing more directly that phototrophic colonization provides a carbon source for subsequent colonization by heterotrophs. The potentially important role that phototrophs could play in providing carbon is suggested by the dominance of carbon-requiring heterotrophs in the 16S rDNA clone libraries from the volcanic substrates. In the case of rocks with basaltic composition, the nature of the weathering generally precludes the establishment of well-defined endolithic zonation of phototrophs; however phototrophs become established on the rock surface and could provide carbon for heterotrophs within. Our data show that phototrophs can become rapidly established on rock surfaces (within the first year) to subsequently colonise the volcanic substrates.

In analogy to volcanically perturbed environments (Carson and Brown 1978; Fermani et al. 2007; Herrera et al. 2009), phototrophs may be some of the first successful colonists of impact-perturbed environments (Cockell et al. 2002; Cockell and Lee 2002), colonizing impact substrates and providing a source of organic carbon for the heterotrophic populations, consistent with their importance today in the shocked rocks in the Haughton structure. The communities that inhabit the shocked gneiss and sandstone rocks today are obviously not immediate post-impact successional communities and the rocks have been subjected to 39 million years of climatic variation since the impact event (Sherlock et al. 2005). However, the bulk chemistry and physical features of the rocks have not been substantially altered by non-impact processes. Therefore, the colonization of the rocks and the effects of impact on them can be considered to provide a faithful insight into the geomicrobiological processes involved in the effects of impact on microbial recolonization processes.

In summary, vesicular rock structures, generated by degassing in the case of volcanism and pressurisation and heating in the case of impact events provide habitats within rocks in which diverse microbial populations become established. The degree of light penetration, which is determined by vesicularity and iron content amongst other factors, will determine the depth to which phototrophs become established. The rate of weathering of the material and its weathering products will influence the diversity of organisms that can be sustained and extent to which the rock substrate provides a stable surface for well-defined endolithic zones to become established. As shown here, investigations on end-member volcanic mineralogies and shocked rocks provide a framework for understanding the patterns of microbial colonization in the post-catastrophe environment and how phototrophs become established to initiate a carbon cycle for heterotrophic organisms.

Acknowledgements I thank the Leverhulme Trust (project number F/00 269/N) and the Royal Society for support for the work on microorganisms on volcanic environments.

References

Abed RMM, Zein B, Al-Thukair A, de Beer D (2007) Phylogenetic diversity and activity of aerobic heterotrophic bacteria from a hypersaline oil-polluted microbial mat. Syst Appl Microbiol 30:319–330

Adamo P, Violante P (1991) Weathering of volcanic rocks from Mt. Vesuvius associated with the lichen *Stereocaulum vesuvianum*. Pedobiologia 35:209–217

Adamo P, Marchetiello A, Violante P (1993) The weathering of mafic rocks by lichens. Lichenologist 25:285–297

Agee JK (1993) Fire ecology of Pacific Northwest forests. Island Press, Washington, DC

Alvarez LW, Alvarez W, Asaro F, Michel HV (1980) Extraterrestrial cause for the Cretaceous-Tertiary extinction – experimental results and theoretical interpretation. Science 208:1095–1108

Atkinson T, Gairns S, Cowan DA, Danson MJ, Hough DW, Johnson DB, Norris PR, Raven N, Robinson C, Robson R, Sharp RJ (2000) A microbiological survey of Montserrat Island hydrothermal biotopes. Extremophiles 4:305–313

Barns SM, Fundyga RE, Jeffries MW, Pace NR (1994) Remarkable archaeal diversity detected in a Yellowstone National Park hot spring environment. Proc Natl Acad Sci USA 91:1609–1613

Bell RA (1993) Cryptoendolithic algae of hot semi-arid lands and deserts. J Phycol 29:133–139

Büdel B, Wessels DCJ (1991) Rock inhabiting blue-green algae/cyanobacteria from hot arid regions. Algologic Stud 64:385–398

Büdel B, Weber B, Kühl M, Pfanz H, Sültemeyer D, Wessels D (2004) Reshaping of sandstone surfaces by cryptoendolithic cyanobacteria: bioalkalization causes chemical weathering in arid landscapes. Geobiology 2:261–268

Carson JL, Brown RM (1978) Studies of Hawaiian freshwater and soil algae. 2. Algal colonization and succession on a dated volcanic substrate. J Phycol 14:171–178

Cockell CS, Lee P (2002) The biology of impact craters – a review. Biol Rev 77:279–310

Cockell CS, Osinski GR (2007) Impact-induced impoverishment and transformation of a sandstone habitat for lithophytic microorganisms. Meteor Planet Sci 42:1985–1993

Cockell CS, Stokes MD (2004) Widespread colonization by polar hypoliths. Nature 431:414

Cockell CS, Stokes MD (2006) Hypolithic colonization of opaque rocks in the Arctic and Antarctic polar desert. Arctic Ant Alpine Res 38:335–342

Cockell CS, Lee P, Osinski G, Horneck G, Broady P (2002) Impact-induced microbial endolithic habitats. Meteor Planet Sci 37:1287–1298

Cockell CS, Osinski GR, Lee P (2003a) The impact crater as a habitat: effects of impact processing of target materials. Astrobiology 3:181–191

Cockell CS, Rettberg P, Horneck G, Scherer K, Stokes DM (2003b) Measurements of microbial protection from ultraviolet radiation in polar terrestrial microhabitats. Polar Biol 26:62–69

Cockell CS, Lee P, Broady P, Lim DSS, Osinski GR, Parnell J, Koeberl C, Pesonen L, Salminen J (2005) Effects of asteroid and comet impacts on habitats for lithophytic organisms – a synthesis. Meteor Planet Sci 40:1901–1914

Cockell CS, Olsson-Francis K, Herrera A, Meunier A (2009a) Alteration textures in terrestrial volcanic glass and the associated bacterial community. Geobiology 7:50–65

Cockell CS, Olsson-Francis K, Knowles F, Kelly L, Herrera A, Thorsteinsson T, Marteinsson V (2009b) Bacteria in weathered basaltic glass, Iceland. Geomicrobiol J 26:491–507

de la Torre JR, Goebel BM, Friedmann EI, Pace NR (2003) Microbial diversity of cryptoendolithic communities from the McMurdo Dry valleys, Antarctica. Appl Environ Microbiol 69:3858–3867

del Pino JSN, Almenar ID, Rivero FN, Rodriguez-Rodriguez A, Rodriguez CA, Herrera CA, Garcia JAG, Hernandez JLM (2007) Temporal evolution of organic carbon and nitrogen forms in volcanic soils under broom scrub affected by a wildfire. Sci Total Environ 378:245–252

Donachie SP, Christenson BW, Kunkel DD, Malahoff A, Alam M (2002) Microbial community in acidic hydrothermal waters of volcanically active White Island, New Zealand. Extremophiles 6:419–425

Eppard M, Krumbein W, Koch C, Rhiel E, Staley J, Stackebrandt E (1996) Morphological, physiological, and molecular characterization of actinomycetes isolated from dry soil, rocks, and monument surfaces. Arch Microbiol 166:12–22

Fermani P, Mataloni G, Van de Vijver B (2007) Soil microalgal communities on an Antarctic active volcano (Deception Iceland, South Shetlands). Polar Biol 30:1381–1393

Fike DA, Cockell CS, Pearce D, Lee P (2003) Heterotrophic microbial colonization of the interior of impact-shocked rocks from Haughton impact structure, Devon Island, Nunavut, Canadian High Arctic. Int J Astrobiol 1:311–323

French BM (2004) The importance of being cratered: the new role of meteorite impact as a normal geological process. Meteor Planet Sci 39:169–197

Friedmann EI (1980) Endolithic microbial life in hot and cold deserts. Origins Life Evol Bios 10:223–235

Friedmann EI (1982) Endolithic microorganisms in the Antarctic cold desert. Science 215:1045–1053

Gaylarde PM, Jungblut A, Gaylarde CC, Neilan BA (2006) Endolithic phototrophs from an active geothermal region in New Zealand. Geomicrobiol J 23:579–587

Gross W, Küver J, Tischendorf G, Bouchaala N, Büsch W (1998) Cryptoendolithic growth of the red alga *Galdieria sulphuraria* in volcanic areas. Eur J Phycol 33:25–31

Gundlapally SR, Garcia-Pichel F (2006) The community and phylogenetic diversity of biological soil crusts in the Colorado Plateau studied by molecular fingerprinting and intensive cultivation. Microb Ecol 52:345–357

Herrera A, Cockell CS (2007) Exploring microbial diversity in volcanic environments: a review of methods in DNA extraction. J Microbiol Meth 70:1–12

Herrera A, Cockell CS, Self S, Blaxter M, Reitner J, Arp G, Dröse W, Tindle AG (2008) Bacterial colonization and weathering of terrestrial obsidian rock in Iceland. Geomicrobiol J 25:25–37

Herrera A, Cockell CS, Self S, Blaxter M, Reitner J, Thorsteinsson T, Arp G, Dröse W, Tindle A (2009) A cryptoendolithic community in volcanic glass. Astrobiology 9:369–381

Hode T, Cady SL, von Dalwigk I, Kristiansson P (2008) Evidence of ancient microbial life in an impact structure and its implications for astrobiology – a case study. In: Seckbach J, Walsh M (eds) From fossils to astrobiology. Springer, Heidelberg, pp 249–273

Jacobsson SP, Gudmundsson MT (2008) Subglacial and intraglacial volcanic formations in Iceland. Jökull 58:179–196

Jezberova J (2006) Phenotypic diversity and phylogeny of picocyanobacteria in mesotrophic and eutrophic freshwater reservoirs investigated by a cultivation-dependent polyphasic approach. PhD Thesis, University of South Bohemia, Czech Republic

Katana A, Kwiatowski JM, Spalik K, Zakrys B, Szalacha E, Szymanska H (2001) Phylogenetic position of *Koliella* (Chlorophyta) as inferred from nuclear and chloroplast SSU rDNA. J Phycol 37:443–451

Kieffer SW, Phakey PP, Christie JM (1976) Shock processes in porous quartzite: transmission electron microscope observations and theory. Contrib Mineral Petrol 59:41–93

Koeberl C, Reimold WU (2004) Post-impact hydrothermal activity in meteorite impact craters and potential opportunities for life. Bioastronomy 2002 Life Stars 213:299–304

Kring DA (1997) Air blast produced by the Meteor Crater impact event and a reconstruction of the affected environment. Meteor Planet Sci 32:517–530

Kring DA (2003) Environmental consequences of impact cratering events as a function of ambient conditions on Earth. Astrobiology 3:133–152

Le Bas MJ, Le Maitre RW, Woolley AR (1992) The construction of the total alkali-silica chemical classification of volcanic rocks. Mineral Petrol 46:1–22

Matthes U, Turner SJ, Larson DW (2001) Light attenuation by limestone rock and its constraint on the depth distribution of endolithic algae and cyanobacteria. Int J Plant Sci 162:263–270

Melosh HJ (1989) Impact cratering: a geologic process. Oxford University Press, Oxford

Metzler A, Ostertag R, Redeker HJ, Stoffler D (1988) Composition of the crystalline basement and shock metamorphism of crystalline and sedimentary target rocks at the Haughton-impact-crater, Devon Island, Canada. Meteoritics 23:197–207

Murray AE, Grzymski JJ (2007) Diversity and genomics of Antarctic marine micro-organisms. Phil Trans R Soc B 362:2259–2271

Nagy ML, Perez A, Garcia-Pichel F (2005) The prokaryotic diversity of biological crusts in the Sonoran desert (Organ Pipe Cactus National Monument, AZ). FEMS Microb Ecol 54:233–245

Naumov MV (2005) Principal features of impact-generated hydrothermal circulation systems: mineralogical and geochemical evidence. Geofluids 5:165–184

Nemergut DR, Anderson SP, Cleveland CC, Martin AP, Miller AE, Seimon A, Schmidt SK (2007) Microbial community succession in an unvegetated, recently deglaciated soil. Microbiol Ecol 53:110–122

Nienow JA, McKay CP, Friedmann EI (1988) The cryptoendolithic microbial environment in the Ross Desert of Antarctica: light in the photosynthetically active region. Microbiol Ecol 16:271–289

Nogales B, Moore ER, Llobet-Brossa E, Rossello-Mora R, Amann R, Timmis KN (2001) Combined use of 16S ribosomal DNA and 16S rRNA to study the bacterial community of polychlorinated biphenyl-polluted soil. Appl Environ Microbiol 67:1874–1884

O'Sullivan LA, Fuller KE, Thomas EM, Turley CM, Fry JC, Weightman AJ (2004) Distribution and culturability of the uncultivated 'AGG58 cluster' of the *Bacteroidetes* phylum in aquatic environments. FEMS Microb Ecol 47:359–370

Omelon CR, Pollard WH, Ferris FG (2006) Chemical and ultrastructural characterization of high arctic cryptoendolithic habitats. Geomicrobiol J 23:189–200

Osinski GR (2007) Impact metamorphism of $CaCO_3$-bearing sandstones at the Haughton structure, Canada. Meteor Planet Sci 42:1945–1960

Osinski GR, Spray JG, Lee P (2001) Impact-induced hydrothermal activity within the Haughton impact structure: generation of a transient, warm, wet oasis. Meteor Planet Sci 36:731–745

Osinski GR, Lee P, Parnell J, Spray JG, Baron MT (2005a) A case study of impact-induced hydrothermal activity: the Haughton impact structure, Devon Island, Canadian high arctic. Meteor Planet Sci 40:1859–1877

Osinski GR, Lee P, Spray JG, Parnell J, Lim DSS, Bunch TE, Cockell CS, Glass B (2005b) Geological overview and cratering model of the Haughton impact structure, Devon Island, Canadian High Arctic. Meteor Planet Sci 40:1759–1776

Pierson BK, Mitchell HK, Ruff-Roberts AL (1993) *Chloroflexus aurantiacus* and ultraviolet radiation: implications for archean shallow-water stromatolites. Origin Life Evol Biosph 23:243–260

Quintern LE, Horneck G, Eschweiler U, Bücker H (1992) A biofilm used as ultraviolet-dosimeter. Photochem Photobiol 55:389–395

Quintern LE, Puskeppeleit M, Rainer P, Weber S, El Naggar S, Eschweiler U, Horneck G (1994) Continuous dosimetry of the biologically harmful UV-radiation in Antarctica with the biofilm technique. J Photochem Photobiol 22:59–66

Saiz-Jimenez C, Garcia-Rowe J, Garcia del Cura MA, Ortega-Calvo JJ, Roekens E, Van Grieken R (1990) Endolithic cyanobacteria in Maastricht Limestone. Sci Total Environ 94:209–220

Schminke HU (2004) Volcanism. Springer, Berlin

Sherlock SC, Kelley SP, Parnell J, Green P, Lee P, Osinski GR, Cockell CS (2005) Re-evaluating the age of the Haughton impact event. Meteor Planet Sci 40:1777–1787

Shivaji S, Reddy GSN, Aduri RP, Kutty R, Ravenschlag K (2004) Bacterial diversity of a soil sample from Schirmacher Oasis, Antarctica. Cell Mol Biol 50:525–536

Stroncik NA, Schminke H (2002) Palagonite – a review. Int J Earth Sci 91:680–697

Stuart JC, Binzel RP (2004) Bias-corrected population, size distribution, and impact hazard for the near-Earth objects. Icarus 170:295–311

Tester PA, Varnam SM, Culver ME, Eslinger DL, Stumpf RP, Swift RN, Yungel JK, Black MD, Litaker RW (2003) Airborne detection of ecosystem responses to an extreme event: phyto-plankton displacement and abundance after hurricane induced flooding in the Pamlico-Albemarle Sound system, North Carolina. Estuaries 26:1353–1364

Thorseth IH, Furnes H, Tumyr O (1991) A textural and chemical study of Icelandic palagonite of varied composition and its bearing on the mechanism of the glass-palagonite transformation. Geochim Cosmochim Acta 55:731–749

Toon OW, Zahnle K, Morrison D, Turco RP, Covey C (1997) Environmental perturbations caused by the impacts of asteroids and comets. Rev Geophys 35:41–78

Turmel M, Ehara M, Otis C, Lemieux C (2002) Phylogenetic relationships among streptophytes as inferred from chloroplast small and large subunit rRNA gene sequences. J Phycol 38:364–375

Turner S (1997) Molecular systematics of oxygenic photosynthetic bacteria. Plant Syst Evol (Suppl) 11:13–52

Ustinova I, Krienitz L, Huss VAR (2001) *Closteriopsis acicularis* (G.M. Smith) Belcher et Swale us a fusiform alga closely related to *Chlorella kessleri* Fott et Nováková (Chlorophyta, Terbouxiophyceae). Eur J Physol 36:341–351

Van Thienen P, Benzerara K, Brueur D, Gillmann C, Labrosse S, Lognonne P, Spohn T (2007) Water, life and planetary habitability. Space Sci Rev 129:167–203

Versh E, Kirsimae K, Joeleht A, Plado J (2005) Cooling of the Kardla impact crater: I. The mineral paragenetic sequence observation. Meteor Planet Sci 40:3–19

Walker JJ, Pace NR (2005) Phylogenetic composition of rocky mountain endolithic microbial ecosystems. Appl Environ Microbiol 73:3497–3504

Walker JJ, Spear JR, Pace NR (2005) Geobiology of a microbial endolithic community in the Yellowstone geothermal environment. Nature 434:1011–1014

Weber B, Wessels DCJ, Büdel B (1996) Biology and ecology of cryptoendolithic cyanobacteria of a sandstone outcrop in the Northern Province, South Africa. Algol Stud 83:565–579

Wignall PB (2001) Large igneous provinces and mass extinctions. Earth Sci Rev 53:1–33

Wolff-Boenisch D, Gíslason SR, Oelkers EH, Putnis CV (2004) The dissolution rates of natural glasses as a function of their composition at pH 4 and 10, and temperatures from 25 to 74°C. Geochim Cosmochim Acta 68:4843–4858

Wolff-Boenisch D, Gíslason SR, Oelkers EH (2006) The effect of crystallinity on dissolution rates and CO_2 consumption capacity of silicates. Geochim Cosmochim Acta 70:858–870

Wu X, Xi WY, Ye WJ, Yang H (2007) Bacterial community composition of a shallow hypertro-phic freshwater lake in China. FEMS Microb Ecol 61:85–96

Chapter 10
Microbial Diversity of Cave Ecosystems

Annette Summers Engel

The formation of natural caves (speleogenesis) is due to any number of processes that result in the hollowing out of rock, including dissolution, mechanical weathering, volcanic activity, or even the melting of glacial ice. Caves are classified based on the solid rock that they developed within, the proximity to the groundwater table (e.g., above, at, or below it), the speleogenetic history of a feature, and the overall passage morphology and organization (e.g., cave length, passage shape, passage arrangement, passage levels) (Fig. 10.1). Caves are one type of feature that characterizes a karst landscape, which develops in soluble rocks (e.g., limestone, dolomite, gypsum, halite) that roughly coincides with the global distribution of carbonate sedimentary rocks of all geologic ages (e.g., Ford and Williams 2007). Although karst comprises ~15–20% of the Earth's ice-free land surface, karst caves are not interconnected, not within the same hydrological drainage basin and definitely not across different drainage basins.

Most karst caves are dissolutional and form by the action of water flowing on and through rocks; these caves can form from epigenic (relating to the surface, as a top-down process) or hypogenic fluids (relating to the subsurface, as a bottom-up process) (e.g., Palmer 2007) (Fig. 10.1). Karst landscapes can link the Earth's surface to the subsurface with entrances and extensive underground water flow systems, but surface karst features do not have to be extensively developed, nor present, for subsurface processes to operate (e.g., Klimchouk 2007). Caves formed in rocks other than carbonates, like basalt or sandstone, are classified as pseudokarst. Erosional caves develop from mechanical scouring or wave action rather than by dissolution. Dissolution caves can transition into erosion caves through time. Sea caves form as erosional caves along sea coasts, and anchialine caves form by dissolution along coasts. Volcanic caves or lava tubes are starkly different than dissolution caves, forming from the cooled crust surrounding flowing lava.

A.S. Engel (✉)
Department of Geology & Geophysics, Louisiana State University, E235 Howe-Russell
Geoscience Complex, Baton Rouge, Louisiana 70803, USA
e-mail: aengel@lsu.edu

L.L. Barton et al. (eds.), *Geomicrobiology: Molecular and Environmental Perspective*,
DOI 10.1007/978-90-481-9204-5_10, © Springer Science+Business Media B.V. 2010

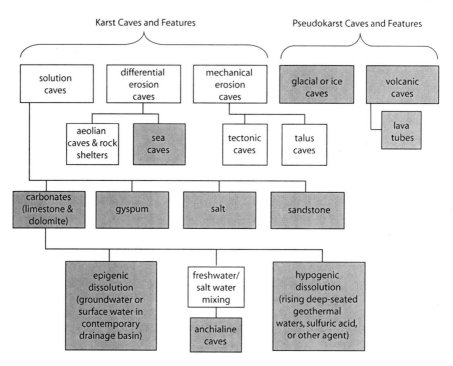

Fig. 10.1 Common types of caves and the processes responsible for forming them, including dissolution and weathering. *Gray boxes* indicate which cave types have been studied to characterize the microbiology, predominately using molecular methods (Modified from White and Culver 2000)

If one considers the extent to which carbonate rocks comprise the rock record, the 100–1,000 m depths to which subsurface carbonates can be karstified, and the longevity of some karst landscapes (e.g., Gale 1992), ranging from several 100,000 to several million years, then it is reasonable to consider that karst caves may serve as a long-term reservoir for microbes in the subsurface. By evaluating the microbial diversity and ecology from different types of caves, a logical set of hypotheses to test may relate to how rock type, fluid availability and geochemistry, and the developmental history of the habitat play a role in community structure and composition. Moreover, because the source(s) of microbes associated with any of the different types of caves or karst features range from soil, water (e.g., groundwater, sea spray, or rainwater), plants or animals, deep-seated fluids circulating in sedimentary basins, or from the rock itself, hypotheses related to surface-subsurface linkages can be tested, as well as hypotheses related to biogeography and endemism. Nevertheless, despite knowing for some time now that the Earth's subsurface contains abundant and active microbial biomass (e.g., Pace 1997; Whitman et al. 1998), our understanding of the microbial diversity in caves, of the factors controlling microbial distribution patterns, and of the role of microbes in cave and karst processes is somewhat incomplete.

This chapter focuses on our current knowledge of the microbial diversity and ecology in karst caves. Because there have been several recent reviews describing the geomicrobiology of caves (e.g., Northup and Lavoie 2001; Barton and Northup 2007), the roles of microbes in geologic and geochemical subsurface karst processes will be briefly highlighted. At this time, and as will become clear, there is insufficient information from all of the different cave types (e.g., volcanic caves, coastal caves, sea caves, anchialine caves and sinkholes, etc.) to make meaningful interpretations. Therefore, this review centers on the microbial diversity of karst caves, from which interesting scientific questions can be addressed (Engel and Northup 2008).

The Karst Habitat

The density, frequency, and number of void spaces in karst are important to the evolution of cave and karst ecosystems. In general, voids within karstifiable rocks form a porosity and permeability continuum, from intergranular spaces, to pores, joints, fractures and fissures, to conduits and cave passages large enough for human entry (e.g., Ford and Williams 2007; Palmer 2007; Bonacci et al. 2009). Voids become active routes for fluid flow that enlarge over time to create a porosity and permeability flow system consisting of matrix permeability, fracture permeability, and conduit permeability, so called the triple porosity or permeability model (e.g., White 2002). The resulting heterogeneous flow system creates three zones of circulation: (1) a region of vertical circulation within the epikarst that is unsaturated (also known as the vadose zone), (2) a zone at the water table that can be saturated, or the epiphreatic zone, and (3) the phreatic zone. Although some caves can extend vertically for thousands of meters into the subsurface, most caves are organized with horizontal passages linked to the surface by flowing (i.e. stream) water and that develop into tiered levels associated with local to regional base-level elevations (e.g., Palmer 2007). Microbial transport from a potential source into and through a karst system will be influenced by vertical and horizontal fluid flow, and may also be affected by sorption onto biofilms forming on surfaces, or retention in zones of slow movement and circulation (e.g., Mahler et al. 2000; White 2002; Pronk et al. 2006; Lehman 2007).

A cave connected to the surface can have three major habitat zones based on light penetration and intensity: entrance, twilight, and dark zone. Each zone has specific physicochemical and nutrient conditions related to geochemical gradients that influence colonization potential and distribution of life (Bonacci et al. 2009), such as metal or dissolved oxygen concentrations, water availability, and even the presence of light, as well as the development of porosity and permeability through time, for example by affecting the saturation state of carbonate minerals in the system. Different parts of a cave can have drastically different physicochemical conditions, from active stream passages subject to flood, to bone-dry passages where water is only available from sporadic dripwater or seasonal sheetflow associated with fractures hydrologically connected to the surface. The cave hygropetric is a habitat associated with flowing water in an otherwise dry cave passage (Sket 2004). In some shallow settings, such as volcanic caves, plant roots may penetrate passages and supply allochthonous organic carbon (e.g., Howarth 1973). Because photosynthesis is not possible in the dark zone, some dark zone ecosystems rely on allochthonous organic material sourced from wind, flowing or dripping water, or guano (e.g., Culver and Pipan 2009), while others use autochthonous organic carbon produced from chemolithoautotrophic primary productivity (e.g., Sarbu et al. 1996; Opsahl and Chanton 2006; Chen et al. 2009; Porter et al. 2009).

Exploration History and Microbiological Methods

The study of biospeleology has only been formalized since the middle of the nineteenth century, even though the pigmentless and eyeless, exotic cave-dwelling animals were the subjects of imagination and scientific inquiry since the sixteenth century in Europe and China (Culver and Pipan 2009; Romero 2009). Obligate cave-adapted animals (troglobites if terrestrial and stygobites if aquatic) have evolved unique characteristics suitable for the subsurface habitat, including adaptations to darkness and extreme environmental conditions spanning nutrient and energy limitation, oxygen deprivation, and geochemically variable solutions. Although there has been a significant increase in biospeleological research worldwide within the past 20 years, our knowledge of cave ecology, population dynamics, nutrient cycling, and life cycles of most cave-adapted animals is still quite limited.

One of the main limitations to studying cave and karst biology, and microbiology, has been obtaining samples and specimens. In any given karst terrain, cave entrances suitable for human entry are rare; more typical are solutionally-enlarged fractures no wider than a few water droplets (e.g., Ford and Williams 2007). Therefore, although samples can be collected directly from natural caves by explorers or from artificial boreholes and wells through pumping or other sampling devices, the limits in exploration have significantly impacted our ability to understand the distribution of life throughout the subsurface, and to appreciate potential risks for contamination. Only recently have some strategies used to explore the bottom of the ocean, specifically the use of autonomous remote-controlled devices, been employed for karst exploration and sampling. DEPTHX is the most well known and highly publicized example (http://www.stoneaerospace.com/products-pages/products-DEPTHX.php), which has been used to examine the sinkholes of Sistema Zacatón, Mexico (Gary and Sharp 2006; Sahl et al. 2009).

The earliest microbiological studies of cave and karst habitats, predominately based on the examination of sediment and water by using microscopy or enrichment and isolation culture-based approaches, identified that microbes were prevalent but not as diverse as in surface habitats like soil (e.g., Caumartin 1963; Gounot 1967). Many cave deposits were considered to have microbiological origins, even if researchers could not prove it. For example, early studies of cave nitrate (saltpeter or saltpetre) deposits (e.g., Hess 1900; Faust 1949, 1968) and carbonate moonmilk and other carbonate speleothem formations (e.g., Went 1969) observed microbes microscopically in the deposits or conducted experiments to show that the deposits could form from byproducts that were due to microbial activity (e.g., Caumartin and Renault 1958). But, most of the experiments did not quantitatively demonstrate that formation was exclusively caused by microbes or a specific set of microbial processes. Even evidence from recent investigations of similar nitrate or carbonate deposits remains, in large part, circumstantial (Taboroši 2006, and references therein). Similarly, iron-oxidizing bacteria were also identified from cave sediments, cave walls, and karst aquifers (e.g., Provencio and Polyak 2001), and recent studies have shown that deposits of ferromanganese minerals do accumulate and change due to microbial activities (Northup et al. 2003; Spilde et al. 2005). Sulfur-oxidizing bacteria were implicated in the sulfuric acid-promoted dissolution of limestone in early research (e.g., Principi 1931; Symk and Drzal 1964; Morehouse 1968; Hubbard et al. 1986), but experimental evidence to support these claims came years later (e.g., Engel et al. 2004b).

Prior to the advent of molecular genetics methods, cave and karst studies relied heavily on microscopy- and culture-based methods developed from medical microbiology. Although many microbes were discovered by using this technique, and for many years this was the only method to obtain information about the microbial diversity and nutrient cycling from karst settings, most uncovered strains were more or less considered affiliated to common soil or aquatic microbes from the surface (e.g., Caumartin 1963; Rusterholtz and Mallory 1994; Mikell et al. 1996; Laiz et al. 1999; Canganella et al. 2002). From these results, some concluded that most of the microbes in caves were translocated soil and surface microbial groups, having been brought into caves from meteoric drip waters, surface streams and air currents, or having been carried into the caves by animals. Most of these microbes were also assumed to be secondary degraders and food sources for higher organisms. Unfortunately, the reaction from these studies was that cave and karst microbes and microbial processes were unimportant from a management perspective and research justification standpoint. Indeed, depending on the type of cave and hydrological connectivity to the surface, some microbes could be sourced from the surface, but there was also mounting evidence that cave microbes were unique from surface soil or aquatic environments (e.g., Vlasceanu et al. 1997; Angert et al. 1998).

The advent of molecular genetics methods provided opportune rescue. The transition came after the 1986 discovery and research of Movile Cave in Romania in the mid-1990s. The ecosystem, which has dozens of animals unique to science, is sustained by chemolithoautotrophic microbial primary productivity (e.g., Sarbu 1990; Sarbu et al. 1996; Chen et al. 2009; Porter et al. 2009). The investigations not only coincided with significant advances in our understanding of the chemosynthetic ecosystems at the deep-sea hydrothermal vents (Jannasch 1985), but also were among the first to apply molecular genetics techniques of the late 1980s and 1990s (e.g., Lane et al. 1985; Woese 1987; Amann et al. 1995; Pace 1997). Since then, and spurred on by the applications of automated DNA sequencing, high-throughput computing analyses of environmental rRNA and functional gene sequence phylogenies, and the development of oligonucleotide rRNA probing, microautoradiography, and stable-isotope probing, the biochemical evaluation of genes, enzymes, and metabolites has been the focus of most cave and karst research. Efforts have gone from answering the basic question "Who's there?" to another simple question that is more difficult to answer, "What are they doing?"

Figure 10.1 shows the types of cave systems that have been investigated using a range of molecular techniques, which, like most microbial ecology studies, predominately center on the use of 16S rRNA gene surveys and phylogenies to provide some general insight into microbial diversity and ecosystem function. Using a combination of culture-dependent and -independent approaches, we know that most microbes colonizing rock surfaces in caves, forming microbial mats and biofilms in cave streams, and inhabiting karst aquifers, are genetically divergent from surface microbial groups and represent novel taxonomic groups (see references in Table 10.1). Taken collectively, these results could indicate that some groups are endemic to karst. To test this hypothesis, however, additional research is needed to culture novel groups to test for possible endemicity experimentally (e.g., Rusterholtz and Mallory 1994; Laiz et al. 2003; Zhou et al. 2007; Snider et al. 2009).

Molecular genetics methods have also provided a way to explore some of the most geochemically or mineralogically extreme cave and karst habitats consisting of extremophilic microbes belonging to oligotrophic, acidophilic, thermophilic, and sulfidophilic communities (Table 10.1). Caves rich in sulfur compounds, whereby hydrogen sulfide enters the cave typically dissolved in cave streams or springs, are referred to as sulfidic caves. These microbial systems are relatively easy to spot, not only because of the dominance of reduced sulfur compounds and sulfate minerals, but because the microbial communities form visually stunning, colorful, and conspicuous microbial mats and biofilms (Engel 2007). Incidentally, the public has been fascinated with these systems, brought on by the introduction of the dark life concept and harrowing exploration stories printed in popular literature (Taylor 1999). Sulfur-based karst microbial geochemistry draws parallels to marine and deep-sea hydrothermal vent systems, as well as provides analogs to possible life on other planetary bodies (Boston et al. 2006). One of the most publicized sites is the Cueva de Villa Luz, Mexico (also known as La Cueva de los Sardinas or Cueva de Azufroso). This is a spectacular system where groundwater springs with elevated concentrations of hydrogen sulfide enter the cave, and the cave walls are covered elemental sulfur, gypsum, and oozing slime, affectionately termed snottites. These formations have pH values averaging <2 (Hose et al. 2000), due in part to the sulfur-oxidizing microbes producing sulfuric acid.

Distribution of Microbial Groups by Cave Type

Since the first published molecular genetics studies from a karst setting in the late 1990s (e.g., Vlasceanu et al. 1997; Angert et al. 1998), the use of culture-independent approaches has recently caused an avalanche of 16S rRNA gene sequence data. In 2009 alone, not only was there a special issue of the *International Journal of*

Table 10.1 Summary of some of the molecular investigations from different types of cave habitats. Many of these studies document the presence of novel species whose similarity to other known species or environmental isolate genetic sequences is relatively low (80–92% sequence identity). Updated from Engel and Northup (2008), and not including all of the contributions in White (2009)

Microbial mats, cave-wall biofilms, cenotes in active sulfidic systems	Iron/Manganese deposits from sediments and cave walls	Oligotrophic, cave-wall and cave pool habitats	Other studies, including from cave paintings and disturbance area (e.g., tourist caves)
Vlasceanu et al. (1997)	Northup et al. (2000)	Northup et al. (2000)	Laiz et al. (1999)
Angert et al. (1998)	Northup et al. (2003)	Holmes et al. (2001)	Schabereiter-Gurtner et al. (2002a)
Humphreys (1999)	Chelius and Moore (2004)	Sanchez-Moral et al. (2003)	Schabereiter-Gurtner et al. (2002b)
Vlasceanu et al. (2000)	Spilde et al. (2005)	Barton et al. (2006)	Laiz et al. (2003)
Hose et al. (2000)		Barton et al. (2007)	Schabereiter-Gurtner et al. (2004)
Engel et al. (2001)		Shabarova and Pernthaler (2009)	Barton et al. (2006)
Canganella et al. (2002)		Snider et al. (2009)	Gonzalez et al. (2006)
Engel et al. (2003)			Ikner et al. (2007)
Engel et al. (2004a)			Weidler et al. (2007)
Hutchens et al. (2004)			Zhou et al. (2007)
Barton and Luiszer (2005)			Kelly et al. (2009)
Macalady et al. (2006)			Pronk et al. (2009)
Macalady et al. (2007)			
Meisinger et al. (2007)			
Spear et al. (2007)			
Porter and Engel (2008)			
Porter et al. (2009)			
Engel et al. (2009)			
Chen et al. (2009)			
Sahl et al. (2009)			

Speleology devoted to cave microbiology research, but work was also presented in a special session on "The Geomicrobiology of Cave and Karst Habitats," held during the International Congress of Speleology (White 2009); this special session doubled the available 16S rRNA gene sequences available in public databases (N. Lee, personal communication 2009). Based on the published studies reporting 16S rRNA gene sequences from cultures or clone libraries from caves and karst (Table 10.1), roughly half of the recognized bacterial phyla have been identified (Fig. 10.2a) and slightly less than half of the recognized archaeal phyla (Fig. 10.2b). A general 18S rRNA gene tree represents the diversity of eukaryotic groups identified from caves, and it is clear that not all of the groups, especially the

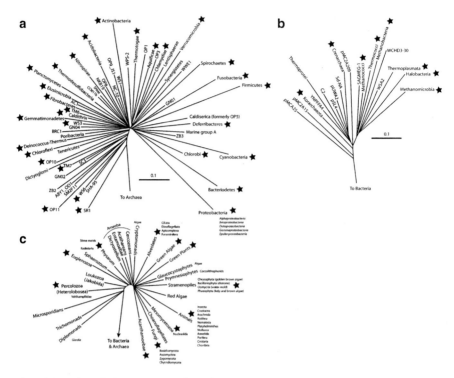

Fig. 10.2 Acknowledged representation of microbes belonging to the domains Bacteria (**a**), Archaea (**b**) and Eukaryota (**c**) identified from cave and karst systems. Divisions detected are indicated by *stars* (★). The schematic topologies for (**a**) and (**b**) are based on phylogenetic trees created using the NAST alignment of 16S rRNA gene sequences from Greengenes (DeSantis et al. 2006) (http://greengenes.lbl.gov). The *bars* in (**a**) and (**b**) indicate changes per nucleotide. The schematic tree of (**c**) is based on the topology from 18S rRNA gene sequence phylogeny of Dawson and Pace (2002)

microeukaryotes, have been recognized (Fig. 10.2c). We know more about general microbial eukaryotic diversity from classical species descriptions rather than from molecular gene sequences (e.g., Dawson and Pace 2002), although this is not a problem restricted to cave and karst settings.

The remainder of the review focuses on the microbial diversity of epigenic and hypogenic cave types (Fig. 10.1), and emphasis is placed on the known bacterial diversity because the number of studies of Archaea from caves and karst is limited (e.g., Vlasceanu 1999; Northup et al. 2003; Chelius and Moore 2004; Barton and Luiszer 2005; Gonzalez et al. 2006) and the diversity of microbial eukaryotes is virtually unknown in nearly all karst habitats. Most of what is known is from studying just a few systems and focusing on specific groups, such as pathogens like *Histoplasmosis*, guanophiles, or organisms associated with deterioration of cave walls, paintings, and speleothems (e.g., Hasenclever et al. 1967; Gittleson and Hoover 1969; Sudzuki and Hosoyama 1991; Coppellotti Krupa and Guidolin 2003; Nieves-Rivera 2003;De Luca et al. 2005; Landolt et al. 2006; Walochnik and Mulec 2009).

Natural and Impact-Associated Communities in Epigenic Caves

Epigenic caves are among the most commonly formed cave systems known (Palmer 2007). These caves form at, or proximal to, the groundwater table, and so changes in regional base level and climate through time affect the location and arrangement of voids (pores through conduits) in the overall karst landscape and drainage basin. Mammoth Cave in Kentucky (USA), with more than 580 km of mapped cave passages, is the most well-known epigenic system worldwide. Hydrological conditions are critical to the development of epigenic systems because carbonate dissolution is influenced by the gravity-driven movement of water from the surface into the subsurface, either as stream flow or meteoric input (e.g., as dripwaters). Terrestrial habitats within epigenic caves include cave-wall surfaces, speleothems, the unsaturated epikarstic zone, hygropetric surfaces, and any other part of the cave system that is at an air-water or air-rock interface. In contrast, saturated aquatic habitats are almost exclusively in the phreatic zone. One hypothesis that has been considered from the emergent 16S rRNA gene sequence datasets is that communities associated with different karst zones are distinct from each other, whereby planktonic microbial communities in the phreatic zone are distinct from the communities forming biofilms on carbonate rock surfaces, or even colonizing the cave-wall surfaces of the epikarst (e.g., Rusterholtz and Mallory 1994; Alfreider et al. 1997; Lehman 2007; Pronk et al. 2009).

Another hypothesis that has been put forth recently relates to endemism and biogeography of karst microbial communities, not just to understand the distribution of microbes in karst settings, but to determine the mechanisms of dispersal that may be related to inoculation source(s) into the subsurface (e.g., Porter and Engel 2008). It should go without saying, but maintaining pristine, natural conditions is difficult when that habitat must be penetrated for study (e.g., Northup et al. 1997; Chelius et al. 2009). Consideration needs to be given to the relative abundances of microbial groups in karst commonly attributed to contamination (e.g., Barton et al. 2006) before assessing whether a specific group's presence is due to natural processes or if the group was introduced from human impact or another type of disturbance (note: it is difficult to determine if a microbial group is truly absent from a community based on the fact that most rRNA-based methods are not exhaustive searches). There have been many studies to characterize the naturally occurring microbial communities in epigenic caves and karst aquifers at all hydrological stages, especially under end-member conditions of base-flow and storm events (e.g., Rusterholtz and Mallory 1994; Mahler et al. 2000; van Beynen and Townsend 2005; Goldscheider et al. 2006; Pronk et al. 2006; Kelly et al. 2009). Storm events, like flash flooding, may pulse diagnostic microbial groups from surface soils, forest litter, or contaminated urban areas into karst, which may alert to changes in water quality. Unfortunately, it has been difficult to make sense of some of the results because insufficient distribution data exist at this time for all cave types, and especially the more common epigenic cave systems (Table 10.1), and we do not understand the roles of most microbes in karst settings to evaluate if detection of a group indicates metabolically active populations in the system (i.e. indigenous), or if the group is passing through or the result of an inoculation event (i.e. potential contaminant, but only detected during or after disturbance events).

For example, comparison of five recent 16S rRNA gene sequence surveys from different caves highlights the complexities that exist in interpreting microbial distribution data in epigenic caves. Some microbial groups, including members of the *Deltaproteobacteria*, *Acidobacteria*, and *Nitrospira*, are abundant and present under almost all hydrological regimes throughout an uncontaminated alpine karst water basin (e.g., Farnleitner et al. 2005; Pronk et al. 2009). But, other groups, including *Firmicutes*, and members of the common soil genera *Bacillus* and *Clostridium,* are identified in the water only after storm events, suggesting that these groups signify some sort of disturbance. However, highly similar 16S rRNA gene sequences (>97% sequence identities) to these alpine karst storm groups were retrieved from pristine to low impact (from tourism) cave sediments (Ikner et al. 2007), pristine cave pool water (Shabarova and Pernthaler 2009), and carbonate moonmilk deposits (Engel, unpublished data). These groups are not the most abundant in the pool water and moonmilk sites, and instead, the most prevalent groups are members of betaproteobacterial Oxalobacteraceae, Methylophilaceae, and Comamonadaceae families. Shabarova and Pernthaler (2009) suggest that these three betaproteobactieral groups point toward a highly specialized endemic subsurface microflora, which is different than the interpretation from Pronk et al. (2009) who suggest that *Deltaproteobacteria*, *Acidobacteria*, and *Nitrospira* are endemic subsurface groups. To complicate matters, the betaproteobacterial groups (with high sequence identities) are abundant in high impact sediments in another cave system with frequent tourists (Ikner et al. 2007). If persistent contamination of nonendemic groups is widespread, then it may be nearly impossible to understand diversity and distribution results. However, almost all of the karst springs in southwestern Illinois (USA) have high counts of enteric bacteria (e.g., *Escherichia coli, Enterococcus* spp., *Staphylococcus* spp.) all year-around (Kelly et al. 2009). In this study, the contamination source can confidently be attributed to domestic (private) wastewater treatment systems.

Some of the earliest work studying the effects of human impact on potentially indigenous microbial communities in karst was associated with expedition camps in large cave systems. In these expeditions, explorers, surveyors, and scientists were required to spend several days to more than a week underground. For long expeditions, explorers acquire water from cave pools, as well as sleep, eat, urinate, and defecate in the caves because it is usually impossible to bring enough water into the cave for each person to use, or to carry liquid waste out of the cave. After *Escherichia coli* was found in the drinking pools from Lechuguilla Cave in New Mexico (USA), the interpretation was that pristine pools were contaminated by *E. coli* from human fecal material (e.g., Northup et al. 1997; Hunter et al. 2004; Lavoie and Northup 2006). The cave explorers were asked to conduct their work in vastly different and somewhat unrealistic ways to minimize the influence of human-associated microbes on the indigenous communities in and around exploration camps. The research findings were met with significant opposition, including concerns about the experimental (e.g., growth medium composition) and analytical (e.g., no genetic identification) results that led to presumptive conclusions that *E. coli* was present in the pool waters, even though *E. coli* had not been previously shown to survive long-term in low-nutrient conditions (Barton and Pace 2005; Hunter et al. 2005). To resolve some of the issues, a broader group of microbes are now considered as "human indicator bacteria" (*E. coli, Staphylococcus aureus*, and high-temperature *Bacillus* spp.) (Lavoie and Northup 2006). From more detailed investigations of different areas of two large caves (Lechuguilla and Mammoth Caves) with low (off-trail and distant from camps) to high human activity (areas with high human activity, like camps and tourist trails), *E. coli* was identified from only 15% of the active urine dumps and urination control areas, *S. aureus* was found in 30% of the high impact and ~3% of the low impact sites, and the abundance of high-temperature *Bacillus* spp. was 8 × higher in high impact than low impact sites (Lavoie and Northup 2006). Experiments to test for persistence reveal that growth of *E. coli* and *S. aureus* decreases significantly to below detection after several weeks, indicating that areas with high impact have the potential to recover to pre-impact levels following human indicator bacterial contamination (Lavoie and Northup 2006). Additional research from tourist (sometimes called "show" or "sacrificial") caves also demonstrate the strong correlation between the number of visitors to that cave and the type and biomass of microbial groups, attributed to the influx of rich organic matter (e.g., Ikner et al. 2007; Chelius et al. 2009). Tourists reverse the concentration and availability of organic carbon in a system by bringing lint, fibers, dander, hair, and trash, as well as human-associated microbes, into a cave system. Where input rates of the unnatural (i.e. impact-derived) organic material are greatest, microbial biomass is higher than in pristine or low impact areas. Interestingly, microbial groups found in high impact zones in modern tourist caves are comparable to communities associated with Paleolithic cave paintings; depending on the types of microbes present, however, the paintings can be damaged (e.g., Laiz et al. 1999; Schabereiter-Gurtner et al. 2002a, b, 2004).

The use of artificial lighting, which might be left on all the time in some show caves, has been one area of research (e.g., Ariño and Saiz-Jimenez 1996; Sanchez-Moral et al. 2003; Mulec and Kosi 2009; Roldán and Hernández-Mariné 2009). Photosynthetic organisms, like algae and plants, have been studied from the entrance zones of caves, but brightly to dimly lit show caves can also have diverse, albeit unnatural, photosynthetic communities (Grobbelaar 2000). The flora associated with cave lighting is termed lampenflora (e.g., Dobat 1998), derived from the French phrase *la maladie verte*, or "green sickness." The diversity of lampenflora has recently been reviewed (Mulec and Kosi 2009), indicating that biodiversity is lower compared to the flora from cave entrances, although there have been no molecular diversity studies to date. Tourists are the main source for spores and material, such that some of the most insidious examples of lampenflora are from caves visited by hundreds of thousands of tourists a year (Grobbelaar 2000; Mulec and Kosi 2009). Mulec et al. (2008) report that the concentration of chlorophyll *a* per lampenflora algae surface area is almost double that of epilithic algae from cave entrances from Slovenian tourist caves. In the Cango Caves (South Africa), which receives over 500,000 visits per year, algal growth significantly decreases with increasing distance from lamps but growth is still detected even with 1 μmol photon/m²/s or less (Grobbelaar 2000). In the Mammoth Cave and tourist caves in Slovenia, increased growth is correlated to warmer temperatures around lights and not to photon flux (Smith and Olson 2007; Mulec and Kosi 2009), although growth is inhibited next to strong lights due to extremely high temperatures. The main concern for the proliferation of lampenflora is the destruction of rocks. Eradication methods of lamenflora and other invasive species has ranged from scrubbing inflicted areas with brushes, using power washers, chemical treatment, turning lights off, or changing light wavelengths or photon flux.

But, even the eradication mechanisms to remove contaminating microbial communities have consequences. Lascaux Cave (France) is the best studied system regarding contamination-impacted and -related microbial communities (e.g., Dupont et al. 2007; Bastian et al. 2009a, b, c). The research provides a mixed success story. Lascaux Cave is filled with thousands of Paleolithic paintings, and the cave was closed to visitors in 1963 to protect the art. Nevertheless, the cave was invaded by *Fusarium solani* species complex in 2001, and the cave walls were treated for 4 years with benzalkonium chloride to remove the fungi (Dupont et al. 2007). In recent diversity studies, >45% of the cave-wall microbial groups are affiliated with known pathogens similar to groups found in contaminated air-conditioning systems and cooling towers in nearby community hospitals and public buildings (Bastian et al. 2009a). Compared to caves with no biocide treatment, the Lascaux Cave communities lacked or had rare *Acidobacteria* and other phyla previously retrieved from cave-wall habitats (e.g., *Actinobacteria*, *Firmicutes, Nitrospira, Bacteroidetes*) (e.g., Zhou et al. 2007), but instead were dominated by *Ralstonia* spp. (*Betaproteobacteria*) and *Pseudomonas* spp. (*Gammaproteobacteria*). The researchers propose that chemical treatment killed typical cave microbial groups and allowed for a less diverse group capable of using the biocide to take over. Rapid fungal colonization could be due to fungal spore dispersal by air movement or the presence of arthropods on the cave walls, including collembola (Bastian et al. 2009c). The animals consume cave-wall microbes living on organic compounds found in the paintings, but also spread the fungal spores on their bodies. The implications of this work are significant for cave management practices; while some might consider removing contaminating microbes acceptable, others likely consider it unacceptable to kill cave animals in the process.

Microbial Communities from Hypogenic Caves

Hypogenic caves and karst commonly form due to the action of rising fluids (water or gases), at or below the water table, but with little to no relationships to surface karst processes and drainage basin evolution. There are different types of hypogenic caves, but most work has focused on caves formed from sulfuric acid. The microbial diversity and associated processes in sulfidic caves have recently been reviewed by Engel (2007), and so this review only highlights comparisons and recent contributions to the subject.

Egemeier (1981) examined the sulfuric acid process in detail from studying hydrogen sulfide (H_2S)-bearing springs and streams, extensive gypsum deposits, and gypsum-replaced carbonate rock walls in Lower Kane Cave in Wyoming (USA). He hypothesized that volatilization of H_2S from the groundwater to the cave atmosphere, and H_2S autoxidation to sulfuric acid on the moist cave walls, caused the replacement of the carbonate with gypsum. The cave streams are undersaturated with respect to gypsum, and so cave passage enlargement ensues if gypsum spalls off into the streams. Nearly all of the subsequent studies from active and ancient caves formed from sulfuric acid dissolution considered that H_2S was predominately oxidized by abiotic processes. However, sulfur-oxidizing bacteria that generate sulfuric acid as a byproduct of their metabolism are more important to carbonate dissolution than abiotic processes alone (e.g., Hose et al. 2000; Engel et al. 2004b; Macalady et al. 2006).

16S rRNA gene sequence surveys have increased our understanding of the microbial diversity from sulfidic systems (Table 10.1). Microbes in hypogenic systems occur as microbial draperies or snottites on cave-wall surfaces, in corrosion residues, and as microbial mats and biofilms in active cave streams and pools. The groups identified from microbial mats in active sulfidic caves have similar microbial community compositions (Engel 2007). *Proteobacteria* are one of the most prevalent groups, and although all classes have been identified, members of the *Epsilonproteobacteria*, *Gammaproteobacteria*, and *Betaproteobacteria* have the highest relative abundances in the mats. Members of these classes are associated with sulfur oxidation and are important to sulfuric acid speleogenesis. The diversity of environmental, uncultivated *Epsilonproteobacteria* has increased significantly by studying the hypogenic cave systems (Campbell et al. 2006; Porter and Engel 2008; Porter et al. 2009). Sequences from many different lineages, including candidate divisions, have also been retrieved recently (e.g., Chen et al. 2009; Engel et al. 2009), possibly indicating a richer diversity in these systems than previously considered.

Snottites on acidic cave-wall surfaces in several active sulfidic caves have been intensely investigated (Hose et al. 2000; Vlasceanu et al. 2000; Engel 2007; Macalady et al. 2007). Acidity is maintained in part because the replacement gypsum physically separates the cave-wall solutions from the underlying carbonate rock that would buffer pH to more neutral conditions, and also because of continued microbial sulfur oxidation. The phylogenetic diversity of these biofilms is similar in all of the studied caves, consisting of extremophiles belonging to the *Proteobacteria* (e.g., *Acidithiobacillus* spp.), *Acidimicrobium*, *Thermoplasmales*, actinomycetes, the bacterial candidate lineage TM6, as well filamentous fungi and protists. Presumably the active snottite and cave-wall microbial communities from active sulfuric acid caves would have some relation to the cave-wall and corrosion residue communities from cave systems that formed in the past from sulfuric acid speleogenesis. But, interestingly, the diversity from cave-wall corrosion residues and ferromanganese deposits from Lechuguilla Cave is different, consisting of diverse bacterial groups, including the proteobacterial genera *Hyphomicrobium*, *Pedomicrobium*, *Leptospirillum*, and *Stenotrophomonas*, and different members of the Crenarchaeota and Euryarchaeota (Northup et al. 2003).

Although the sulfur cycle has been the focus of many studies from hypogenic caves, methane cycling is also beginning to be of interest to karst researchers. Methanotrophic and methanogenic microbial groups, while expected in some groundwater systems with high concentrations of dissolved methane (Opsahl and Chanton 2006), have almost exclusively been studied in detail from Movile Cave using enrichment and isolation techniques, $^{13}CH_4$-labelling incorporation studies, and ^{13}C-DNA analysis (Vlasceanu 1999; Hutchens et al. 2004; Chen et al. 2009). The contribution, if any, of methanogenic or methanotrophic activity to speleogenesis is unknown, although methane (and other one-carbon compounds) is potentially significant to the ecosystem development and primary productivity (Opsahl and Chanton 2006; Chen et al. 2009).

The Future of Cave Microbial Diversity Research

Caves are important and relatively accessible habitats to study subsurface microbial diversity. Although the important advances in our understanding of the geomicrobiological and biogeochemical roles of cave and karst microbes in metal and nutrient cycles, including carbonate dissolution and precipitation, have been made, many challenges still lie before us. One research direction will undoubtedly continue, and that is related to determining overall habitat biodiversity. This research direction is a common thread in most microbial ecology and environmental microbiology studies. However, future research needs to transcend simply answering "Who's there?" and begin to address "What are they doing?" Another set of questions relate to "Where did they come from?" It is becoming clear that more baseline data are needed to test hypotheses related to the distribution, dispersal, and reservoir size of the different, and possibly distinctive, microbial groups in these subsurface habitats. Do surface-derived microorganisms die if transported into caves and become food for higher level organisms, or can surface-derived microbes function in the restricted (i.e. based on physicochemical conditions) habitat and subsequently colonize caves? If the latter is possible, then how do the microbes adapt to the more stable conditions of karst? At what genetic divergence level can subsurface microbial groups be distinguished from their surface, or other habitat, counterparts and instead be considered indigenous to the cave? One can hypothesize that, in the more stable, low-nutrient (i.e. low organic carbon) habitats, selection may lead to genome streamlining and size reduction. Are there metabolic functions that signify endemism? Unlike the possible transport via dripwaters, to what extent do air currents and groundwater bring microbes into the cave environment? For this question to be answered, the hydrogeological connectivity that exists, and has existed, between surface and subsurface cave and karst environments needs to be better characterized. Once these research directions take shape, the importance of understanding karst processes and the microbial contributions to these processes will increase, and we will be able to better protect and conserve the natural habitats through time.

Acknowledgements I appreciate the stimulating conversations and ideas to improve this review that came from M.L. Porter, J. Mulec, and N. Lee. Support for this review was provided by the United States National Science Foundation (DEB-0640835), the Louisiana Board of Regents (LEQSF [2006-09]-RD-A-03 and NSF/LEQSF(2005)-Pfund-04), and Louisiana State University.

References

Alfreider A, Krossbacher M, Psenner R (1997) Groundwater samples do not reflect bacterial densities and activity in subsurface systems. Water Res 31:832–840

Amann RI, Ludwig W, Schleifer KH (1995) Phylogenetic identification and *in situ* detection of individual microbial cells without cultivation. Microbiol Rev 59:143–169

Angert ER, Northup DE, Reysenbach AL, Peek AS, Goebel BM, Pace NR (1998) Molecular phylogenetic analysis of a bacterial community in Sulphur river, Parker cave, Kentucky. Am Mineral 83:1583–1592

Ariño X, Saiz-Jimenez C (1996) Biological diversity and cultural heritage. Aerobiologia 12:279–282

Barton HA, Luiszer F (2005) Microbial metabolic structure in a sulfidic cave hot spring: potential mechanisms of biospeleogenesis. J Cave Karst Stud 67:28–38

Barton HA, Northup DE (2007) Geomicrobiology in cave environments: past, current and future perspectives. J Cave Karst Stud 69:163–178

Barton HA, Pace NR (2005) Discussion: Persistent coliform contamination in Lechuguilla Cave pools. J Cave Karst Stud 67:55–57

Barton HA, Taylor MR, Lubbers BR, Pemberton AC (2006) DNA extraction from low biomass carbonate rock: an improved method with reduce contamination and the low biomass contaminant database. J Microbiol Methods 66:21–31

Barton HA, Taylor NM, Kreate MP, Springer AC, Oehrle SA, Bertog JL (2007) The impact of host rock geomicrobiology on bacterial community structure in oligotrophic cave environments. Int J Speleol 36:93–104

Bastian F, Alabouvette C, Jurado V, Saiz-Jimenez C (2009a) Impact of biocide treatments on the bacterial communities of the Lascaux cave. Naturwissenschaften 96:863–868

Bastian F, Alabouvette C, Saiz-Jimenez C (2009b) Bacteria and free-living amoeba in the Lascaux cave. Res Microbiol 160:38–40

Bastian F, Alabouvette C, Saiz-Jimenez C (2009c) The impact of arthropods on fungal community structure in Lascaux cave. J Appl Microbiol 106:1456–1462

Bonacci O, Pipan T, Culver DC (2009) A framework for karst ecohydrology. Environ Geol 56:891–900

Boston PJ, Hose LD, Northup DE, Spilde MN (2006) The microbial communities of sulfur caves: a newly appreciated geologically driven system on Earth and potential model for Mars. In: Harmon RS, Wicks C (eds) Perspectives on karst geomorphology, hydrology, and geochemistry – a tribute volume to Derek C. Ford and William B. White. Geological Society of America Special Paper 404, pp 331–344

Campbell BJ, Engel AS, Porter ML, Takai K (2006) The versatile *ε-proteobacteria*: key players in the sulphur cycle. Nat Rev Microbiol 4:458–468

Canganella F, Bianconi G, Gambacorta A, Kato C, Uematsu K (2002) Characterisation of heterotrophic microorganisms isolated from the "Grotta Azzura" of Cape Palinuro (Salerno, Italy). Mar Ecol 23:1–10

Caumartin V (1963) Review of the microbiology of underground environments. Bull Nat Speleol
    Soc 25:1–14
Caumartin V, Renault P (1958) La corrosion biochimique dans un reseau karstique et al genèse du
    mondmilch. Notes Biospéleologiques 13:87–109
Chelius MK, Moore JC (2004) Molecular phylogenetic analysis of Archaea and Bacteria in Wind
    Cave, South Dakota. Geomicrobiol J 21:123–134
Chelius MK, Beresford G, Horton H, Quirk M, Selby G, Simpson RT, Horrocks R, Moore JC
    (2009) Impacts of alterations of organic inputs on the bacterial community within the sedi-
    ments of Wind Cave, South Dakota, USA. Int J Speleol 38:1–10
Chen Y, Wu L, Boden R, Hillebrand A, Kumaresan D, Moussard H, Baciu M, Lu Y, Murrell JC
    (2009) Life without light: microbial diversity and evidence of sulfur- and ammonium-based
    chemolithotrophy in Movile Cave. ISME J 3:1093–1104
Coppellotti Krupa O, Guidolin L (2003) Taxonomy and ecology of ciliate fauna (Protozoa,
    Ciliophora) from karst caves in North East Italy. Subterranean Biol 1:3–11
Culver DC, Pipan T (2009) The biology of caves and other subterranean habitats. Oxford
    University Press, Oxford
Dawson SC, Pace NR (2002) Novel kingdom-level eukaryotic diversity in anoxic environments.
    PNAS 99:8324–8329
De Luca E, Toniello V, Coppellotti O (2005) Protozoi de acque carsiche in un'area nord orientale
    della provincial di Treviso. Speleol Venta 13:124–131
DeSantis TZ, Hugenholtz P, Larsen N et al (2006) greengenes, a chimera-checked 16S rRNA gene
    database and workbench compatible with ARB. Appl Environ Microbiol 72:5069–72
Dobat K (1998) Flore de la luminére artificiélle (lampenflora-maladie verte). In: Juberthie C,
    Decu V (eds) Encyclopaedia Biospeleologica, Tome 2. Société Biospéologie, Moulis-Bucarest,
    pp 1325–1335
Dupont J, Jacquet C, Dennetière B, Lacoste S, Bousta F, Orial G, Cruaud C, Couloux A,
    Roquebert MF (2007) Invasion of the French Palaeolithic painted cave of Lascaux by members
    of the *Fusarium solani* species complex. Mycologia 99:526–533
Egemeier S (1981) Cave development by thermal waters. Bull Nat Speleol Soc 43:31–51
Engel AS (2007) Observations on the biodiversity of sulfidic karst habitats. J Cave Karst Stud
    69:187–206
Engel AS, Northup DE (2008) Caves and karst as model systems for advancing the microbial
    sciences. In: Martin J, White WB (eds) Frontiers in Karst Research. Karst Waters Institute
    Special Publication 13, LeesburgVirginia, pp 37–48
Engel AS, Porter ML, Kinkle BK, Kane TC (2001) Ecological assessment and geological signifi-
    cance of microbial communities from Cesspool Cave, Virginia. Geomicrobiol J 18:259–274
Engel AS, Lee N, Porter ML, Stern LA, Bennett PC, Wagner M (2003) Filamentous
    "*Epsilonproteobacteria*" dominate microbial mats in sulfidic caves. Appl Environ Microbiol
    69:5503–5511
Engel AS, Porter ML, Stern LA, Quinlan S, Bennett PC (2004a) Bacterial diversity and ecosystem
    function of filamentous microbial mats from aphotic (cave) sulfidic springs dominated by
    chemolithoautotrophic "*Epsilonproteobacteria*.. FEMS Microbiol Ecol 51:31–53
Engel AS, Stern LA, Bennett PC (2004b) Microbial contributions to cave formation: new insights
    into sulfuric acid speleogenesis. Geology 32:369–372
Engel AS, Meisinger DB, Porter ML, Payn RA, Schmid M, Stern LA, Schleifer KH, Lee NM (2009)
    Linking phylogenetic and functional diversity to nutrient spiraling in microbial mats from Lower
    Kane Cave (USA). ISME J 4:98–110. doi:10.1038/ismej.2009.91
Farnleitner AH, Wilhartitz I, Ryzinska G et al (2005) Bacterial dynamics in spring water of alpine
    karst aquifers indicates the presence of stable autochthonous microbial endokarst communi-
    ties. Environ Microbiol 7:1248–1259
Faust B (1949) The formation of saltpeter in caves. Bull Nat Speleol Soc 11:17–23
Faust B (1968) Notes on the subterranean accumulation of saltpetre. J Spelean Hist 1:3–11
Ford D, Williams P (2007) Karst hydrogeology and geomorphology. Wiley, West Sussex

Gale SJ (1992) Long-term landscape evolution in Australia. Earth Surf Proc Land 17:323–343

Gary MO, Sharp JM Jr (2006) Volcanogenic karstification of Sistema Zacatón. In: Harmon RS, Wicks C (eds) Perspectives on Karst Geomorphology, Hydrology, and Geochemistry – A tribute volume to Derek C. Ford and William B. White. Geological Society of America Special Paper 404, pp 79–89

Gittleson SM, Hoover RL (1969) Cavernicolous protozoa: review of the literature and new studies in Mammoth Cave, Kentucky. Annales Spéléol 24:737–776

Goldscheider N, Hunkeler D, Rossi P (2006) Review: microbial biocenoses in pristine aquifers and an assessment of investigative methods. Hydrogeol J 14:926–941

Gonzalez JM, Portillo MC, Saiz-Jimenez C (2006) Metabolically active Crenarchaeota in Altamira cave. Naturwissenschaften 93:42–45

Gounot AM (1967) Le microflore des limons argileaux souterains: son activité productrice dans la biocénose cavernicole. Annal Spéléol 22:23–143

Grobbelaar JU (2000) Lithophytic algae: a major threat to the karst formation of show caves. J Appl Phycol 12:309–315

Hasenclever HF, Shacklette MH, Young RV, Gelderman GA (1967) The natural occurrence of *Histoplasma capsulatum* in a cave: 1. epidemiological aspects. Am J Epidemiol 86:238–245

Hess WH (1900) The origin of nitrates in cavern earths. J Geol 8:129–134

Holmes AJ, Tujula NA, Holley M, Contos A, James JM, Rogers P, Gillings MR (2001) Phylogenetic structure of unusual aquatic microbial formations in Nullarbor caves, Australia. Environ Microbiol 3:256–264

Hose LD, Palmer AN, Palmer MV, Northup DE, Boston PJ, DuChene HR (2000) Microbiology and geochemistry in a hydrogen-sulphide rich karst environment. Chem Geol 169:399–423

Howarth FG (1973) The cavernicolous fauna of Hawaiian lava tubes, 1. Introduction. Pac Insect 15:139–151

Hubbard DA, Herman JS, Bell PE (1986) The role of sulfide oxidation in the genesis of Cesspool Cave, Virginia, USA. In: 9th International Congress of Speleology, vol 1. Barcelona, Spain, pp 255–257

Humphreys WF (1999) Physico-chemical profile and energy fixation in Bundera Sinkhole, an anchialine remiped habitat in north-western Australia. J R Soc West Aust 82:89–98

Hunter AJ, Northup DE, Dahm CN, Boston PJ (2004) Persistent coliform contamination in Lechuguilla Cave pools. J Cave Karst Stud 66:102–110

Hunter AJ, Northup DE, Dahm CN, Boston PJ (2005) Persistent coliform contamination in Lechuguilla Cave pools, response: Barton and Pace discussion. J Cave Karst Stud 67:133–135

Hutchens E, Radajewski S, Dumont MG, McDonald IR, Murrell JC (2004) Analysis of metha-notrophic bacteria in Movile Cave by stable isotope probing. Environ Microbiol 6:111–120

Ikner LA, Toomey RS, Nolan G, Neilson JW, Pryor BM, Maier R (2007) Culturable microbial diveristy and the impact of tourism in Kartchner Caverns, Arizona. Microb Ecol 53:30–42

Jannasch HW (1985) Review lecture: the chemosynthetic support of life and the microbial diversity at deep-sea hydrothermal vents. Proc R Soc Lond Ser B Biol Sci 225:277–297

Kelly WR, Panno SV, Hackley KC, Martinsek AT, Krapac IG, Weibel CP, Storment EC (2009) Bacteria contamination of groundwater in a mixed land-use karst region. Water Qual Expo Health 1:69–78

Klimchouk AB (2007) Hypogene speleogenesis: hydrogeological and morphogenetic perspective. Special Paper No. 1. National Cave and Karst Research Institute, Carlsbad, New Mexico

Laiz L, Groth I, Gonzalez I, Saiz-Jimenez C (1999) Microbiological study of the dripping waters in Altamira cave (Santillana del Mar, Spain). J Microbiol Methods 36:129–138

Laiz L, Gonzalez-Delvalle M, Hermosin B, Ortiz-Martinez A, Saiz-Jimenez C (2003) Isolation of cave bacteria and substrate utilization at different temperatures. Geomicrobiol J 20:479–489

Landolt JC, Stephenson SL, Slay ME (2006) Dictyostelid cellular slime moulds from caves. J Cave Karst Stud 68:22–26

Lane DJ, Pace B, Olsen GJ, Stahl DA, Sogin ML, Pace NR (1985) Rapid determination of 16S ribosomal RNA sequences for phylogenetic analyses. PNAS 82:6955–6959

Lavoie KH, Northup DE (2006) Bacteria as indicators of human impact in caves. In: Rea GT (ed) 7th National Cave and Karst Management Symposium, Proceedings. NICKMS Steering Committee, Albany, NY, pp 40–47

Lehman RM (2007) Understanding of aquifer microbiology is tightly linked to sampling approach. Geomicrobiol J 24:331–341

Macalady JL, Lyon EH, Koffman B, Albertson LK, Meyer K, Galdenzi S, Mariani S (2006) Dominant microbial populations in limestone-corroding stream biofilms, Frasassi Cave system, Italy. Appl Environ Microbiol 72:5596–5609

Macalady JL, Jones DS, Lyon EH (2007) Extremely acidic, pendulous cave wall biofilms from the Frasassi Cave system, Italy. Environ Microbiol 9:1402–1414

Mahler BJ, Personne JC, Lods GF, Drogue C (2000) Transport of free and particulate-associated bacteria in karst. J Hydrol 238:179–193

Meisinger DB, Zimmermann J, Ludwig W, Schleifer KH, Wanner G, Schmid M, Bennett PC, Engel AS, Lee NM (2007) In situ detection of novel *Acidobacteria* in microbial mats from a chemolithoautotrophically based cave ecosystem (Lower Kane Cave, WY, USA). Environ Microbiol 9:1523–1534

Mikell AT, Smith CL, Richardson JC (1996) Evaluation of media and techniques to enumerate heterotrophic microbes from karst and sand aquifer springs. Microb Ecol 31:115–124

Morehouse DF (1968) Cave development via the sulfuric acid reaction. Bull Nat Speleol Soc 30:1–10

Mulec J, Kosi G (2009) Lampenflora algae and methods of growth control. J Cave Karst Stud 71:109–115

Mulec J, Kosi G, Vrhovšek D (2008) Characterization of cave aerophytic algal communities and effects of irradiance levels on production of pigments. J Cave Karst Stud 70:3–12

Nieves-Rivera ÁM (2003) Mycological survey of Río Camuy Caves Park, Puerto Rico. J Cave Karst Stud 65:23–29

Northup DE, Lavoie K (2001) Geomicrobiology of caves: a review. Geomicrobiol J 18:199–222

Northup DE, Beck KM, Mallory LM (1997) Human impact on the microbial communities of Lechuguilla Cave: is protection possible during active exploration? J Cave Karst Stud 59:166

Northup DE, Dahm CN, Melim LA et al (2000) Evidence for geomicrobiological interactions in Guadalupe Caves. J Cave Karst Stud 62:80–90

Northup DE, Barns SM, Yu LE et al (2003) Diverse microbial communities inhabiting ferroman-ganese deposits in Lechuguilla and Spider caves. Environ Microbiol 5:1071–1086

Opsahl SP, Chanton JP (2006) Isotopic evidence for methane-based chemosynthesis in the Upper Floridan aquifer food web. Oecologia 150:89–96

Pace NR (1997) A molecular view of microbial diversity and the biosphere. Science 276:734–740

Palmer AN (2007) Cave Geology. Cave Books, Dayton, OH

Porter ML, Engel AS (2008) Diversity of uncultured *Epsilonproteobacteria* from terrestrial sulfidic caves and springs. Appl Environ Microbiol 74:4973–4977

Porter ML, Engel AS, Kinkle B, Kane TC (2009) Productivity-diversity relationships from chemolithoautotrophically based sulfidic karst systems. Int J Speleol 38:27–40

Principi P (1931) Fenomeni di idrologia sotterranea nei dintorni di Triponzo (Umbria). Grotte d'Ital 5:1–4

Pronk M, Goldscheider N, Zopfi J (2006) Dynamics and interaction of organic carbon, turbidity and bacteria in a karst aquifer system. Hydrogeol J 14:473–484

Pronk M, Goldscheider N, Zopfi J (2009) Microbial communities in karst groundwater and their potential use for biomonitoring. Hydrogeol J 17:37–48

Provencio PP, Polyak VJ (2001) Iron oxide-rich filaments: possible fossil bacteria in Lechuguilla Cave, New Mexico. Geomicrobiol J 18:297–309

Roldán M, Hernández-Mariné M (2009) Exploring the secrets of the three-dimensional architec-ture of phototrophic biofilms in caves. Int J Speleol 38:41–53

Romero A (2009) Cave Biology: life in darkness. Cambridge University Press, Cambridge
Rusterholtz KJ, Mallory LM (1994) Density, activity, and diversity of bacteria indigenous to a karstic aquifer. Microb Ecol 28:79–99
Sahl JW, Fairfield N, Harris JK, Wettergreen D, Stone WC, Spear JR (2010) Novel microbial diversity retrieved from an autonomous robotic exploration of the world's deepest vertical phreatic sinkhole. Astrobiology 10:201–213
Sanchez-Moral S, Cañaveras JC, Laiz L, Saiz-Jimenez C, Bedoya J, Luque L (2003) Biomediated precipitation of calcium carbonate metastable phases in hypogean environments. Geomicrobiol J 20:491–500
Sarbu SM (1990) The unusual fauna of a cave with thermomineral waters containing H_2S from Southern Dobrogea, Romania. Mém Biospéol 17:191–195
Sarbu SM, Kane TC, Kinkle BK (1996) A chemoautotrophically based cave ecosystem. Science 272:1953–1955
Schabereiter-Gurtner C, Saiz-Jimenez C, Piñar G, Lubitz W, Rölleke S (2002a) Altamira Cave Paleolithic paintings harbour partly unknown bacterial communities. FEMS Microbiol Lett 211:7–11
Schabereiter-Gurtner C, Saiz-Jimenez C, Piñar G, Lubitz W, Rölleke S (2002b) Phylogenetic 16S rRNA analysis reveals the presence of complex and partly unknown bacterial communities in Tito Bustillo Cave, Spain, and on its Palaeolithic paintings. Environ Microbiol 4:392–400
Schabereiter-Gurtner C, Saiz-Jimenez C, Piñar G, Lubitz W, Rölleke S (2004) Phylogenetic diversity of bacteria associated with Paleolithic paintings and surrounding rock walls in two Spanish caves (Llonín and La Garma). FEMS Microbiol Ecol 47:235–247
Shabarova T, Pernthaler J (2009) Investigation of bacterioplankton communities in aquatic karst pools in Bärenschacht Cave of Bernese Oberland. In: White WB (ed) 15th International Congress of Speleology, Proceedings, vol 1, Symposia Part 1. National Speleological Society, Huntsville, Alabama, USA, pp 416–421
Sket B (2004) The cave hygropetric – a little known habitat and its inhabitants. Arch Hydrobiol 160:413–425
Smith T, Olson R (2007) A taxonomic survey of lamp flora (algae and cyanobacteria) in electrically lit passages within Mammoth Cave National Park, Kentucky. Int J Speleol 36:105–114
Snider JR, Goin C, Miller RV, Boston PJ, Northup DE (2009) Ultraviolet radiation sensitivity in cave bacteria: evidence of adaptation to the subsurface? Int J Speleol 38:11–22
Spear JR, Baton HA, Robertson CE, Francis CA, Pace NR (2007) Microbial community biofabrics in a geothermal mine adit. Appl Environ Microbiol 73:6172–6180
Spilde MN, Northup DE, Boston PJ, Schelble RT, Dano KE, Crossey LJ, Dahm CN (2005) Geomicrobiology of cave ferromanganese deposits: a field and laboratory investigation. Geomicrobiol J 22:99–116
Sudzuki M, Hosoyama Y (1991) Microscopic animals from Gyoku-sen-dô Cave and its water quality. J Speleol Soc Japan 16:38–44
Symk B, Drzal M (1964) Research on the influence of microorganisms on the development of karst phenomena. Geog Pol 2:57–60
Taboroši D (2006) Biologically influenced carbonate speleothhems. In: Harmon RS, Wicks C (eds) Perspectives on Karst Geomorphology, Hydrology, and Geochemistry – A tribute volume to Derek C. Ford and William B. White. Geological Society of America Special Paper 404, pp 307–317
Taylor MR (1999) Dark Life: Martian Nanobacteria, Rock-eating Cave Bugs, and Other Extreme Organisms of Inner earth and Outer Space. Scribner, New York
van Beynen P, Townsend K (2005) A disturbance index for karst environments. Environ Manage 36:101–116
Vlasceanu L (1999) Thriving in the dark: the microbiology of two chemoautotrophically-based groundwater ecosystems. PhD Dissertation, University of Cincinnati, Cincinnati, OH
Vlasceanu L, Popa R, Kinkle BK (1997) Characterization of Thiobacillus thioparus LV43 and its distribution in a chemoautotrophically based groundwater ecosystem. Appl Environ Microbiol 63:3123–3127

Vlasceanu L, Sarbu SM, Engel AS, Kinkle BK (2000) Acidic cave-wall biofilms located in the Frasassi Gorge, Italy. Geomicrobiol J 17:125–139

Walochnik J, Mulec J (2009) Free-living amoebae in carbonate precipitating microhabitats of karst caves and a new vahlkampfiid amoeba, *Allovahlkampfia spelaea* gen. nov., sp. nov. Acta Protozool 48:25–33

Weidler GW, Dornmayr-Pfaffenhuemer M, Gerbl FW, Heinen W, Stan-Lotter H (2007) Communities of Archaea and Bacteria in a subsurface radioactive thermal spring in the Austrian Central Alps, and evidence of ammonia-oxidizing *Crenarchaeota* source. Appl Environ Microbiol 73:259–270

Went FW (1969) Fungi associated with stalactite growth. Science 16:385–386

White WB, Culver DC (2000) Cave, definition of. In: Culver DC, White WB (eds). Encyclopedia of Caves. Elsevier Academic Press, Burlington, MA, pp. 81–85

White WB (2002) Karst hydrology: recent developments and open questions. Eng Geol 65:85–105

White WB (2009) Proceedings, International Congress of Speleology, Kerrville, Texas

Whitman WB, Coleman DC, Wiebe WJ (1998) Prokaryotes: the unseen majority. PNAS 95:6578–6583

Woese CR (1987) Bacterial evolution. Microbiol Rev 51:221–271

Zhou JP, Gu YQ, Zou CS, Mo MH (2007) Phylogenetic diversity of bacteria in an earth-cave in Guizhou Province, Southwest of China. J Microbiol 45:105–112

Chapter 11
Statistical Evaluation of Bacterial 16S rRNA Gene Sequences in Relation to Travertine Mineral Precipitation and Water Chemistry at Mammoth Hot Springs, Yellowstone National Park, USA

Héctor García Martín, John Veysey, George T. Bonheyo, Nigel Goldenfeld, and Bruce W. Fouke

It is possible that common earth-surface geological features can arise as a result of bacteria interacting with purely physical and chemical processes. The ability to distinguish ancient and modern mineral deposits that are biologically influenced from those that are purely abiotic in origin will advance our ability to interpret microbial evolution from the ancient rock record on earth and potentially other planets. As a step toward deciphering biotic from abiotic processes, we have combined

H.G. Martín, J. Veysey, and N. Goldenfeld
Department of Physics, University of Illinois at Urbana-Champaign, 1110 West Green Street, Urbana, IL 61801-3080, USA

G.T. Bonheyo and B.W. Fouke (✉)
Department of Geology, University of Illinois at Urbana-Champaign, 1301 West Green Street, Urbana, IL 61801-2938, USA
e-mail: fouke@illinois.edu

B.W. Fouke
Department of Microbiology, University of Illinois Urbana-Champaign, 601 S. Goodwin Avenue, Urbana, IL 61801, USA

N. Goldenfeld and B.W. Fouke
University of Illinois Urbana-Champaign, Institute for Genomic Biology, 1206 W. Gregory Drive, Urbana, IL 61801, USA

H.G. Martín
Joint BioEnergy Institute, Emeryville, CA 94608, USA
and
Physical Biosciences Division, Lawrence Berkeley National Laboratory, Berkeley, CA 94710, USA

G.T. Bonheyo
Pacific Northwest Laboratory, Marine Sciences Laboratory, 1529 W. Sequim Bay Rd, Sequim, WA 98382, USA

L.L. Barton et al. (eds.), *Geomicrobiology: Molecular and Environmental Perspective*,
DOI 10.1007/978-90-481-9204-5_11, © Springer Science+Business Media B.V. 2010

carbonate mineralogical and geochemical analyses together with community-based microbial genetic analyses in hot spring drainage systems at Mammoth Hot Springs in Yellowstone National Park. Previously (Fouke et al. 2000, 2003), we reported the shape and chemistry of carbonate mineral deposits (*travertine*), which have formed along the hot spring outflow. This travertine exhibits five distinct ecological zonations (termed sedimentary depositional *facies*) even though most physical and chemical attributes of the spring water change smoothly and continuously over the course of the drainage outflow path. Here, we document an unexpectedly sharp correlation between microbial phylogenetic diversity and travertine facies, which suggests that changes in bacterial community composition are a sensitive indicator of environmental conditions along the spring outflow. These results provide an environmental context for constraining abiotic and biotic theories for the origin of distinct crystalline structures and chemistries formed during hot spring travertine precipitation.

In order to quantitatively track and identify the mechanisms and products of microbial fossilization during calcium carbonate precipitation, we have initiated a biocomplexity study of microbe-mineral-environmental interactions at Mammoth Hot Springs in Yellowstone National Park. The goal is to determine whether microbial community structure and activity directly influence the precipitation of high-temperature terrestrial calcium carbonate mineral deposits, called *travertine* (Ford and Pedley 1996; Pentecost 2005). Initially, we determined that the Mammoth spring drainage systems are composed of five ecological partitions (called *sedimentary facies*), an analysis based solely on the shape, structure and chemical composition of the travertine mineral deposits (Fouke et al. 2000). Analyses of spring water chemistry were then integrated with the travertine facies along the drainage system to establish water-mineral precipitation baselines defining the bulk system-level chemical evolution of the spring outflow system (Fouke et al. 2000). The next phase of the research was to spatially map the ecological distribution of bacteria with respect to the travertine facies in which they live, using what have now become standard culture-independent molecular analyses of 16S rRNA gene sequences (Fouke et al. 2003).

The present study is a rigorous statistical evaluation of these bacterial 16S rRNA gene sequences, which is the first such analysis completed within a natural environmental setting. We had not yet developed nor applied these statistical approaches at the time of the publication of the original clone libraries (Fouke et al. 2003). However, the results presented in the present paper are essential in that they provide the first quantitative validation of: (1) the completeness and randomness of these types of molecular microbial analyses in the environment; and (2) the system-scale correlation of microbial community composition with calcium carbonate mineral precipitation, which is the critical ecological relationship required to begin to identify the dynamics of microbial influence on carbonate mineral precipitation. These statistical approaches are universally applicable to other molecular studies of microbial ecology and will therefore be a valuable tool for studying biocomplex dynamics in the environment.

Geological Setting of Mammoth Hot Springs

Subsurface waters erupt at Mammoth Hot Springs to precipitate terraced crystalline deposits, called travertine, which is composed of aragonite and calcite (Friedman 1970; Sorey 1991). Our studies were conducted at Spring AT-1 (Fouke et al. 2000), located on Angel Terrace, in the upper terrace region of the Mammoth Hot Springs complex. Spring AT-1 is typical of the hot springs found at Mammoth Hot Springs, in that as the spring water flows away from the subsurface vent, the water cools, degases CO_2, increases in pH, and precipitates travertine that steadily changes composition from nearly 100% aragonite to nearly 100% calcite. Precipitation rates are rapid (Kandianis et al. 2008) and can reach 5 mm/day. The rapid precipitation partially seals the vents and reroutes surface outflow, causing the spring flow path to regularly change in direction and intensity, which in turn influences subsequent travertine precipitation. The dynamical interplay between fluid flow and travertine precipitation, be it primarily biotic or abiotic in origin is complex and not yet understood. The hot springs harbor diverse communities of microorganisms, representing at least 21 divisions of bacteria (Fouke et al. 2003).

In order to analyze the physical, geological, and biological aspects of this rapidly changing hydrothermal system, we first subdivided the spring drainage system into a series of recognizable sub-environments or ecological partitions along the flow path. These sub-environments are known as sedimentary depositional *facies*. A facies is defined as a sedimentary rock deposit that represents the sum total of physical, chemical, geological, and biological processes active in a natural environment of sediment deposition and mineral accumulation. Each facies has its own distinct mineralogical and hydrological features and may therefore be readily identified, even if the overall drainage system significantly changes and migrates. Our previous work defined a five-component travertine facies model for Spring AT-1. This is based on physical and chemical characteristics of the spring water (temperature, pH, elemental and isotopic chemistry) and associated travertine (crystalline growth form and fabric, mineralogy, elemental and isotopic chemistry), quantitative modeling of this aqueous and solid data, and limited microscopic observations of the microbiology (Fig. 11.1). The following is a brief summary of these five facies, called the vent, apron and channel, pond, proximal slope, and distal slope (Fouke et al. 2000).

The facies model allows equivalent ecological locations in the spring drainage systems to be analyzed over time, despite nearly constant changes in the rate or direction of spring flow, and thus allows comparisons to be made between springs in different geographic locations and of different geological ages. Remarkably, we find that the physical structures characteristic of each facies develop sharp boundaries instead of gradual transitional zones (Fouke et al. 2000). Although any given travertine facies may be as much as tens of meters long and cover hundreds of square meters in area, the boundary between facies is relatively abrupt, occurring over as little as 1 cm in distance between the pond (1–3 m in length along the spring flowpath) and proximal slope (10–15 m in length) or up to 10 cm between the proximal slope and distal slope (10–15 m in length).

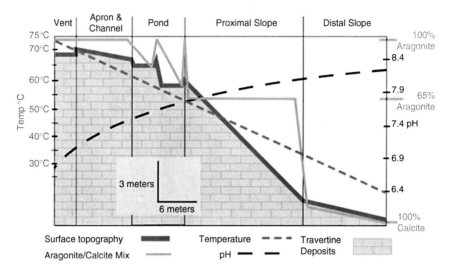

Fig. 11.1 Facies model. Cross-sectional view of Spring AT-1 with 2 × vertical exaggeration to highlight the topography of the spring features. Trends in pH, temperature and travertine aragonite/calcite mineralogical ratios overlay the structural representation to show how these attributes change with increasing distance from the spring outflow source vent

The aqueous chemistry of the hot spring drainage system is dominated by CO_2 degassing and dropping temperature as proven by Rayleigh-type fractionation calculations of spring water dissolved inorganic carbon (DIC) and its associated $\delta^{13}C$ versus (Fouke et al. 2000). While these physical factors help drive the rapid precipitation of carbonate crystals to deposit travertine at rates as high as 5 mm/day at the pond lip, they are not the exclusive controls on precipitation. Significant biological controls on travertine crystal form and isotope chemistry have been identified where travertine crystals entomb and preserve the shape of filamentous *Aquificales* bacteria, and by quantitative subtraction of degassing and temperature effects on $\delta^{13}C$ and $\delta^{18}O$ isotopic fractionation in the spring water and the travertine. These robust disequilibrium signatures may be biologically mediated and systematically increase in magnitude from the high (73°C) to the low (≤25°C) temperature portions of the Spring AT-1 outflow.

We then conducted a culture-independent molecular survey of the bacterial communities, which are distinctly partitioned between travertine depositional facies in the surface drainage system of Spring AT-1 (Fouke et al. 2003). PCR amplification and sequencing of 16S rRNA genes with universally conserved bacterial primers has identified over 553 unique partial and 104 complete gene sequences (derived from more than 14,000 clones) affiliated with 221 unique species that represent 21 bacterial divisions. These sequences exhibited less than 12% similarity in bacterial community composition between each of the travertine depositional facies. This implies that relatively little downstream bacterial transport and colonization take place despite the rapid and continuous flow of spring water from the high-temperature to low-temperature facies. These results suggest that travertine depositional facies, which are independently determined by the physical and chemical conditions of the hot spring drainage system, effectively predict bacterial community composition as well as the morphology and chemistry of travertine precipitation.

Materials and Methods

Field work and sample collection. We collected multiple samples from within each of the five travertine facies at Spring AT-1 for the purpose of conducting the first direct correlation of bacterial 16S rRNA gene sequence identifications with travertine mineral precipitation in the context of sedimentary depositional facies (Fouke et al. 2003). Field photographs and detailed diagrams depicting aerial and cross-sectional views of Spring AT-1, and sampling positions, have previously been published (Fouke et al. 2000, 2003). As a brief summary, samples were collected from the interior of each of the five facies, with each sample occurring within the continuous primary flow path of the primary hot spring drainage outflow (Veysey et al. 2008). The sampling strategy for the present study was to conduct an initial characterization of the microbial communities inhabiting each travertine facies. Therefore, each sample was collected from the middle of each facies and was thus laterally separated from the next sample by as much as a few meters. With the results presented in this study and our previous work (Fouke et al. 2003), our ongoing microbiological analyses of Spring AT-1 is currently focused on detailed mm-scale sampling across the boundaries between facies, as well as a correlation of specific crystal morphologies and chemistries with microbial phylogenetic and functional diversity. However, these next progressive and strategic stages of our analysis of Spring AT-1 would not be possible without the synthesis of the data presented in the present paper.

DNA extraction, PCR amplification, cloning and sequencing. The DNA extraction protocols and 16S rRNA gene sequence PCR amplification protocols employed have been optimized to avoid biases and have previously been described (Frias-Lopez et al. 2002; Fouke et al. 2003). DNA amplified using universal bacterial primers was then cloned in order to isolate the individual 16S rRNA gene sequences. To maximize the number of unique sequences identified (thus better characterizing the total diversity of the spring system) we chose to avoid sequencing identical clones derived from a single PCR reaction. Of the greater than 14,000 clones generated, approximately 5,000 clones were screened by RFLP analyses and 1,050 potentially unique clones were selected for sequencing. Ultimately, 657 partial 16S rRNA gene sequences were obtained, and 108 of these were sequenced as contigs to completion.

Nucleotide sequence accession numbers. The GenBank accession numbers for the 16S rRNA gene sequences analyzed in this study have previously been reported (Fouke et al. 2003).

Statistical analyses. We analyzed our sequences using three Operational Taxonomic Unit (OTU) definitions, defined by sequence differences of 0.5%, 1%, and 3%, to determine whether our interpretations of environmental partitioning could be affected by such variation. The lower bound is due to our PCR and sequence derived error rate (Tindall and Kunkel 1988; Barnes 1992) and the 3% difference is a typical OTU definition (Stackebrandt and Goebel 1994). In our accumulation curves, a straight line would indicate that we have sampled only a small subset of the total biodiversity: new OTUs are found at a constant rate with each additional new sample analyzed. If a facies is well sampled, however, the curve will flatten asymptotically when the number of samples, n, is large, because novel OTU sequences are detected with decreasing frequency.

To quantitatively estimate how well each facies has been sampled, accumulation curves were fitted to analytical curves obtained by modeling the sampling process. We assume that in each environmental sample collected, there is a maximum of N possible bacterial cells that could be detected, and that each of these cells would be present and detected in the sample with a probability p, regardless of the cell's identity. The factor p includes the combined probability of the cell being captured and detected through the process of DNA extraction and amplification of the 16S rRNA gene sequences via PCR. Thus, we use multiple methods of DNA extraction to eliminate cell durability biases and amplify the 16S rRNA gene via PCR. Finally, we screen the resultant clone library in an attempt to sequence only unique clones within that sample, as opposed to repeatedly sequencing identical clones. In this manner we increase the likelihood that an OTU will be detected even if it is not numerically dominant in the clone library (which may be due to extraction, amplification, and cloning biases rather than environmental population abundance).

The likelihood that each sequence we analyze will represent a new OTU is approximated as $(1-S/S_o)$, where S is the number of different OTUs already identified and S_o is the total number of different OTUs present in the environment. For each sequence, the probability that the number of different OTUs will increase is $p(1-S/S_o)$. This leads to an accumulation curve of the type $S = S_m (1-exp(-Kt))$, where t is the maximum number of individuals that would be found if $p = 1$ and K is a constant related to the sampling procedure. This is not quite what was represented in the accumulation curves, since we only have information about samples rather than individuals, as explained above. Nonetheless, the number of samples n is simply $n = t/N$, so $S = S_m (1-exp(-Kn))$. The parameters K and S_m were determined from a linear fit of $log(dS/dn)$ versus $-n$. Estimates through other methods were also attempted: fits to hyperbolic accumulation curves (Colwell and Coddington 1994) were not convincing and non-parametric methods (Krebs 1989; Chao and Lee 1994) yielded variances that were too large to be trustworthy.

Results

We identified 193 OTUs using the 3% cutoff and found that 90% of these could be identified in only one of the facies (partitioned between facies). There were 237 OTUs using the 1% cutoff and 331 OTUs using the 0.5% cutoff with 91%

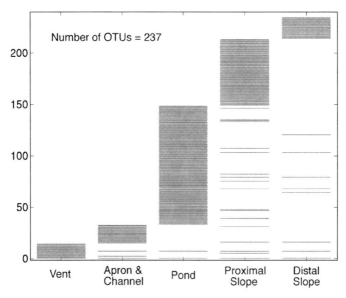

Fig. 11.2 Species present in each facies: 1% OTU definition. Each OTU is numbered sequentially, starting with OTUs that first appear in the Vent facies, followed by OTUs that first appear in the Apron and Channel, then the Pond, the Proximal Slope, and lastly the Distal Slope facies. The figure provides a graphical representation of where each OTU (y-axis) is found (x-axis)

and 93% (respectively) of the sequences partitioning to a single facies. Figure 11.2 graphically represents the distribution of sequences between the five facies using a 1% OTU definition. The plots for the 3% and 0.5% definitions are similar in appearance; however, under the 3% rule, two sequences may be found in all five facies (20). Finally, the total number of sequences that can be found in more than one facies remains low under each OTU definition: 19 OTUs under the 3% definition, 20 OTUs under the 1% definition, and 24 OTUs under the 0.5% definition.

Accumulation curves were generated for the three different OTU definitions (3%, 1% and 0.5%) for each facies and the results for the pond facies are shown in Fig. 11.3b. The curve from each OTU definition collapses into the same curve, giving some confidence in the robustness of the sampling procedure and the validity of the assumption of random sampling used to derive the exponential accumulation curve. We see this pattern no matter which OTU definition is used. In the model above, all of the OTUs were assumed equally likely to appear (hence the factor $1-S/S_o$). In a more realistic model the probability of finding each OTU should be proportional to its abundance. However, the approximations used above describe the data well and provide a tractable expression for the accumulation curve.

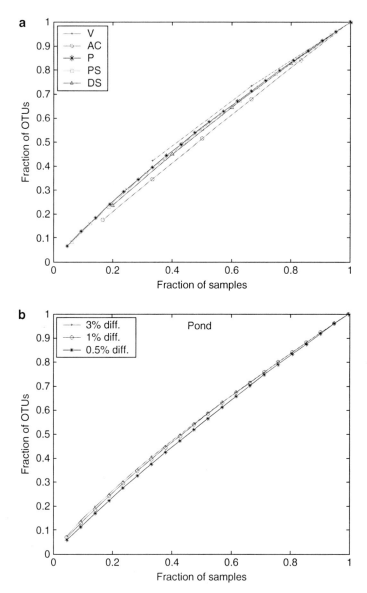

Fig. 11.3 Accumulation curves and exponential fits. (**a**) Accumulation curve generated for each facies using a 1% OTU definition. (**b**) Accumulation curves generated for the pond facies using a 3%, 1%, and 0.5% OTU definitions. Accumulation curves for different OTU definitions collapse into the same curve when the x and y-axis are properly scaled by the total number of OTUs and samples, respectively

Discussion

Although different microbial species have specific growth requirements and preferred temperature and pH ranges, the tight partitioning with respect to the travertine facies is nonetheless remarkable. First, it is surprising that very few of the upstream sequences were not also detected downstream. We initially expected that the rapid flow of the spring would result in downstream transport of microbial cells, and thus we thought that many sequences would also being identified downstream of their point of initial detection. Consequently, we performed most of our analyses on the first four facies extending from the vent. Surprisingly, the sequences detected in the water column of one facies, which are presumably most susceptible to being flushed downstream, were not typically detected downstream of their original facies. Secondly, because bacterial species have a preferred range of environmental growth conditions, we expected that many sequences would be found across facies boundaries, coinciding with gradual temperature and pH changes. However, the facies boundaries proved to be nearly absolute boundaries with respect to detected bacterial 16S rRNA gene sequences. Although we observed particular sequences over a range of conditions within each travertine facies, with very few exceptions, OTUs were not found to traverse the facies boundaries.

Inferred metabolic activity of the identified bacteria, derived from comparison of our sequences to GenBank, indicates that the bacterial communities found in the spring drainage system change from primarily chemolithotrophic in the vent facies, to photoautotrophic and ultimately to heterotrophic in the distal slope facies. Associated with this transition is an observed increase in the total number of OTUs and their associated bacterial divisions from the vent to the pond facies. The number of OTUs decreases, however, with down flow progression into the proximal slope and distal slope facies. These trends in our data can be interpreted as follows: fewer OTUs and bacterial divisions would be expected at the upper temperature limits of the spring where little organic matter is available for heterotrophy and the temperature is at the upper limit for photosynthesis (Miller and Castenholz 2000). Although the pond, proximal-slope and distal-slope facies have temperature profiles that would support both autotrophic and heterotrophic lifestyles (Fouke et al. 2000), we actually find a reduction in the number of species represented in the proximal-slope and distal-slope facies. This variation results from differences in the environmental stability of each facies with regards to temperature, pH, and water flow (Veysey et al. 2008). Ponds, for example, have the widest temperature and pH range of any facies and show greater fluctuations in flow direction and intensity.

To validate our interpretation of facies partitioning, we need to determine what proportion of the total community in each facies we have identified. Severe undersampling might prevent us from identifying OTUs that actually do occur in multiple facies. Estimates for the total number of OTUs in each facies are made using an exponential fit to the accumulation curve in Fig. 11.3. The accumulation curve plots the number of different OTUs, S, found in a given number of samples versus this number of samples, n. Since all of the samples are assumed to be equivalent, this graph is an average over all possible permutations of these samples. Accumulation curves are traditionally made using the number of individuals as the x-axis instead of the number of samples (Hughes et al. 2001). However, our samples amalgamate large numbers of individuals: we have information regarding which OTUs are present in each sample, but not the OTU identity for every individual in the sample. The abundance of unique gene sequences in the clone libraries are not representative of the abundances in the environmental sample due to the inherent DNA extraction and PCR biases. Therefore, the clone library data cannot be used to make accumulation curves.

Accumulation curves were generated based on the three different OTU definitions (3%, 1% and 0.5%) for each facies, with the results for the pond facies shown as an example in Fig. 11.3b. The curve from each OTU definition collapses into the same curve, giving confidence in the robustness of the sampling procedure and the validity of the assumption of random sampling used to derive the exponential accumulation curve. Thus, since all of the individual cells are captured with equal probability, we expect that the observed OTUs represent the most numerically abundant bacteria in each facies. Consequently, we conclude that these species (and therefore most of the bacterial consortia) are partitioned according to the travertine facies model. This finding constrains abiotic and biotic theories for the origin of travertine terraces (Kandianis et al. 2008; Veysey and Goldenfeld 2008).

Acknowledgements This research was supported by the National Science Foundation Biocomplexity in the Environment Coupled Biogeochemical Cycles Program (EAR 0221743), the National Science Foundation Geosciences Postdoctoral Research Fellowship Program (EAR-0000501), the Petroleum Research Fund of the American Chemical Society Starter Grant Program (34549-G2), and the University of Illinois Urbana-Champaign Critical Research Initiative. This work was completed under National Park Service research permit number 3060R. Conclusions in this study are those of the authors and do not necessarily reflect those of the funding or permitting agencies. Discussions with A. Salyers and C. Woese are gratefully acknowledged.

References

Barnes WM (1992) The fidelity of Taq polymerase catalyzing PCR is improved by an N-terminal deletion. Gene 112:29–35
Chao A, Lee S-M (1994) Estimating population size via sample coverage for closed capture-recapture models. Biometrics 50:88
Colwell RK, Coddington A (1994) Estimating terrestrial biodiversity through extrapolation. Phil Trans R Soc Lond Ser B 345:101–108
Ford TD, Pedley HM (1996) A review of tufa and travertine deposits of the world. Earth Sci Rev 41:117–175
Fouke BW, Farmer JD, Des Marais D, Pratt L, Sturchio NC, Burns PC, Discipulo MK (2000) Depositional facies and aqueous-solid geochemistry of travertine-depositing hot springs (Angel Terrace, Mammoth Hot Springs, Yellowstone National Park, USA). J Sediment Res 70:265–285
Fouke BW, Bonheyo GT, Sanzenbache BL, Frias-Lopez J (2003) Partitioning of bacterial communities between travertine depositional facies at Mammoth Hot Springs, Yellowstone National Park, USA. Can J Earth Sci 40:1531–1548

Frias-Lopez J, Zerkle AL, Bonheyo GT, Fouke BW (2002) Cyanobacteria diversity associated with coral black band disease in Caribbean and Indo-Pacific reefs. Appl Environ Microbiol 69:2409–2413

Friedman I (1970) Some investigations of the deposition of travertine from hot springs: I. The isotope chemistry of a travertine-depositing spring. Geochim Cosmochim Acta 34:1303–1315

Hughes JB, Hellmann JJ, Ricketts TH, Bohannan BJM (2001) Counting the uncountable: statistical approaches to estimating microbial diversity. Appl Environ Microbiol 67:4399–4406

Kandianis MT, Fouke BW, Veysey J, Johnson RW, Inskeep W (2008) Microbial biomass: a catalyst for $CaCO_3$ precipitation in advection-dominated transport regimes. GSA Bull 120:442–450

Krebs C (1989) Ecological methodology. Harper & Row, New York

Miller SR, Castenholz RW (2000) Evolution of thermotolerance in hot spring cyanobacteria of the genus *Synechococcus*. Appl Environ Microbiol 66:4222–4229

Pentecost A (2005) Travertine. Springer, Heidelberg

Sorey ML (1991) Effects of potential geothermal development in the Corwin Springs known geothermal resources area, Montana, on the thermal features of Yellowstone National Park. Water Resources Investigations Report No. 91-4052. US Geological Survey

Stackebrandt E, Goebel B (1994) Taxonomic Note: a place for DNA-DNA reassociation and 16S rRNA sequence analysis in the present species definition in bacteriology. Int J Syst Bacteriol 44:846–849

Tindall KR, Kunkel TA (1988) Fidelity of DNA synthesis by the Thermus aquaticus DNA polymerase. Biochemistry 27:6008–6013

Veysey J, Goldenfeld N (2008) Watching rocks grow. Nat Phys 3:1–5

Veysey J, Fouke BW, Kandianis MT, Schickel TJ, Johnson RW, Goldenfeld N (2008) Reconstruction of water temperature, pH, and flux of ancient hot springs from travertine depositional facies. J Sediment Res 78:69–76

Chapter 12
Compositional, Physiological and Metabolic Variability in Microbial Communities Associated with Geochemically Diverse, Deep-Sea Hydrothermal Vent Fluids

Ken Takai and Kentaro Nakamura

Introduction

Deep-sea hydrothermal vent environments represent one of the most physically and chemically diverse biomes in Earth. The chemical and thermal gradients (e.g., >350°C across distances as small as several centimeters in active chimneys) provide a wide range of niches for microbial communities living there (Huber and Holden 2008; Nakagawa and Takai 2008; Reysenbach et al. 2000; Takai et al. 2006a). Psychrophiles, mesophiles, thermophiles and hyperthermophiles (organisms growing best from 4°C to above 80°C) thrive by chemolithoautotrophy or heterotrophy, utilizing abundant available inorganic and organic chemical energy, carbon and other element sources. They reside as free-living forms in the rocky and sedimentary mixing interfaces between hot, highly reductive hydrothermal fluids (high temperatures of endmember hydrothermal fluids) and ambient seawaters beneath and at the seafloor (along subseafloor hydrothermal fluid paths, and in and on chimneys and sediments) and within lower temperatures of diffuse fluids mainly resulting from subseafloor mixing between endmember hydrothermal fluids and ambient seawaters, and as facultative or obligate symbionts on and within invertebrate hosts (Takai et al. 2006a).

Over the past 2 decades, microbiologists have gained significant insights into the compositional and functional diversity of microbial communities in these unique ecosystems. In particular, the microbial communities have been extensively explored in the chimney structures hosting high temperatures of endmem-

K. Takai (✉)
Subsurface Geobiology Advanced Research (SUGAR) Project, Japan Agency for Marine-Earth Science and Technology (JAMSTEC), 2-15 Natsushima-cho, Yokosuka 237-0061, Japan
and
Precambrian Ecosystem Laboratory, Japan Agency for Marine-Earth Science and Technology (JAMSTEC), 2-15 Natsushima-cho, Yokosuka 237-0061, Japan
e-mail: kent@jamstec.go.jp

K. Nakamura
Precambrian Ecosystem Laboratory, Japan Agency for Marine-Earth Science and Technology (JAMSTEC), 2-15 Natsushima-cho, Yokosuka 237-0061, Japan

L.L. Barton et al. (eds.), *Geomicrobiology: Molecular and Environmental Perspective*, DOI 10.1007/978-90-481-9204-5_12, © Springer Science+Business Media B.V. 2010
251

ber-like hydrothermal fluid emissions (McCliment et al. 2006; Nakagawa and Takai 2006; Nakagawa et al. 2006; Nunoura and Takai 2009; Pagé et al. 2008; Perner et al. 2007a, b; Takai et al. 2008a, 2009; and see references before 2006 in Takai et al. 2006a) and in the relatively lower temperatures of diffuse flow fluids (see references in Huber and Holden 2008). Much of the early research was conducted in the representative deep-sea hydrothermal systems in the Mid Ocean Ridges (MORs), however recently, as the greater heterogeneity of the physical and chemical characteristics of hydrothermal fluids in the Volcanic Arcs (VAs) and Backarc Basins (BABs) and even of hydrothermal fluids in the ultramafic rocks-associated systems recently found in the MORs, several studies have demonstrated the great inter-fields heterogeneity of the microbial communities, particularly present in the chimney structures the most adjacently influenced by the hydrothermal fluids, among a variety of fields (Brazelton et al. 2006; Nunoura and Takai 2009; Takai et al. 2006a, 2008a, 2009). These results have suggested the impact of endmember-like hydrothermal fluid chemistry on the compositions and functions of proximal microbial communities in the subseafloor as well as at the seafloor habitats. Of course, it is still not entirely clear how the hydrothermal fluid chemistry control the compositional and functional diversity and extent of the adjacent microbial communities. However, several disciplines of approaches begin to clarify the geochemical-microbial interrelation based on the thermodynamic modeling of microbially available energy potentials by various chemolithotrophic metabolisms in the hydrothermal mixing zones and the inter-fields comparisons of functionally active microbial communities coupled with hydrothermal inputs of energy sources (McCollom 2007; McCollom and Shock 1997; Shock and Holland 2004; Takai et al. 2008a, 2009).

In this chapter, we first overview the physical and chemical characteristics of endmember-like hydrothermal fluids and the compositions and functions of the proximal microbial communities in several well-studied deep-sea hydrothermal fields, representing the distinct geographical and geological settings. As it has been already pointed out (Takai et al. 2008a, 2009), the inter-fields comparison highlights a concept that different physical and chemical characteristics of hydrothermal fluids would control the compositional, physiological and metabolic variability in the surrounding microbial communities. In addition, to find the key for the geochemical-microbial relevance, the energetic potentials of indigenous chemolithotrophic metabolisms of the microbial communities are thermodynamically modeled under given conditions of habitats in the hydrothermal mixing zones and compared among the different hydrothermal fields. The thermodynamic energy state of potentially available energy metabolisms in each of the hydrothermal fields is highly associated with the emerging pattern of constituent microbial populations showing specific energy metabolisms in each of the microbial communities characterized. Hence, the chemolithotrophic energy potentials may be one of the significant geochemical aspects to control the compositional and functional variability of microbial ecosystems in the deep-sea hydrothermal environments.

Method for Thermodynamic Calculation of Available Energy Metabolisms

In this chapter, the physical and chemical characteristics of endmember hydrothermal fluids in the representative fields (Kairei field in the Central Indian Ridge, Mariner field in the Lau Basin, Iheya North field in the Okinawa Trough and TOTO caldera field in the Mariana Arc) are compiled from the previously reported literatures (Gamo et al. 2004; Kumagai et al. 2008; Nakagawa et al. 2005, 2006; Takai et al. 2004a, 2008a) (Table 12.1). All the hydrothermal fields overviewed in this chapter more or less have the physical and chemical variations of endmember hydrothermal fluids within the field, which is driven by the subseafloor phase-separation and -partition processes. Thus, several endmembers of phase-separation-influenced hydrothermal fluids are listed in Table 12.1. In addition, not all the physical and chemical properties have been described and then, the previously unpublished data are added in Table 12.1. The other blank values are tentatively calculated according to a simple phase-separation ratio that is linearly reproduced by the deviation of chloride concentrations in the hydrothermal fluids, or are tentatively substituted by the values of the physically similar hydrothermal fluids in the similar hydrothermal fields. The corresponding values in the ambient seawater (mixing counterpart) are taken from the data previously reported (Charlou et al. 1996; McCollom 2007).

For this study, a thermodynamic reaction-path model was employed to calculate the fluid composition during mixing between seawater and hydrothermal fluids. The estimated composition was then used to determine the available energy yields of a variety of potential metabolic reactions as a function of temperature. The thermodynamic model employed followed the method developed by McCollom and Shock (1997), Shock and Holland (2004), and McCollom (2007). The model calculation began with 1 kg of vent fluid and continued with successive increments of seawater until a seawater-to-vent fluid mixing ratio of 1,000:1 was reached. In the mixing calculations, we assumed that all the redox reactions and the mineral precipitations were kinetically inhibited at the timescale of fluid mixing. In addition, the temperature was used as a direct function of mixing ratio, ignoring conductive cooling and/or heating, so that the addition of each increment of seawater lowered the temperature of the mixed fluid.

The amounts of chemical energy potentially available for chemolithotrophic metabolisms in the mixed fluid (bioavailable energy) were determined by calculating the Gibbs free energy of each of the metabolic reactions. The chemolithotrophic energy metabolisms applied to thermodynamic modeling are listed in Table 12.2. The overall Gibbs free energy of reaction can be calculated using the equation:

$$\Delta Gr = \Delta Gr° + RT \ln Q \tag{12.1}$$

where ΔGr is the Gibbs free energy of reaction, $\Delta Gr°$ is the standard state Gibbs free energy of reaction, R is the universal gas constant, T is the temperature in Kelvin, and Q is the activity quotient of the compounds involved in the reaction. The Q term takes into account the contribution of the fluid composition to the Gibbs

Table 12.1 Physical and chemical characteristics of hydrothermal fluids in different deep-sea hydrothermal fields

Field	Kairei in CIR		Mariner in LB		Iheya North in OT			TOTO caldera in MVA	Rainbow in MAR	Ambient seawater
Vent site	Kali	Fugen	SC and MC	CRC	NBC	CBC	E18	Sulfur chimney		
Temp. (°C)	360	360	365	365	311	311	311	103	365	2
Depth (m)	2,450	2,420	1,908	1,921	980	990	1,000	2,951	2,300	
pH	3.4	3.4	2.4	2.66	4.8	5.3	4.6[a]	5.2	2.8	7.8
H_2 (mM)	3.7	3	0.096	0.045	0.229[b]	0.135	0.346	0.001[a]	16	0.0000004
H_2S (mM)	4.8	6.4	6.9	9	4.5[b]	2.7	6.8	14.6	1.2	0
CH_4 (mM)	0.123	0.2	0.007	0.008	6.9[b]	4.1	10.4	0.002[a]	2.5	0.0000003
CO_2 (mM)	9.5	10	41	72	230[b]	136	347	59[a]	16	2.3
N_2 (mM)	1.43[b]	1.43[b]	0.458[a]	1.1[a]	0.0346[b]	0.0189	0.0487	1[a]	1.8	0.59
CO (mM)	0.0005	0.0007	0.0003[b]	0.0004[b]	<0.0001	<0.0001	<0.0001	0.001[b]	0.005	0
SO_4 (mM)	0	0	0	0	0	0	0	25.3	0	27.9
NH_4 (mM)	0.0274	0.01	0.0161[a]	0.011[a]	2.1[b]	1.85[b]	2.97[b]	<0.01	0.01	0.00005
Na (mM)	518	489	454	387	405	745	288	432	553	464
Cl (mM)	623	606	597	531	511	864	338	502	750	546
Ca (mM)	27.2	26.7	45.2	40.6	18.1	19.9	11.9	11.6	66.6	10.2
Mg (mM)	0	0	0	0	0	0	0	51.9	0	52.7
K (mM)	13.8	12.7	31.5	26.4	73	79	56.2	9.7	20.4	9.8
Fe (mM)	5.4	3.9	12.8	11.1	0.251[b]	0.424	0.167	0.5[b]	24.1	0.0000015
Mn (mM)	0.828	0.84	6	4.9	0.669	0.46	0.445	0.018[b]	2.25	0
NO_3 (mM)	0	0	0	0	0	0	0	0.04[b]	0	0.03
Si (mM)	16	16.8	15.6	15.2	10.5	11.6	6.9[b]	3.4	6.9	0.16
δD (H_2)	−368[b]	−391[b]	−340[b]	−374[b]	−430[b]	n.d.	n.d.	n.d.	n.d.	n.d.

$\delta^{13}C$ (CO_2)	-5.3	-5.1	-0.1	-0.8	-9.7[b]	n.d.	n.d.	-1.5	
$\delta^{13}C$ (CH_4)	-9.8	-18.0	-3.4	+3.7	-53.8[b]	n.d.	n.d.	-38.1	
$\delta^{34}S$ (H_2S)	n.d.[c]	n.d.	+6.4	+6.9	+12[b]	n.d.	n.d.	n.d.	
Reference	Kumagai et al. 2008; Takai et al. 2004a	Kumagai et al. 2008; Takai et al. 2004a	Takai et al. 2008a	Takai et al. 2008a	Nakagawa et al. 2005	Nakagawa et al. 2005	Nakagawa et al. 2005	Gamo et al. 2004; Nakagawa et al. 2006	Charlou et al. 2002	Charlou et al. 1996; McCollom 2007

[a]Values taken from a similar hydrothermal fluid: for Mariner field from Brothers caldera NW field (unpublished); for E18 site in Iheya North field from a gas-rich fluid in the Yonaguni Knoll IV field (unpublished); for TOTO caldera field from Nikko seamount (Lupton et al. 2008)

[b]Unpublished data

[c]n.d.: not determined

Values calculated according to a simple phase-separation ratio that is linearly reproduced by the deviation of chloride concentrations in the hydrothermal fluids

Table 12.2 Potential microbial chemolithotrophic metabolisms in deep-sea hydrothermal mixing zones

Energy metabolism	Overall chemical reaction	Identified (I)/ Cultured (C)[a]	Remark[b]
Aerobic methanotrophy	$CH_4 + 2O_2 = CO_2 + 2H_2O$	I but not C	A
Anoxic methanotrophy with SO_4-reduction	$CH_4 + SO_4^{2-} = HCO_{3-} + HS^- + H_2O$	I but not C	A
Anoxic methanotrophy with NO_3-reduction	$CH_4 + 2NO_3^- = HCO_3^- + 3OH^- + N_2$	Not I and C	A
Hydrogenotrophic methanogenesis	$H_2 + 1/4CO_2 = 1/4CH_4 + 1/2H_2O$	I and C	A
Homo-acetogenesis	$H_2 + 1/4CO_2 = 1/8CH_3COOH + 1/4H_2O$	Not I and C	C
Hydrogenogenic carboxydotrophy	$CO + H_2O = H_2 + CO_2$	I and C	A
Hydrogenotrophic SO_4-reduction	$H_2 + SO_4^{2-} + 2H^+ = 1/4H_2S + H_2O$	I and C	A
Hydrogenotrophic S-reduction	$H_2 + S^0 = H_2S$	I and C	A
Thiotrophic (H_2S-oxidizing) O_2-reduction	$H_2S + 2O_2 = SO_4^{2-} + 2H^+$	I and C	A
Thiotrophic (S-oxidizing) O_2-reduction	$S^0 + H_2O + 3/2O_2 = SO_4^{2-} + 2H^+$	I and C	B
Thiotrophic (S_2O_3-oxidizing) O_2-reduction	$S_2O_3^{2-} + 5/2O_2 = 2SO_4^{2-}$	I and C	C
Thiotrophic (S_2O_3-oxidizing) denitrification	$S_2O_3^{2-} + 5/3NO_3^- = 2SO_4^{2-} + 5/6N_2$	I and C	C
Thiotrophic (S_2O_3-oxidizing) ammonification	$S_2O_3^{2-} + 2H_2O + NO_3^- = 2SO_4^{2-} + NH_4^+$	I and C	C
Thiotrophic (S-oxidizing) denitrification	$S^0 + 2/5H_2O + 6/5NO_3^- = SO_4^{2-} + 4/5H^+ + 3/5N_2$	I and C	B
Thiotrophic (S-oxidizing) ammonification	$S^0 + 3/4NO_3^- + 7/4H_2O = SO_4^{2-} + 3/4NH_4^+ + 1/2H^+$	I and C	B
Thiotrophic (H_2S-oxidizing) denitrification	$H_2S + 2H^+ + 2NO_3^- = SO_4^{2-} + N_2 + 2H_2O$	I and C	A
Hydrogenotrophic O_2-reduction	$H_2 + 1/2O_2 = H_2O$	I and C	A
Hydrogenotrophic Fe(III)-reduction	$H_2 + 2Fe^{3+} = 2Fe^{2+} + 2H^+$	I and C	A
Fe(II)-oxidizing O_2-reduction	$Fe^{2+} + 1/4O_2 + H^+ = Fe^{3+} + 1/2H_2O$	I but not C	A
Fe(II)-oxidizing denitrification	$Fe^{2+} + 1/5NO_3^- + 2/5H_2O + 1/5H^+ = 1/10N_2 + Fe^{3+} + OH^-$	I and C	A
Mn(II)-oxidizing O_2-reduction	$Mn^{2+} + 1/2O_2 + H_2O = MnO_2(s) + 2H^+$	I and C	A
Mn(II)-oxidizing denitrification	$Mn^{2+} + 1/5NO_3^- + 6/5H^+ = 1/10N_2 + Mn^{4+} + 3/5H_2O$	Not I and C	D
Nitrification (NO_2-oxidazing O_2-reduction)	$NO_2^- + 1/2O_2 + 2OH^- + 2H^+ = NO_3^- + 2H_2O$	Not I and C	C
Nitrification (NH_4-oxidazing O_2-reduction)	$NH_4^+ + 2O_2 = NO_3^- + H_2O + H^+$	I and C	A

Hydrogenotrophic denitrification	$H_2 + 2/5NO_3^- + 2/5H^+ = 1/5N_2 + 6/5H_2O$	I and C	A
Hydrogenotrophic ammonification	$H_2 + 1/4NO_3^- + 1/2H^+ = 1/4NH_4^+ + 3/4H_2O$	I and C	A
Anoxic ammonia-oxidation (ANAMOX)	$NH_4^+ + NO_2^- = N_2 + 2H_2O$	I but not C	A
S-disproportionation	$S^0 + H_2O = 1/4SO_4^{2-} + 3/4H_2S + 1/2H^+$	Not yet I and C	B
S_2O_3-disproportionation	$S_2O_3^{2-} + H_2O = SO_4^{2-} + H_2S$	Not yet I and C	C

[a] This means whether a microbial component representing the chemolithotrophic metabolism is identified or cultivated in deep-sea hydrothermal environments

[b] Remark means the explanation of each of the metabolisms as follows: A; fully used for the thermodynamic calculation but not all the results shown, B; used for the thermodynamic calculation on an assumption that the maximum amount of available S^0 (polysulfide) is equal to that of H_2S, C; not used for the thermodynamic calculation due to the lacks of compositional data, D; not used for the thermodynamic calculation due to the lack of reliable thermodynamic data for Mn^{4+}

energy of each reaction, which was determined based on the chemical composition of the mixed fluid estimated by the reaction-path calculations. The Gibbs energy of the metabolic reactions involving elemental sulfur (S^0) in this study was estimated with the assumptions that the maximum amount of available S^0 (practically polysulfide) was equal to that of H_2S and that the thiotrophic energy metabolisms utilized S^0 (practically polysulfide) but not H_2S. In the following sections, for the sake of convenience, the maximum ΔGr values in two hypothetical mixing water having temperature ranges between 4°C and 50°C and between 50°C and 125°C represent the chemolithotrophic metabolisms of the psychlophilic to mesophilic and thermophilic to hyperthermophilic microbial components, respectively.

The mixing calculations were performed with the aid of the computer program EQ3/6, Version 8.0 (Wolery and Jarek 2003). The thermodynamic database necessary for the EQ3/6 operations, as well as the values of the standard Gibbs energy for the chemolithotrophic metabolic reactions ($\Delta Gr°$), were generated using the SUPCRT92 code and database (Johnson et al. 1992). The core of the SUPCRT92 database is the thermodynamic data set for aqueous species and complexes from Shock and Helgeson (1988, 1990) and Shock et al. (1997). The database used here has more recent upgrades of the slop98.dat and spec02.dat databases (see Wolery and Jove-Colon 2004), as well as OBIGT.dat database (see also http://www.predcent.org/).

Case Studies

Kairei Hydrothermal Field in the Central Indian Ridge (CIR)

Geological Settings and Physical–Chemical Characteristics of Hydrothermal Fluids

The Kairei hydrothermal field is located in the first segment of the Central Indian Ridge (CIR-S1), which was the first deep-sea hydrothermal field discovered in the Indian Ocean (Hashimoto et al. 2001). The local bathymetric topography and the dive surveys using *DSRV Shinkai 6500* revealed that the hydrothermal activities of the Kairei field were distributed along the basaltic lava flow (Kumagai et al. 2008). Thus, it was initially believed that the Kairei field was a typical basalt-hosted hydrothermal system (Van Dover et al. 2001; Gallant and Von Damm 2006). However, the hydrothermal fluid chemistry of the Kairei field was known to be atypical as compared to other basalt-hosted hydrothermal systems and even any other previously known hydrothermal systems in the MORs (Gallant and Von Damm 2006). The endmember hydrothermal fluids are characterized by a very high H_2 concentration, a relatively high Si concentration and a remarkably low CH_4/H_2 ratio (Gallant and Von Damm 2006; Kumagai et al. 2008) (Table 12.1).

The H_2-enriched high-temperature hydrothermal vent fluids have been often found in several hydrothermal fields located along the Mid Atlantic Ridge (MAR) (Charlou et al. 2002; Melchert et al. 2008). These MAR H_2-rich hydrothermal fluids are generally attributed to the serpentinization of abyssal peridotite tectonically uplifted and emplaced near the hydrothermal fields (Charlou et al. 2002). However, these MAR H_2-rich hydrothermal fluids usually exhibit higher CH_4 and lower Si concentrations than the CIR Kairei hydrothermal fluids (Charlou et al. 2002). In fact, the CH_4 and Si concentrations in the Kairei hydrothermal fluids are similar to those in the typical basalt-hosted hydrothermal fluids (Gallant and Von Damm 2006). These results have puzzled geochemists since its discovery how the Kairei hydrothermal fluids are generated. Recently, the enigmatic hydrothermal fluid chemistry of the Kairei field has been unraveled by the discovery of specific geologic structure mainly consisting of lower crustal ultramafic-like rocks such as troctolites using *DSRV Shinkai 6500* surveys and by the petrological characterization of the highly serpentinized minerals in the rocks (Nakamura et al. 2009). The theoretical model calculations have predicted that all the previously enigmatic chemical features (high H_2 and Si concentrations and low CH_4/H_2 ratio) in the Kairei hydrothermal fluids can be explained by the serpentinization of troctolites and the subsequent hydrothermal reactions with basaltic lava near the Kairei field. Thus, it is now recognized that the Kairei field is a both basalt- and ultramafic rocks-associated hydrothermal system.

The Kairei hydrothermal fluids represent slightly different endmembers influenced by the subseafloor phase-separation (Gallant and Von Damm 2006). The Kali vent site is a principal hydrothermal vent site hosting slightly brine-rich fluid and the Fugen chimney site has a relatively gas-rich endmember fluid (Table 12.1). The Fugen chimney is located at the easternmost part of the hydrothermal field and is likely derived from the longest branched hydrothermal fluid path in the predicted subseafloor hydrogeologic structure (Gallant and Von Damm 2006). Although the Fugen chimney fluid is generally enriched with various gas components, only H_2 in the Fugen chimney fluid is less than in the Kali vent fluid (Takai et al. 2004a) (Table 12.1). The stable isotopic signatures of $D(H_2)$, $\delta^{13}C(CO_2)$ and $\delta^{13}C(CH_4)$ vary between the Kali and Fugen hydrothermal fluids (Table 12.1). Together with the concentration variations, the lower $D(H_2)$ value in the Fugen chimney fluid should show a higher extent of lower temperatures of isotopic equilibrium (the most likely promoted by microbial H_2 consumption) (unpublished data), the higher $\delta^{13}C(CO_2)$ value in the Fugen chimney fluid may be explained by the isotopic fractionation during the microbial consumption (Takai et al. 2004a) and the significantly lower $\delta^{13}C(CH_4)$ in the Fugen fluid can be attained only by the microbial methanogenesis (Takai et al. 2004a, 2008b). All these geochemical results are well consistent with the composition and metabolic activity of the subseafloor microbial communities (occurrence of HyperSLiME) delineated from the microbiological characterization of the components entrained from the hydrothermal fluids (Takai et al. 2004a).

Microbial Communities in the Chimney Structures and the Fluids

The microbial communities in the hydrothermal fluids and the chimney structures of the Kali and Fugen sites were characterized by using the quantitative cultivation technique, the 16S rRNA gene clone analysis and the FISH analysis (Takai et al. 2004a). The microbial components present in high temperature hydrothermal fluids, most of which were entrained by the hydrothermal fluid flows from the subseafloor habitats and were thermally dead but small populations were still considered to survive at the time of recovery, were trapped with the in situ colonization systems (ISCS) deployed in the hydrothermal emissions (Takai et al. 2004a). In this chapter, the viable cell count of each of the cultivated microbial populations (roughly classified into the genus to order level by the 16S rRNA gene sequence similarity) is summarized in Table 12.3 and the cultivated microbial population structures are mainly discussed as common inventories for the functionally active microbial community structures in the habitats among different hydrothermal fields.

From the microbial communities trapped from the Kali vent and Fugen chimney fluids (the ISCS communities), members of *Thermococcus* or *Pyrococcus* were the most abundant populations (Table 12.3). These *Thermococcales* species are usually recognized to be chemoorganotrophs and represent thermophilic to hyperthermophilic consumers and decomposers in the trophic state of microbial communities. However, it has become evident that some of the *Thermococcus* members can chemolithoautotrophically grow by hydrogenogenic carboxydotrophy using CO as the energy and carbon sources (Lee et al. 2008; Sokolova et al. 2004). Thus, it seems likely that some of the *Thermococcus* and *Pyrococcus* populations contribute the primary production of microbial communities associated with the hydrothermal fluids by their possible carboxydotrophy. The second abundant cultivated population was thermophilic and hyperthermophilic *Methanococcales* members, all of which represented hydrogenotorphic (H_2-trophic), methanogenic chemolithoautotrophs (Table 12.3). The other cultivated microbial components were thermophilic and hyperthermophilic members of *Aquifex* and *Persephonella* (both H_2-trophic and/or thiotrophic chemolithoautotrophs), *Archaeoglobus* (Organotrophic and H_2-trophic SO_4-reducing chemolithotrophs). Although our previous microbiological characterization did not obtain the culture (Takai et al. 2004a), *Thermodesulfatator* species were known to be the dominating, H_2-trophic SO_4-reducing chemolithoautotrophic bacteria in the chimney microbial communities of the Kairei field (Moussard et al. 2004). In our recent characterization of microbial communities in the Kairei field using the samples obtained again in 2006, the *Thermodesulfatator* population in the Kali vent fluid was as much as the *Methanococcales* population (unpublished data). Thus, the H_2-trophic SO_4-reducing chemolithoautotrophs such as *Archaeoglobus* and *Thermodesulfatator* members potentially dominate the microbial communities in the Kali vent and Fugen chimney fluids. In addition, certain populations of mesophilic and thermophilic *Epsilonproteobacteria* such as *Hydrogenimonas* and

Table 12.3 Viable cell count of each microbial population in hydrothermal fluid and chimney microbial communities among different hydrothermal fields

Population density (cells/g wet)	Kairei Kali ISCS	Kairei Fugen chimney & ISCS	Mariner SC & MC chimney & ISCS	Mariner CRC chimney	Iheya North NBC chimney & ISCS	Iheya North CBC chimney & ISCS	Iheya North E18 ISCS	TOTO caldera Sulfur chimney chimney & ISCS
Thermococcus	3×10^5–1.5×10^6	1×10^6–5×10^6	10^5–10^6	10^6–10^7	5×10^6–3.5×10^7	2.5×10^5–1.8×10^6	100–720	10^6–10^7
Pyrococcus	<10	5×10^6–2.5×10^7	<10	<10	20–140	2–15	3–20	<10
Methanococcales	4000–2×10^4	2×10^4–1×10^5	<10	<10	90–650	10–70	3800–2.6×10^4	<10
Archaeoglobus	100–500	<10	<10	<10	1.2×10^5–8×10^7	<10	<15	<10
Aquifex	100–500	5–25	1000–10^4	10^6–10^7	230–1600	<10	<15	<10
Persephonella	100–500	100–500	1000–10^4	10^6–10^7	1.6×10^6–12×10^7	7–50	<15	<10
Desulfurobacteria	<10	<10	<10	<10	<4	6–38	3–20	<10
Thermaceae	<10	<10	100–1000	<10		500–3500	<15	<10
Deferribacter	<10	<10	1000–10^4	1000–10^4	<4	<10	<15	<10
Rhodothermus	<10	<10	10^4–10^5	10^5–10^6	<4	<10	<15	<10
Tepidibacter	<10	<10	<10	<10	25–180	<10	<15	<10
Nautiliales	<10	<10	<10	<10	8×10^4–5.6×10^5	<10	<15	300–3×10^3
Hydrogenimonas	<10	1000–5000	<10	<10	<4	25–180	<15	<10
Nitratiruptor	<10	<10	<10	<10	2.2×10^5–1.5×10^6	37–260	<15	<10
Nitratifractor	<10	<10	<10	<10	<4	33–230	<15	<10
Sulfurovum	<10	<10	<10	<10	1.4×10^4–7.3×10^5	7–50	<15	<10
Sulfurimonas	<10	100–500	10^5–10^6	10^5–10^6	5.6×10^5–3.9×10^6	1800–1.3×10^4	2500–1.7×10^4	3000–3×10^4
Thioreductor	<10	<10	<10	<10	<4	<10	600–4300	<10
Thiomicrospira	<10	<10	<10	<10	<4	<10	<15	2000–2×10^4
heterotrophic *Gammaproteobacteria*	<10	<10	5×10^6–2.5×10^7	5×10^6–2.5×10^7	n.d	n.d	n.d	10^5–10^6
Others	<10	<10	100–1000	<10	n.d	n.d	n.d	n.d
Reference	Takai et al. 2004a	Takai et al. 2004a	Takai et al. 2008a	Takai et al. 2008a	Nakagawa et al. 2005	Nakagawa et al. 2005	Nakagawa et al. 2005	Nakagawa et al. 2006

Sulfurimonas were found in the Fugen chimney fluids (Table 12.3). These *Epsilonproteobacteria* are also H_2-trophic and thiotrophic chemolithoautotrophs. The microbial communities in the Fugen chimney habitats were very similar with the community in the Fugen chimney fluid (Takai et al. 2004a).

The culture-independent molecular analyses (16S rRNA gene clone and FISH analyses) generally show the similar compositions of microbial communities in the hydrothermal fluids and chimney (Takai et al. 2004a). However, the FISH analysis clearly demonstrated the numerical predominance of *Methanococcales* populations in the microbial communities in the Kali and Fugen fluids and strongly suggested that the hyperthermophilic H_2-trophic methanogens would sustain the microbial communities as the primary producers (Takai et al. 2004a). Thus, both the geochemical and microbiological characterizations pointed to the possible occurrence of hyperthermophilic subsurface lithotrophic microbial ecosystem (HyperSLiME) beneath the Kairei hydrothermal field (Takai et al. 2004a).

Thermodynamic Potentials of Various Chemolithotrophic Energy Metabolisms

As shown in the hydrothermal fluid chemistry data, two endmember hydrothermal fluids are chemically indistinct (Table 12.1). Thus, although the biomass and composition of whole viable populations in the microbial communities are considerably different between in the Kali and Fugen site fluids, the calculation of Gibbs free energy changes of potential chemolithorophic metabolisms does not give a significant variation (Fig. 12.1). The calculated pattern in potential energy yields of chemolithotrophic metabolisms shows that a variety of energy metabolisms are available in the lower and higher temperatures of mixing zones in the Kairei field. The H_2-trophic methanogenesis, SO_4-, NO_3- and O_2-reductions, and the thiotrophic NO_3- and O_2-reductions are all thermodynamically predominant energy metabolisms in the microbial communities (Fig. 12.1), and are also detected as functionally active microbial components such as *Methanocaccales*, *Archaeoglobus* and *Thermodesulfatator*, *Aquifex*, *Persephonella*, *Hydrogenimonas* and *Sulfrumonas* species in the communities (Table 12.1). Interestingly, the H_2-trophic methanogenesis, SO_4- and S-reductions are energetically much advantageous for hyperthermophiles. This may explain why the thermophilic to hyperthermophilic H_2-trophic chemolithoautotrophs dominate the microbial communities in the Kairei hydrothermal fluids and support the occurrence of HyperSLiME. In addition, the calculated energy metabolism potentials suggest that there should be left the previously and presently unexplored chemolithotrophic components such as aerobic and anaerobic methanotrophs in the in situ microbial communities (Fig. 12.1). As compared to other deep-sea hydrothermal fields described below, the Kairei hydrothermal field can host the most energetically diversified metabolisms. Probably the diversity of potential microbial energy metabolisms is sustained by abundant H_2 concentrations in the hydrothermal fluids. Since the high H_2 concentrations in the hydrothermal

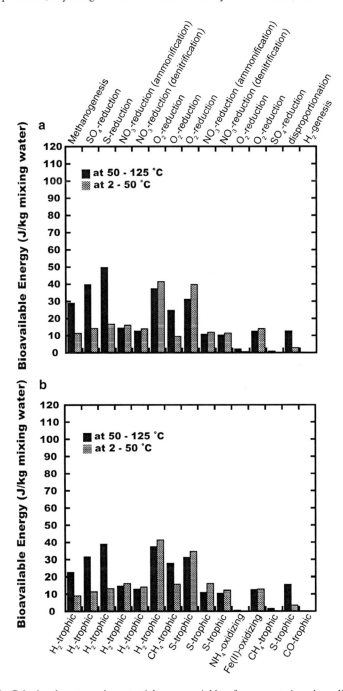

Fig. 12.1 Calculated patterns in potential energy yields of representative chemolithotrophic metabolisms in lower and higher temperatures of mixing zones in the Kairei field. (**a**) The pattern in the hydrothermal mixing zones in the Kali vent site, (**b**) the pattern in the hydrothermal mixing zones in the Fugen chimney site

fluids are supplied from a geologic (geochemical) process, the serpentinization of ultramafic rocks, it may be further implied that the ultramafic rocks-associated hydrothermal systems host a greater diversity of chemolithotrophic microbial communities than the other types of hydrothermal systems.

Mariner Hydrothermal Field in the Lau Basin (LB)

Geological Settings and Physical–Chemical Characteristics of Hydrothermal Fluids

The Mariner field is one of the hydrothermal systems recently discovered in the Lau Basin (LB) (Takai et al. 2008a). It is located at the northern end of the central Valu Fa Ridge and is hosted by basaltic lava (Escrig et al. 2009). However, the high resolution of bathymetric survey and the *DSRV Shinkai6500* dive survey suggested that the Mariner hydrothermal activity was little associated with local faults and fissures and occurred at small volcanic crater-like structures (Ferrini et al. 2008; Takai et al. 2008a). Thus, the Mariner field represents a basalt-hosted hydrothermal system but is considerably influenced by the local magmatic activity (Ferrini et al. 2008; Takai et al. 2008a). The magmatic input was evidently identified in the hydrothermal fluid chemistry in the Mariner field (Ishibashi et al. 2006; Takai et al. 2008a).

There are found numerous hydrothermal vent sites in the Mariner field (Takai et al. 2008a). The general characteristics of the Mariner hydrothermal fluids are the strong acidity, with pH as low as 2.4 and the high concentrations of metallic elements such as Fe and Mn (Table 12.1), both of which may be due to a significant input of magmatic volatiles such as SO_2 (Takai et al. 2008a). The slightly distinct, endmember hydrothermal fluids were found in the Mariner field (Table 12.1). The vapor-lost (Cl-enriched) hydrothermal fluids are predominant, and the Snow Chimney (SC) site, a center of hydrothermal activity for the Mariner field, and the Monk Chimney (MC) site host this type of endmember hydrothermal fluid (Table 12.1) (Takai et al. 2008a). In contrast, from the Crab Restaurant Chimney (CRC) site, slightly vapor-gained (Cl-depleted) hydrothermal fluid is obtained (Table 12.1). The variability in the hydrothermal fluid chemistry is derived from the subseafloor hydrothermal fluid boiling (phase-separation) and has an impact on the compositions and functions of the adjacent subseafloor and seafloor microbial communities (Takai et al. 2008a).

The stable isotopic analyses of gas components in the hydrothermal fluids demonstrated novel isotopic characteristics of the hydrothermal volatiles (Table 12.1). The stable carbon isotopic compositions of CO_2 in both the endmember fluid were the most ^{13}C-enriched among the previously reported values. This is consistent with the relatively high magmatic input shown by other chemical characteristics and may due to the quite high temperatures of isotopic equilibrium between CO_2 and CH_4 in the source magma and relatively shallow occurrence of the hydrothermal reaction zones beneath the Mariner field (Takai et al. 2008a). In addition, as discussed in the hydrothermal fluids of the Kairei field, the lower $D(H_2)$ value in the CRC fluid shows a higher extent of lower temperatures of isotopic equilibrium (probably derived by more active H_2-trophic and H_2-genic metabolisms of the subseafloor microbial communities) (Takai et al. 2008a) and the significantly higher $\delta^{13}C(CH_4)$ in the CRC fluid can be explained only by the microbial methanotrophy in the subseafloor habitats (Takai et al. 2008a).

Microbial Communities in the Chimney Structures

The microbial communities in the chimney structures of the SC, MC and CRC sites were characterized by using the quantitative cultivation technique and the 16S rRNA gene clone analysis (Takai et al. 2008a). From all the chimney habitats studied, facultatively anaerobic heterotrophic *Pseudomonas*, *Idiomarina* and *Pseudoalteromonas* species within the *Gammaproteobacteria* were detected as the predominant cultivable populations (Table 12.3). Since these bacterial species are the heterotrophic bacterial components often detected in various deep-sea, non-hydrothermal environments, it is still uncertain whether they may represent opportunistic or indigenous cultivable bacterial consumers and decomposers of the chimney microbial communities in the Mariner field. Other than mesophilic heterotrophs, *Thermococcus* (potentially chemoorganotrophs) and *Sulfurimonas* (H_2-trophic and/or thiotrophic chemolithoautotrophs) species were cultivated as the predominant cultivated populations commonly found in the chimney habitats (Table 12.3). In the CRC chimney hosting the more gas-enriched hydrothermal fluid, thermophilic H_2-trophic and/or thiotrophic chemolithoautotrophs such as *Aquifex* and *Persephonella* populations dominated the cultivated microbial community as much as *Thermococcus* members (Table 12.3). In addition, a sizable population of *Deferribacter* (thermophilic organotrophic and H_2-trophic, NO_3^--, S- and Fe(III)-reducers) and *Rhodothermus* (aerobic thermophilic heterotorophs) members was identified in the Mariner field chimneys.

Variability in the cultivated microbial community was also evident in the comparison between the chimney habitats within the Mariner field (Table 12.3). The viable cell counts of *Thermococcus*, *Aquifex* and *Persephonella* species was more than 10^6 cells/g in the CRC site, one order magnitude higher than the populations of *Thermococcus* and *Sulfurimonas* species in the SC and MC sites. In addition, an increased population of thermophilic, H_2-trophic and thiotrophic chemolithoautotrophs such as *Aquifex* and *Persephonella* members was found in the chimney associated with the gas-enriched hydrothermal fluid (Tables 12.1 and 12.3). The variability in hydrothermal fluid chemistry among the chimneys may have an impact on the cultivable microbial community structures in the chimney habitats and even in the potential subseafloor habitats (Takai et al. 2008a). The culture-independent 16S rRNA gene clone analysis also supports in part the abundant occurrence of *Thermococcus* and *Sulfurimonas* phylotypes in the CSC and MC chimney habitats although the bacterial and archaeal phylotypes that are not detected by the cultivation technique dominate the 16S rRNA gene libraries (Takai et al. 2008a).

Thermodynamic Potentials of Various Chemolithotrophic Energy Metabolisms

The culture-dependent microbiological characterization demonstrated that the biomass and composition of whole viable populations differed between in the chimney structures hosing the gas-enriched and -depleted hydrothermal fluids. Nevertheless, since the endmember hydrothermal fluids in the Marine field are not so chemically distinct (Table 12.1), the calculation of Gibbs free energy changes of potential chemolithorophic metabolisms provides a quite similar result (Fig. 12.2). In both lower and higher temperatures of mixing zones in the Mariner fields, two potential chemolithotrophic metabolisms are able to supply the abundant energy for the primary production of the microbial communities; thiotrophic O_2-reduction and Fe-oxidizing O_2-reduction (Fig. 12.2). Among the microbial components detected by the cultivation-dependent characterization, the *Aquifex*, *Persephonella* and *Sulfurimonas* populations are the candidates of thiotrophic O_2-reducers (Table 12.1), and indeed represented the practical capability of the thiotrophic O_2-reducing growth in the cultivation experiments although most of them were also capable of hydrogenotrophic NO_3- and O_2-reduction, and thiotrophic NO_3-reduction (Takai et al. 2008a). As the evident Fe-oxidizing O_2-reducing chemolithotrophs so far identified in the deep-sea hydrothermal environments, only a zetaproteobacterium of *Mariprofundus ferooxydans* and several *Alphaproteobacteria* strains are known (Edwards et al. 2003; Emerson et al. 2007). However, none of the cultures and the 16S rRNA gene sequences phylogenetically associated with these previously known Fe-oxidizing O_2-reducing chemolithotrophs was detected in any of the chimney structures in the Mariner field.

In the predominant bacterial and archaeal phylotypes detected in the chimney habitats, the phylotypes of the *Nitrospirae* and Marine Benthic Euryarcheaota Group E were known to dominate the prokaryotic rRNA gene communities specifically in the iron (hydr)oxides-rich habitats (Suzuki et al. 2004; Takai et al. 2008a). The physiological characteristics and energy metabolisms of these previously uncultivated bacterial and archaeal components are completely unclear, but the ecotope preference of these phylotypes has suggested that such previously uncultivated bacteria and archaea represent potential Fe-oxidizing chemolithotrophs (Suzuki et al. 2004; Takai et al. 2008a). If so, the calculated pattern in potential energy yields of chemolithotrophic metabolisms may thus point to the compositional and metabolic traits of the microbial communities in the hydrothermal mixing zones in the Mariner field.

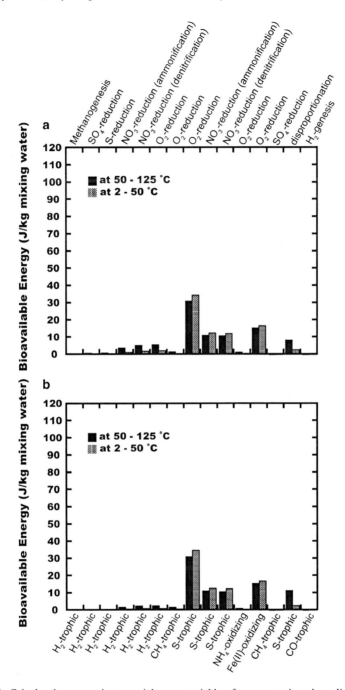

Fig. 12.2 Calculated patterns in potential energy yields of representative chemolithotrophic metabolisms in lower and higher temperatures of mixing zones in the Mariner field. (**a**) The pattern in the hydrothermal mixing zones in the Snow chimney (SC) and Monk chimney (MC) sites, (**b**) the pattern in the hydrothermal mixing zones in the Crab Restaurant chimney (CRC) site

Iheya North Hydrothermal Field in the Okinawa Trough (OT)

Geological Settings and Physical–Chemical Characteristics of Hydrothermal Fluids

The Okinawa Trough (OT) is a "rifting phase" Backarc Basin covered with thick terrigenous sediments (Letouzey and Kimura 1986) and hosts a total of six active hydrothermal fields known so far (Glasby and Notsu 2003). The Iheya North hydrothermal field is the most extensively studied field among the Okinawa Trough hydrothermal systems and is located in the central depression of the Iheya North Knoll (Takai et al. 2006a). Since the Iheya North Knoll consists for the most part of pumice and dacite at the seafloor, the Iheya North field represents a sediments- and felsic rocks-associated hydrothermal activity (Ishibashi and Urabe 1995). The hydrothermal activities in the Okinawa Trough are commonly characterized by the fluids containing extraordinary high concentrations of gaseous carbon compounds such as CO_2 and CH_4 (Konno et al. 2006) and by the liquid CO_2 emissions and the CO_2 hydrate formation found in the sediments near the fields (Sakai et al. 1990; Inagaki et al. 2006). In the Iheya Noth field, six large hydrothermal mounds and accompanying hydrothermal fluid emissions such as North Edge Chimney (NEC), Event 18 (E18), North Big Chimney (NBC), Central Big Chimney (CBC), High Radioactivity Vent (HRV) and South Big Chimney (SBC) are identified (Takai et al. 2006a). The highest fluid temperature and the highest flow rate have consistently been recorded at the NBC mound, indicating that the NBC is generated on the main conduit of hydrothermal fluid flow in the Iheya North field.

The fluid temperatures and flow rates systematically decrease with increasing distance from the NBC. Additionally, the hydrothermal fluid chemistry changes significantly between the different hydrothermal mounds (Table 12.1) (Takai et al. 2006a). The chemical composition of the NBC endmember fluid is relatively stable through the long term of chemical monitoring (about 10 years) and the chlorinity ranges between 75% and 100% of seawater chlorinitiy (Table 12.1). This indicates that the hydrothermal fluid from NBC experiences the moderate phase-separation and is a slightly gas-enriched fluid (Table 12.1). In contrast, the CBC hydrothermal fluid is brine-enriched and the E18 fluid with lower chlorinity is considerably influenced by a vapor phase input (Table 12.1). The variation in the endmember hydrothermal fluid chemistry is highly influenced by the subseafloor phase-separation and –partition and the mixing extent of different phases is probably controlled by the complex subseafloor hydrogeologic structures (Takai et al. 2006a). As compared to the hydrothermal fluids listed in Table 12.1, the Iheya North hydrothermal fluids are notable for the extremely high concentrations of CH_4, CO_2 and NH_4, all of which are important energy, carbon and nitrogen sources for the microbial activities (Table 12.1). In contrast, the hydrothermal fluids are highly depleted with Fe (Table 12.1) and are thus clear hydrothermal emissions. At the present, the stable isotopic characterization is not in detail determined (Table 12.1). However, the presence of isotopically very light H_2 and CH_4 strongly suggest that functionally active subseafloor microbial communities should occur in the whole hydrothermal circulation pathway and should affect the biogeochemical processes (e.g., microbial H_2 generation and consumption, and methanogenesis) associated with the subseafloor hydrothermal circulation. The microbial-function-impacted and phase-separation-induced variability in these volatile components of the hydrothermal fluids has a great impact on the compositions and functions of the microbial communities occurring near the hydrothermal fluid discharges.

Microbial Communities in the Fluids and Chimney Structures

The microbial communities present in the representative hydrothermal fluids and chimney structures (NBC, CBC and E18 sites) were studied by cultivation-dependent and -independent techniques in the Iheya North field (Nakagawa et al. 2005). A great variety of cultivated microbial components were obtained from the NBC and CBC chimney habitats and the in situ colonization systems (ISCS) deployed in the NBC, CBC and E18 hydrothermal fluids (Table 12.3). In the NBC fluid and chimney structure, members of *Thermococcus* or *Pyrococcus* (potential chemoorganotrophs) were the most abundant populations (Table 12.3). Other predominant cultivated populations were members of *Persephonella* (H_2-trophic and/or thiotrophic S-, NO_3- and O_2-reducers), *Nitratiruptor* (H_2-trophic NO_3-reducers), *Sulfurimonas* (H_2-trophic and/or thiotrophic S-, NO_3- and O_2-reducers), *Archaeoglobus* (Organotrophic and H_2-trophic SO_4-reducers), *Sulfurovum* and *Nitratifractor* (both H_2-trophic and/or thiotrophic S-, NO_3- and O_2-reducers) and *Nautiliales* (H_2-trophic S- and NO_3-reducers), representing >10^5 cells/g of populations (Table 12.3) (Nakagawa et al. 2005). Considerably lower populations of *Aquifex* (H_2-trophic and/or thiotrophic S-, NO_3- and O_2-reducers), *Methanococcales* (H_2-trophic methanogens) and *Tepidibacter* (chemoorganotrophs) species were also detected in the chimney habitats of NBC site (Table 12.3) (Nakagawa et al. 2005). In the CBC fluid and chimney (brine-enriched fluid site), the whole viable biomass was relatively lower than in the NBC site. The most abundant cultivated component was *Thermococcales* members while the viable counts of *Sulfurimonas, Persephonella, Sulfurovum* and *Methanococcales* members were considerably reduced (Table 12.3) (Nakagawa et al. 2005). Members of *Desulfurobacteria* (H_2-trophic S- and NO_3-reducers) and *Thermaceae* (aerobic heterotrophs) were only cultivated in the CBC chimney and ISCS (Table 12.3). Interestingly, in the E18 ISCS microbial community obtained from the gas-enriched hydrothermal fluid, the most abundant cultivated population was the members of *Methanococcales* (Table 12.3). In addition, the *Sulfurimonas and Thioreductor* species (H_2-trophic S-reducers) were more abundantly cultivated than the *Thermococcales* members (Table 12.3). As far as we know, this is the only the chimney habitat providing the larger viable counts of the chemolithoautotrophs than that of the *Thermococcus* species in the global deep-sea hydrothermal fields studied so far.

The variability in the microbial communities of different hydrothermal vent sites was generally supported by the results from the culture-independent, 16S rRNA gene clone analysis (Nakagawa et al. 2005). As previously discussed (Nakagawa et al. 2005), the significant intra-field variation in the compositions and metabolisms of the microbial communities may be affected by the physical and chemical variation of the hydrothermal fluids in the Iheya North field. Probably, the intra-field heterogeneity of the hydrothermal fluid chemistry, particularly of the volatile species, is the most significantly generated by the subcritical phase-separation and -partition processes of the hydrothermal fluids. The subcritical phase-separation of hydrothermal fluid occurs below the critical point of the seawater (406°C and ~30 MPa) at a depth of <3,000 m (Bischoff and Rosenbauer 1985) and acts to provide the two types of hydrothermal fluids; one is incessantly being enriched with the vapor (gas) phase and the other is also losing the vapor (gas) phase but the salinity is slightly affected. Thus, the hydrothermal systems occurring at relatively shallower depths of ocean floors and crusts in the Volcanic Arcs and Backarc Basins may generate the greater habitational, compositional and functional diversity of the microbial communities together with the greater heterogeneity of the hydrothermal fluid chemistry than those in the MORs.

Thermodynamic Potentials of Various Chemolithotrophic Energy Metabolisms

The calculation of Gibbs free energy changes of potential chemolithotrophic metabolisms also provides somewhat different patterns in energy state of the microbial communities in the different hydrothermal vent sites (Fig. 12.3). In the Iheya North field, relatively diverse chemolithotrophic energy metabolisms are energetically favorable as in the CIR Kairei field (Fig. 12.3). However, due to the moderate concentration of H_2 in the Iheya North hydrothermal fluids, the H_2-trophic methanogenesis, S-reduction and SO_4-reduction are not advantageous even in the E18 site hosting the gas-enriched hydrothermal fluid. Nevertheless, the H_2-trophic NO_3- and O_2-reductions can provide considerable energy for the primary production (Fig. 12.3). Instead, the aerobic methanotrophy, thiotrophic O_2-reduction and aerobic ammonia-oxidation have the abundant potentials for the indigenous chemolithotophic metabolisms in the mixing zones of the Iheya North field (Fig. 12.3). In particular, the aerobic methanotrophy and ammonia-oxidation are based on the high inputs of CH_4 and NH_4 from the Iheya North hydrothermal fluids and may result in the occurrence of unique microbial communities in the Okinawa Trough hydrothermal systems. However, the thermodynamic calculation of the energy yields demonstrates that the most energetically dominating chemolithotrophic metabolism in the Iheya North field should be thermophilic to hyperthermophilic anoxic methane oxidation (AMO) coupled with SO_4-reduction (Fig. 12.3). This is very surprising because there has been unknown any of the microorganism capable of thermophilic to hyperthermophilic SO_4-reducing AMO although many psychrophilic to mesophilic AMO consortia consisting of ANME archaea and SO_4-reducing bacteria have been identified in the marine hydrothermal and non-hydrothermal sedimentary environments (Jørgensen and Boetius 2007). However, the thermophilic to hyperthermophilic SO_4-reducing AMO metabolism is energetically possible and the lipid biomarker signatures for the previously uncultivated (hyper)thermophilic AMO archaea were obtained from a Guaymas Basin hydrothermal field that had similar geologic settings with the Okinawa Trough systems (Schouten et al. 2003). In other

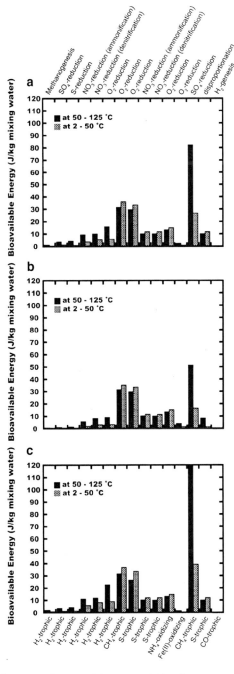

Fig. 12.3 Calculated patterns in potential energy yields of representative chemolithotrophic metabolisms in lower and higher temperatures of mixing zones in the Iheya North field. (**a**) The pattern in the hydrothermal mixing zones in the North Big Chimney (NBC) sites, (**b**) the pattern in the hydrothermal mixing zones in the Central Big Chimney (CBC) site, (**c**) the pattern in the hydrothermal mixing zones in the E18 site

words, the thermodynamic calculation of potential chemolithotrophic metabolisms may point to the significance of previously unidentified chemolithotrophic metabolisms and components for the indigenous microbial community productions in the Iheya North field and the similar hydrothermal systems.

In addition, the cultivation-dependent microbiological characterization does not detect the microbial populations potentially driving the aerobic methanotrophy and ammonia-oxidation (Table 12.3). As the 16S rRNA gene clone analyses of the chimney habitats and other mixing zones have clearly indicated that the potentially methanotrophic gamma-proteobacterial and ammonia-oxidizing Marine Group I crenarchaeal phylotypes dominate the microbial rRNA gene community structures (Nakagawa et al. 2005; Takai et al. 2004b), such cultivation-resistant microbial components likely catalyze the chemolithotrophic metabolisms in situ. Many of the cultivated microbial components can operate the H_2-trophic NO_3- and O_2-reductions and the thiotrophic O_2-reduction predicted as the predominant chemolithotrophic metabolisms by the thermodynamic calculation. However, the relatively large populations of the H_2-trophic methanogens (*Methanococcales* members), SO_4- (*Archaeglobus* spp.) and S-reducers (*Desulfurobactria, Nautiliales* and *Thioreductor* members, etc.) are a little contradictory to the thermodynamic model.

TOTO Caldera Field in the Mariana Volcanic Arc (MVA)

Geological Settings and Physical–Chemical Characteristics of Hydrothermal Fluids

The TOTO caldera is an off-ridge volcano located in the southernmost Mariana Trough and in an extension of Mariana Volcanic Arc (Nakagawa et al. 2006). The TOTO caldera has an asymmetric caldera depression between the southwestern and the northeastern halves and a large lava dome in the northeastern half. There has been no rock samples obtained from the volcano, but based on the petrological characteristics of the rocks recovered from the adjacent seamounts (Fryer et al. 1998), the whole volcano seems to consist mainly of basaltic lava highly influenced by the arc magma input. The hydrothermal activity in the TOTO caldera is located along a fissure in the lava dome, and numerous clear diffusing flows and accompanying faunal communities are found (Nakagawa et al. 2006). The temperatures of the diffuse fluids range from ~10°C to ~100°C (Nakagawa et al. 2006). In the southernmost part of the hydrothermal field, clear diffuse fluids emit from yellow-colored elemental sulfur chimneys and a white smoker vent site is near the sulfur chimneys. The hydrothermal fluid and the microbial communities in the sulfur chimney site were intensively studied (Nakagawa et al. 2006).

The TOTO caldera hydrothermal fluids represent an extremely strong acidity, a pH value as low as pH 1.6 in the white smoker fluid (Nakagawa et al. 2006). This is still the lowest pH value of deep-sea hydrothermal fluid ever reported. The white smoker fluid shows the highest temperature of 170°C among the all the hydrothermal fluids measured in the TOTO caldera while the Mg concentrations of the TOTO caldera fluids, even in the white smoker fluid, are similar with that in the seawater (Nakagawa et al. 2006). In addition, the white smoker fluid has a similar chemical composition of major elements other than H_2S and SO_4 with the seawater (Nakagawa et al. 2006). With other geochemical characteristics of the fluids, it was suggested that the TOTO caldera field would represent a novel hydrothermal system driven by the rapid mixing between the oxygenated seawater and the superheated magmatic vapor in the sub-seafloor environment (Nakagawa et al. 2006). This type of hydrothermal activities was first described in the DESMOS field in the Manus Basin (Gamo et al. 1997) and has been recently discovered in many of the submarine volcanoes in the Mariana Volcanic Arc (Lupton et al. 2008; Resing et al. 2007). Thus, the physical and chemical characteristics of the diffuse hydrothermal fluid in the sulfur chimney listed in Table 12.1 are not the endmember values but the measured values.

The notable chemical characteristics of the TOTO caldera diffuse fluid are the high concentrations of H_2S and SO_4 highly inputted from the magmatic volatile (Table 12.1). In addition, as the hydrothermal fluid is acidic, the hydrothermal leaching of basaltic lava enriches the metallic ions such as Fe and Mn as compared to the seawater. Since the hydrothermal fluid does not experience the high temperatures of hydrothermal rock-water reaction, the redox potential of the hydrothermal fluid is likely oxidative and seems to be insufficient for the growth of strictly anaerobic methanogens. Nevertheless, the stable carbon isotopic composition of CH_4 shows the highly ^{13}C-depleted value (Table 12.1). Based on the isotopic signature, Gamo et al. (2004) speculated the possible subseafloor microbial methanogenesis in the TOTO caldera field. However, none of the microbiological data supporting the speculation has been yet obtained as described below (Nakagawa et al. 2006).

Microbial Community in the Sulfur Chimney

The microbial community in the sulfur chimney was characterized by quantitative cultivation, 16S rRNA gene clone analysis and FISH analysis (Nakagawa et al. 2006). As compared to the microbial communities found in other deep-sea hydrothermal fields, a relatively simple composition of microbial community was identified (Table 12.1) (Nakagawa et al. 2006). The most abundant cultivated microbial components were the members of *Thermococcus* and *Shewanella* (potential chemoorganotrophs), while their growth was strongly associated with S-reduction (Nakagawa et al. 2006). In addition, an aerobic mesophilic heterotroph *Marinobacter* sp. was cultivated as the abundant microbial component (Table 12.1). Other than these consumers and decomposers in the microbial community, the members of *Lebetimonas* (*Nautiliales*) (H_2-trophic S-reducer), *Sulfurimonas* (H_2-trophic S-reducer) and *Thiomicrospira* (thiotrophic O_2-reducer) were detected in the sulfur chimney (Table 12.1). Not only the compositional diversity but also the metabolic versatility of these microbial components was quite simple (Nakagawa et al. 2006). Although the viable populations of these epsilonproteobacterial chemolithotrophs were rather low, the FISH analysis clearly demonstrated that more than 80% of cells in the sulfur chimney were *Epsilonproteobacteria* (Nakagawa et al. 2006). Thus, it was delineated that the H_2-trophic S-reducing and/or thiotrophic O_2-reducing *Epsilonproeobacteria* would sustain the microbial communities in the diffuse fluids and the sulfur chimneys as the primary producers (Nakagawa et al. 2006).

Thermodynamic Potentials of Various Chemolithotrophic Energy Metabolisms

As the microbiological characterization revealed, the calculation of Gibbs free energy changes of chemolithotrophic metabolisms gives a very simple pattern of potential energy metabolisms of the microbial communities in the TOTO caldera hydrothermal fluid and the mixing zones (Fig. 12.4). The microbially available energy metabolisms are only the thiotrophic O_2-reduction and the Fe(II)-oxidizing O_2-reduction (Fig. 12.4). The microbiological characterizations detected none of the potential methanotrophic

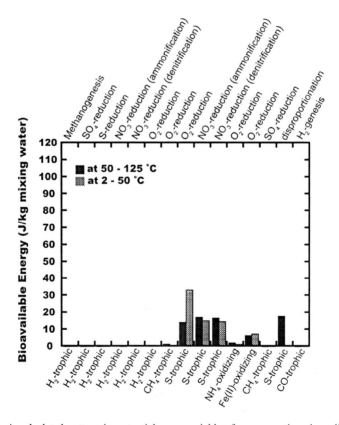

Fig. 12.4 A calculated pattern in potential energy yields of representative chemolithotrophic metabolisms in lower and higher temperatures of mixing zones in the TOTO caldera sulfur chimney site

microbial components but the predominant epsilonproteobacterial H_2-trophic S-reducers or thiotrophic O_2-reducers. Only based on the microbiological characterization, it is difficult to predict whether the predominant epsilonproteobacterial primary producers adopt H_2-trophic S-reduction or thiotrophic O_2-reduction as the operative energy metabolism. However, together with the thermodynamic calculation data of the potential energy state, it seems very likely that the primary production by the *Epsilonproteobacteria* and *Thiomicrospira* members is sustained by the aerobic oxidation of the reduced sulfur compounds plentifully supplied from the magmatic volatiles by means of diffuse fluids in the TOTO caldera.

Inter-Fields Comparison of Variability in Thermodynamic Energy State and Microbial Community

It is a well-known fact that microbiological characterization, any of the cultivation-dependent or -independent techniques, has substantial methodological bias, and even if several different disciplines of approaches are combined, it does not completely resolve the in situ composition and function of indigenous microbial community. In addition, if the geochemical environments in the hydrothermal mixing zones can be theoretically predicted according to a given mixing pattern based on the physical and chemical characteristics of the endmember hydrothermal fluids and the ambient seawaters, the in situ mixing behaviors are spatially and physically variable and the physical and chemical characteristics may be not uniformly comparable in many cases. Thus, it is practically very difficult to compare the geochemical environments and the compositions and functions of microbial communities in the hydrothermal mixing zones among the different hydrothermal fields and even within a field. However, the systematically accumulated and elaborated data for the hydrothermal fluid chemistry and the indigenous microbial community would reveal significant pieces of clues to understand the compositional, physiological and metabolic variability of the microbial communities in the deep-sea hydrothermal habitats. Probably, the phase-separation-controlled variability in the hydrothermal fluid chemistry and the adjacent microbial community identified within several hydrothermal fields represents the excellent examples (Nakagawa et al. 2005; Takai et al. 2008a, 2009). In this chapter, however, the thermodynamic modeling of microbially available energy potentials does not fully reveal the intra-field variability in the interrelation between the chemolithotrophic metabolic potentials and the compositions of the microbial communities as described above.

On the other hand, as described above, it is intuitively justified that the compositional and metabolic diversity of the microbial communities dwelling in the mixing zones is strongly associated with the physical and chemical variation of the proximal hydrothermal fluids. It is also implied that the potential geochemical-microbial interrelation stands on the thermodynamic energy state of chemolithotrophic metabolisms supporting the primary production of microbial communities. In the last section of this chapter, we attempt to examine how the hydrothermal fluid chemistry energetically constraints the primary production of microbial communities in the hydrothermal mixing zones.

First, correlation between the available energy yield of specific chemolithotrophic metabolisms and the viable population of the microbial components capable of the specific metabolisms is examined (Fig. 12.5). The thermophilic to hyperthermophilic H_2-trophic *Methanococcales* members, the thermophilic to hyperthermophilic *Aquificales* and *Desulfurobacteria* members and the *Epsilonproteobacteria* with the versatile energy metabolisms are chosen as the representative primary producers that are frequently detected in different deep-sea hydrothermal fields (Fig. 12.5). Among these chemolithoautotrophic components, the cultivated population density of the thermophilic to hyperthermophilic H_2-trophic methanogens (*Methanococcales* members) is moderately correlated with the calculated energy potential of methanogenesis reaction ($4H_2 + CO_2 \rightarrow CH_4 + 2H_2O$) in the geochemical environment (Fig. 12.5a), whereas no correlation is found between the cultivated population density of the *Aquificales* and *Desulfurobacteria* members or the *Epsilonproteobacteria* and the calculated energy potentials of their possible metabolisms (H_2-trophic S-reduction, NO_3-reduction and O_2-reduction and thiotrophic NO_3-reduction and O_2-reduction) (Fig. 12.5b and c). The correlation between the thermodynamic energy potential of methanogenesis and the viable count of methanogens implies that the thermodynamic energy state of chemolithotrophic metabolisms could in part control the primary production potential in the deep-sea hydrothermal microbial ecosystem, at least the H_2-trophic methanogens' contribution. In a similar manner, direct correlation between the hydrothermal fluid input of principal energy sources such as H_2 and H_2S and the primary production potential of representative chemolithoautotrophs in the proximal microbial community is tested (Fig. 12.6). In this case, the H_2 concentration in the hydrothermal fluid is also correlated with the cultivated population density of the thermophilic to hyperthermophilic H_2-trophic methanogens (Fig. 12.6a), but the cultivated biomass of the *Aquificales* and *Desulfurobacteria* members or the *Epsilonproteobacteria* has no correlation with the H_2 and/or H_2S concentrations in the hydrothermal fluid (Fig. 12.6b and c). These results strongly suggest that the H_2 concentration in the hydrothermal fluid is the primary chemical factor directly to control the thermodynamic energy potential and the existent biomass and production of the thermophilic to hyperthermophilic H_2-trophic methanogens in the hydrothermal mixing zones. Indeed, a similar idea was hypothesized at the discovery of HyperSLiME in the Kairei hydrothermal field (Takai et al. 2004a). The abundant H_2, exceeding a

Fig. 12.5 (continued) between them (*dot line*). A *open square* indicates the presumed value that will be obtained from the MAR Rainbow hydrothermal field of which the thermodynamic energy potential was previously calculated by McCollom (2007). (**b**) The thermodynamic energy potential of a total of possible metabolisms (H_2-trophic S-reduction, NO_3-reduction and O_2-reduction and thiotrophic NO_3-reduction and O_2-reduction) and the cultivated population of the (hyper)thermophilic *Aquificales* and *Desulfurobacteria* members. (**c**) The thermodynamic energy potential of a total of possible metabolisms (H_2-trophic S-reduction, NO_3-reduction and O_2-reduction and thiotrophic NO_3-reduction and O_2-reduction) and the cultivated population of the psychlophilic to mesophilic *Epsilonproteobacteria*

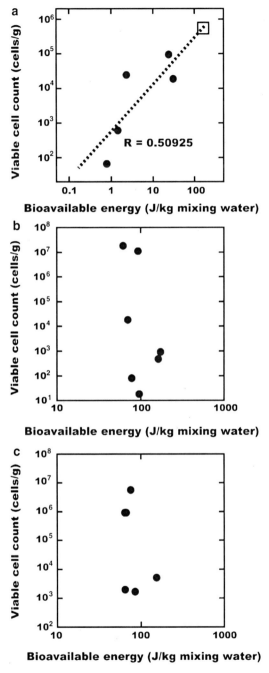

Fig. 12.5 Correlation between the available energy yield of specific chemolithotrophic metabolisms and the viable population of the microbial components capable of the specific metabolisms. (**a**) The thermodynamic energy potential of the (hyper)thermophilic H_2-trophic methanogenesis and the cultivated population of the (hyper)thermophilic H_2-trophic methanogens. A moderate correlation is found

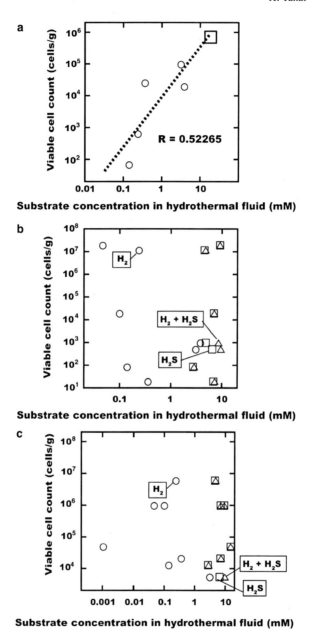

Fig. 12.6 Correlation between the hydrothermal fluid input of principal energy sources and the viable population of the microbial components capable of the specific metabolisms. (**a**) The H$_2$ concentration in the hydrothermal fluids and the cultivated population of the (hyper)thermophilic H$_2$-trophic methanogens. A moderate correlation is found between them (*dot line*). A *open square* indicates the presumed value from the MAR Rainbow hydrothermal field. (**b**) The H$_2$ and/or H$_2$S concentrations in the hydrothermal fluids and the cultivated population of the (hyper)thermophilic *Aquificales* and *Desulfurobacteria* members. (**c**) The H$_2$ and/or H$_2$S concentrations in the hydrothermal fluids and the cultivated population of the psychlophilic to mesophilic *Epsilonproteobacteria*

certain threshold concentration, in the hydrothermal fluid would prepare the prosperous occurrence of (hyper)thermophilic H_2-trophic methanogens in the proximal microbial community (Takai et al. 2004a). Furthermore, since the hydrothermal serpentinization of ultramafic rocks is the most ubiquitous and efficient process to provide the extraordinary high H_2 concentration to the hydrothermal fluid through the history of Earth, the Ultramafics-Hydrothermalism-Hydrogenogenesis-HyperSLiME (UltraH^3) linkage was predicted as a substantial interrelation between the geological settings, geochemical environments and microbial community function supporting the possible occurrence of the most ancient living ecosystem (Takai et al. 2006b). If the correlation between the thermodynamic energy potential of methanogenesis (H_2 concentration in the hydrothermal fluid) and the existent activity of (hyper)thermophilic H_2-trophic methanogens is completely proved, such a previously proposed hypothesis would be justified as the thermodynamic energy basis of interrelation between the geological settings, geochemical environments and microbial community function. As shown in Figs. 12.5a and 12.6a, the microbiological characterization of microbial communities in the most H_2-enriched, peridotite-hosted Rainbow hydrothermal field (Charlou et al. 2002) would provide a key insight into our discussion in the future.

Conclusions and Perspectives

In this chapter, we overview the compositional, physiological and metabolic variability in microbial communities associated with the hydrothermal fluid chemistry in the four well-studied fields. The deep-sea hydrothermal fields represent the different geographical locations and the typical geological settings. However, the examples of the physical and chemical characteristics of hydrothermal fluids and the compositions and functions of the microbial communities are quite limited and insufficient for the inter-fields comparison. The quantitative and qualitative limitation is probably found in both the geochemical and microbiological characterizations. However, the biggest problem is the absence of uniformly comparable (less biased) quantitative data for the compositions and activities of the existent microbial communities in the deep-sea hydrothermal environments. It is shown in this chapter that the results obtained from quantitative cultivation technique are generally helpful to estimate the compositional, physiological and metabolic characteristics of the microbial communities. However, the results may be strongly influenced by the methodological and technical variance among the investigations and researchers. Together with the cultivation-dependent analyses, the culture-independent but the function-associated, quantitative molecular techniques should be incorporated into the future investigation of clarifying the geochemical-microbial interactions in the deep-sea hydrothermal environments. In addition, it is also true that the exploration of new deep-sea hydrothermal fields has frequently led to the discovery of previously unexpected geochemical processes and of the associating microbial and macrofaunal community compositions and functions. Thus, we need not only to intensify the interdisciplinary investigation to each of the presently known deep-sea hydrothermal systems but also to direct the exploratory investigation into the yet unknown deep-sea hydrothermal ecosystems.

References

Bischoff JL, Rosenbauer RJ (1985) An empirical equation of state for hydrothermal seawater (3.2 percent NaCl). Am J Sci 285:725–763

Brazelton WJ, Schrenk MO, Kelley DS, Baross JA (2006) Methane- and sulfur-metabolizing microbial communities dominate the Lost City hydrothermal field ecosystem. Appl Environ Microbiol 72:6257–6270

Charlou JL, Fouquet Y, Donval JP, Auzende JM, Jean-Baptiste P, Stievenard M, Michel S (1996) Mineral and gas geochemistry of hydrothermal fluids on an ultrafast spreading ridge: East Pacific Rise, 17° to 19°S (Naudur cruise, 1993)-phase separation processes controlled by volcanic and tectonic activity. J Geophys Res 101:15899–15919

Charlou JL, Donval JP, Fouquet Y, Jean-Baptiste P, Holm N (2002) Geochemistry of high H_2 and CH_4 vent fluids issuing from ultramafic rocks at the Rainbow hydrothermal field (36j14VN, MAR). Chem Geol 191:345–359

Edwards KJ, Rogers DR, Wirsen CO, McCollom TM (2003) Isolation and characterization of novel psychrophilic, neutrophilic, Fe-oxidizing, chemolithoautotrophic alpha- and gamma-proteobacteria from the deep sea. Appl Environ Microbiol 69:2906–2913

Emerson D, Rentz JA, Lilburn TG, Davis RE, Aldrich H, Chan C, Moyer CL (2007) A novel lineage of proteobacteria involved in formation of marine fe-oxidizing microbial mat communities. PLoS One 2:e667

Escrig S, Bézos A, Goldstein SL, Langmuir CH, Michael PJ (2009) Mantle source variations beneath the Eastern Lau Spreading Center and the nature of subduction components in the Lau basin–Tonga arc system. Geochem Geophys Geosyst 10:Q04014. doi:10.1029/2008GC002281

Ferrini VL, Tivey MK, Carbotte SM, Martinez F, Roman C (2008) Variable morphologic expression of volcanic, tectonic, and hydrothermal processes at six hydrothermal vent fields in the Lau back-arc basin. Geochem Geophys Geosyst 9:Q07022. doi:10.1029/2008GC002047

Fryer P, Sujimoto H, Sekine M, Johnson LE, Kasahara J, Masuda H, Gamo T, Ishi T, Ariyoshi M, Fujioka K (1998) Volcanoes of the southwestern extension of the active Mariana island arc: new swath-mapping and geochemical studies. Island Arc A:596–607

Gallant RM, Von Damm KL (2006) Geochemical controls on hydrothermal fluids from the Kairei and Edmond Vent Fields, 23°–25°S, Central Indian Ridge. Geochem Geophys Geosyst 7:Q06018. doi:10.1029/2005GC001067

Gamo T, Okamura K, Charlou JL, Urabe T, Auzende JM, Ishibashi J, Shitashima K, Kodama Y, Shipboard scientific party of the ManusFlux Cruise (1997) Acidic and sulfate-rich hydrothermal fluid from the Manus basin, Papua New Guinea. Geology 25:139–142

Gamo T, Masuda H, Yamanaka T et al (2004) Discovery of a hydrothermal venting site in the southernmost Mariana Arc: Al-rich hydrothermal plumes and white smoker activity associated with biogenic methane. Geochem J 38:527–534

Glasby GP, Notsu K (2003) Submarine hydrothermal mineralization in the Okinawa Trough, SW of Japan: an overview. Ore Geol Rev 23:299–339

Hashimoto J, Ohta S, Gamo T, Chiba H, Yamaguchi T, Tsuchida S, Okudaira T, Watabe H, Yamanaka T, Kitazawa M (2001) First hydrothermal vent communities from the Indian Ocean discovered. Zool Sci 18:717–721

Huber JA, Holden JF (2008) Modeling the impact of diffuse vent microorganisms along Mid-Ocean ridges and flanks. In: Lowell RP, Seewald JS, Metaxas A, Perfit MR (eds) Magma to microbe: modeling hydrothermal processes at oceanic spreading centers, vol 178, Geophysical Monograph Series. American Geophysical Union, Washington, DC, pp 215–231

Inagaki F, Kuypers MM, Tsunogai U et al (2006) Microbial community in a sediment-hosted CO2 lake of the southern Okinawa Trough hydrothermal system. Proc Natl Acad Sci USA 103:14164–14169

Ishibashi J, Urabe T (1995) Hydrothermal activity related to Arc-Backarc magamtism in the Western Pacific. In: Taylor B (ed) Backarc Basins: tectonics and magmatism. Plenum Press, New York, pp 451–495

Ishibashi J, Lupton JE, Yamaguchi T, Querellou J, Nunoura T, Takai K (2006) Expedition reveals changes in Lau Basin hydrothermal system, Eos Trans. AGU 87:13–17

Johnson JW, Oelkers EH, Helgeson HC (1992) SUPCRT92: a software package for calculating the standard molal thermodynamic properties of minerals, gases, aqueous species, and reactions from 1 to 5000 bar and 0 to 1000°C. Comput Geosci 18:899–947

Jørgensen BB, Boetius A (2007) Feast and famine–microbial life in the deep-sea bed. Nat Rev Microbiol 5:770–781

Konno U, Tsunogai U, Nakagawa F, Nakaseama N, Ishibashi J, Nunoura T, Nakamura K (2006) Liquid CO_2 venting on the seafloor: Yonaguni Knoll IV hydrothermal system, Okinawa Trough. Geophys Res Lett 33:L16607. doi:10.1029/2006GL026115

Kumagai H, Nakamura K, Toki T et al (2008) Geological background of the Kairei and Edmond hydrothermal fields along the Central Indian Ridge: implications of their vent fluids' distinct chemistry. Geofluids 8:239–251

Lee HS, Kang SG, Bae SS et al (2008) The complete genome sequence of *Thermococcus onnurineus* NA1 reveals a mixed heterotrophic and carboxydotrophic metabolism. J Bacteriol 190:7491–7499

Letouzey J, Kimura M (1986) Okinawa Trough: genesis of a back-arc basin developing along a continental margin. Tectonophysics 125:209–230

Lupton J, Lilley M, Butterfield D, Evans L, Embley R, Massoth G, Christenson B, Nakamura K, Schmidt M (2008) Venting of a separate CO_2-rich gas phase from submarine arc volcanoes: examples from the Mariana and Tonga-Kermadec arcs. J Geophys Res 113:B08S12. doi:10.1029/2007JB005467

McCliment EA, Voglesonger KM, O'Day PA, Dunn EE, Holloway JR, Cary SC (2006) Colonization of nascent, deep-sea hydrothermal vents by a novel archaeal and nanoarchaeal assemblage. Environ Microbiol 8:114–125

McCollom TM (2007) Geochemical constraints on sources of metabolic energy for chemolithoautotrophy in ultramafic-hosted deep-sea hydrothermal systems. Astrobiology 7:933–950

McCollom TM, Shock EL (1997) Geochemical constraints on chemolithoautotrophic metabolism by microorganisms in seafloor hydrothermal systems. Geochim Cosmochim Acta 61:4375–4391

Melchert B, Devey CW, German CR, Lackschewitz KS, Seifert R, Walter M, Mertens C, Yoerger DR, Baker ET, Paulick H, Nakamura K (2008) First evidence for high-temperature off-axis venting of deep crustal/mantle heat: the Nibelungen hydrothermal field, southern Mid-Atlantic Ridge. Earth Planet Sci Lett 275:61–69

Moussard H, L'Haridon S, Tindall BJ, Banta A, Schumann P, Stackebrandt E, Reysenbach AL, Jeanthon C (2004) *Thermodesulfatator indicus* gen. nov., sp. nov., a novel thermophilic chemolithoautotrophic sulfate-reducing bacterium isolated from the Central Indian Ridge. Int J Syst Evol Microbiol 54:227–233

Nakagawa S, Takai K (2006) The isolation of thermophiles from deep-sea hydrothermal environments. In: Rainey FA, Oren A (eds) Extremophiles. Methods in microbiology (vol 35). Elsevier, London, pp 57–91

Nakagawa S, Takai K (2008) Deep-sea vent chemoautotrophs: diversity, biochemistry and ecological significance. FEMS Microbiol Ecol 65:1–14

Nakagawa S, Takai K, Inagaki F, Chiba H, Ishibashi J, Kataoka S, Hirayama H, Nunoura T, Horikoshi K, Sako Y (2005) Variability in microbial community and venting chemistry in a sediment-hosted backarc hydrothermal system: impacts of subseafloor phase-separation. FEMS Microbiol Ecol 54:141–155

Nakagawa T, Takai K, Suzuki Y, Hirayama H, Konno U, Tsunogai U, Horikoshi K (2006) Geomicrobiological exploration and characterization of a novel deep-sea hydrothermal system at the TOTO caldera in the Mariana Volcanic Arc. Environ Microbiol 8:37–49

Nakamura K, Morishita T, Bach W, Klein F, Hara K, Okino K, Takai K, Kumagai H (2009) Serpentinized troctolites exposed near the Kairei hydrothermal field, Central Indian Ridge: insights into the origin of the Kairei hydrothermal fluid supporting a unique microbial ecosystem. Earth Planet Sci Lett 280:128–136

Nunoura T, Takai K (2009) Comparison of microbial communities associated with phase-separation-induced hydrothermal fluids at the Yonaguni Knoll IV hydrothermal field, the Southern Okinawa Trough. FEMS Microbiol Ecol 67:351–370

Pagé A, Tivey MK, Stakes DS, Reysenbach AL (2008) Temporal and spatial archaeal colonization of hydrothermal vent deposits. Environ Microbiol 10:874–884

Perner M, Kuever J, Seifert R, Pape T, Koschinsky A, Schmidt K, Strauss H, Imhoff JF (2007a) The influence of ultramafic rocks on microbial communities at the Logatchev hydrothermal field, located 15 degrees N on the Mid-Atlantic Ridge. FEMS Microbiol Ecol 61:97–109

Perner M, Seifert R, Weber S, Koschinsky A, Schmidt K, Strauss H, Peters M, Haase K, Imhoff JF (2007b) Microbial CO_2 fixation and sulfur cycling associated with low-temperature emissions at the Lilliput hydrothermal field, southern Mid-Atlantic Ridge (9 degrees S). Environ Microbiol 9:1186–1201

Resing JA, Lebon G, Baker ET, Lupton JE, Embley RW, Massoth GJ, Chadwick WW, de Ronde CEJ (2007) Venting of acid-sulfate fluids in a high-sulfidation setting at NW rota-1 submarine volcano on the Mariana Arc. Econ Geol 102:1047–1061

Reysenbach AL, Banta AB, Boone DR, Cary SC, Luther GW (2000) Microbial essentials at hydrothermal vents. Nature 404:835–845

Sakai H, Gamo T, Kim ES et al (1990) Unique chemistry of the hydrothermal solution in the mid-Okinawa Trough backarc basin. Geophys Res Lett 17:2133–2136

Schouten S, Wakeham SG, Hopmans EC, Sinninghe Damsté JS (2003) Biogeochemical evidence that thermophilic archaea mediate the anaerobic oxidation of methane. Appl Environ Microbiol 69:1680–1686

Shock EL, Helgeson HC (1988) Calculation of the thermodynamic and transport properties of aqueous species at high pressures and temperatures: correlation algorithms for ionic species and equation of state predictions to 5 kb and 1000°C. Geochim Cosmochim Acta 52:2009–2036

Shock EL, Helgeson HC (1990) Calculation of the thermodynamic and transport properties of aqueous species at high pressures and temperatures: standard partial molal properties of organic species. Geochim Cosmochim Acta 54:915–945

Shock EL, Holland ME (2004) Geochemical energy sources that support the subsurface biosphere. In: Wilcock WSD, Delong EF, Kelley DS, Baross JA, Cary SC (eds) The subseafloor biosphere at Mid-Ocean Ridges, vol 144, Geophysical Monograph Series. AGU, Washington, DC, pp 153–165

Shock EL, Sassani DC, Willis M, Sverjensky DA (1997) Inorganic species in geologic fluids: correlations among standard molal thermodynamic properties of aqueous ions and hydroxide complexes. Geochim Cosmochim Acta 61:907–950

Sokolova TG, Jeanthon C, Kostrikina NA, Chernyh NA, Lebedinsky AV, Stackebrandt E, Bonch-Osmolovskaya EA (2004) The first evidence of anaerobic CO oxidation coupled with H_2 production by a hyperthermophilic archaeon isolated from a deep-sea hydrothermal vent. Extremophiles 8:317–323

Suzuki Y, Inagaki F, Takai K, Nealson KH, Horikoshi K (2004) Microbial diversity in inactive chimney structures from deep-sea hydrothermal systems. Microb Ecol 47:186–196

Takai K, Gamo T, Tsunogai U, Nakayama N, Hirayama H, Nealson KH, Horikoshi K (2004a) Geochemical and microbiological evidence for a hydrogen-based, hyperthermophilic subsurface lithoautotrophic microbial ecosystem (HyperSLiME) beneath an active deep-sea hydrothermal field. Extremophiles 8:269–282

Takai K, Oida H, Suzuki Y, Hirayama H, Nakagawa S, Nunoura T, Inagaki F, Nealson KH, Horikoshi K (2004b) Spatial distribution of Marine Crenarchaeota Group I in the vicinity of deep-sea hydrothermal systems. Appl Environ Microbiol 70:2404–2413

Takai K, Nakagawa S, Reysenbach AL, Hoek J (2006a) Microbial ecology of Mid-Ocean Ridges and Back-Arc Basins. In: Christie DM, Fisher CR, Lee SM, Givens S (eds) Back-Arc spreading systems: geological, biological, chemical and physical interactions, vol 166, Geophysical Monograph Series. AGU, Washington, DC, pp 185–213

Takai K, Nakamura K, Suzuki K, Inagaki F, Nealson KH, Kumagai H (2006b) Ultramafics-Hydrothermalism-Hydrogenesis-HyperSLiME (UltraH³) linkage: a key insight into early microbial ecosystem in the Archean deep-sea hydrothermal systems. Paleontol Res 10:269–282

Takai K, Nunoura T, Ishibashi J et al (2008a) Variability in the microbial communities and hydro-thermal fluid chemistry at the newly-discovered Mariner hydrothermal field, southern Lau Basin. J Geophys Res 113:G02031. doi:10.1029/2007JG000636

Takai K, Nakamura K, Toki T, Tsunogai U, Miyazaki M, Miyazaki J, Hirayama H, Nakagawa S, Nunoura T, Horikoshi K (2008b) Cell proliferation at 122 degrees C and isotopically heavy CH_4 production by a hyperthermophilic methanogen under high-pressure cultivation. Proc Natl Acad Sci USA 105:10949–10954

Takai K, Nunoura T, Suzuki Y et al (2009) Variability in microbial communities in black smoker chimneys at the NW caldera vent field, Brothers volcano, Kermadec arc. Geomicrobiol J. 26:552–569

Van Dover CL, Humphris SE, Fornari D et al (2001) Biogeography and ecological setting of Indian Ocean hydrothermal vents. Science 294:818–823

Wolery TW, Jarek RL (2003) Software user's manual. EQ3/6, Version 8.0. U.S. Dept. of Energy Report, 10813-UM-8.0-00. Sandia National Laboratories, Albuquerque, New Mexico, p 376

Wolery TJ, Jove-Colon CF (2004) Qualification of thermodynamic data for geochemical model-ing of mineral–water interactions in dilute systems. U.S. Dept. of Energy Report, ANL-WIS-GS-000003 REV00. Bechtel SAIC Company, LLC, Las Vegas, Nevada, p 212

Chapter 13
The Molecular Geomicrobiology of Bacterial Manganese(II) Oxidation

Bradley M. Tebo, Kati Geszvain, and Sung-Woo Lee

Introduction

Manganese is the second most abundant transition metal found in the Earth's crust. It has a significant biological role as it is a cofactor of enzymes such as superoxide dismutase and is the key metal in the reaction center of photosystem II. In the environment, manganese is mostly found in three different oxidation states: II, III, and IV. Mn(II), primarily occurring as the soluble Mn^{2+} species, is the thermodynamically favored state at low pH and Eh while insoluble Mn(III) and Mn(IV) oxides are favored at high pH and Eh. Thus, studies of Mn in the environment have almost always employed this paradigm for defining different Mn phases based on operational definitions: Mn that passes through a 0.2 or 0.4 μm filter is defined as soluble Mn(II) while Mn that is trapped by the filter are the solid phase Mn(III,IV) oxides. Soluble Mn species other than Mn(II) were thought not to be important because Mn(III) ions are not stable in solution and rapidly disproportionate to Mn(II) and Mn(IV). However, recent work on the mechanism of bacterial Mn(II) oxidation has demonstrated that Mn(III) occurs as an intermediate in the oxidation of Mn(II) to Mn(IV) oxides (Webb et al. 2005b; Parker et al. 2007; Anderson et al. 2009b) and that a variety of inorganic and organic ligands can complex Mn(III) and render it relatively stable in solution. In this article we review these new insights into the molecular mechanism of bacterial Mn(II) oxidation and recent advances in our understanding of Mn(II) oxidation in the environment. The study of the importance of Mn in the environment needs to employ the new paradigm for Mn cycling which takes into account the role of soluble Mn(III) species.

B.M. Tebo (✉), K. Geszvain, and S.-W. Lee
Division of Environmental and Biomolecular Systems, Oregon Health & Science University,
20000 NW Walker Road, Beaverton, OR 97006, USA
e-mail: tebo@ebs.ogi.edu

L.L. Barton et al. (eds.), *Geomicrobiology: Molecular and Environmental Perspective*,
DOI 10.1007/978-90-481-9204-5_13, © Springer Science+Business Media B.V. 2010

Manganese(III,IV) Oxide Formation

Mn(II) can be oxidized to insoluble Mn(III,IV) oxides through both abiotic and biotic processes in the environment. However, the abiotic manganese oxidation process is much slower than biological processes (Hastings and Emerson 1986), which are carried out by a variety of different bacteria and fungi. Therefore, in many environments, microbial processes are considered to be primarily responsible for the formation of manganese oxides (Emerson et al. 1982; Tebo et al. 1984; Tebo and Emerson 1985; Clement et al. 2009; Dick et al. 2009).

Most known bacteriogenic manganese oxides are layer type with a high percentage of vacancies in their mineral lattices (Bargar et al. 2000; Villalobos et al. 2003; Jürgensen et al. 2004; Bargar et al. 2005; Saratovsky et al. 2006; Villalobos et al. 2006) and a relatively large surface area (Nelson et al. 1999; Villalobos et al. 2003). Because of these features, bacteriogenic manganese oxides are capable of oxidation or sorption of transition metals and other elements (Tebo et al. 2004). Additionally, manganese oxides can oxidize a variety of different organic compounds (Stone and Morgan 1984; Stone 1987; Tebo et al. 2004) including the oxidation of humic substances to low-molecular-weight organic acids (Sunda and Kieber 1994). Therefore, it is important to understand how manganese oxides are formed as they control the fate of other various elements, both inorganic and organic. Fortunately, the manganese oxides produced by model microbial cultures in the lab under environmentally relevant conditions are very similar if not the same as those produced in the environment (Bargar et al. 2009; Dick et al. 2009). Describing the recent studies of the properties of the biogenic oxides is beyond the scope of this review; the reader is referred to a number of other reviews and recent articles (Villalobos et al. 2003; Tebo et al. 2004; Bargar et al. 2005; Webb et al. 2005a, c; Villalobos et al. 2006; Spiro et al. 2009).

Bacterial Manganese(II) Oxidation

The ability to oxidize Mn(II) to either Mn(III) or Mn(IV) has been found in a diverse group of bacteria including α-, β-, γ-Proteobacteria, Actinobacteria, and Firmicutes (Tebo et al. 2004, 2005) and from diverse environments (Francis et al. 2001; Takeda et al. 2002; Templeton et al. 2005; Dick et al. 2006; Anderson et al. 2009a; Cahyani et al. 2009). Although bacteria capable of oxidizing Mn(II) have been isolated from diverse phyla, the physiological role of Mn oxidation is still unclear. As oxidation of Mn(II) is an energetically favorable process, it is possible that some bacteria could gain energy out of the process as suggested by some studies (Kepkay and Nealson 1987; Ehrlich and Salerno 1990). However, it is still not conclusive that the cells indeed benefit from this process. Also, manganese oxides could be increasing the availability of carbon sources that were previously non-available as they are capable of breaking down natural organic matter (Sunda and Kieber 1994). Or, as the bacterium tends to get encrusted by the manganese oxides it produces, forming manganese oxides could be its way of protecting itself from toxic compounds in the environment (Ghiorse 1984). A number of other functions for Mn(II) oxidation have also been proposed (Tebo et al. 2004).

The pathways for biotic manganese oxidation are now fairly well known (Fig. 13.1). Bacterial manganese oxidation occurs in the presence of oxygen according to the following reaction:

$$Mn^2 + 1/2\,O_2 + H_2O \leftrightarrow MnO_2 + 2H^+$$

It has been proposed that manganese oxidation in the environment could also be coupled to nitrate reduction (Murray et al. 1995; Luther et al. 1997), however, no bacteria have been isolated that are capable of coupling the two processes. Furthermore, recent studies showed that alternative electron acceptors, such as nitrate, nitrite and nitrous oxide, were not able to stimulate manganese oxidation while O_2 could, leading the authors to conclude that microbial manganese oxide formation is not coupled to an alternate electron acceptor (Schippers et al. 2005; Clement et al. 2009).

The chemical mechanism of Mn(II) oxidation to MnO_2 has been the subject of several recent investigations. Although early work had suggested that insoluble Mn(III)-bearing oxides were formed as an intermediate in bacterial formation of MnO_2 (Murray et al. 1985; Hastings and Emerson 1986; Mann et al. 1988;

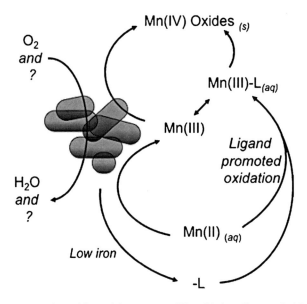

Fig. 13.1 The current view of bacterial manganese(II) oxidation. Enzymatic Mn(II) oxidation proceeds via two one-electron steps, oxidation of Mn(II) to Mn(III) and Mn(III) to Mn(IV). In the presence of complexing organic or inorganic ligands (L), the Mn(III) intermediate can form a soluble Mn(III) complex (Mn(III)–L) which can be either stable in solution or undergo oxidation or disproportionation to Mn(IV) oxides. Under low iron conditions, some Mn(II)-oxidizing bacteria and many non-Mn(II)-oxidizing bacteria may produce organic ligands that can abiotically promote the oxidation of Mn(II) to Mn(III)-L

Mandernack et al. 1995), in situ x-ray absorption spectroscopic measurements have now conclusively shown that solid phase Mn(III) is not a major product or intermediate in Mn(II) to Mn(IV) oxidation under environmentally realistic conditions of Mn(II) concentration, temperature and pH (Bargar et al. 2000).

The molecular mechanism of Mn(II) oxidation has been speculated on theoretical grounds to occur via two sequential one electron transfer reactions (Luther 2005). However, this has been experimentally difficult to demonstrate because soluble Mn(III) is unstable and may occur only transiently (i.e., never accumulate in detectable concentrations). Using scanning transmission x-ray microscopy (STXM), a significant amount of Mn(III) was observed in clumps of Mn(II)-oxidizing bacterial cells, which was used to infer that bacterial Mn(II) oxidation involves one electron transfer processes (Pecher et al. 2003). However, this work was inconclusive because Mn(III) oxides are known to be formed by bacteria when the Mn(II) concentrations are unrealistically high (millimolar and greater as was used in these preparations; Hastings and Emerson 1986; Mandernack et al. 1995; Bargar et al. 2005) and STXM cannot distinguish between different forms of Mn(III) (soluble versus mineral).

There are several ligands known to complex and stabilize Mn(III) in solution. For example, pyrophosphate forms a colored Mn(III)-pyrophosphate complex (Mn(III)-PP) which can be readily measured using a UV-Vis spectrometer (Kostka et al. 1995; Webb et al. 2005b). Recent studies using the model Mn(II)-oxidizing bacteria, *Aurantimonas manganoxydans* SI85-9A1 and spores of *Bacillus* sp. strain SG-1 (see below), have shown that pyrophosphate can be used as an agent to trap and measure Mn(III)-PP products (Webb et al. 2005b; Anderson et al. 2009b). This Mn(III)-PP, however, only occurs as an intermediate since with time, the Mn(III)-PP is converted to insoluble brown-black MnO_2 particles. When Mn(III)-PP is the initial substrate for MnO_2 formation by *Bacillus* SG-1 spores, the rate of Mn(III)-PP disappearance is the same as the rate of disappearance of Mn(III)-PP detected when Mn(II) was the starting substrate (Webb et al. 2005b). Based on these observations, it is likely that the bacterial Mn(II) oxidation process is a two-step, one electron transfer process with Mn(III) occurring as an intermediate. This process is dependent on O_2 and, at least in *Bacillus* sp. SG-1, both the Mn(II) to Mn(III) and the Mn(III) to Mn(IV) oxidation steps are catalyzed by the same enzyme as neither reaction occurs in a Mn oxidase mutant (Webb et al. 2005b). The potential for complexed Mn(III) to occur as an intermediate in bacterial Mn(II) oxidation has important implications with respect to Mn oxide mineral formation and nutrient Fe acquisition. In other model Mn(II)-oxidizing bacteria, *Pseudomonas putida* strains MnB1 and GB-1, Fe limitation leads to the production of fluorescent siderophores (pyoverdines; PVDs), Fe(III)-chelating compounds produced by microorganisms under low iron conditions to acquire Fe from the environment. The pyoverdines complex Mn(III) even more tightly than Fe(III) (Parker et al. 2004) and prevent the formation of Mn oxides as long as the PVD concentration is in excess of the total [Mn + Fe] concentration (Parker et al. 2007). Thus, Fe exerts a major control on Mn oxide formation in *P. putida* and probably in other Mn(II)-oxidizing bacteria as well. The ability of siderophores to form Mn(III)-siderophore complexes appears to be a general characteristic of siderophores. Desferrioxamine B (DFOB), for example, also binds Mn(III) with a high stability constant comparable to that for Fe(III) (Faulkner et al. 1994; Parker et al. 2004; Duckworth and Sposito 2005a).

Not only do siderophores (PVD and DFOB) complex Mn(III), but they also promote oxidation of Mn(II) (i.e., ligand-promoted Mn(II) oxidation; Fig. 16.1) (Duckworth and Sposito 2005b; Parker et al. 2007). It had been previously thought that Mn(II) oxidation was a process that was primarily carried out by either enzymes or catalyzed by oxide surfaces. However, it has been shown that in the presence of oxygen, ligand-Mn(II) can be air-oxidized to ligand-Mn(III) (Faulkner et al. 1994; Klewicki and Morgan 1998; Duckworth and Sposito 2005b; Parker et al. 2007). Specifically, siderophores could therefore have a more significant role in manganese oxidation, especially in iron-limiting environments.

These studies have suggested a role for Mn(III) in the environment. If a ligand (e.g., pyrophosphate or a siderophore) is present during the manganese oxidation process, then the intermediate Mn(III) can be complexed with the ligand to form a stable, soluble Mn(III) which will diffuse into the environment. This Mn(III)-siderophore complex could then be available for Mn(III) oxidation or reduction resulting in the formation of Mn(IV) oxides or Mn(II), respectively. Recent studies have shown that siderophores can be found in pM to nM levels in various environments including in seawater (Mawji et al. 2008), in fresh water (Duckworth et al. 2009b), and in terrestrial soils (Holmstrom et al. 2004; Essen et al. 2006). Therefore, it is possible that siderophore-promoted Mn(II) oxidation in the environment could have more significance than expected. Another source of soluble Mn(III) is through the dissolution of Mn(III) bearing minerals. Various ligands have been shown to be able to promote the formation of Mn(III) complexes which would then be available for further redox processes (Klewicki and Morgan 1999; Duckworth and Sposito 2005a; Peña et al. 2007; Wang and Stone 2008). Therefore, soluble Mn(III)-complexes are not just a lab artifact but a type of Mn species that needs more consideration in the environment. For a more detailed description of the role of siderophores on biogeochemical cycling of manganese and iron the reader is referred to Duckworth et al. (2009a).

The Molecular Microbiology of Manganese(II) Oxidation

Much progress has been made recently in understanding the biochemical mechanisms and enzymology of bacterial Mn(II) oxidation. In the following, a critical review of what is known about the molecular mechanisms of bacterial Mn(II) oxidation using different model organisms is provided.

Mn(II) Oxidation by Bacillus sp. Strain SG-1 Spores

While the ability to oxidize Mn(II) has been identified in a wide variety of bacteria, only in a fraction of these bacteria have the enzymes responsible been in any way characterized. One of the best understood systems is the oxidation of Mn(II) by *Bacillus* spores. The marine Gram positive bacterium *Bacillus* sp. SG-1 has been studied for many years as a model system for bacterially-mediated Mn(II) oxidation (Emerson et al. 1982). In this species, oxidation only occurs in mature spores by an enzyme located in the spore coat and results in the spore being coated in an insoluble Mn(IV) oxide shell (Emerson et al. 1982; de Vrind et al. 1986). In bacilli, spore formation can be triggered by starvation (Piggot 1996) and renders the cell resistant to various environmental threats, such as heat, desiccation, radiation, oxidants, and proteases (Nicholson et al. 2000). Mature spores are metabolically dormant and are thought to have little impact on their environment. However, *Bacillus* species which form Mn(II)-oxidizing spores are diverse and found in many environments, such as hydrothermal plumes and coastal marine sediments (Francis and Tebo 2002; Dick et al. 2006). Thus the impact of spores on the biogeochemical cycling of elements in the environment is likely underestimated.

In the first genetic analysis of bacterial Mn(II) oxidation, random transposon mutagenesis of *Bacillus* sp. SG-1 was employed to identify mutants that retained the ability to sporulate but failed to oxidize Mn(II). This screen identified two chromosomal regions, designated MnxI and MnxII, as involved in oxidation (van Waasbergen et al. 1993). Subsequent mapping of the transposon insertion sites revealed that the two regions are closely linked and encompass a region encoding seven genes that are thought to form an operon (the *mnx* operon, Fig. 13.2)

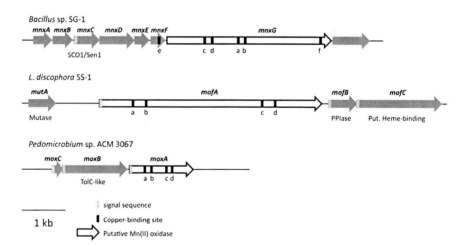

Fig. 13.2 MCO genes identified as encoding the putative Mn(II) oxidase. Depicted as *blue arrows* are the genes that encode the putative Mn(II) oxidase in *Bacillus* sp. SG-1, *L. discophora* SS-1 and *Pedomicrobium* sp. ACM 3067, while neighboring genes are *grey arrows*. Gene names, when available, are listed above the genes and the putative functions of the non-MCO genes are below. Mutase = phosphoglycerate mutase; PPIase = peptidyl-prolyl-*cis-trans*-isomerase. Cu^{2+} binding sites are marked with orange rectangles and are lettered according to sequence homology. For references, see text

(van Waasbergen et al. 1996). Most of the seven *mnx* genes have no sequence similarity to characterized proteins so their function remains unknown. One exception is *mnxG*, which encodes a protein with homology to multicopper oxidases (MCO, described further below). The *mnxC* gene encodes a protein with homology to SCO1/SenC proteins (van Waasbergen et al. 1996). This class of proteins is involved in the synthesis of cytochrome C oxidase, and in some cases these proteins function as copper chaperones that deliver Cu to the enzyme active site (Frangipani and Haas 2009). Thus MnxC may be required for delivery of Cu to the MCO protein MnxG. The significance of MnxC to Mn(II) oxidation is unclear, however, since the *mnxC* gene is not present in the *mnx* operons of other Mn(II)-oxidizing *Bacilli* (Dick et al. 2008b).

Biochemical characterization of the Mn(II) oxidase from several related marine *Bacillus* species confirmed a role of MnxG in oxidation. The oxidase activity, localized in the exosporium (Francis and Tebo 2002), was partially purified from *Bacillus* sp. PL-12 and MB-7 as a band that formed Mn(IV) oxides in an in-gel activity assay and identified by tandem mass spectrometry (Dick et al. 2008b). Most of the peptides recovered were revealed to be unique to MnxG, while one peptide was from a small protein encoded by the *mnxF* gene, which is located upstream of *mnxG* in the putative *mnx* operon (Fig. 16.2) (van Waasbergen et al. 1996). Thus, the MCO protein MnxG has been identified as required for Mn(II) oxidation through both genetics and biochemistry. Further support of the identification of an MCO as the Mn(II) oxidase is the fact that addition of Cu(II) to *Bacillus* sp. SG-1 spores stimulated Mn(II) oxidation (van Waasbergen et al. 1996).

The MCO family of enzymes is found in plants, animals, fungi and bacteria and includes enzymes such as ceruloplasmin, laccase and ascorbate oxidase (Solomon et al. 1996). MCOs are characterized by conserved Cu-binding motifs (Ryden and Hunt 1993), with most containing four Cu cofactors; human ceruloplasmin is unusual in that it has six Cu ions (Solomon et al. 1996). In the best characterized MCO enzymes, such as ceruloplasmin and laccase, an electron is removed from the substrate and transferred to a type 1 Cu site. From there the electron is shuttled to a trinuclear cluster consisting of one type 2 and two type 3 Cu sites, where four electrons (from four substrate molecules) ultimately reduce O_2 to $2H_2O$ (Solomon et al. 1996). The substrates for these enzymes are varied, ranging from metal ions to organic compounds (Solomon et al. 1996). The fungal laccases are involved in degradation of lignin and humic substances (Baldrian 2006) but have also been shown to oxidize Mn(II) to Mn(III) (Hofer and Schlosser 1999). In the case of human ceruloplasmin and yeast Fet3p, the substrate is Fe(II) (Sakurai and Kataoka 2007). Thus, there is a precedent supporting a role for an MCO in metal oxidation.

The oxidation of Mn(II) to Mn(IV) requires the transfer of two electrons, while MCO enzymes catalyze transfer of one electron. Thus it would be expected that MnxG would catalyze formation of Mn(III) from Mn(II) or Mn(IV) from Mn(III) but not the complete reaction. Evidence from spectroscopic and kinetic experiments with *Bacillus* sp. SG-1 suggests that bacterial Mn(II) oxidation proceeds through a transient Mn(III) intermediate before formation of Mn(IV) (see above). Each step is enzymatically catalyzed, with the Mn(II) to Mn(III) step being rate-limiting

(Webb et al. 2005b). In the purification of Mn(II) oxidase activity from *Bacillus* sp. PL-12 and MB-7, mass spectrometry primarily identified MnxG in the active fraction (Dick et al. 2008b) and spores generated from a *mnxG* mutant strain were defective in both formation of the trapped Mn(III)-PP intermediate and decay of Mn(III)-PP (Webb et al. 2005b). Thus it appears that MnxG alone may catalyze the two electron oxidation reaction required to oxidize Mn(II) to Mn(IV). MnxG is rare among MCO proteins in having five predicted Cu binding motifs (Dick et al. 2008b). The additional Cu-binding site in MnxG may reflect the need for additional Cu cofactors to accomplish the two electron transfer required for Mn(II) oxidation to Mn(IV). MnxF, which co-purified with MnxG in *Bacillus* sp. MB-7, also contains a predicted Cu-binding site and thus may play a catalytic role (Dick et al. 2008b). Alternatively, MnxF may shuttle Cu ions to MnxG (O'Halloran and Culotta 2000), or it may have a structural function and help to anchor the oxidase to the exosporium (Dick et al. 2008b).

MCO Enzymes Identified as the Mn(II) Oxidase in Other Genera

An MCO enzyme has also been identified as the Mn(II) oxidase in the β-proteobacteria (Corstjens et al. 1997). The freshwater β-proteobacterium *Leptothrix discophora* is able to oxidize both iron and manganese (van Veen et al. 1978). Similar to *Bacillus* spores, oxidation by *L. discophora* occurs extracellularly, in association with polymers that commonly form a structured sheath (Ghiorse 1984). The model organism for studying Mn(II) oxidation in the β-proteobacteria is the strain *L. discophora* SS1. This strain lost its ability to form structured sheaths after isolation but retains the ability to oxidize Mn(II) (Adams and Ghiorse 1985) via a secreted macromolecule. The highest level of oxidation activity is found in stationary phase cultures (de Vrind-de Jong et al. 1990). Mn(II) oxidation by *L. discophora* SS1 is affected by iron concentration (El Gheriany et al. 2009) and stimulated by Cu(II) (Brouwers et al. 2000).

The putative Mn(II) oxidase was identified in *L. discophora* SS1 through a combination of biochemical, immunological and genetic approaches. The Mn(II) Oxidizing Factor (MOF) was purified from spent culture medium and used to produce antibodies. The αMOF antibodies were then used to screen an expression library for genomic DNA clones that code for expression of MOF. This screen identified the gene *mofA*, which encodes a putative MCO of 1,662 amino acids, with a putative signal sequence at the N-terminus for export of MofA from the cell (Corstjens et al. 1997). The *mofA* gene has been detected in other Mn(II)-oxidizing *Leptothrix* sp. strains but not in other Mn(II)-oxidizing genera (Siering and Ghiorse 1997). While Mn(II) oxidation by *L. discophora* is affected by iron concentration, the level of *mofA* transcript is not (El Gheriany et al. 2009). Also, the levels of *mofA* transcript during the cell cycle do not mirror the levels of Mn(II) oxidase activity (El Gheriany et al. 2009). This suggests that Fe may affect MofA post-transcriptionally or that Fe is required for a cofactor of MofA or an accessory protein.

The gene *mofC*, located downstream of *mofA* in a putative operon, encodes a protein with a putative heme-binding motif specific to c-type cytochromes (Brouwers et al. 2000) so MofC may be involved in Mn(II) oxidation and its activity may be affected by Fe concentration.

An MCO has also been implicated in Mn(II) oxidation by the α-proteobacterium *Pedomicrobium* sp. ACM 3067 (Ridge et al. 2007). The *Pedomicrobium* genus is made up of budding-hyphal bacteria found in aquatic environments (Cox and Sly 1997); most species in this genus are capable of oxidizing Mn(II) (Gebers and Beese 1988). Much like *L. discophora*, the solid Mn(IV) oxides generated by *Pedomicrobium* sp. accumulate on extracellular polysaccharides on the cell surface (Sly et al. 1990). Oxidation occurs in early to mid exponential phase and is inhibited by positively charged antibiotics such as gentamicin and by poly-L-lysine (Larsen et al. 1999). The ability to oxidize Mn(II) requires the presence of copper ions, implying that the oxidase is a copper-dependent enzyme (Larsen et al. 1999). Using degenerate primers specific for the copper binding motifs of MCO proteins, the gene *moxA* was identified in *Pedomicrobium* sp. ACM 3067 (Ridge et al. 2007). *moxA* encodes a putative MCO of 476 amino acids, with a signal sequence for export. The gene was disrupted by plasmid integration and, of the 12 potential plasmid integrants isolated, one (ML2) was shown to be defective for Mn(II) oxidation and to have its *moxA* gene disrupted. Therefore, MoxA was identified as the Mn(II) oxidase enzyme in *Pedomicrobium* sp. ACM 3067. However, it is not clear what the oxidation phenotypes of the other 11 plasmid integrants were or where the plasmid integrated in these strains (Ridge et al. 2007). Thus, it is possible that the defect in oxidation shown by ML2 is due to a second mutation in the strain, not the plasmid integrated into *moxA*. Unambiguous identification of the Mn(II) oxidase from *Pedomicrobium* sp. ACM 3067 awaits biochemical purification of the activity.

A Calcium-Binding Heme Peroxidase Involved in Mn(II) Oxidation

Many members of the globally-dispersed *Aurantimonadaceae* family of the α-proteobacteria are capable of oxidizing Mn(II) (Anderson et al. 2009a). Oxidation by the strain *A. manganoxydans* SI85-9A1 has been studied in the most detail; like other oxidizers, it precipitates Mn(IV) oxides on its cell surface (Caspi et al. 1996). Unlike most other oxidizing species, *A. manganoxydans* SI85-9A1 oxidizes Mn(II) early in growth (Dick et al. 2008a), making its oxidation behavior more similar to *Pedomicrobium* sp. ACM 3067 (Larsen et al. 1999). Mn(II) oxidation by *A. manganoxydans* SI85-9A1 occurs during growth on organic acids like formate and on glycerol but not during growth on glucose, even though growth rate is enhanced on this substrate (Dick et al. 2008a). Growth on glycerol is very slow but is substantially increased by the addition of $MnCl_2$. Except for a moderate effect in *Erythrobacter* sp. SD-21 (see below), no other characterized Mn(II) oxidizing organism displays this effect of Mn(II) on growth (Dick et al. 2008a).

In the genome sequence of *A. manganoxydans* SI85-9A1, there are two homologs to *moxA,* the putative MCO Mn(II) oxidase from *Pedomicrobium* sp. ACM 3067, which have been named *moxA1* and *moxA2* (Dick et al. 2008a). Based on this, it was expected that the Mn(II) oxidase enzyme from this species would be an MCO. However, purification of the oxidizing activity from the loosely bound outer membrane fraction of *A. manganoxydans* SI85-9A1 identified a hemolysin-type Ca^{2+}-binding peroxidase as the oxidase (Anderson et al. 2009b). Heme-containing manganese peroxidase enzymes (MnPs) have previously been identified in eukaryotes (Camarero et al. 2000; Palma et al. 2000); in particular, MnPs are important for the degradation of lignin by fungi (Reddy and D'Souza 1994). In the well-studied MnP from *Phanerochaete chrysosporium,* one H_2O_2 equivalent is consumed to produce two Mn(III) equivalents (Wariishi et al. 1992). Mn(II) oxidation by *A. manganoxydans* SI85-9A1 appears to proceed through an Mn(III) intermediate and is stimulated by the addition of H_2O_2 to the isolated loosely bound outer membrane fraction (Anderson et al. 2009b). Furthermore, using a heme-staining in-gel assay, a heme-binding protein was seen to co-localize with the Mn(II) oxidase and Mn(II) oxidation was stimulated by the addition of Ca^{2+} (Anderson et al. 2009b). Taken together, these data support the identification of a Ca^{2+} binding heme peroxidase as the Mn(II) oxidase in this species. Consequently, the protein was named MopA, for manganese-oxidizing peroxidase.

In the marine α-proteobacterium *Erythrobacter* sp. SD-21, Mn(II) oxidation begins in stationary phase, with a slower apparent rate of oxidation in cultures grown in light as opposed to those grown in dark (Francis et al. 2001). With the addition of $MnCl_2$ to the growth medium, the bacteria grow to a higher cell density but the rate of growth is not detectably affected (Francis et al. 2001). In an early indication that *Erythrobacter* sp. SD-21 did not employ an MCO for Mn(II) oxidation, addition of Cu^{2+} to cell-free extracts did not increase Mn(II) oxidation activity, while addition of Ca^{2+}, NAD^+ and pyrroloquinoline quinone (PQQ) did (Johnson and Tebo 2008). Subsequently, a Ca^{2+} binding heme peroxidase similar to MopA was isolated from this strain (Anderson et al. 2009b). The stimulatory effect of Ca^{2+}, then, is likely due to its effect on MopA. The effect of PQQ and NAD^+ is proposed to be through reaction of PQQ and NADH to produce peroxide, which could then be used by the peroxidase to oxidize Mn(II) (Anderson et al. 2009b). The addition of PQQ and NAD^+ together also slightly enhanced oxidation by *A. manganoxydans* SI85-9A1 (Anderson et al. 2009b).

The product of Mn(II) oxidation by cell-free extracts from *Erythrobacter* sp. SD-21 is soluble and does not have the characteristic brown color of Mn(IV) oxides. On plates, the oxides are present in a halo surrounding the colonies, which would be consistent with a soluble Mn(III) complex diffusing through the agar then disproportionating to Mn(II) and Mn(IV) (Johnson and Tebo 2008). Thus the product of Mn(II) oxidation in this species may be Mn(III), not Mn(IV). In agreement with this, an Mn(III)-PP complex that was stable for up to 30 h could be detected through in vitro oxidation assays with cell-free extracts (Johnson and Tebo 2008). In the fungal systems, MnPs oxidize Mn(II) to Mn(III) in the presence of chelators; the chelated Mn(III) then oxidizes lignin and the other substrates of this enzyme

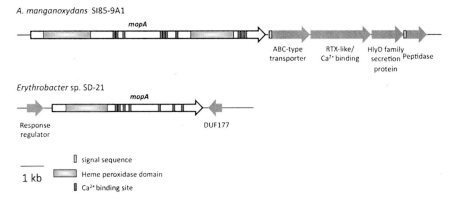

Fig. 13.3 Peroxidase genes identified as encoding the putative Mn(II) oxidase. Depicted as *blue arrows* are the genes that encode the putative Mn(II) oxidase in *A. manganoxydans* SI85-9A1 and *Erythrobacter* sp. SD-21, while neighboring genes are *grey arrows*. Gene names, when available, are listed above the genes and the putative functions of the non-peroxidase genes are below. For references, see text

(Glenn et al. 1986; Perez and Jeffries 1992). In the absence of chelators the Mn(III) is unstable and disproportionates into Mn(IV) and Mn(II) (Glenn et al. 1986). Thus, Mn(III) would be the expected product of a Mn(II) oxidase like MopA. However, *A. manganoxydans* SI85-9A1 forms Mn(IV) oxides (Caspi et al. 1996; Anderson et al. 2009b); this suggests its MopA enzyme is capable of two electron transfers to produce Mn(IV) or that a second enzyme present in the purified active fraction carries out the second electron transfer. Of possible significance is the observation that MopA from *A. manganoxydans* SI85-9A1 is predicted to have two heme peroxidase domains while the MopA from *Erythrobacter* sp. SD-21 is only predicted to have one (Fig. 13.3) (Anderson et al. 2009b); perhaps the second peroxidase domain is required for a second electron transfer to oxidize Mn(III) to Mn(IV).

The Mn(II) Oxidase of Pseudomonas putida *GB-1 Remains Elusive*

The bulk of the genetic experiments investigating bacterial Mn(II) oxidation have been performed in the γ-proteobacterium *Pseudomonas putida* GB-1 (and the closely related strain MnB1; (Schweisfurth 1973; Brouwers et al. 1997; Caspi et al. 1998), due to its genetic tractability and the ease with which it can be grown and maintained in the lab environment. Like the other characterized bacterial Mn(II) oxidizers, *P. putida* GB-1 precipitates Mn(IV) oxides on its cell surface (Okazaki et al. 1997). A genetic screen for *P. putida* GB-1 mutants that failed to oxidize Mn(II) identified genes with homology to components of the general secretory pathway; transposon insertions in these genes resulted in mutants that produced the oxidase enzyme but failed to produce Mn(IV) oxides on their surface (de Vrind et al. 2003). A similar secretion pathway may be employed by other Mn(II)-oxidizing bacteria. In *P. putida* GB-1 Mn(II) oxidation begins at the onset of stationary phase (Jung and Schweisfurth 1979), requires oxygen (Okazaki et al. 1997) and iron (Murray et al. 2005; Parker et al. 2007), and is inhibited by the presence of some antibiotics (Rosson et al. 1984; Caspi et al. 1998).

Several rounds of random transposon mutagenesis have been undertaken to identify the Mn(II) oxidase of *P. putida* GB-1. In one of these experiments, a mutant strain was isolated (GB-1-007) that lost the ability to oxidize Mn(II) and that had a transposon insertion in a MCO-encoding gene. This gene was named *cumA* and was proposed to encode the Mn(II) oxidase of *P. putida* GB-1 (Brouwers et al. 1999). Recent work has shown, however, that in-frame deletion of *cumA* has no effect on Mn(II) oxidation and that the mutation responsible for the oxidation defect of GB-1-007 was a frame-shift mutation in a sensor histidine kinase gene (*mnxS1*). The *mnxS1* gene is located immediately upstream of a second sensor histidine kinase gene (*mnxS2*) and a response regulator gene (*mnxR*), each of which are required for Mn(II) oxidation in *P. putida* GB-1 (Geszvain and Tebo, 2010). Thus the GB-1-007 mutant revealed not the Mn(II) oxidase enzyme itself but a two-component regulatory pathway essential for oxidation.

Other factors required for Mn(II) oxidation in *P. putida* GB-1 and *P. putida* MnB1 have been identified through random transposon mutagenesis screens. For example, transposon insertions within the c-type cytochrome maturation operon *ccm* resulted in a loss of Mn(II) oxidation (Caspi et al. 1998; de Vrind et al. 1998). This result suggests that cytochrome c is involved in the transfer of electrons from Mn(II) to oxygen or that a c-type cytochrome is part of the Mn(II) oxidase enzyme. Interestingly, two mutants with transposon insertions in *ccmF* also exhibited dere-pressed pyoverdine synthesis and secreted porphyrins into the medium (de Vrind et al. 1998). Thus, the lack of Mn(IV) oxide formation seen with these mutants may be due to trapping of the oxidation intermediate Mn(III) by the excess pyoverdine (Murray et al. 2005; Parker et al. 2007). However, the fact that whole cell lysates from the *ccmF* mutants also failed to oxidize Mn(II) (de Vrind et al. 1998) suggests secreted pyoverdines are not the cause of the defect, since they would be washed away during preparation of the lysates. Another class of mutants defective in Mn(II) oxidation is those with transposon insertions within genes encoding components of the Krebs cycle: the succinate dehydrogenase (*sdh*) complex genes, the lipoate acetyltransferase subunit of pyruvate dehydrogenase (*aceA*), and two isoci-trate dehydrogenase genes (*icd*) (Caspi et al. 1998). These mutants also were defective in cytochrome c oxidase activity and thus may affect Mn(II) oxidation by the same mechanism as the *ccm* operon mutants. On the other hand, these mutants also exhibited a growth defect that could be responsible for the inability to oxidize Mn(II). Mn(II) oxidation by *P. putida* MnB1 was also disrupted by transposon insertions in *trpE*, the gene which encodes the α subunit of the tryptophan biosyn-thesis enzyme anthranilate synthetase (Caspi et al. 1998). Oxidation by this mutant could be restored by supplementation of the growth medium with tryptophan, but it could also be restored by addition of PQQ (Johnson and Tebo 2008).

Thus far, the *P. putida* GB-1 Mn(II) oxidase enzyme has not been identified genetically. There is some evidence to suggest that the addition of copper enhances Mn(II) oxidation by *P. putida* GB-1 (Brouwers et al. 1999) so it is possible that an MCO protein other than CumA is involved in Mn(II) oxidation in this species. Mn(II) oxidation by a cell-free extract from *P. putida* MnB1 could be substantially stimulated by addition of PQQ and moderately enhanced by NAD^+ (Johnson and Tebo 2008). In this way, *P. putida* is similar to *A. manganoxydans* SI85-9A1 and *Erythrobacter* sp. SD-21 and thus it may also employ a peroxidase enzyme for Mn(II) oxidation. Partial biochemical purification of the oxidase activity suggested the presence of two oxidizing factors of 250 and 180 kDa (Okazaki et al. 1997). The possible presence of multiple Mn(II) oxidase enzymes may explain the difficulty in genetically identifying the enzyme since it would be difficult to simultaneously disrupt the genes for each enzyme. In the basidiomycete *Stropharia rugosoannulata*, a laccase (a type of MCO) and an MnP have been shown to work together, with the MnP using H_2O_2 produced via oxidation of Mn(II) to Mn(III) by the laccase (Schlosser and Hofer 2002). Therefore, it is possible that the two previously identified factors in *P. putida* may be an MCO and a peroxidase.

The Molecular Complexity of Bacterial Mn Oxidation

The three MCOs identified as the putative Mn(II) oxidase show little similarity to one another outside their copper-binding motifs. At the gene level, while *mnxG* is the last gene of a putative seven gene operon (van Waasbergen et al. 1996) and *moxA* the last gene of a three gene operon (Ridge et al. 2007), *mofA* is predicted to have its own promoter (Corstjens et al. 1997) and fall at the beginning of a three gene operon (Brouwers et al. 2000) (Fig. 13.2). The proteins are also quite different. MnxG and MofA are both large proteins of over one thousand amino acids (van Waasbergen et al. 1996; Corstjens et al. 1997), while MoxA is less than 500 amino acids (Ridge et al. 2007). MnxG is the only Mn(II)-oxidizing MCO predicted to have more than four copper-binding sites (Dick et al. 2008b). The order of the Cu-binding sites in MnxG is also unusual (Fig. 16.2) (Dick et al. 2008b). Thus, while it was possible to identify a putative Mn(II)-oxidizing MCO from *Pedomicrobium* using just sequences conserved among MCOs, there are no unifying characteristics among the known Mn(II) oxidases that would make it possible to identify which MCO in the genome of a Mn(II)-oxidizing organism is the oxidase. As a further complication, *A. manganoxydans* SI85-9A1 and *Erythrobacter* sp. SD-21 employ MopA, a peroxidase enzyme, to oxidize Mn(II) and their two MopA proteins also vary quite a bit in size, domain organization, and genomic context (Fig. 13.3).

Aside from the extracellular location of the oxidase activity, there are few common themes among the characterized Mn(II) oxidizing strains (Table 13.1). In most cases the final product is Mn(IV) oxide, although *Erythrobacter* sp. SD-21 may stop at Mn(III). Oxidation occurs in spores or stationary phase cultures, except for

Table 13.1 Comparison of the model Mn(II) oxidizing bacteria

Strain	Class	Growth effect[a]	Time of oxidation	Putative Mn(II) oxidase	Rxn product[b]	Mn(III)?[c]	Stimulant[d]
A. manganoxydans SI85-9A1	Alpha[e]	Stimulate growth rate	Exponential	Peroxidase (MopA)	Mn(IV)	Yes	Ca^{2+}, H_2O_2, PQQ, NAD^+
Erythrobacter sp. SD-21	Alpha[e]	Stimulate growth yield	Stationary	Peroxidase (MopA)	Mn(III)?	Yes	Ca^{2+}, PQQ, NAD^+
Pedomicrobium sp. ACM 3067	Alpha[e]	None	Exponential	MCO (MoxA)	?	?	Cu^{2+}
L. discophora SS-1	Beta[e]	None	Stationary	MCO (MofA)	?	?	Fe(II), Cu^{2+}
P. putida GB-1	Gamma[e]	None	Stationary	?	Mn(IV)	Yes	PQQ, NAD^+, Cu^{2+}
Bacillus sp. SG-1	Bacilli	None	Mature spore	MCO (MnxG)	Mn(IV)	Yes	Cu^{2+}

[a]The effect on growth of the addition of Mn(II) to media
[b]The end product of Mn(II) oxidation
[c]Is there detectable Mn(III) produced during oxidation?
[d]Listed are compounds that increase Mn(II) oxidation by cultures or cell extracts of the respective strains
[e]Classes within the phylum Proteobacteria

A. manganoxydans SI85-9A1 and *Pedomicrobium* sp. ACM 3067 and the addition of Mn(II) only appears to affect the growth of *A. manganoxydans* SI85-9A1 and *Erythrobacter* sp. SD-21. This variability in the physiology of Mn(II) oxidation may reflect variability in the role(s) oxidation plays in each organism.

Environmental Perspectives

The advances in our understanding of bacterial Mn(II) oxidation have provided an impetus for more mechanistic studies of Mn cycling in the environment. In this section we briefly summarize two environments that reinforce the notion that the model laboratory systems currently under study are relevant to aquatic systems.

Suboxic Zones of Anoxic and Seasonally Anoxic Basins: Mn(III) Is Important!

In redox-stratified marine and estuarine environments, stable suboxic zones can form as a transition zone between the overlying oxic waters and the deeper hydrogen sulfide-containing waters and defined as those zones where $O_2 < 5$ µM and H_2S is undetectable. In the suboxic zones the O_2 gradient is negligible, oxidized forms of nitrogen (NO_2^-, NO_3^-) become depleted, and particulate Mn oxides have a maximum (Fig. 13.4). Mn cycling (oxidation and reduction) is particularly pronounced in these suboxic zones and contributes to their maintenance (Tebo 1991; Oguz et al. 2001; Konovalov et al. 2003, 2004; Yakushev et al. 2007).

The Black Sea, the world's largest anoxic basin, has served as a model for studies of Mn cycling in the suboxic zone. One of the major discoveries during the 1988 Black Sea expedition aboard the R/V Knorr was the geographically extensive, broad suboxic zone (Fig. 13.4). At the time, measurements of the Mn(II) oxidation rates indicated very rapid O_2-dependent microbially-catalyzed oxidation and suggested that this process was responsible for maintenance of the suboxic zone, i.e., that the microbially-produced Mn oxides capped the vertical flux of hydrogen sulfide. Subsequent modeling studies of the extensive 1988 data set also pointed to the Mn oxides as being important; however, because the flux of O_2 wasn't sufficient to account for all of the H_2S being consumed in the suboxic zone, Mn(II) oxidation coupled to nitrate reduction was proposed to occur (Murray et al. 1995). This hypothesis has been the subject of further investigation by German and U.S. scientists during the last decade. Direct measurements using radiotracer techniques did not provide evidence for any electron acceptor for Mn oxide formation other than O_2, although O_2 was found to be required in only very small amounts (Schippers et al. 2005; Clement et al. 2009). Based on the laboratory results showing that soluble Mn(III) was important

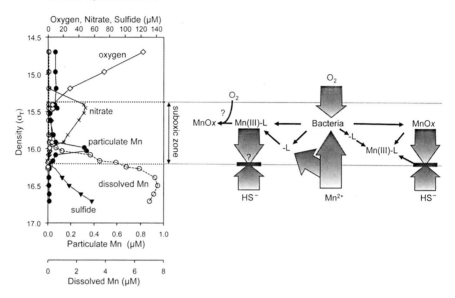

Fig. 13.4 A model of Mn cycling in the suboxic zone of the Black Sea taking into account the formation of Mn(III) complexes with organic or inorganic ligands (L). On the *left* depth profiles of important redox chemicals are shown plotted versus the density (σ_T) of the water column. The suboxic zone is defined as the region in which the gradients of oxygen and sulfide are very small or negligible. MnO_x refers to Mn(IV) oxides

(Webb et al. 2005b; Parker et al. 2007), a concerted effort to develop a new method to measure Mn(III) complexes and apply it to field measurements was made (Trouwborst et al. 2006). Direct measurement of Mn(III) in water samples from the suboxic zone of the Black Sea, and the seasonally anoxic Chesapeake Bay indicated that Mn(III) concentrations could be substantial (as high as 4–5 μM in the Black Sea) and account for up to 100% of the dissolved Mn in the suboxic zone (Trouwborst et al. 2006). Although the natural ligand(s) for Mn(III) complexation have not been determined, Mn(II)-oxidizing bacteria isolated from the Black Sea have been shown to be abundant and to produce Mn(III) complexing siderophores (Fig. 13.4) and preliminary evidence exists for the in situ production of Mn(III) (Clement 2006). Modeling studies, using the measured values of Mn(III) in the Black Sea, can now account for removal of all of the H_2S flux into the suboxic zone without requiring the contribution of other electron acceptors (Yakushev et al. 2007).

The importance of Mn(III) in the environment is not limited to large anoxic basins. Mn(III) was also found in the seasonally suboxic zone in the Chesapeake Bay (Trouwborst et al. 2006) and in sediments of the St. Laurentian estuary (A. Madison, G.W. Luther and B.M. Tebo, unpublished). The occurrence of soluble Mn(III) complexes as important redox chemical species in the environment is anticipated to be a widespread phenomenon, whose implications may be significant.

The Columbia River Estuary: Identification of the Environmental Mn Oxidase

While it's been extremely challenging to identify the enzymes that are involved in Mn(II) oxidation in the model organisms in the laboratory, the more difficult question is whether the same type of enzymes are involved in bacterial Mn(II) oxidation in the environment. Do our model systems have any relevance in the field?

The Columbia River is the largest river draining the Pacific Northwest region of the United States and is a major source of dissolved nutrients and trace metals, including Mn, to the California Current (Barnes et al. 1972; Aguilar-Islas and Bruland 2006; Bruland et al. 2008). The natural tidal cycles and high river flow create a dynamic estuary with high turbidity due to sediment load in the river water and resuspension during mixing of the oceanic and fresh waters, particularly during spring tides when Mn concentrations are particularly high and resuspension of Mn oxides occurs. During this period the Mn cycle is undoubtedly very active, with both Mn(II) oxidation and Mn oxide reduction expected to be occurring to various extents in different parts of the estuary and the estuarine plume that extends into the coastal ocean.

In an attempt to identify the enzymes active in Mn(II) oxidation in the Columbia River estuary and plume, Anderson et al. (in preparation) collected large volume water samples, concentrated them using tangential flow filtration and assayed for Mn(II) oxidation activity. Three different samples containing the highest Mn(II) oxidation activity were used for proteomic analysis. The samples were fractionated into five fractions: secreted, loosely bound outer membrane (LBOM), soluble (periplasmic and cytoplasmic), inner membrane, and outer membrane (OM). Mn(II) oxidation assays revealed activity only in the LBOM and OM fractions, consistent with localization studies done with the model Mn oxidizers *A. manganoxydans* SI85-9A1 and *Erythrobacter* sp. SD-21 (Johnson and Tebo 2008; Anderson et al. 2009b). The proteins in these fractions were separated using size exclusion chromatography and the active fractions were concentrated and analyzed using tandem mass spectrometry. A total of 105 protein homologs were identified in GenBank with a large percentage showing homology to *A. manganoxydans* SI85-9A1 based on the genome sequence. Many of the proteins identified were associated with motility, receptor or transport functions. However, of particular interest was the identification of both a putative hemolysin-type Ca-binding heme peroxidase and a putative MCO suggesting that both types of enzymes could be involved in Mn(II) oxidation in the environment. Q-PCR primers were designed for heme peroxidase genes and used to interrogate genomic DNA extracted from the same samples as the proteins extractions as well as from other samples. The heme peroxidase gene copy number ranged between 3.7×10^4 and 2.6×10^5 in the estuarine samples. Although we don't know the proportion of these heme peroxidases that oxidize Mn(II), these results are consistent with a heme peroxidase mechanism of Mn(II) oxidation in the estuary.

Implications and Future Directions

Recent progress in the laboratory in understanding the molecular underpinnings of bacterial Mn(II) oxidation has required a revision in the way we think about Mn(II) oxidation in the environment. Clearly, the old paradigm of dissolved and particulate Mn (defined by what goes through or gets trapped by a 0.2 µm filter), no longer applies: redox active Mn(III) may exist as soluble organic or inorganic complexes in the environment.

The new Mn paradigm requires us to reconsider the Mn cycle. Since Mn(III) may compete for the same ligands (siderophores) as Fe(III), the Fe and Mn cycles are even more intimately intertwined than previously thought. In major regions of the ocean, primary productivity is Fe-limited and >99% of the available Fe is complexed with organic ligands believed to be siderophores or their breakdown products. Since Mn is in greater abundance in surface seawater, Mn(III) may complex to the same ligands as those that bind Fe(III) and indirectly affect Fe availability to phytoplankton. This may also partially account for the surface maximum in dissolved Mn, previously thought to be due entirely to photoreduction of Mn oxides (Sunda et al. 1983; Sunda and Huntsman 1988). Perhaps Mn(II)-oxidizing bacteria are able to overcome Fe limitation by oxidizing Mn(II) to Mn(III) which would exchange with the Fe(III) in the complexed aquatic pool and the Fe(III) would in turn become available to the bacteria (Parker et al. 2004, 2007). The knowledge that Mn(III) can occur as stable soluble complexes with siderophores invites further speculation that, by analogy to the case for Fe, this is the reason for the sustained Mn concentrations in the deep ocean (Johnson 2006).

Additionally, from laboratory investigations we now have a better sense that abiotic reactions driven indirectly by Mn(II)-oxidizing bacteria may at times be extremely important, e.g., ligand-promoted oxidation of Mn(II) to Mn(III) by siderophores or the autooxidation of Mn(II) on the surface of extremely reactive preformed bacteriogenic Mn oxides. Understanding the global biogeochemical cycle of Mn will require us to be able to distinguish and quantify these processes.

Acknowledgements We gratefully acknowledge funding for our research from the National Science Foundation (MCB-0630355 and OCE-0221500), the NSF Science and Technology Center for Coastal Margins Observation and Prediction (CMOP; OCE 0424602) and grant number P42ES010337 from the National Institute of Environmental Health Sciences (NIEHS). The contents of this publication are solely the responsibility of the authors and do not necessarily represent the official views of the NSF or NIEHS.

References

Adams LF, Ghiorse WC (1985) Influence of Manganese on Growth of a Sheathless Strain of *Leptothrix discophora*. Appl Environ Microbiol 49:556–562
Aguilar-Islas AM, Bruland KW (2006) Dissolved manganese and silicic acid in the Columbia River plume: a major source to the California current and coastal waters off Washington and Oregon. Mar Chem 101:233–247

Anderson CR, Dick GJ, Chu ML, Cho JC, Davis RE, Brauer SL, Tebo BM (2009a) *Aurantimonas manganoxydans*, sp. nov. and *Aurantimonas litoralis*, sp. nov.: Mn(II) oxidizing representatives of a globally distributed clade of alpha-Proteobacteria from the order Rhizobiales. Geomicrobiol J 26:189–198

Anderson CR, Johnson HA, Caputo N, Davis RE, Torpey JW, Tebo BM (2009b) Mn(II) oxidation is catalyzed by heme peroxidases in "*Aurantimonas manganoxydans*" strain SI85-9A1 and *Erythrobacter* sp. strain SD-21. Appl Environ Microbiol 75:4130–4138

Baldrian P (2006) Fungal laccases – occurrence and properties. FEMS Microbiol Rev 30:215–242

Bargar JR, Tebo BM, Villinski JE (2000) In situ characterization of Mn(II) oxidation by spores of the marine *Bacillus* sp strain SG-1. Geochim Cosmochim Acta 64:2775–2778

Bargar JR, Tebo BM, Bergmann U, Webb SM, Glatzel P, Chiu VQ, Villalobos M (2005) Biotic and abiotic products of Mn(II) oxidation by spores of the marine *Bacillus* sp. strain SG-1. Am Mineral 90:143–154

Bargar JR, Fuller CC, Marcus MA, Brearley AJ, Perez De la Rosa M, Webb SM, Caldwell WA (2009) Structural characterization of terrestrial microbial Mn oxides from Pinal Creek, AZ. Geochim Cosmochim Acta 73:889–910

Barnes CA, Duxbury AC, Morse BA (1972) Circulation and selected properties of the Columbia river effluent at sea. In: Pruter AT, Alverson DL (eds) The Columbia river estuary and adjacent ocean waters. University of Washington Press, Seattle, pp 41–80

Brouwers G-J, de Vrind JPM, Corstjens PLAM, Westbroek P, de Vrind-de Jong EW (1997) *Pseudomonas* genes involved in the oxidation of manganese. Abstracts, Geological Society of America Annual Meeting, Salt Lake City, UT

Brouwers GJ, de Vrind JP, Corstjens PL, Cornelis P, Baysse C, de Vrind-de Jong EW (1999) *cumA*, a gene encoding a multicopper oxidase, is involved in Mn^{2+} oxidation in *Pseudomonas putida* GB-1. Appl Environ Microbiol 65:1762–1768

Brouwers GJ, Vijgenboom E, Corstjens PLAM, de Vrind JPM, de Vrind-de Jong EW (2000) Bacterial Mn^{2+} oxidizing systems and multicopper oxidases: an overview of mechanisms and functions. Geomicrobiol J 17:1–24

Bruland KW, Lohan MC, Aguilar-Islas AM, Smith GJ, Sohst B, Baptista A (2008) Factors influencing the chemistry of the near-field Columbia river plume: nitrate, silicic acid, dissolved Fe, and dissolved Mn. J Geophys Res Oceans 113:C00B02

Cahyani VR, Murase J, Ishibashi E, Asakawa S, Kimura M (2009) Phylogenetic positions of Mn^{2+}-oxidizing bacteria and fungi isolated from Mn nodules in rice field subsoils. Biol Fertil Soils 45:337–346

Camarero S, Ruiz-Duenas FJ, Sarkar S, Martinez MJ, Martinez AT (2000) The cloning of a new peroxidase found in lignocellulose cultures of *Pleurotus eryngii* and sequence comparison with other fungal peroxidases. FEMS Microbiol Lett 191:37–43

Caspi R, Haygood MG, Tebo BM (1996) Unusual ribulose-1, 5-bisphosphate carboxylase/oxygenase genes from a marine manganese-oxidizing bacterium. Microbiology 142: 2549–2559

Caspi R, Tebo BM, Haygood MG (1998) c-type cytochromes and manganese oxidation in *Pseudomonas putida* MnB1. Appl Environ Microbiol 64:3549–3555

Clement BG (2006) Biological Mn(II) oxidation in freshwater and marine systems: new perspectives on reactants, mechanisms and microbial catalysts of Mn cycling in the environment. University of California San Diego, La Jolla, CA

Clement BG, Luther GW III, Tebo BM (2009) Rapid, oxygen-dependent microbial Mn(II) oxidation kinetics at sub-micromolar oxygen concentrations in the Black Sea suboxic zone. Geochim Cosmochim Acta 73:1878–1889

Corstjens PLAM, de Vrind JPM, Goosen T, de Vrind-de Jong EW (1997) Identification and molecular analysis of the *Leptothrix discophora* SS-1 *mofA* gene, a gene putatively encoding a manganese-oxidizing protein with copper domains. Geomicrobiol J 14:91–108

Cox TL, Sly LI (1997) Phylogenetic relationships and uncertain taxonomy of *Pedomicrobium* species. Int J Syst Bacteriol 47:377–380 de Vrind JP, de Vrind-de Jong EW, de Voogt JW, Westbroek P, Boogerd FC, Rosson RA (1986) Manganese oxidation by spores and spore coats of a marine *Bacillus* species. Appl Environ Microbiol 52:1096–1100

de Vrind JP, Brouwers GJ, Corstjens PL, den Dulk J, de Vrind-de Jong EW (1998) The cytochrome c maturation operon is involved in manganese oxidation in *Pseudomonas putida* GB-1. Appl Environ Microbiol 64:3556–3562

de Vrind J, de Groot A, Brouwers GJ, Tommassen J, de Vrind-de Jong E (2003) Identification of a novel Gsp-related pathway required for secretion of the manganese-oxidizing factor of *Pseudomonas putida* strain GB-1. Mol Microbiol 47:993–1006

de Vrind-de Jong EW, Corstjens PL, Kempers ES, Westbroek P, de Vrind JP (1990) Oxidation of manganese and iron by Leptothrix discophora: Use of N, N, N', N'-Tetramethyl-p-Phenylenediamine as an indicator of metal oxidation. Appl Environ Microbiol 56:3458–3462

Dick GJ, Lee YE, Tebo BM (2006) Manganese(II)-oxidizing bacillus spores in Guaymas basin hydrothermal sediments and plumes. Appl Environ Microbiol 72:3184–3190

Dick GJ, Podell S, Johnson HA et al (2008a) Genomic insights into Mn(II) oxidation by the marine alphaproteobacterium *Aurantimonas* sp. strain SI85-9A1. Appl Environ Microbiol 74:2646–2658

Dick GJ, Torpey JW, Beveridge TJ, Tebo BM (2008b) Direct identification of a bacterial manganese(II) oxidase, the multicopper oxidase MnxG, from spores of several different marine *Bacillus* species. Appl Environ Microbiol 74:1527–1534

Dick GJ, Clement BG, Webb SM, Fodrie FJ, Bargar JR, Tebo BM (2009) Enzymatic microbial Mn(II) oxidation and Mn biooxide production in the Guaymas basin deep-sea hydrothermal plume. Geochim Cosmochim Acta 73:6517–6530

Duckworth OW, Sposito G (2005a) Siderophore-manganese(III) interactions II. Manganite dissolution promoted by desferrioxamine B. Environ Sci Technol 39:6045–6051

Duckworth OW, Sposito G (2005b) Siderophore-manganese(III) interactions. I. Air-oxidation of manganese(II) promoted by desferrioxamine B. Environ Sci Technol 39:6037–6044

Duckworth OW, Bargar JR, Sposito G (2009a) Coupled biogeochemical cycling of iron and manganese as mediated by microbial siderophores. Biometals 22:605–613

Duckworth OW, Holmstrom SJM, Pena J, Sposito G (2009b) Biogeochemistry of iron oxidation in a circumneutral freshwater habitat. Chem Geol 260:149–158

Ehrlich HL, Salerno JC (1990) Energy coupling in Mn^{2+} oxidation by a marine bacterium. Arch Microbiol 154:12–17

El Gheriany IA, Bocioaga D, Hay AG, Ghiorse WC, Shuler ML, Lion LW (2009) Iron requirement for Mn(II) oxidation by *Leptothrix discophora* SS-1. Appl Environ Microbiol 75:1229–1235

Emerson S, Kalhorn S, Jacobs L, Tebo BM, Nealson KH, Rosson RA (1982) Environmental oxidation rate of manganese(II): bacterial catalysis. Geochim Cosmochim Acta 46:1073–1079

Essen SA, Bylund D, Holmstrom SJM, Moberg M, Lundstrom US (2006) Quantification of hydroxamate siderophores in soil solutions of podzolic soil profiles in Sweden. Biometals 19:269–282

Faulkner KM, Stevens RD, Fridovich I (1994) Characterization of Mn(III) complexes of linear and cyclic desferrioxamines as mimics of superoxide dismutase activity. Arch Biochem Biophys 310:341–346

Francis CA, Tebo BM (2002) Enzymatic manganese(II) oxidation by metabolically dormant spores of diverse *Bacillus* species. Appl Environ Microbiol 68:874–880

Francis CA, Co EM, Tebo BM (2001) Enzymatic manganese(II) oxidation by a marine α-proteobacterium. Appl Environ Microbiol 67:4024–4029

Frangipani E, Haas D (2009) Copper acquisition by the SenC protein regulates aerobic respiration in *Pseudomonas aeruginosa* sp. PAO1. FEMS Microbiol Lett 298:234–240

Gebers R, Beese M (1988) *Pedomicrobium americanum* sp. nov. and *Pedomicrobium australicum* sp. nov. from Aquatic Habitats, *Pedomicrobium* gen. emend., and *Pedomicrobium ferrugineum* sp. emend. Int J Syst Bacteriol 38:303–315

Geszvain K, Tebo BM (2010) Identification of a two-component regulatory pathway essential for Mn(II) oxidation in *Pseudomonas putida* GB-1. Appl Environ Microbiol 76:1224–1231

Ghiorse WC (1984) Biology of iron- and manganese-depositing bacteria. Annu Rev Microbiol 38:515–550

Glenn JK, Akileswaran L, Gold MH (1986) Mn(II) oxidation is the principal function of the extracellular Mn-peroxidase from *Phanerochaete chrysosporium*. Arch Biochem Biophys 251:688–696

Hastings D, Emerson S (1986) Oxidation of manganese by spores of a marine *Bacillus* – kinetic and thermodynamic considerations. Geochim Cosmochim Acta 50:1819–1824

Hofer C, Schlosser D (1999) Novel enzymatic oxidation of Mn^{2+} to Mn^{3+} catalyzed by a fungal laccase. FEBS Lett 451:186–190

Holmstrom SJM, Lundstrom US, Finlay RD, van Hees PAW (2004) Siderophores in forest soil solution. Biogeochemistry 71:247–258

Johnson KS (2006) Manganese redox chemistry revisited. Science 313:1896–1897

Johnson HA, Tebo BM (2008) In vitro studies indicate a quinone is involved in bacterial Mn(II) oxidation. Arch Microbiol 189:59–69

Jung WK, Schweisfurth R (1979) Manganese oxidation by an intracellular protein of a *Pseudomonas* species. Z Allg Mikrobiol 19:107–115

Jürgensen A, Widmeyer JR, Gordon RA, Bendell-Young LI, Moore MM, Crozier ED (2004) The structure of the manganese oxide on the sheath of the bacterium *Leptothrix discophora*: an XAFS study. Am Mineral 89:1110–1118

Kepkay PE, Nealson KH (1987) Growth of a manganese oxidizing *Pseudomonas* sp in continuous culture. Arch Microbiol 148:63–67

Klewicki JK, Morgan JJ (1998) Kinetic behavior of Mn(III) complexes of pyrophosphate, EDTA, and citrate. Environ Sci Technol 32:2916–2922

Klewicki JK, Morgan JJ (1999) Dissolution of β-MnOOH particles by ligands: pyrophosphate, ethylenediaminetetraacetate, and citrate. Geochim Cosmochim Acta 63:3017–3024

Konovalov SK, Luther GW, Friederich GE et al (2003) Lateral injection of oxygen with the Bosporus plume – fingers of oxidizing potential in the Black Sea. Limnol Oceanogr 48:2369–2376

Konovalov S, Samodurov A, Oguz T, Ivanov L (2004) Parameterization of iron and manganese cycling in the Black Sea suboxic and anoxic environment. Deep-Sea Res Pt I 51:2027–2045

Kostka JE, Luther III GW, Nealson KH (1995) Chemical and biological reduction of Mn(III)-pyrophosphate complexes – potential importance of dissolved Mn(III) as an environmental oxidant. Geochim Cosmochim Acta 59:885–894

Larsen EI, Sly LI, McEwan AG (1999) Manganese(II) adsorption and oxidation by whole cells and a membrane fraction of *Pedomicrobium* sp. ACM 3067. Arch Microbiol 171:257–264

Luther GW (2005) Manganese(II) oxidation and Mn(IV) reduction in the environment – two one-electron transfer steps versus a single two-electron step. Geomicrobiol J 22:195–203

Luther GW, Sundby B, Lewis BL, Brendel PJ, Silverberg N (1997) Interactions of manganese with the nitrogen cycle: alternative pathways to dinitrogen. Geochim Cosmochim Acta 61:4043–4052

Mandernack KW, Post J, Tebo BM (1995) Manganese mineral formation by bacterial spores of a marine *Bacillus*, strain SG-1: evidence for the direct oxidation of Mn(II) to Mn(IV). Geochim Cosmochim Acta 59:4393–4408

Mann S, Sparks NHC, Scott GHE, de Vrind-de Jong EW (1988) Oxidation of manganese and formation of Mn_3O_4 (hausmannite) by spore coats of a marine *Bacillus* sp. Appl Environ Microbiol 54:2140–2143

Mawji E, Gledhill M, Milton JA, Tarran GA, Ussher S, Thompson A, Wolff GA, Worsfold PJ, Achterberg EP (2008) Hydroxamate siderophores: occurrence and importance in the Atlantic ocean. Environ Sci Technol 42:8675–8680

Murray JW, Dillard JG, Giovanoli R, Moers H, Stumm W (1985) Oxidation of Mn(II): initial mineralogy, oxidation state and aging. Geochim Cosmochim Acta 49:463–470

Murray JW, Codispoti LA, Friederich GE (1995) Oxidation-reduction environments: the suboxic zone in the Black Sea. In: Huang CP, O'Melia CR, Morgan JJ (eds) Aquatic chemistry: interfacial and interspecies processes. American Chemical Society, Washington, DC, pp 157–176

Murray K, Mozafarzadeh M, Tebo B (2005) Cr(III) oxidation and Cr toxicity in cultures of the manganese(II)-oxidizing *Pseudomonas putida* strain GB-1. Geomicrobiol J 22:151–159

Nelson YM, Lion LW, Ghiorse WC, Shuler ML (1999) Production of biogenic Mn oxides by *Leptothrix discophora* SS-1 in a chemically defined growth medium and evaluation of their Pb adsorption characteristics. Appl Environ Microbiol 65:175–180

Nicholson WL, Munakata N, Horneck G, Melosh HJ, Setlow P (2000) Resistance of *Bacillus* endospores to extreme terrestrial and extraterrestrial environments. Microbiol Mol Biol Rev 64:548–572

O'Halloran TV, Culotta VC (2000) Metallochaperones, an intracellular shuttle service for metal ions. J Biol Chem 275:25057–25060

Oguz T, Murray JW, Callahan AE (2001) Modeling redox cycling across the suboxic-anoxic interface zone in the Black Sea. Deep-Sea Res Pt I 48:761–787

Okazaki M, Sugita T, Shimizu M, Ohode Y, Iwamoto K, de Vrind-de Jong EW, de Vrind JP, Corstjens PL (1997) Partial purification and characterization of manganese-oxidizing factors of *Pseudomonas fluorescens* GB-1. Appl Environ Microbiol 63:4793–4799

Palma C, Martinez AT, Lema JM, Martinez MJ (2000) Different fungal manganese-oxidizing peroxidases: a comparison between *Bjerkandera* sp. and *Phanerochaete chrysosporium*. J Biotechnol 77:235–245

Parker DL, Sposito G, Tebo BM (2004) Manganese(III) binding to a pyoverdine siderophore produced by a manganese(II)-oxidizing bacterium. Geochim Cosmochim Acta 68: 4809–4820

Parker DL, Morita T, Mozafarzadeh ML, Verity R, McCarthy JK, Tebo BM (2007) Inter-relationships of MnO_2 precipitation, siderophore-Mn(III) complex formation, siderophore degradation, and iron limitation in Mn(II) oxidizing bacterial cultures. Geochim Cosmochim Acta 71:5672–5683

Pecher K, McCubbery D, Kneedler E, Rothe J, Bargar J, Meigs G, Cox L, Nealson K, Tonner B (2003) Quantitative charge state analysis of manganese biominerals in aqueous suspension using Scanning Transmission X-ray Microscopy (STXM). Geochim Cosmochim Acta 67:1089–1098

Peña J, Duckworth OW, Bargar JR, Sposito G (2007) Dissolution of hausmannite (Mn_3O_4) in the presence of the trihydroxamate siderophore desferrioxamine B. Geochim Cosmochim Acta 71:5661–5671

Perez J, Jeffries TW (1992) Roles of manganese and organic acid chelators in regulating lignin degradation and biosynthesis of peroxidases by *Phanerochaete chrysosporium*. Appl Environ Microbiol 58:2402–2409

Piggot PJ (1996) Spore development in *Bacillus subtilis*. Curr Opin Genet Dev 6:531–537

Reddy CA, D'Souza TM (1994) Physiology and molecular biology of the lignin peroxidases of *Phanerochaete chrysosporium*. FEMS Microbiol Rev 13:137–152

Ridge JP, Lin M, Larsen EI, Fegan M, McEwan AG, Sly LI (2007) A multicopper oxidase is essential for manganese oxidation and laccase-like activity in *Pedomicrobium* sp. ACM 3067. Environ Microbiol 9:944–953

Rosson RA, Tebo BM, Nealson KH (1984) Use of poisons in the determination of microbial manganese binding rates in seawater. Appl Environ Microbiol 47:740–745

Ryden LG, Hunt LT (1993) Evolution of protein complexity: the blue copper-containing oxidases and related proteins. J Mol Evol 36:41–66

Sakurai T, Kataoka K (2007) Structure and function of type I copper in multicopper oxidases. Cell Mol Life Sci 64:2642–2656

Saratovsky I, Wightman PG, Pasten PA, Gaillard J-F, Poeppelmeier KR (2006) Manganese oxides: parallels between abiotic and biotic structures. J Am Chem Soc 128:11188–11198

Schippers A, Neretin LN, Lavik G, Leipe T, Pollehne F (2005) Manganese(II) oxidation driven by lateral oxygen intrusions in the western Black Sea. Geochim Cosmochim Acta 69:2241–2252

Schlosser D, Hofer C (2002) Laccase-catalyzed oxidation of Mn^{2+} in the presence of natural Mn^{3+} chelators as a novel source of extracellular H_2O_2 production and its impact on manganese peroxidase. Appl Environ Microbiol 68:3514–3521

Schweisfurth R (1973) Manganoxydierende bakterien I. Isolierung und bestimmung einiger stamme von manganbakterien. Z Allg Mikrobiol 13:341–347

Siering PL, Ghiorse WC (1997) PCR detection of a putative manganese oxidation gene *mofA* in environmental samples and assessment of *mofA* gene homology among diverse manganese-oxidizing bacteria. Geomicrobiol J 14:109–125

Sly LI, Arunpairojana V, Dixon DR (1990) Binding of colloidal MnO_2 by extracellular polysaccharides of *Pedomicrobium manganicum*. Appl Environ Microbiol 56:2791–2794

Solomon EI, Sundaram UM, Machonkin TE (1996) Multicopper oxidases and oxygenases. Chem Rev 96:2563–2606

Spiro TG, Bargar JR, Sposito G, Tebo BM (2010) Bacteriogenic manganese oxides. Acc Chem Res 43:2–9

Stone AT (1987) Reductive dissolution of manganese(III/IV) oxides by substituted phenols. Environ Sci Technol 21:979–988

Stone AT, Morgan JJ (1984) Reduction and dissolution of manganese(III) and manganese(IV) oxides by organics. 2. Survey of the reactivity of organics. Environ Sci Technol 18:617–624

Sunda WG, Huntsman SA (1988) Effect of sunlight on redox cycles of manganese in the southwestern Sargasso Sea. Deep-Sea Res 35:1297–1317

Sunda WG, Kieber DJ (1994) Oxidation of humic substances by manganese oxides yields low-molecular-weight organic substrates. Nature 367:62–64

Sunda WG, Huntsman SA, Harvey GR (1983) Photoreduction of manganese oxides in seawater and its geochemical and biological implications. Nature 301:234–236

Takeda M, Kamagata Y, Ghiorse WC, Hanada S, Koizumi J (2002) *Caldimonas manganoxidans* gen. nov., sp. nov., a poly(3-hydroxybutyrate)-degrading, manganese-oxidizing thermophile. Int J Syst Evol Microbiol 52:895–900

Tebo BM (1991) Manganese(II) oxidation in the suboxic zone of the Black Sea. Deep-Sea Res 38:S883–S905

Tebo BM, Emerson S (1985) Effect of oxygen tension, Mn(II) concentration, and temperature on the microbially catalyzed Mn(II) oxidation rate in a marine fjord. Appl Environ Microbiol 50:1268–1273

Tebo BM, Nealson KH, Emerson S, Jacobs L (1984) Microbial mediation of Mn(II) and Co(II) precipitation at the O_2/H_2S interfaces in 2 anoxic fjords. Limnol Oceanogr 29:1247–1258

Tebo BM, Bargar JR, Clement B, Dick G, Murray KJ, Parker D, Verity R, Webb SM (2004) Biogenic manganese oxides: properties and mechanisms of formation. Annu Rev Earth Planet Sci 32:287–328

Tebo BM, Johnson HA, McCarthy JK, Templeton AS (2005) Geomicrobiology of manganese(II) oxidation. Trends Microbiol 13:421–428

Templeton AS, Staudigel H, Tebo BM (2005) Diverse Mn(II)-oxidizing bacteria isolated from submarine basalts at Loihi Seamount. Geomicrobiol J 22:127–139

Trouwborst RE, Clement BG, Tebo BM, Glazer BT, Luther GW III (2006) Soluble Mn(III) in suboxic zones. Science 313:1955–1957

van Veen WL, Mulder EG, Deinema MH (1978) The *Sphaerotilus-Leptothrix* group of bacteria. Microbiol Rev 42:329–356

van Waasbergen LG, Hoch JA, Tebo BM (1993) Genetic analysis of the marine manganese-oxidizing *Bacillus* sp. strain SG-1: protoplast transformation, Tn917 mutagenesis, and identification of chromosomal loci involved in manganese oxidation. J Bacteriol 175:7594–7603

van Waasbergen LG, Hildebrand M, Tebo BM (1996) Identification and characterization of a gene cluster involved in manganese oxidation by spores of the marine *Bacillus* sp. strain SG-1. J Bacteriol 178:3517–3530

Villalobos M, Toner B, Bargar J, Sposito G (2003) Characterization of the manganese oxide produced by *Pseudomonas putida* strain MnB1. Geochim Cosmochim Acta 67:2649–2662

Villalobos M, Lanson B, Manceau A, Toner B, Sposito G (2006) Structural model for the biogenic Mn oxide produced by *Pseudomonas putida*. Am Mineral 91:489–502

Wang Y, Stone AT (2008) Phosphonate- and carhoxylate-based chelating agents that solubilize (hydr)oxide-bound Mn-III. Environ Sci Technol 42:4397–4403

Wariishi H, Valli K, Gold MH (1992) Manganese(II) oxidation by manganese peroxidase from the basidiomycete *Phanerochaete chrysosporium*. Kinetic mechanism and role of chelators. J Biol Chem 267:23688–23695

Webb SM, Bargar JR, Tebo BM (2005a) Structural Characterization of biogenic manganese oxides produced in sea water by the marine *Bacillus* sp. strain SG-1. Am Mineral 90:1342–1357

Webb SM, Dick GJ, Bargar JR, Tebo BM (2005b) Evidence for the presence of Mn(III) intermediates in the bacterial oxidation of Mn(II). Proc Natl Acad Sci USA 102:5558–5563

Webb SM, Tebo BM, Bargar JR (2005c) Structural influences of sodium and calcium ions on the biogenic manganese oxides produced by the marine *Bacillus* sp., strain SG-1. Geomicrobiol J 22:181–193

Yakushev EV, Pollehne F, Jost G, Kuznetsov I, Schneider B, Umlauf L (2007) Analysis of the water column oxic/anoxic interface in the Black and Baltic seas with a numerical model. Mar Chem 107:388–310

Chapter 14
Role of Microorganisms in Banded Iron Formations

Inga Koehler, Kurt Konhauser, and Andreas Kappler

Banded iron formations (BIF) represent the largest source of iron in the world. They formed throughout the Precambrian, and today are globally distributed on the remnants of the ancient cratons. The first BIF dates back to at least 3.9–3.8 billion years. Little is known about this early period in earth's history, in particular about the presence of molecular oxygen, O_2, and therefore also about the deposition mechanisms of BIF at that time.

Composition, Occurance, and Spatial/Temporal Distribution of BIF'S

Mineralogy

The composition of BIF is dominated by silica (~40–50%) and iron (~20–40%). They are considered to be of sedimentary origin, but always display a diagenetic and metamorphic overprint which sometimes significantly altered the original sediment in terms of its composition and mineralogy. Therefore, the main minerals phases now found in BIF, such as hematite ($Fe_2^{III}O_3$), magnetite ($Fe_2^{III}Fe^{II}O_4$), chert (SiO_2) and stilpnomelane ($K(Fe^{II}Mg,Fe^{III})_8(Si,Al)_{12}(O,OH)_{27}$) are actually of secondary origin. Proposed primary minerals are ferric hydroxide ($Fe(OH)_3$), siderite ($Fe^{II}(CO_3)$) (partially secondary), greenalite ($(Fe)_3Si_2O_5(OH)_4$) and amorphous silica (Klein 2005). The iron in BIF originated as dissolved Fe(II) from submarine hydrothermal vents and was subsequently transformed to dissolved Fe(III) in the upper water column by either

I. Koehler and A. Kappler
Center for Applied Geosciences, University of Tübingen,
Sigwartstrasse 10, D-72076 Tübingen, Germany

K. Konhauser (✉)
Department of Earth and Atmospheric Sciences, University of Alberta,
Edmonton, Alberta, T6G 2E3, Canada
e-mail: kurtk@ualberta.ca

L.L. Barton et al. (eds.), *Geomicrobiology: Molecular and Environmental Perspective*,
DOI 10.1007/978-90-481-9204-5_14, © Springer Science+Business Media B.V. 2010

abiological or biological oxidation. The ferric iron then hydrolyzed rapidly to ferric hydroxide and settled to the sea floor where further transformations ensued.

An early BIF categorization was done by James (1954, 1966) who classified BIF with regards to their mineralogy. Carbonate dominated BIF usually contain alternating chert- and inorganic carbon-rich mineral layers, the latter composed of ankerite ($Ca(Fe, Mg, Mn)(CO_3)_2$) and siderite. Those with a high amount of hematite and magnetite are classified as oxide rich BIF, but they may also contain subsidiary amounts of siderite and iron silicates. Silica rich BIF are dominated by chert, a variety of silicate minerals, such as stilpnomelane, minnesotaite ($(Fe, Mg)_3Si_4O_{10}(OH)_2$), greenalite, and carbonates.

Tectonic Setting

BIF have been broadly classified according to tectonic setting, size and lithology as either Algoma or Superior type (Gross 1965). The Superior type (S-type) first appears around 3 billion years ago (Ga) during the Archean and extends to around 1.7 Ga (in the Proterozoic) (Fig. 14.1). Stratigraphically, the S-type BIF occur in almost coeval, large pericratonic basins (Kholodov 2008). They are dominated by

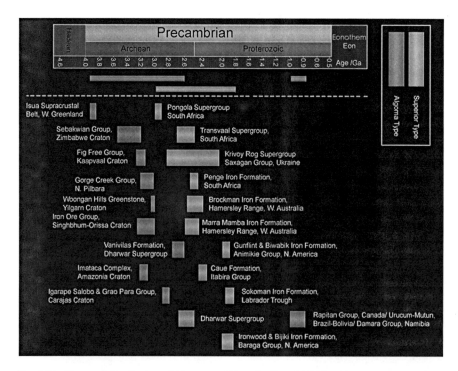

Fig. 14.1 Temporal distribution of Algoma and Superior Type BIF. The Algoma type occur from 3.9 to 2.6 Ga and then again from 1.0 to 0.85 Ga. While the Superior type commonly occur between 1.7 and 3.0 Ga

hematite, magnetite, siderite, ankerite and greenalite. There is usually no direct association between these BIF and volcanic rocks.

In contrast, Algoma type BIF is often replaced by volcanic rock sequences. Their thickness and lateral expansion is smaller than those of the Superior type BIFs. They usually appear in late Archean, but there are also Algoma type BIF found in late Proterozoic (Fig. 14.1).

Spatial and Temporal Distribution

The oldest BIF are those of the Isua Supracrustal Belt in Western Greenland with an approximate age of ~3.8–3.9 Ga (Mojzsis et al. 1996). Further Archean BIF includes the Sebakwian Group in Zimbabwe with an age of ~3.2–3.6 Ga and the late Archean Vanivilas Formation in the Dharwar Supergroup in India with an age of ~2.6–2.8 Ga. The transition zone between the Archean and the Proterozoic is marked by BIF in the Transvaal Supergroup, South Africa, with an age of ~2.7–2.5 Ga and the Brockman Iron Formation at the Hamersley Range in Western Australia with an age of ~2.6–2.4 Ga. The Gunflint & Biwabik Iron Formation in Northern America can be assigned to the Proterozoic BIF with an age of ~2.0–1.9 Ga. BIF then disappear from the rock record until around 750 million years ago with their re-emergence in the Rapitan Group in Canada with an age of ~750 Ma. These latter BIF are associated with Snowball Earth glacial events (Hoffman and Schrag 2000).

Microbial and Chemical Processes Generating BIF Source Sediment

The time of BIF deposition spans major evolutionary changes in the Earth's surface composition, from an early anoxic atmosphere dominated by CO_2 and CH_4 to an atmosphere that became partially oxygenated. Therefore, it is likely that BIF formed via different mechanisms throughout the Precambrian. These mechanisms are briefly discussed below.

Oxidation of Fe(II) by Cyanobacterial O_2

The traditional model of BIF precipitation assumes the oxidation of hydrothermal Fe(II), either via abiotic oxidation by cyanobacterially-produced O_2 (Cloud 1973; Klein and Beukes 1989) and/or biotic oxidation by chemolithotrophic bacteria (aerobic Fe(II)-oxidizers) (Fig. 14.2). Both models suggest the presence of free molecular oxygen in the Precambrian ocean and therefore require the presence of oxygenic photosynthesis at that time of Earth history. This raises the question of when oxygen first appeared and became relevant for Fe(II) oxidation in the ancient

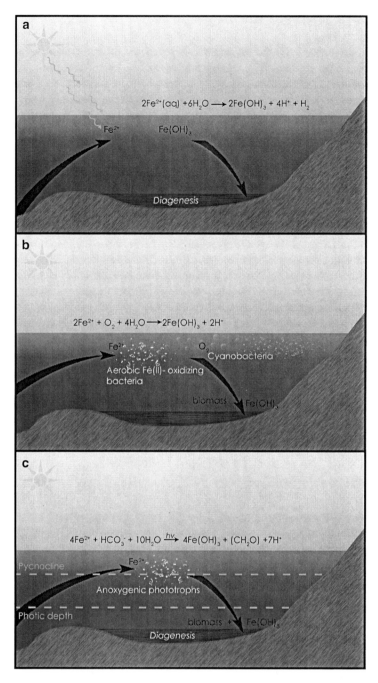

Fig. 14.2 Summary of Fe(II) oxidation processes potentially involved in BIF deposition. (a) Abiotic photooxidation of hydrothermal Fe(II) by UV light with sedimentation of iron minerals. (b) Chemical oxidation of hydrothermal Fe(II) by cyanobacterial produced O_2 and Fe(II) oxidation by aerobic Fe(II)-oxidizing bacteria with sedimentation of either iron minerals only (chemical oxidation) or sedimentation of joint biomass and iron minerals (aerobic Fe(II)-oxidizing bacteria). (c) Direct biological oxidation of Fe(II) by anoxygenic phototrophic Fe(II)-oxidizing bacteria with joint sedimentation of biomass and iron minerals

oceans. Previously, microfossils found in the 3.45 billion year old Apex chert of Warrawoona in Australia were suggested as evidence for the existence of cyanobacteria at that time (Schopf 1993), although the biogenicity of those fossils has since been questioned (Brasier et al. 2002). Much younger evidence comes from the 2.7 Ga stromatolitic assemblages of Tumbiana Formation in Western Australia, which suggests that the primary microbial mat community was comprised of cyanobacteria (Buick 1992). Additionally, the analyses of biomarkers, which are organic remains of biological molecules, can be used to identify the former presence of life in rocks. Cyanobacteria display a variety of potential organic biomarker molecules which can serve as a fingerprint. An example are 2α-methylhopanes, organic molecules that are present in membranes of modern cyanobacteria, which were also extracted from bitumen of the 2.6 Ga Marra Mamba Iron Formation and the 2.5 Gyr Mt. McRae Shale of the Hamersley Group, Western Australia (Brocks et al. 1999; Summons et al. 1999). However, such hopanes have been found in anoxygenic phototrophic Fe(II)-oxidizers as well (Rashby et al. 2007). In addition to hopanes, steranes can be used to track oxygenic photosynthesis. They give direct indication for the existence of oxygen since the biosynthesis of the sterols (precursor to steranes) requires oxygen (Runnegar 1991). These compounds were found in the 2.7 billion-years-old shales of the Jeerinah Formation, Hamersley Group (Brocks et al. 1999, Fig. 14.3).

Further evidence for the presence of oxygen comes from the composition of rocks themselves. For instance, the observation of mass-independent fractionation (MIF) of sulfur isotopes in rocks older than 2.4 Ga, but not younger than 2.1 Ga points to changes in the overall sulfur cycle during that time. Specifically, the loss of the MIF signal is directly attributed to the concentration of atmospheric oxygen, and it is widely accepted that the change in sulfur isotopes indicates a shift from anoxic photochemical reactions to an atmosphere with some free oxygen (Farquhar et al. 2000; Mojzsis 2003). Concentrations of more than 10^{-5} PAL would have oxidized sulfur to sulfate preventing the preservation of a MIF signal (Pavlov and Kasting 2002) via initiation of bacterial sulfate reduction. Additionally, the presence of oxygen prior to 2.4 Ga would have had an oxidative effect on siderite, detrital uraninite, and pyrite. But since these minerals have been found in rocks such as the siliciclastic sediments of the Pilbara Craton, it has to be assumed that oxygen was not available prior to this point in early earth atmosphere (Rasmussen and Buick 1999). Recent analyses of chromium isotopes in Banded Iron Formations also yield a hint for the rise of atmospheric oxygen around 2.45–2.2 Ga (Frei et al. 2009). Cr is sensitive to the redox state of the environment and therefore provides an excellent oxygen indicator in the rock record. As soon as atmospheric oxygen became available, oxidative weathering set in producing the oxidized and more mobile hexavalent [Cr(VI)], a process that prefers the heavier Cr isotope. The rise in O_2 in the atmosphere can therefore be followed by Cr isotope analysis in sedimentary deposits such as BIF rocks. The mobility of molybdenum (Mo) is also sensitive to the redox status of the environment. In the absence of molecular oxygen, Mo is preserved in crustal sulfide minerals leading to low Mo concentrations in the oceans and in the sediment. A shift in the concentrations of Mo to higher values in the rock record therefore point to an increase of oxygen in the atmosphere leading to an oxidative weathering of the Mo-bearing

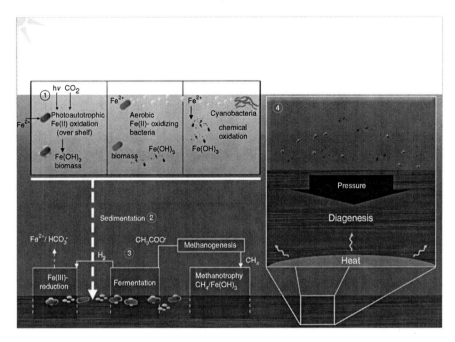

Fig. 14.3 Model summarizing potential biological and chemical processes during BIF deposition. (1) Hydrothermal Fe(II) is oxidized by photoautotrophic anoxygenic Fe(II)-oxidizing bacteria, aerobic Fe(II)- oxidizing bacteria or via chemical oxidation by cyanobacterially-produced O_2. (2) Biomass and Fe(III) sediment to ocean floor as cell-mineral aggregates. (3) After sedimentation metabolically driven redox processes by fermenters and Fe(III) reducers take place, possibly also involving methanogens and methanotrophs. (4) Pressure and temperature alter the source sediment and cause induced diagenetic/metamorphic overprint

sulfide rocks. Mo analyses in the late Archean Mount McRae Shale in Western Australia carried out by Anbar et al. (2007) point to the presence of small amounts of O_2 in the environment more than 50 million years before the start of the Great Oxidation Event. They assigned the rise of oxygen at around 2.5 Ga +/−8 Ma.

Collectively, these findings raise the question of how Fe(III) minerals in BIF older than 2.5–2.7 Ga were possible without the presence of oxygen? It is possible that while other processes account for the deposition of major parts of the early BIF, cyanobacteria still could have played a minor role in their formation, but establishing the oxic environment later than their "first" appearance. As mentioned above, biomarkers like steranes, and the fossil record provide evidence for the presence of cyanobacteria around 2.6–2.7 Ga.

UV-Photooxidation Model

A model which offers a possible explanation for the formation of ferric iron oxides without oxygen involves photooxidation of ferrous iron by UV radiation (Cairns-Smith 1978; Francois 1986). This model is based on the high levels of ultraviolet radiation on early Earth due to the lack of an ozone layer at that time. Under such conditions, dissolved ferrous iron species, such as Fe^{2+} or $Fe(OH)^+$, absorb radiation in the 200–400 nm range, leading to the formation of dissolved ferric iron [reaction 14.1], which in turn, hydrolyzes to form ferric hydroxide at circumneutral pH (Cairns-Smith 1978; Braterman et al. 1983)

$$2Fe^{2+}_{(aq)} + 2H^+ + h\nu \rightarrow 2Fe^{3+}_{(aq)} + H_2 \uparrow \qquad (14.1)$$

However the photochemical oxidation has mainly been observed in simple aqueous systems and the UV light probably was absorbed to a large extent within the ocean surface layer. Also it has been demonstrated that the precipitation of amorphous ferrous silicates and carbonates would have occurred faster than photochemical oxidation (Konhauser et al. 2007). Therefore UV photooxidation can probably be ruled out as primary mechanism to explain the formation of Fe oxide minerals in BIF prior to 2.7–2.5 Ga.

Direct Biological Oxidation of Fe^{2+} by Anoxygenic Phototrophic Fe(II)-Oxidizing Bacteria

Phototrophic Fe(II)-oxidizing bacteria were discovered about 2 decades ago and might represent an alternative hypothesis to early BIF formation in the absence of O_2 (Widdel et al. 1993; Heising et al. 1999; Straub et al. 1999; Posth et al.2010a). These bacteria could have oxidized ferrous iron to ferric iron within the photic zone of the oceans through their photosynthetic process that involves light-energy fuelled CO_2 fixation coupled to the microbial oxidation of Fe^{2+} (Konhauser et al. 2002). The attractiveness of this concept is that it explains BIF deposition in the absence of molecular oxygen using the abundant availability of Fe^{2+}, light and CO_2 at that time. It has even been demonstrated by eco-physiological lab experiments in combination with modeling that these phototropic bacteria would have been capable of oxidizing enough Fe(II) to explain the large expansion of BIF deposits (Kappler et al. 2005). Growth experiments have even shown that the phototrophs can effectively oxidize Fe(II) up to a few 100 m of water depth (Kappler et al. 2005).

An additional argument in favor of the presence of these organisms in the Archean is the presence of enormous amounts of Fe^{2+} that could have served at that time as their electron donors. Interestingly, it has recently been demonstrated that two Fe(II)-rich lakes (Lake Matano in Indonesia and Lake LaCruz in Spain) indeed harbor phototrophic Fe(II)-oxidizing bacteria in the photic zone of the water column (Crowe et al. 2008; Walter et al. 2009). Additionally, phylogenetic analysis of the enzymes involved in the (bacterio-)chlorophyll biosynthesis shows that anoxygenic photosynthetic lineages are more deeply rooted than the oxygenic cyanobacterial lineages (Xiong 2006). However, it has to be noted that to date there is no actual physical or chemical evidence for existence of Fe(II)-oxidizing phototrophs in the Archean (Posth et al. 2010b). A possible approach for such evidence could be finding organic biomarkers that are unique for these bacteria, for example unique pigments involved in photosynthesis and radical scavenging (radicals forming during Fe-Fenton reactions) – both processes highly relevant in systems where photosynthetic Fe(II)-oxidizers are present.

Post-Depositional Processes (Including Microbial Activity) in BIF

The question of what minerals in BIF are of primary or secondary origin has been widely discussed. The formation of quartz from siliceous gelatinous precipitate is mostly accepted at this point, while thermodynamic calculations by Berner (1969) show that hematite can form as a result of the dehydration of either goethite or ferrihydrite under diagenetic conditions. The presence of a number of reduced iron phases in BIF complicates the picture because some are considered to represent primary precipitates that formed within the anoxic water column (e.g., spheroidal siderite), when concentrations of ferrous iron and bicarbonate [reaction 14.2], originating from a combination of hydrothermal sources and microbial respiration of sedimented organic carbon, exceeded siderite supersaturation (Tice and Lowe 2004). Other ferrous iron minerals, including magnetite, rhombic siderite, ferrosilicates (stilpnomelane, chlorite), ankerite, and pyrite, formed during diagenesis and metamorphism (e.g., Ayres 1972; Perry et al. 1973; McConchie 1987). In terms of magnetite, a number of petrographic studies have described the secondary origins of the magnetite, including (1) disseminated grains within but obscuring sedimentary laminae, (2) laminated beds that clearly truncate sedimentary layering, (3) layer-discordant veins, and (4) cleavage fills (Han 1978; McConchie 1987; Morris 1993; Krapež et al. 2003). Much of the magnetite likely formed when Fe^{2+}, formed via microbial Fe(III) reduction (see below), reacted with the initial ferric hydroxide precursors [reaction 14.3]. However, the fact that magnetite frequently appears in association with siderite-rich bands also supports the notion that magnetite could have formed via oxidation of siderite (only when O_2 was available) [reaction 14.4] or by reaction with hematite [reaction 14.5] (Figs. 14.4 and 14.5).

$$8Fe^{2+} + 16HCO_3^- \rightarrow 8FeCO_3 + 8H_2O + 8CO_2 \tag{14.2}$$

$$8Fe^{2+} + 16Fe(OH)_3 + 16OH^- \rightarrow 8Fe_3O_4 + 32H_2O \tag{14.3}$$

$$3FeCO_3 + \tfrac{1}{2}O_2 \rightarrow Fe_3O_4 + 3CO_2 \tag{14.4}$$

$$Fe_2O_3 + FeCO_3 \rightarrow 4Fe_3O_4 + CO_2 \tag{14.5}$$

BIF also show an abundance of light carbon isotope signatures within the carbon layers (Garrels et al. 1973; Baur et al. 1985) consistent with the initial presence of microbial biomass during sedimentation. In addition, highly negative $\delta^{56}Fe$ values in 2.9 Ga old magnetite (Yamaguchi et al. 2005), with comparable negative

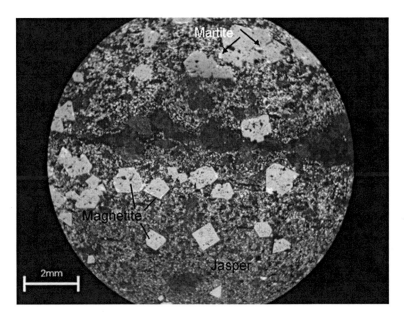

Fig. 14.4 BIF *thin section* M1, Mamatwan Manganese Mine, Kuruman, North Cape Province, South Africa. The sample is approximately 2.20–2.22 billion years old. Sample shows white magnetite crystals in a jasperous fine grained matrix. The single magnetite crystals are subhedral to euhedral and their sizes ranges from 1.0 to 0.5 mm. In the *upper part* of the section magnetite forms aggregates with a size of up to 2 mm. Several crystals display narrow bright white rims identified as martite (see *arrows*). The *red color* in the jasperous matrix comes from fine grained hematite particles

fractionations as observed in experimental culture with dissimilatory Fe(III)-reducing bacteria (Johnson et al. 2003), point towards the antiquity of such an anaerobic respiratory pathway. Moreover, the recognition that a variety of deeply-branching (and presumed very ancient) hyperthermophilic *Bacteria* and *Archaea* can reduce Fe(III) to Fe(II) (Vargas et al. 1998), strengthens the likelihood that such a metabolism occurred very early on Earth.

Assuming that sedimentation of biomass and Fe(III) took place during anoxygenic phototrophic or aerobic Fe(II) oxidation in the Archean, it could be argued that the electrons present in the biomass associated with the sedimented Fe(III) minerals could theoretically re-reduce all Fe(III) leaving no remains for a later formation of hematite (Konhauser et al. 2005). However, since the Fe(III) oxide hematite and also the mixed Fe(II)-Fe(III) oxide magnetite are major constituents of many BIF it can be assumed that only minor amount of organic carbon were initially deposited with the Fe(III)-rich sediment. Alternatively, the biomass and electrons could have been removed from the sediment by Fe(III)-independent redox processes, e.g. by fermentation or methanogenesis. Calculations by Konhauser et al. (2005) showed that only about 3% of the biomass reached final burial. Indeed, Konhauser et al. (2005) further suggested that some of the initially produced biomass was transformed by hydrolysis and fermentation leading to a possible

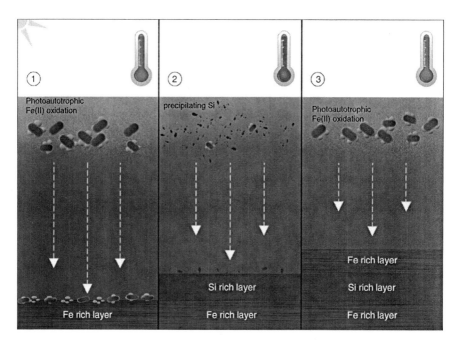

Fig. 14.5 Possible deposition of alternating iron and silicate mineral layers triggered by temperature fluctuations in the ocean water: (1) and (3) Moderate/higher temperatures yield relatively high photoautotrophic bacteria oxidation rates and thus iron(III) mineral formation. Therefore, biomass and Fe(III) settle together to the seafloor. (2) With decreasing temperatures photoautotrophic oxidation rates slow down and at the same time lower temperatures initiate abiotic Si precipitation from Si oversaturated ocean water. Si minerals then settle to the seafloor

removal of electrons in form of reduced compounds (e.g. H_2) away from the sediment that otherwise would have been available to reduce Fe(III). Furthermore, some of the H_2 or organic compounds could have been used by methanogens. The highly negative $\delta^{13}C$ values of around −40‰ to −60‰ in 2.8–2.6 Ga old kerogens point to intensive methane formation at that time (Hayes 1983).

Limitation of Microbial Processes in BIF's by Nutrients and Trace Metals

Phosphate is essential to life due to its important role in many biomolecules such as DNA, RNA and ATP, as well as a constituent in cellular membranes. It was suggested that the strong affinity of phosphate to Fe(III) hydroxides could have depleted the Archean ocean of phosphate leading to a "phosphate crisis", i.e. constraining microbial activity due to limiting bioavailable concentrations of phosphate in the water column (Bjerrum and Canfield 2002). However, since Archean oceans had high concentrations of dissolved and amorphous silica, the silica would have competed with phosphate for sorption sites at the ferric hydroxide particle surface lowering sorption of phosphate compared to silica-free systems. Additionally, the co-precipitation of iron and silica changes the particle's point of zero charge (PZNC) also lowering phosphate binding and phosphate removal from the ocean water (Konhauser et al. 2007). Overall, it was therefore suggested that phosphate in the Archean was probably not limiting.

Besides nutrients such as phosphate, trace metals are important for microorganisms due to their role as metal-cofactors. Although the microorganisms usually need only small quantities, trace metals such as nickel, copper, and cobalt can represent a limiting factor for microbial growth. A possible approach to constrain the plausibility of a certain bacterial activity at the time of BIF formation is therefore to estimate concentrations of trace metals required by a certain physiological group of microbes in the Archean ocean from the abundance of these trace metals in BIF. However, to use this approach to constrain certain microbial processes, several questions have to be addressed.

1. Is the trace metal distribution in BIF representative of the composition of the Archean ocean or has there been major remobilization due to diagenesis and/or metamorphism?
2. Do certain trace metals show different sorption behavior to abiogenic versus biogenic Fe(III) minerals and is there a difference in co-precipitation during abiotic versus biotic Fe(II) oxidation?
3. Is there an influence by silica on the sorption and co-precipitation behavior of trace metals?

Recently, analysis of Ni concentrations in BIF, in combination with quantification of Ni sorption behavior to Fe(III) (hydr)oxides (similar to the ones assumed to be the primary minerals of BIF), suggested a decrease in the amount of bioavailable Ni around 2.7 Ga ago probably due to a cooling of the mantle. As a consequence, microbial methane production could have slowed since Ni is essential for these microbes ((Jaun and Thauer 2007), potentially initiating the so-called Great Oxidation Event around 2.4 Ga (Konhauser et al. 2009). Another potentially interesting trace metal is cobalt, which has been identified as being involved in Fe(II) oxidation in phototrophic Fe(II)-oxidizing bacteria and thus possibly limited photoferrotrophic activity in the Archean (Jiao et al. 2005). However, sorption behaviour of Co to abiogenic and biogenic Fe(III) (hydr)oxides has not been correlated with BIF Co concentrations thus far.

Mechanisms of Altering Iron and Silica Mineral Layering – A Potential Role of Microorganisms?

BIF show a characteristic layering of Fe-rich and silica/carbonate-rich bands. The thickness of the layers varies between micro-scale to meter-thick units (Trendall 1968; Beukes and Klein 1992; Klein 2005) and some BIF show a wide lateral continuity of up to hundreds of kilometers. This raises the question of whether the alternating banding has a unifying trigger (Posth et al. 2008), and whether the precipitation of either iron or silica somehow influences the formation of the other. It has been suggested that the layering is due to seasonal stratification or yearly climate cycles (Morris 1993), in which periodic upwelling or hydrothermal pulsation of Fe(II)-rich waters was punctuated by seasonal evaporation of silica (Holland 1973; Garrels 1987; Jacobsen and Pimental-Klose 1988; Siever 1992).

However, these models have been questioned since they do not explain the lateral banding of both Fe(III) and silica minerals. Additionally, they do not explain why iron and silica do not co-precipitate to a large extent (Trendall 1968; Garrels 1987; Morris 1993). Accordingly, a new model has been suggested that links temperature-induced bacterial Fe(II)-oxidation to abiotic silicification as a way of explaining the banding (Posth et al. 2008). The activity of phototrophic Fe(II)-oxidizing bacteria (Fe(II) oxidation rates) shows a strong temperature dependency (Hegler et al. 2008; Posth et al. 2008). When the bacteria were incubated under changing temperature conditions and in the presence of silica, at water temperatures of 25–30°C, the Fe(II) oxidation rate and the ferric hydroxide precipitation was high. Below and above these temperatures, Fe(II) oxidation and the precipitation of the biogenic iron(III) minerals decreased. At lower temperatures when Fe(III) mineral precipitation decreased, however, the precipitation of amorphous silica was triggered. As temperatures increased to 25–30°C, Fe(III) minerals were again precipitated by the Fe(II)-oxidizing phototrophs, effectively allowing the alternating layering of silica and Fe minerals. Interestingly, the rate of Fe(II) oxidation was not affected by the silica and Fe(III) precipitation seems to be decoupled from silica precipitation. This is probably due to the change in Fe(III) mineral surface charge from positive to negative due to the sorption of cell organic matter preventing silica sorption (Posth et al. 2008). This decoupling suggests that a separate deposition of Fe-rich and silica-rich layers in BIF could have been triggered by one unifying parameter, temperature.

These findings raise the question of whether ocean temperature cycles in the Archean existed and were strong enough to trigger the banding. Estimates of water temperature in the Archean are poorly constrained, and range from 10°C to 85°C (Knauth and Lowe 2003; Knauth 2005; Robert and Chaussidon 2006; Kasting et al. 2006; Jaffrés et al. 2007; Shields and Kasting 2007). An additional difficulty in constraining these past temperature is the high potential for diagenetic overprinting, which might distort the data. Based on the most recent interpretations, the general climate in the Archean was around 10–33°C (Kasting et al. 2006) and consequently, mesophilic Fe(II)-oxidizing phototrophs, as used in the experiments by Posth et al. (2008), would prosper in such an environment.

Nevertheless, two independent lines of evidence suggest that such temperature fluctuations took place in the ancient ocean. First, modern ocean temperature cycles depend on incoming currents and seasonal variations and show temperature changes as needed for the effects observed by Posth et al. (2008). Second, a tilt in Earth's axis (obliquity) could have caused seasonal temperature changes (Laskar and Robutel 1993).

Conclusions

Banded iron formations not only serve as possible archives for the Precambrian atmosphere and hydrosphere but they shed insights into the biotic processes occurring at that time. It is generally believed that the large amounts of oxidized Fe present in the BIFs were formed by oxidation of hydrothermal Fe(II). The oxidation of the Fe(II) after 2.7–2.5 Ga ago could be due to cyanobacterial O_2, however, the formation of BIF older than 2.7–2.5 Ga requires an anoxic mechanism for Fe(II) oxidation. Currently, anoxygenic photosynthetic bacteria provide the most plausible explanation of Fe(III) mineral deposition in the anoxic Archean. The role of microorganisms in BIF deposition also goes beyond providing the O_2 for Fe(II) oxidation or direct Fe(II) oxidation. Biomass and ferric hydroxides deposited at the sea floor provide the necessary substrates for Fe(III)-reducing bacteria, fermenters and methanogens. These microorganisms probably reworked the sediments to a significant extent before ultimate lithification.

Acknowledgements This work was supported by research grants from the German Research Foundation (DFG) made to AK (KA 1736/2-1, 2-2, 4-1, and 12-1), funding from the DFG and the University of Tuebingen to IK, and the Natural Sciences and Engineering Research Council of Canada to KK. We would also like to thank Nicole Posth and Merle Eickhoff for helpful comments.

References

Anbar AD, Duan Y, Lyons TW, Arnold GL, Kendall B, Creaser RA, Kaufman AJ (2007) A whiff of oxygen before the great oxidation event? Science 317:1903–1906

Ayres DE (1972) Genesis of iron-bearing minerals in banded iron formation mesobands in the Dales Gorge member, Hamersley Group, Western Australia. Econ Geol 67:1214–1233

Baur ME, Hayes JM, Studley SA, Walter MR (1985) Millimeter-scale variations of stable isotope abundances in carbonates from banded iron formations in the Hamersley Group of Western Australia. Economic Geol 80:270–282

Berner RA (1969) Goethite stability and the origin of red beds. Geochim Cosmochim Acta 33:267–273

Beukes NJ, Klein C (1992) Models for iron-formation deposition. In: Schopf JW, Klein C (eds) The proterozoic biosphere: a multidisciplinary study. University of Cambridge Press, Cambridge, UK, pp 147–151

Bjerrum CJ, Canfield DE (2002) Ocean productivity before about 1.9 Gyr ago limited by phosphorus adsorption onto iron oxides. Nature 417:159–162

Brasier MD, Green OR, Jephcoat AP, Kleppe AK, Van Kranendonk MJ, Lindsay JF, Steele A, Grassineau NV (2002) Questioning the evidence for Earth's oldest fossils. Nature 416:76–81

Braterman PS, Cairns-Smith AG, Sloper RW (1983) Photo-oxidation of hydrated Fe^{2+} – significance for banded iron formations. Nature 303:163–164

Brocks JJ, Logan GA, Buick R, Summons RE (1999) Archean molecular fossils and the early rise of eukaryotes. Science 285:1033–1036

Buick R (1992) The antiquity of oxygenic photosynthesis: evidence for stromatolites in sulphate-deficient Archaean lakes. Science 255:74–77

Cairns-Smith AG (1978) Precambrian solution photochemistry, inverse segregation, and banded iron formations. Nature 276:807–808

Cloud P (1973) Paleoecological significance of the banded iron-formation. Econ Geol 68:1135–1143

Crowe SA, Jones C, Katsev S et al (2008) Photoferrotrophs thrive in an Archean Ocean analogue. Proc Natl Acad Sci USA 105:15938–15943

Farquhar J, Bao H, Thiemens M (2000) Atmospheric influence of Earth's earliest sulfur cycle. Science 289:756–758

Francois LM (1986) Extensive deposition of banded iron formations was possible without photosynthesis. Nature 320:352–354

Frei R, Gaucher C, Poulton SW, Canfield DE (2009) Fluctuations in Precambrian atmospheric oxygenation recorded by chromium isotopes. Nature 461:250–253

Garrels RM, Perry EA Jr, MacKenzie FT (1973) Genesis of Precambrian iron-formations and the development of atmospheric oxygen. Econ Geol 68:1173–1179

Garrels RM (1987) A Model for the deposition of the microbanded Precambrian iron formations. American Journal of Science 287:81–106

Gross GA (1965) Geology of iron deposits in Canada, Volume 1. General geology and evaluation of iron deposits, Geological Survey of Canada Economic Report, 22

Han T-M (1978) Microstructures of magnetite as guides to its origin in some Precambrian iron-formations. Fortschr Mineral 56:105–142

Hayes JM (1983) Geochemical evidence bearing on the origin of aerobiosis, a speculative hypothesis. In: Schopf JW, Klein C (eds) Earth's earliest biosphere, its origins and evolution. Princeton University Press, Princeton, NJ, pp 291–301

Hegler F, Posth NR, Jiang J, Kappler A (2008) Physiology of phototrophic iron(II)- oxidizing bacteria-implications for modern and ancient environments. FEMS Microbiol Ecol 66:250–260

Heising S, Richter L, Ludwig W, Schink B (1999) *Chlorobium ferrooxidans* sp. nov., a phototrophic green sulfur bacterium that oxidizes ferrous iron in coculture with a *Geospirillum* sp. strain. Arch Microbiol 172:116–124

Hoffman PF, Schrag DP (2000) Snowball Earth. Sci Am 282(January):68–75

Holland HD (1973) The oceans: a possible source of iron in iron-formations. Econ Geol 68:1169–1172

Jacobsen SB, Pimentel-Klose MR (1988) A Nd isotopic study of the Hamersley and Michipicoten banded iron formations: the source of REE and Fe in Archean oceans. Earth Planet Sci Lett 87:29–44

Jaffrés JBD, Shields GA, Wallmann K (2007) The oxygen isotope evolution of seawater: a critical review of a long-standing controversy and an improved geological water cycle model for the past 3.4 billion years. Earth Sci Rev 83:83–122

James HL (1954) Sedimentary facies of iron-formation. Econ Geol 49:236–294

James HL (1966) Chemistry of the iron-rich sedimentary rocks. In: Fleischer M (ed) Data of geochemistry, 6th edn. Paper 440-W. US Govt. Printing Office, Washington, DC

Jaun B, Thauer RK (2007) Nickel and its surprising impact in nature. In: Sigel A, Sigel H, Sigel RKO (eds) Metal ions in life sciences, vol 2. Wiley, Chichester, UK, pp 323–356

Jiao Y, Kappler A, Croal LR, Newman DK (2005) Isolation and characterization of a genetically tractable photoautotrophic Fe(II)-oxidizing bacterium, *Rhodopseudomonas palustris* Strain TIE-1. Appl Environ Microbiol 71:1–10

Johnson CM, Beard BL, Beukes NJ, Klein C, O'Leary JM (2003) Ancient geochemical cycling in the Earth as inferred from Fe isotope studies of banded iron formations from the Transvaal craton. Contrib Mineral Petrol 144:523–547

Kappler A, Pasquero C, Konhauser KO, Newman DK (2005) Deposition of banded iron formations by anoxygenic phototrophic Fe(II)-oxidizing bacteria. Geology 33:865–868

Kasting JF, Howard MT, Wallmann K, Veizer J, Shields G, Jaffrés J (2006) Paleoclimates, ocean depth, and the oxygen isotopic composition of seawater. Earth Planet Sci Lett 252:82–93

Kholodov VN (2008) Siderite formation and evolution of sedimentary iron ore deposition in the Earth's history. Geol Ore Deposits 50:299–319

Klein C (2005) Some Precambrian Banded Iron Formations (BIFs) from around the world: their age, geologic setting, mineralogy, metamorphism, geochemistry, and origin. Am Mineral 90:1473–1499

Klein C, Beukes NJ (1989) Geochemistry and sedimentology of a facies transition from limestone to iron-formation deposition in the Early Proterozoic Transvaal Supergroup, South Africa. Econ Geol 84:1733–1774

Knauth LP (2005) Temperature and salinity history of the Precambrian Ocean: implications for the course of microbial evolution. Palaeogeogr Palaeoclimatol Palaeoecol 219:53–69

Knauth PL, Lowe DR (2003) High Archaen climatic temperature inferred from oxygen isotope geochemistry of cherts in the 3.5 Ga Swaziland Supergroup, South Africa. Geol Soc Am Bull 115:566–580

Konhauser KO, Hamade T, Raiswell R, Morris RC, Ferris FG, Southam G, Canfield DE (2002) Could bacteria have formed the Precambrian banded iron formations? Geology 30:1079–1082

Konhauser KO, Newman DK, Kappler A (2005) The potential significance of microbial Fe(III) reduction during deposition of Precambrian banded iron formations. Geobiology 3:167–177

Konhauser KO, Amskold L, Lalonde SV, Posth NR, Kappler A, Anbar A (2007) Decoupling photochemical Fe(II) oxidation from shallow-water deposition. Earth Planet Sci Lett 258:87–100

Konhauser KO, Pecoits E, Lalonde SV, Papineau D, Nisbet EG, Barley ME, Arndt NT, Zahnle K, Kamber BS (2009) Oceanic nickel depletion and a methanogen famine before the great oxidation event. Nature 458:750–754

Krapež B, Barley ME, Pickard AL (2003) Hydrothermal and resedimented origins of the precursor sediments to banded iron formation: sedimentological evidence from the Early Paleoproterozoic Brockman supersequence of Western Australia. Sedimentology 50:979–1011

Laskar J, Robutel P (1993) The chaotic obliquity of the planets. Nature 361:608–612

McConchie D (1987) The geology and geochemistry of the Joffre and Whaleback Shale members of the Brockman iron formation, Western Australia. In: Appel PWU, LaBerge GL (eds) Precambrian iron-formations. Theophrastus, Athens

Mojzsis SJ (2003) Probing early atmospheres. Nature 425:249–251

Mojzsis SJ, Arrhenius G, McKeegan KD, Harrison TM, Nutman AP, Friend CRL (1996) Evidence for life on Earth before 3, 800 million years ago. Nature 384:55–59

Morris RC (1993) Genetic modelling for banded iron-formation of the Hamersley Group, Pilbara Craton, Western Australia. Precambrian Res 60:243–286

Pavlov AA, Kasting JF (2002) Mass-independent fractionation of sulfur isotopes in Archean sediments: strong evidence for an anoxic Archean atmosphere. Astrobiology 2:27–41

Perry EC, Tan FC, Morey GB (1973) Geology and stable isotope geochemistry of the Biwabik iron formation, northern Minnesota. Econ Geol 68:1110–1125

Posth NR, Hegler F, Konhauser KO, Kappler A (2008) Alternating Si and Fe deposition caused by temperature fluctuations in Precambrian oceans. Nat Geosci 10:703–708

Posth NR, Konhauser KO, Kappler A (2010a) Microbiological processes in BIF deposition. In: Glenn C, Jarvis I (eds) Authigenic minerals: sedimentology, geochemistry, origins, distribution and applications. Journal of Sedimentology IAS Special Publication Series (in press)

Posth NR, Konhauser KO, Kappler A (2010b) Banded iron formations. In: Thiel V, Reitner J (eds) Encyclopedia of geobiology. Springer, Hiedelberg (in press)

Rashby SE, Sessions AL, Summons RE, Newman DK (2007) Biosynthesis of 2-ethylbacteriohopanepolyols by an anoxygenic phototroph. Proc Natl Acad Sci USA 104:15099–15104

Rasmussen B, Buick R (1999) Redox state of the Archean atmosphere: evidence from detrital heavy metals in ca. 3250–2750 Ma sandstones from the Pilbara Craton. Aust Geol 27:115–118

Robert F, Chaussidon M (2006) A Paleotemperature curve for the Precambrian oceans based on silicon isotopes in cherts. Nature 443:969

Runnegar B (1991) Precambrian oxygen levels estimated from the biochemistry and physiology of early eukaryotes. Palaeogeogr Palaeoclimatol Palaeoecol 71:97–111

Schopf JW (1993) Microfossils of the early Archean Apex Chert: new evidence of the antiquity of life. Science 260:640–646

Shields GA, Kasting JF (2007) Palaeoclimatology: evidence for hot early oceans? Nature 447:E1

Siever R (1992) The silica cycle in the Precambrian. Geochim Cosmochim Acta 56:3265–3272

Straub KL, Rainey FR, Widdel F (1999) *Rhodovulum iodosum* sp. nov. and *Rhodovulum robiginosum* sp. nov., two new marine phototrophic ferrous-iron-oxidizing purple bacteria. Int J Syst Bacteriol 49:729–735

Summons RE, Jahnke LL, Hope JM, Logan GA (1999) 2-Methylhopanoids as biomarkers for cyanobacterial oxygenic photosynthesis. Nature 400:554–557

Tice MM, Lowe DR (2004) Photosynthetic microbial mats in the 3,416 Myr old ocean. Nature 431:549–552

Trendall AF (1968) Three Great Basins of Precambrian banded iron formation deposition: a systematic comparison. Geol Soc Am Bull 79:1527–1544

Vargas M, Kashefi K, Blunt-Harris EL, Lovely DR (1998) Microbiological evidence for Fe(III) reduction on early Earth. Nature 395:65–67

Walter XA, Picazo A, Miracle RM, Vicente E, Camacho A, Aragno M, Zopfi J (2009) Anaerobic microbial iron oxidation in an iron-meromictic lake. Geochim Cosmochim Acta 73(13):A1405

Widdel F, Schnell S, Heising S, Ehrenreich A, Assmus B, Schink B (1993) Ferrous iron oxidation by anoxygenic phototrophic bacteria. Nature 362:834–836

Xiong J (2006) Photosynthesis: what color was its origin? Genome Biol 7:245

Yamaguchi KE, Johnson CM, Beard BL, Ohmoto H (2005) Biogeochemical cycling of iron in the Archean Paleoproterozoic Earth: constraints from iron isotope variations in sedimentary rocks from the Kaapvaal and Pilbara Cratons. Chem Geol 218:135–169

Chapter 15
Synergistic Roles of Microorganisms in Mineral Precipitates Associated with Deep Sea Methane Seeps

Huifang Xu

The discovery of huge deep sea methane reservoirs in the form of ice-like solid crystals of methane clathrate hydrates (methane hydrates or gas hydrate for short) and carbonate deposits at the Hydrate Ridge in the Cascadia subduction zone (Suess et al. 1985) has stimulated global exploration and extensive studies of cold methane seeps and vents associated with the hydrate deposits (Paull and Dillon 2001; Trehu et al. 2004). Many deep sea methane hydrate deposits have been discovered around the world, for instance, in the Gulf of Mexico (Pohlman et al. 2008; Sassen et al. 1998, 2004), Monterey Bay of California (Gieskes et al. 2005; Lorenson et al. 2002; Stakes et al. 1999), Black Sea (Peckmann et al. 2001), Sea of Okhotsk, Eastern Siberia (Greinert and Derkachev 2004), the Gulf of Cadiz (Stadnitskaia et al. 2008), the Kuroshima Knoll of southern part of the Ryukyu Arc (Takeuchi et al. 2007), and South China Sea (Han et al. 2008; Lu 2007). It has been estimated that there are about 1,000 ~24,000 Gt of carbon in global methane hydrate zones (Dickens et al. 1997; Harvey and Huang 1995; Kvenvolden 1988; MacDonald 1990; Makogon and Makogon 1997). Methane hydrates have the potential to be a future energy source however methane is also a greenhouse gas. The release of methane from the methane hydrate can have a profound effect on the global climate (Kvenvolden 1998). In geological record, the rapid global temperature change at the Paleocene-Eocene boundary at ~55 millions of years ago, the Paleocene–Eocene Thermal Maximum (PETM), may be related to catastrophic methane release from sea sediments (Dickens et al. 1995; Katz et al. 2001; Zachos et al. 2005). At the interface between uprising methane from dissociation of methane hydrate and sulfate from sea water, a distinct microbial consortium mediates anaerobic oxidation of methane (AOM) through a net reaction of

$$CH_4 + SO_4^{2-} \rightarrow HS^- + HCO_3^- + H_2O. \tag{15.1}$$

H. Xu (✉)
Department of Geoscience, and NASA Astrobiology Institute,
University of Wisconsin-Madison, 1215 West Dayton Street, Madison, WI 53706, USA
e-mail: hfxu@geology.wisc.edu

L.L. Barton et al. (eds.), *Geomicrobiology: Molecular and Environmental Perspective*,
DOI 10.1007/978-90-481-9204-5_15, © Springer Science+Business Media B.V. 2010

The interaction results in a unique microbial ecosystem associated with AOM in deep sea environments (Jørgensen and Boetius 2007). The methane-fueled microbial communities in anoxic sediments above methane hydrates have the highest biomass in known marine ecosystems (Boetius et al. 2000; Michaelis et al. 2002). The interactions among pore fluids, ambient sediments and the metabolic products from coexisting syntrophic microorganisms result in distinctive mineral precipitations and mineral assemblages with Fe-sulfides, graphitic carbon, and variety of carbonate minerals.

In many cold methane seep and vent sites, chemosynthetic communities that derive energy from H_2S and methane oxidation have developed at the sea floor (Fig. 15.1) (Levin et al. 2000; Sassen et al. 1998; Treude et al. 2003). *Beggiatoa* spp, hydrogen sulfide-oxidizing organisms of tube worms, methane-utilizing clams (*Calyptogena*) and bivalves (*Acharax*) are distributed in the seep and vent sites according to the fluxes of up-rising hydrogen sulfide and methane (Bohrmann et al. 2002; Treude et al. 2003). The bacteria mats of *Beggiatoa* are associated with high uprising hydrogen sulfide and methane areas at the sea floor.

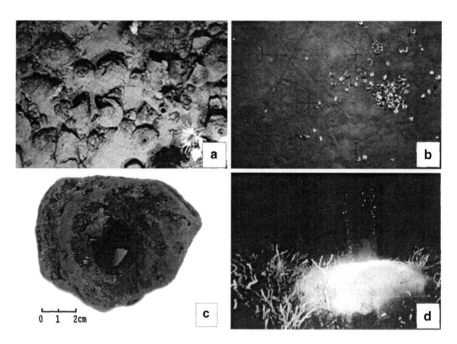

Fig. 15.1 (**a**) A photo showing carbonate deposit and carbonate chimneys at a seep site in the South China Sea. (**b**) A photo showing bio-mats of Beggiatoa (slightly yellowish) and bivalves on a carbonate deposit at a seep site in the South China Sea. (**c**) Hand specimen of a carbonate chimney (Lu 2007, used by permission). (**d**) A photo showing a typical methane hydrate mound (~2 m across) at the Bush Hill site with vents and a nearby chemosynthetic community of tube worms and Beggiatoa covering the mound (Sassen et al. 1998, used by permission)

Methane Production and Methane Hydrate Formation

There are three distinct processes that produce methane: biogenic through methanogenesis, thermogenic (thermal cracking at high temperature), and geothermal (or, abiotic) through serpentinization of Fe-bearing olivine. The methane-dominated gases produced through these three methane production processes have different isotopic signatures and chemical compositions. The source of the methane or methane production process can be determined using the ratio of methane to the sum of ethane and propane $[C_1/(C_2 + C_3)]$ and carbon isotopic fractionation. The carbon fractionation is measured with respect to the standard Pee Dee Belemnite (PDB). The ratio difference ($\delta^{13}C$) of ^{13}C and ^{12}C is defined by:

$$\delta^{13}C = [(^{13}C/^{12}C)_{Sample} / (^{13}C/^{12}C)_{PDB} - 1] \times 1000. \qquad (15.2)$$

The factor 1,000 yields results expressed in thousandths, or *per mil* (‰), not %.

In similar way, the ratio (δD) of D (deuterium, ^2H) and ^1H in methane, and the ratio ($\delta^{18}O$) of ^{18}O and ^{16}O in carbonate can be obtained relative to their standards of standard mean ocean water (SMOW), or Vienna standard mean ocean water (VSMOW). Two diagrams in Fig. 15.2 illustrate the regions for different methane sources.

Biogenic Methane Production through Methanogenesis

Methanogenesis is the final step in the anaerobic degradation of organic carbon (Megonigal et al. 2003). Methanogens use H_2 and acetate generated by fermenting bacteria to produce methane through the following reactions:

$$CO_2 + H_2 \rightarrow CH_4 \qquad (15.3)$$

$$CH_3COOH \rightarrow CO_2 + CH_4. \qquad (15.4a)$$

or,

$$CH_3COO^- + H_2O \rightarrow HCO_3^- + CH_4. \qquad (15.4b)$$

It is reported that 45% of all H_2-consuming (hydrogenotrophic) methanogens can substitute formate for H_2 in the reaction (15.3) (Garcia et al. 2000). The reaction (15.4) is known as acetate fermentation or acetoclastic methanogenesis and is restricted to two genera of *Methanosarcina* and *Methanosaeta* (formally *Methanothrix*) (Megonigal et al. 2003).

The reaction (15.3) is for H_2-consuming (or H_2-scavenging) methanogens, and the reaction (15.4) is for acetate-utilizing methanogens. The reaction (15.4) will result in carbon isotope fractionation between CO_2 and CH_4. The CO_2 generated in

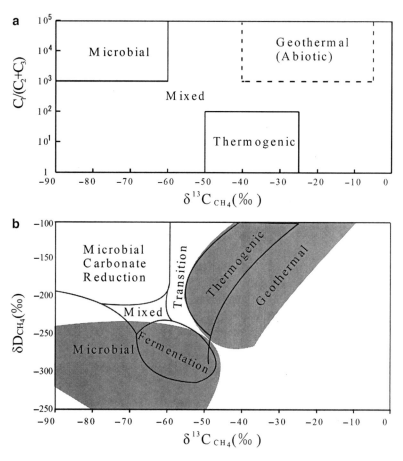

Fig. 15.2 (a) A plot of composition of C_1 (methane) and $C_1/(C_2 + C_3)$ ratio to distinguish biogenic, thermogenic, and geothermal (abiotic) methane. Modified from Claypool and Kvenvolden (Claypool and Kvenvolden 1983). (**b**): A plot of hydrogen and carbon isotopes to distinguish biogenic, thermogenic, and geothermal (abiotic) methane (Modified from Whitaker 1994; Lorenson et al. 2002)

pore fluids will be enriched in ^{13}C, whereas the methane will be depleted in ^{13}C. Thus carbonate minerals formed in such kinds of environments will have high $\delta^{13}C$ values (Greinert et al. 2001; Mazzullo 2000). Consumption of H_2 by H_2-scavenging methanogens promotes fermentation reactions with H_2 as part of the reaction products (Bryant et al. 1967). The two metabolically different types of micro-organisms (syntrophs) cooperate together and form a mutually dependent consortium and unique ecosystem. Because the fermenting bacteria and methanogens are in a symbiotic relationship, a general net reaction for decomposing bio-organics into methane may be written as:

$$2CH_2O + 2H_2 \rightarrow CH_4 + CO_2 + 2H_2O. \tag{15.5}$$

Thermogenic Methane Production

Thermogenic degradation of pre-existing or buried organics releases CH_4 and other hydrocarbons when microbes are no longer active at high temperatures (T > 80°C). For instance, significant thermogenic gas vents and seeps into the water column occur at the Bush Hill site in the Gulf of Mexico (Sassen et al. 1998). This site is characterized by gas hydrate mounds that outcrop on the seafloor.

The thermogenic process is basically a thermal cracking of organics and oils breaking larger molecules into smaller ones. For instance, methane can be generated by breaking of C–C bond in ethane and through reactions of:

$$C H_3CH_3 \rightarrow 2CH_3 \bullet (radicals). \tag{15.6}$$

$$CH_3 \bullet (radical) + CH_3CH_3 \rightarrow CH_4 + CH_3CH_2 \bullet (radical). \tag{15.7}$$

Thermogenic methane forms at a deeper depth relative to the biogenic methane produced through methanogenesis. Thermogenic methane-dominated gas generally contains a small amount of wet gas components ethane, propane and C_{4+} -hydrocarbons, whereas the biogenic methane gas is very dry and pure and is dominated by methane. Both chemical and isotopic compositions can be used to differentiate between these processes (Fig. 15.2). The uprising thermogenic methane through faults can directly contribute to deep sea methane seeps and vents (Greinert et al. 2001; Sassen et al. 2004).

Geothermal Methane Production Through Serpentinization of Ultramafic Rocks

Both hydrogen and methane enriched fluids were observed in mid-ocean ridge hydrothermal systems and nearby the ridge systems (e.g., the recently discovered Lost City hydrothermal vent system) (Kelley et al. 2001, 2005). Circulation of ocean waters through the hydrothermal system results in a redox reaction between water and Fe (II) in olivine in an oxygen free environment which produces magnetite and hydrogen (Charlou et al. 1998). Although real olivine contains about 20 mole% of the fayalite component, and serpentine may contain small amount of Fe (II), the redox reaction at high temperatures can be simplified as (Reeburgh 2007):

$$6(Mg_{1.5},Fe_{0.5})SiO_4 + 7H_2O \rightarrow 3Mg_3Si_2O_5 (OH)_4 + Fe_3O_4 + H_2 \tag{15.8}$$

$$\text{Olivine} \qquad\qquad \text{Serpentine magnetite}$$

The hydrogen gas produced will react with dissolved carbon dioxide to form methane through the Fischer – Tropsch reaction in the presence of metal oxides and metal as catalysts through reaction (15.3).

This type of methane will have higher $\delta^{13}C$ values due to its carbon source from dissolved inorganic carbon in ocean water. The $\delta^{13}C$ values range from -8.8 to -13.6 in the Lost City fluids (Keir et al. 2005).

In low temperature hydrothermal systems like diffuse seeps away from mid-ocean ridges (German and Von Damm 2004), the dissolved hydrogen may also be used by methanogens to produce methane through the reaction (15.3). In the Lost City hydrothermal filed ecosystem, 16S rRNA gene sequence corresponding to the *Methanosarcinales* phylotype were found in high temperature chimneys, while a phylotype of anaerobic methanotrophic Archaea (ANME-1) was restricted to low-temperature, less vigorously venting sites (Brazelton et al. 2006). This type of methane in not important in the methane seeps, but it could be very important in the early anoxic Earth ~2 billion years ago.

Methane Hydrate Formation and Dissociation

Both thermogenic and methanogenic methane gas are the main sources of the deep sea methane hydrate. Based on the methane hydrate phase diagram, at low temperature ($<4°C$) and elevated pressure (>60 bar, or more than 600 m below sea level), methane and water or pore fluids form ice-like solid phases of methane clathrate hydrate, or methane hydrate $(CH_4)(H_2O)_6$ in sediments (Kvenvolden 1988; Sloan 1998).

$$CH_4 + 6H_2O = (CH_4)(H_2O)_6. \tag{15.9}$$

The oxygen isotope fractionation of reaction (15.9) will result in heavy oxygen in the solid methane hydrate with respect to sea-water and pore fluid. There are two major types of gas hydrate phases (structure I and structure II) with methane in large cage sites (Jeffrey and McMullan 1967). Neutral molecules of CO_2, H_2S and ethane may also occupy the methane site in the structure of the methane hydrate crystals. The solid methane hydrate crystals in sea sediments can also serve a stable source for methane and carbon dioxide.

$$(1-x-y)CH_4 + xCO_2 + yH_2S + 6H_2O = [(CH_4)_{1-x},(CO_2)_x,(H_2S)_y](H_2O)_6. \tag{15.10}$$

When temperate increases and/or overlain pressure decreases due to local structure change (faults, slide, and turbidite cuts), or large scale tectonic, climate changes, the solid methane hydrate becomes thermodynamically unstable and dissociates into water and methane. The process could be slow and gentle, or rapid and even catastrophic.

$$(CH_4)(H_2O)_6 = CH_4 + 6H_2O. \tag{15.11}$$

The methane generated through the slow dissociation process results in methane seeps and vents. In an ambient air environment, dissociation of 1 m^3 of methane hydrate will produce 164 m³ of methane gas and 0.8 m³ of liquid water. Water from the dissociation of methane hydrate may also contribute to cold methane seeps.

Anaerobic Oxidation of Methane (AOM) at the Sulfate-Methane Transition (SMT) Zone

Anaerobic oxidation of methane (AOM) occurs in the marine environment is one of the global major methane sinkers (Reeburgh 2007). A net reaction for the anaerobic oxidation of methane at SMT zone can be written as reaction (15.1).

The reaction will increase the alkalinity, but also slightly lowers the pH of sea water-dominated pore fluids (Mazzullo 2000; Pohlman et al. 2008). A linear relationship between alkalinity and HS⁻ concentration was observed in fluids from methane hydrate seep sites (Gieskes et al. 2005). The reactions result in carbonate precipitation and pyrite formation, if there is Fe(II) from surrounding and ambient sediments and solutions. The coupled methane oxidation and sulfate reduction at the sulfate-methane transition (SMT) zone results in sharp decreases in sulfate and methane concentrations in sediment depth profiles (Fig. 15.3). The reaction rate at the SMT zone is very high and the reaction rate drops dramatically away from the SMT zone (Dale et al. 2008; Knab et al. 2008). AOM for general marine environments has been reviewed recently by several authors (Kasten and Jørgensen 2000; Megonigal et al. 2003; Reeburgh 2007). In most marine AOM, methane in the pore fluid above SMT zone is nearly depleted, and concentration of dissolved sulfides drops to zero very fast (Kasten et al. 1998; Niewöhner et al. 1998). In the AOM in methane seep and vent sites, a high biomass above the methane hydrate (may reach ~10¹² cells per cm³) (Boetius et al. 2000; Treude et al. 2007) results in surplus hydrogen sulfide and bisulfide in pore fluids and seep and vent fluids

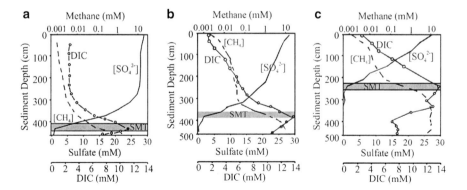

Fig. 15.3 Pore fluid chemistry profiles of the biogeochemically active pore fluids from seep sits in Keathley Canyon in the northern Gulf of Mexico. (**a**) An extreme concavity site (KC03-05), (**b**) an intermediate concavity site (KC03-07), (**c**) a slight concavity site (KC03-19). All figures are modified from Pohlman et al. (2008). DIC = dissolved inorganic carbon

(Pohlman et al. 2008). Both dissolved sulfides and methane in seep and vent fluids provide nutrients for the chemosynthetic communities (like sulfide-oxidizing bacteria, tube worms and methane-consuming mussels and clams) at seep and vent sites (Figs. 15.1 and 15.3).

Recent studies indicate that there are two types of microorganisms (anaerobic methane oxidizers and sulfate-reducing bacteria (SRB) partnering together in an obligate syntrophic relationship to carry out reaction (15.1). Results from 16S rRNA gene sequence and carbon isotope signatures from the microorganisms collected from methane seep sites indicate that *Archaea* related to *Methanosarcinales* are the dominant anaerobic methanotrophs that produce acetate and hydrogen for the associated SRB (Boetius et al. 2000; Hinrichs et al. 1999; Orphan et al. 2001a). Two groups of methanogens, specifically archaeal genes, ANME-1 and ANME-2 were identified in the anaerobic methane oxidation Archean- SRB consortium in seep site SMT zones (Boetius et al. 2000; Hinrichs et al. 1999; Orphan et al. 2001a, b). Three possible reverse methanogenesis processes were proposed for AOM at the SMT zone for methane oxidation (MO) and sulfate reduction (SR) (Hoehler et al. 1994; Megonigal et al. 2003; Reeburgh 2007). The first hypothesis is an analog to reverse hydrogenotrophic methanogenesis:

$$2CH_4 + 2H_2O \rightarrow CO_2 + 4H_2 . (MO) \qquad (15.12)$$

$$4H_2 + SO_4^{2-} + H^+ \rightarrow HS^- + 4H_2O . (SR) \qquad (15.13)$$

The second hypothesis includes reactions of:

$$CH_4 + HCO_3^- \rightarrow CH_3COO^- + 4H_2O . (MO) \qquad (15.14)$$

$$CH_3COO^- + SO_4^{2-} + 4H_2O \rightarrow HS^- + 2HCO_3^- . (SR) \qquad (15.15)$$

The third hypothesis includes reactions of:

$$2CH_4 + 2H_2O \rightarrow CH_3COOH + 4H_2 . (MO) \qquad (15.16)$$

$$4H_2 + SO_4^{2-} + H^+ \rightarrow HS^- + 4H_2O . (SR) \qquad (15.17)$$

$$CH_3COOH + SO_4^{2-} + 4H_2O \rightarrow HS^- + 2HCO_3^- + H^+ . (SR) \qquad (15.18)$$

The third mechanism may be more realistic because it is more thermodynamically favorable (Valentine and Reeburgh 2000). It was reported that *Archaea* of the ANME-2 group are associated with high methane flux seeps and vents, whereas, ANME-1 group are associated with low flux methane seeps (Stadnitskaia et al. 2008). The overall redox reactions involved in AOM can be expressed in Fig. 15.4.

The interaction between uprising methane and dissolved sulfate in pore fluids and sea water produces unique ecosystems. Amount of dissolved methane and sulfide varies at different sites and different levels. The interactions among pore

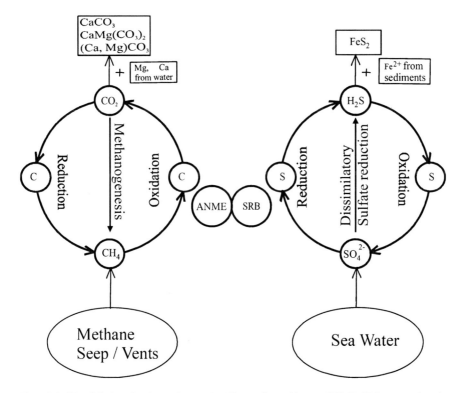

Fig. 15.4 Simplified cycles for carbon and sulfur performed by an ANME-SRB consortium in seep and vent sites. C and S mean elemental C and S or organic C and S, respectively

fluid and sea water, products from microbial metabolisms and ambient sediments result in distinctive mineral precipitates of carbonates and sulfides associated with methane hydrates and seeps vents.

Mineral Precipitates

Precipitation of Pyrite and Graphitic Carbon

Framboidal pyrite is very common in the sediments associated with methane hydrates through AOM. Precipitation of pyrite and other sulfide minerals is controlled by HS^- from the dissimilatory sulfate reduction of reaction (15.1) and Fe in the sediments. The dissimilatory sulfate reduction in marine environments is a very important process in controlling sulfur and carbon cycling (Jørgensen 2000). The process accounts more than half of total organic carbon oxidation (Canfield et al. 1993; Jørgensen 1982). At SMT zone, the SRB-generated H_2S (dominated by HS^- in circumneutral pH condition) (Stumm and Morgan 1996) will react with Fe(II) and Fe(III) from the sediments and precipitates Fe-sulfides. The metastable Fe-monosulfide, FeS, will react with intermediate sulfur species (elemental S^0 and poly-sulfides) and transforms into pyrite as a final product (Berner 1984). Intermediate sulfur species could form by HS^- oxidation (Jørgensen 2000). The reaction may be simplified as:

$$Fe(II) + HS^- \rightarrow FeS \qquad (15.19)$$

and

$$FeS + S^0 \rightarrow FeS_2. \qquad (15.20)$$

Intermediate sulfide structures like greigite (Fe_3S_4) and pyrrhotite ($Fe_{1-x}S$) may also occur in low concentration gradients of methane near disseminated methane hydrate (Larrasoana et al. 2007). The filamentous structure of pyrite framboid pseudomorphs after *Beggiatoa* was reported in a gas hydrate mound at Bush Hill, Gulf of Mexico (Sassen et al. 2004). This type of pyrite may be formed through this process at the sea floor because the presence of internal sulfur granules in *Beggiatoa* is commonly seen.

Another proposed mechanism for pyrite formation is the reaction between FeS and H_2S with hydrogen as a by product (Rickard 1997; Rickard and Luther 1997; Theberge and Luther 1997), although it is very difficult to simulate the process in the laboratory due to the very low hydrogen concentration. The reactions may be written as:

$$FeS + H_2S \rightarrow [FeS - SH_2] \qquad (15.21)$$

$$[FeS - SH_2] \rightarrow [FeS_2 \cdot H_2] \qquad (15.22)$$

$$[FeS_2 \cdot H_2] \rightarrow FeS_2 + H_2. \qquad (15.23)$$

A net reaction is:

$$FeS + H_2S \rightarrow FeS_2 + H_2. \qquad (15.24)$$

A third mechanism for the formation of pyrite is by direct precipitation through reactions of Fe(II), sulfur and polysulfide when the solution is under-saturated with respect to amorphous FeS (Giblin and Howarth 1984; Howarth 1979; Luther 1991).

SEM images show framboidal pyrite clusters associated with carbonate (protod-olomite) (Fig. 15.5a) from site 1 of South China Sea, and a pyrite micro-chimney, or micro-chimney-like pyrite from site 2 of South China Sea (Fig. 15.5b–e). SEM images shows framboidal pyrite core surrounded by elongated pyrite crystals that form pyrite spherules (Fig. 15.5d). High-resolution TEM image shows planar defects in the pyrite spherules (Fig. 15.5e). The pyrite (P) spherules contain marcasite (M)

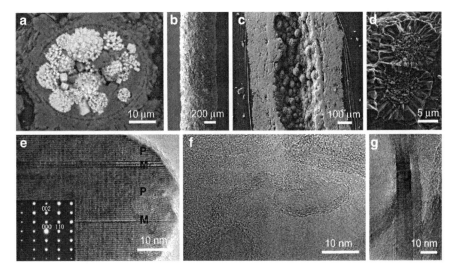

Fig. 15.5 SEM image showing pyrite framboids and protodolomite filling chamber of a foraminifer from site 1 of the South China Sea (**a**). (**b**) A pyrite micro-chimney composed of many spherical aggregates. (**c**) SEM image showing the inner wall of the chimney pyrite. (**d**) High-magnification SEM image showing spherical pyrite aggregates consist of a framboidal core with pyrite spherules. (**e**) High-resolution TEM image showing a pyrite spherule contains marcasite lamellar domains. (**f**) High-resolution TEM images showing a nanotube-like carbon, and (**g**) a graphitic nano-plate

domains along (001) of pyrite. The marcasite and pyrite host keep a strict orientation relationship: marcasite (101)//(001) of pyrite, and [010] direction of marcasite//[100] of pyrite. This is the least misfit plane between pyrite and marcasite. A similar microstructure was observed in a hydrothermal pyrite spherule (Dodony et al. 1996).

Both pyrite and marcasite have the same composition but different crystal structures. Pyrite is the stable polymorph of FeS_2. In general, the presence of neutral H_2S species (i.e., relatively lower pH) is required in order to precipitate marcasite (Murowchick et al. 1986; Schoonen and Barnes 1991a, b). The observed texture of a spherule pyrite rim with marcasite domains surrounding a framboidal pyrite core (Fig. 15.5d) indicates different solution conditions during the pyrite precipitation. The chimney-like pyrite precipitates in worm tubes serve as channels for methane and HS-bearing seep fluids. The reaction (15.1) results in an increase in alkalinity and slightly lowers the pH if there is no pyrite precipitation. In circumneutral (lightly basic) solutions, pyrite precipitates and forms pyrite framboids.

Graphitic nano-crystals and nano-tube-like carbon are closely associated with the pyrite spherules (Fig. 15.5f and g). This type of elemental carbon (instead of by the complete oxidation of methane) may indicate gas-phase methane in the fluid besides dissolved methane. Carbon nanotubes can also be synthesized in aqueous solutions with solid catalysts (Gogotsi et al. 2001; Hata et al. 2004). In the methane seeps areas, there is methane in dissolved and gas phase states even at the SMT zone. The uprising methane and dissolved sulfides will interact with the pyrite surface. The close relationship between nano-tube-like carbon and pyrite spherules indicate that pyrite, a semiconductor, serves as a catalyst during the reaction from methane to elemental carbon. Partial oxidation of methane occurs through sorption of methane molecules on pyrite surface (notation: >S) and pre-formed C on pyrite surface (notation: >S–C_n) followed by electron donation to pyrite crystals:

$$CH_4 \rightarrow > S - CH_4 + e^- \rightarrow > S - C + e^- \text{ (in pyrite)} + H^+ \text{ (in chimney solution)}.$$

and,

$$> S - C_n + CH_4 \rightarrow > S - C_{n+1} + e^- \text{ (in pyrite)} + H^+ \text{ (in chimney solution)}.$$

Whereas, ambient electron acceptors like Fe(III), Mn(IV) and oxygen gain electrons from the neighboring pyrite. The pyrite-catalyzed net reaction may be expressed as:

$$CH_4 (g) + O_2 (g) \rightarrow C + 2H_2O \qquad (15.25)$$

The above pyrite-catalyzed reaction process results in acidification of the solution in the pyrite-chimney chamber which will increase the ratio between neutral [H_2S] species and bisulfide [HS^-]. Lower pH and increasing [H_2S] will result in marcasite growth on pyrite substrate (Murowchick et al. 1986; Schoonen and Barnes 1991a, b) and formation of pyrite spherules as showed in Fig. 15.5d and e. The coupled methane oxidation on pyrite surface, pH change in the chimney solution, and the growth of marcasite and pyrite spherules can be expressed as:

$$\text{Methane in fluid} \rightarrow \text{methane on pyrite/or pre} - \text{formed carbon}$$
$$\rightarrow C + H^+ \text{ (lower pH)};$$
$$HS^- + H_2S \text{ in fluid} \rightarrow H_2S + HS^- \text{ (lower pH)}$$
$$\rightarrow \text{spherule pyrite} + \text{marcasite growth}.$$

There is no carbonate in all the observed pyrite chimneys that also indicates a low pH condition during the spherule pyrite and nanotube-like carbon formation.

Precipitations of Carbonate Minerals

AOM will result in alkalinity increase of pore fluids and carbonate precipitation (Gieskes et al. 2005). Stoichiometric dolomite $CaMg(CO_3)_2$ should precipitate from sea-water and sea-water-derived pore fluids according to thermodynamic equilibria (Hardie 1987). However, aragonite precipitates from modern sea water because solvent water that has large dipole moment forms a strongly hydrated Mg^{2+} surface complex and inhibits dolomite and calcite nucleation and growth (Lippmann 1973). Minerals of aragonite, calcite, low-magnesian calcite (LMC), high-magnesian calcite

(HMC), protodolomite (proto), disordered dolomite, and dolomite (actually Ca-rich dolomite, not stoichiometric dolomite) occur in the carbonate precipitates associated with methane seeps and vents (Greinert et al. 2001; Naehr et al. 2007; Takeuchi et al. 2007). The carbonate minerals may form tube-like chimneys that are channels for seeping and venting fluids (Fig. 15.1c). Framboidal pyrite may coexist with HMC, dolomite, or protodolomite (Fig. 15.5a). Calcite and LMC can be biogenic or bioclasts of coccoliths and foraminifera fragments. The boundary between LMC and HMC ranges from 8 mole % of $MgCO_3$ to 11 mole % of $MgCO_3$. In sea-water, most biogenic magnesian calcite contains ~11 mole % of $MgCO_3$ that may have same solubility of aragonite. Figure 15.6a shows X-ray diffraction patterns of carbonate deposits from three different methane seep sits in South China Sea. The carbonate could be dominated by dolomite, mixture of HMC and protodolomite, and mixture of aragonite and HMC. Protodolomite with a very broad (104) diffraction peak indicates both structural (Mg–Ca disordering) and compositional heterogeneity. High-resolution TEM image and Fourier transform patterns from different areas show disordered dolomite (without super lattice reflections) and weakly ordered dolomite (protodolomite) (Fig. 15.7).

Recent studies in some marine AOM areas without obvious methane seeps and vents indicate dolomite and protodolomite are closely related to the methanogenesis zone and lower part of the SMT zone (Mazzullo 2000). The concentration of methane in pore fluids above SMT zone is about zero. However, protodolomite (~40–44 mole % of $MgCO_3$) and HMC can precipitate above SMT zones in methane

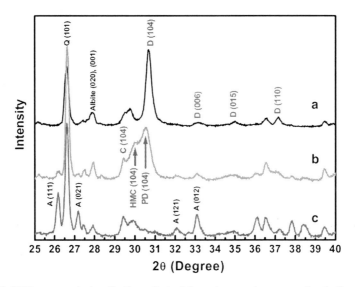

Fig. 15.6 XRD patterns (using Cu Kα radiation) from three methane seep sites in South China Sea. *Top pattern*: dolomite is the major minerals (~45 mole % of $MgCO_3$); *Middle pattern*: protodolomite (~40–44 mole % of $MgCO_3$) and HMC are the major minerals; *Bottom pattern*: aragonite and HMC are the major minerals. Very broad (104) diffraction peak characterizes disordered dolomite and protodolomite (PD) with composition range of ~40 to 44 mole% of $MgCO_3$. C = calcite, Qtz = quartz, Ab = albite

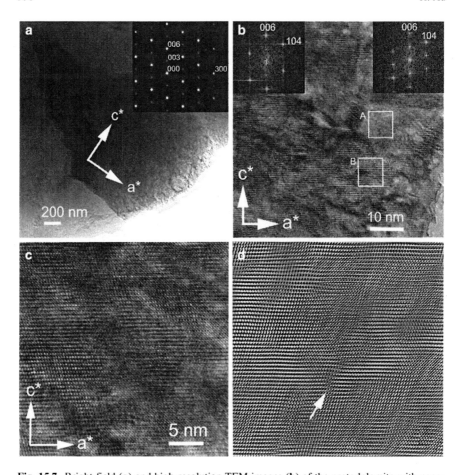

Fig. 15.7 Bright-field (**a**) and high-resolution TEM images (**b**) of the protodolomite with nano-domains and low-angle boundaries among the neighboring crystals. An inserted electron diffraction pattern shows weak supper lattice reflections characterizing poorly ordered dolomite structure. There are domains with dolomite and calcite structures (see inserted Fourier transform patterns from two selected areas). Some areas show anti-phase-domain-like boundaries (**c, d**) that result from Mg–Ca ordering-induced phase transition. Image D is a filtered inverse Fourier transform of image C, and clearly shows the domain boundary, i.e., off-set of (001) dolomite lattice fringes (indicated by *arrow*) between neighboring domains

seep and vent areas (Takeuchi et al. 2007). The dehydration and desolvation of the hydrated Mg^{2+} ion and dolomite surface Mg^{2+}–water complex is the key to the dolomite formation. Both dissolved methane and hydrogen sulfide have low molecular dipole moments and serve as catalysts for lowering dehydration energy of Mg^{2+}–water surface complex. Our synthesis experiments indicate that concentration of $MgCO_3$ in HMC and dolomite is related to molecular dipole moments (μ) of dissolved molecules that serve as catalysts at a given temperature (Fig. 15.8). Increasing temperature will enhance Mg incorporation into HMC and dolomite

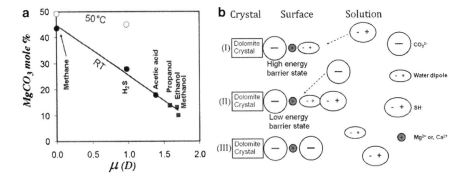

Fig. 15.8 (a) Diagram showing relationship between molecular dipole moment (μ) of added catalysts and mole % of $MgCO_3$ in the crystallized calcite or disordered dolomite at about room temperature (RT). All the solutions are with 5:1 Mg/Ca ratio. Data for methanol, ethanol, and propanol are from published data (Falini et al. 1996). Molecular dipole moments are from Lide (2004). The unit for dipole moment is debye (D). Two 50°C data of disordered dolomite are plotted in *circles*. (b) Diagram showing 1-dimensional crystals diagrams showing catalytic role of dissolved bisulfide on lowering dehydration energy of Mg^{2+} on calcite or dolomite surface. (I) Water dipole bonding with the surface Mg^{2+} due to strong hydration of Mg^{2+}; (II) Interaction between dissolved bisulfide with the surface water results in orientation/distance change of the water dipole, and weakening bonding between the Mg^{2+} and surface water; (III) Aqueous carbonate can repel the water dipole and form bonds with the surface Mg^{2+}. The calcite crystal grows through this process

dramatically in the presence of the catalysts. The presence of dissolve sulfides result in precipitation of HMC at low temperatures and disordered dolomite at temperatures of 40°C or higher. Presence of dissolved methane results in the formation of protodolomite or disordered dolomite (due to fast precipitation and growth). In solutions without methane and dissolved sulfides, no dolomite precipitates. Laboratory synthesis experiments also show that aragonite precipitates if elemental sulfur also exists with the catalysts in forms of colloids. The mechanism by which elemental sulfur inhibits HMC and dolomite precipitation is still not clear. Observed carbonate deposits with HMC and dolomite chimneys above SMT zone, and aragonite at sea floor level in the Kuroshima Knoll of southern part of the Ryukyu Arc can be explained based on newly obtained experimental results (Fig. 15.8). HMC and dolomite chimneys can precipitate in sediments above the SMT zone due to the presence of both methane and sulfide in pore fluids. Only aragonite precipitates at the sea floor (above the carbonate chimney layer) because of oxidation of sulfide and/or formation of elemental sulfur.

The dolomite or protodolomite formed in the methanogenesis zone will have high $\delta^{13}C$ values (Greinert et al. 2001; Mazzullo 2000). The carbon isotope fractionation will enrich ^{13}C in bicarbonate and deplete ^{13}C in methane. The dolomite formed in this environment will have high $\delta^{13}C$ values, i.e., group A carbonate (Figs. 15.9 and 15.10). The dolomite or calcian dolomite formed in these areas generally contains small amounts of Fe(II) or $FeCO_3$. In extreme cases, even

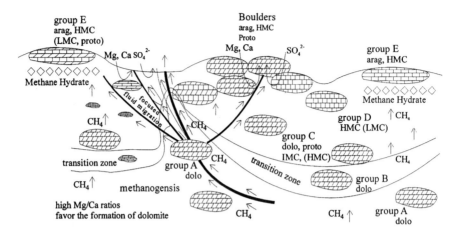

Fig. 15.9 Schematic diagram showing the origin of petrographically and isotopically different carbonate rocks relative to the controlling diagenetic environments. Group A carbonates are restricted to the deep methanogenic zone resulting in the formation of dolomite. Group B carbonates, also dolomite dominated, are representative of the transition between sulfate-reduction and methanogenesis. Group C carbonate, with typical methane-derived values but protodolomite and dolomite as major carbonate phases have formed in the sulfate-reduction zone. Group D carbonates without obvious dolomite contain carbon from mixture of degraded organic matter and methane oxidation and are generated close to surface. Group E are aragonite-rich carbonates formed at the surface or near surface environments, with carbon from methane oxidation (Modified from Greinert et al. 2001)

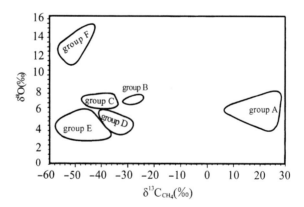

Fig. 15.10 Carbon and oxygen isotope values of carbonate from the northern summit of the Hydrate Ridge. Five distinctive six groups (group A to F) can be distinguished in their dominate carbonate cement phases as well as their carbon source sources and formation conditions illustrated in Fig. 16.9. The carbonate directly associated with methane hydrate falls in group F. Its carbon isotopes are typical methane-generated carbon source, whereas their extremely high $\delta^{13}C$ value is still unsolved (Greinert et al. 2001)

siderite can precipitate. The existence of Fe^{2+} in dolomite also indicates a reducing dolomitizing fluid that contains dissolved methane and other organics that are responsible for the dolomite precipitation. It has been shown that fermenting bacteria can transfer electrons to Fe(III) in sediments (Lovley 2000). The cooperative metabolism between methanogens and fermenting bacteria results in CO_2 or dissolved HCO_3^- as the common product of all these reactions which can result in Fe^{2+}-bearing carbonate (Coleman and Raiswell 1993) through a general net reaction of,

$$2Fe_2O_3 \text{ (sediments)} + 7CH_2O \text{ (organic)} = 4FeCO_3 \text{ (in dolomite)} \\ + 3CH_4 + H_2O. \qquad (15.26)$$

The produced methane in the above reaction serves as a catalyst for dolomite formation.

If the methane is produced mainly through the reaction (15.26), the dolomite precipitated in this zone will have high $\delta^{13}C$ values (Fig. 15.10) (Greinert et al. 2001; Mazzullo 2000). However, the dolomite formed in SMT zone will be depleted in ^{13}C and Fe due to AOM (Mazzullo 2000). HMC and dolomite associated with methane seeps/vents in and above SMT zone are also depleted in ^{13}C and Fe (Greinert et al. 2001; Takeuchi et al. 2007). The AOM of reaction (15.1) will result in ^{13}C–depleted dissolved inorganic carbon. Precipitation of pyrite results in Fe-depleted HMC and dolomite.

Geological Records and Biosignatures

Reported methane seeps and vents in the rock record are based on large carbon isotope excursions, tepee like structures in rocks, ^{13}C-depleted carbonate, associated pyrite framboids, and lipids of microorganisms preserved in carbonate rocks. One well studied site is a Neoproterozoic postglacial cap carbonate (Doushantuo cap carbonate) in Yangtze platform of south China (Jiang et al. 2003, 2006). The cap carbonate formed during destabilization of methane hydrate during the deglaciation at ~635 Ma (millions of years ago). The cap carbonate consisting of disrupted limestone and/or dolomite (with micrite, dolomicrite, pyrite framboids, and barite) (layer 1), and laminated limestone and/or dolomite with tepee-like structures (layer 2). The mineral assemblage and texture is similar to some modern methane seep and vent-induced carbonate deposits. The cap carbonate overlies the glaciogenic diamictite (poorly or non-sorted conglomerate with a wide range of clasts and clay minerals) (Jiang et al. 2003). The tepee-like structures indicate paleo-methane vents. A huge carbon isotope excursion ($\delta^{13}C$ ranges from +5‰ to −40‰) was observed in a thin layer right above the tepee-like layer. This ^{13}C-depleted layer was formed during AOM of uprising methane from dissociation of methane hydrate. Global deglaciation (or the end of Snowball Earth) resulted in a massive dissociation of methane hydrate (Jiang et al. 2003, 2006).

Another case is the Late Cretaceous (Campanian) Tepee Buttes (limestone) in the Western Interior Seaway, USA (Birgel et al. 2006, 2008; Birgel and Peckmann 2008). Highly ^{13}C–depleted archaeal lipids (δ^{13}C ranges from $-102\%_o$ to $-118\%_o$) and bacterial biomarkers (δ^{13}C ranges from $-51\%_o$ to $-73\%_o$), corroborating a syntrophic consortium of *archaea* and bacteria mediating AOM were reported in the limestone Tepee Buttes (Birgel et al. 2006). In the relatively recent geologic past, it is proposed that the sudden release and oxidation of methane hydrate at the Paleocene-Eocene boundary contributed to a rapid global temperature change at the Paleocene–Eocene Thermal Maximum (PETM) (Dickens et al. 1995; Katz et al. 2001).

Acknowledgement Author thanks NASA Astrobiology Institute (N07-5489), National Science Foundation (EAR-0810150, 0824890, 0958000), and Office of Basic Energy Science of U.S. Department of Energy for supporting my research. Author also thanks helps from Dr. Hiromi Konishi, Fangfu Zhang, Yubing Sun, and Jason Huberty of University of Wisconsin, and Dr. Hongfeng Lu and Prof. Xiaoming Sun of Sun Yat-Sen University. I also acknowledge journal of "Geology" for permitting me to use one photo (Fig. 15.1d), and Dr. Lu for using three photos (Fig. 15.1a–c) from his PhD dissertation.

References

Berner RA (1984) Sedimentary pyrite formation – an update. Geochim Cosmochim Acta 48:605–615

Birgel D, Peckmann J (2008) Aerobic methanotrophy at ancient marine methane seeps: a synthesis. Org Geochem 39:1659–1667

Birgel D, Peckmann J, Klautzsch S, Thiel V, Reitner J (2006) Anaerobic and aerobic oxidation of methane at Late Cretaceous seeps in the Western Interior Seaway, USA. Geomicrobiol J 23:565–577

Birgel D, Himmler T, Freiwald A, Peckmann J (2008) A new constraint on the antiquity of anaerobic oxidation of methane: Late Pennsylvanian seep limestones from southern Namibia. Geology 36:543–546

Boetius A, Ravenschlag K, Schubert CJ, Rickert D, Widdel F, Gieseke A, Amann R, Jørgensen BB, Witte U, Pfannkuche O (2000) A marine microbial consortium apparently mediating anaerobic oxidation of methane. Nature 407:623–626

Bohrmann G, Suess E, Greinert J (2002) Gas hydrate carbonates from Hydrate Ridge, Cascadia convergent margin: indicators of near-seafloor clathrate deposits. Proceedings of the fourth international conference on gas hydrates, Yokohama, Japan, pp 102–107

Brazelton WJ, Schrenk MO, Kelley DS, Baross JA (2006) Methane- and sulfur-metabolizing microbial communities dominate the Lost City hydrothermal field ecosystem. Appl Environ Microbiol 72:6257–6270

Bryant MP, Wolin EA, Wolin MJ, Wolfe RS (1967) *Methanobacillus Omelianskii* a symbiotic association of 2 species of bacteria. Arch Mikrobiol 59:20–31

Canfield DE, Jørgensen BB, Fossing H, Glud R, Gundersen J, Ramsing NB, Thamdrup B, Hansen JW, Nielsen LP, Hall POJ (1993) Pathways of organic-carbon oxidation in 3 continental-margin sediments. Mar Geol 113:27–40

Charlou JL, Fouquet Y, Bougault H, Donval JP, Etoubleau J, Jean-Baptiste P, Dapoigny A, Appriou P, Rona PA (1998) Intense CH4 plumes generated by serpentinization of ultramafic rocks at the intersection of the 15d20'N fracture zone and the Mid-Atlantic Ridge. Geochim Cosmochim Acta 62:2323–2333

Claypool GE, Kvenvolden KA (1983) Methane and other hydrocarbon gases in marine sediment. Ann Rev Earth Planet Sci 11:299–327

Coleman ML, Raiswell R (1993) Microbial mineralization of organic-matter-mechanisms of self-organization and inferred rates of precipitiation of diagenetic minerals. Phil Trans R Soc Lond Ser A Math Phys Eng Sci 344:69–87

Dale AW, Regnier P, Knab NJ, Jorgensen BB, Van Cappellen P (2008) Anaerobic oxidation of methane (AOM) in marine sediments from the Skagerrak (Denmark): II. Reaction-transport modeling. Geochim Cosmochim Acta 72:2880–2894

Dickens GR, Oneil JR, Rea DK, Owen RM (1995) Dissociation of oceanic methane hydrate as a cause of the carbon-isotope excursion at the end of the Paleocene. Paleoceanography 10:965–971

Dickens GR, Paull CK, Wallace P (1997) Direct measurement of in situ methane quantities in a large gas-hydrate reservoir. Nature 385:426–428

Dodony I, Posfai M, Buseck PR (1996) Structural relationship between pyrite and marcasite. Am Mineralog 81:119–125

Falini G, Gazzano M, Ripamonti A (1996) Magnesium calcite crystallization from water-alcohol mixtures. Chem Commun 9:1037–1038

Garcia JL, Patel BKC, Ollivier B (2000) Taxonomic phylogenetic and ecological diversity of methanogenic Archaea. Anaerobe 6:205–226

German CR, Von Damm KL (2004) Hydrothermal processes. In: Elderfield H (ed) Tretise on geochemistry, volume 6: the oceans and marine geochemistry. Elsevier, New York, pp 181–222

Giblin AE, Howarth RW (1984) Porewater evidence for a dynamic sedimentary iron cycle in salt marshes. Limnol Oceanogr 29:47–63

Gieskes J, Mahn C, Day S, Martin JB, Greinert J, Rathburn T, McAdoo B (2005) A study of the chemistry of pore fluids and authigenic carbonates in methane seep environments: Kodiak Trench, Hydrate Ridge, Monterey Bay, and Eel River Basin. Chem Geol 220:329–345

Gogotsi Y, Libera JA, Guvenc-Yazicioglu A, Megaridis CM (2001) In situ multiphase fluid experiments in hydrothermal carbon nanotubes. Appl Phys Lett 79:1021–1023

Greinert J, Derkachev A (2004) Glendonites and methane-derived Mg-calcites in the Sea of Okhotsk, Eastern Siberia: implications of a venting-related ikaite/glendonite formation. Mar Geol 204:129–144

Greinert J, Bohrmann G, Suess E (2001) Gas hydrate-associated carbonates and methane-venting at hydrate ridge: classification, distribution, and origin of authigenic lithologies. In: Paull CK, Dillon WP (eds) Natural gas hydrates: occurrence, distribution, and detection. American Geophysical Union, Washington, DC, pp 99–113

Han XQ, Suess E, Huang YY, Wu NY, Bohrrnann G, Su X, Eisenhauer A, Rehder G, Fang YX (2008) Jiulong methane reef: microbial mediation of seep carbonates in the South China Sea. Mar Geol 249:243–256

Hardie LA (1987) Dolomitization – a critical view of some current views. J Sediment Petrol 57:166–183

Harvey LDD, Huang Z (1995) Evaluation of the potential impact of methane clathrate destabilization on future global warming. J Geophys Res Atmosph 100(D2):2905–2926

Hata K, Futaba DN, Mizuno K, Namai T, Yumura M, Iijima S (2004) Water-assisted highly efficient synthesis of impurity-free single-waited carbon nanotubes. Science 306:1362–1364

Hinrichs KU, Hayes JM, Sylva SP, Brewer PG, DeLong EF (1999) Methane-consuming archaebacteria in marine sediments. Nature 398:802–805

Hoehler TM, Alperin MJ, Albert DB, Martens CS (1994) Field and laboratory studies of methane oxidation in an anoxic marine sediment – evidence for a methanogen-sulfate reducer consortium. Glob Biogeochem Cycles 8:451–463

Howarth RW (1979) Pyrite – its rapid formation in a salt-marsh and its importance in ecosystem metabolism. Science 203:49–51

Jeffrey GA, McMullan RK (1967) Clathrate hydrates progress in inorganic chemistry, vol 8. Wiley, New York, pp 43–108

Jiang GQ, Kennedy MJ, Christie-Blick N (2003) Stable isotopic evidence for methane seeps in Neoproterozoic postglacial cap carbonates. Nature 426:822–826

Jiang GQ, Kennedy MJ, Christie-Blick N, Wu HC, Zhang SH (2006) Stratigraphy, sedimentary structures, and textures of the late Neoproterozoic doushantuo cap carbonate in south China. J Sediment Res 76:978–995

Jørgensen BB (1982) Mineralization of organic-matter in the sea bed – the role of sulfate reduction. Nature 296:643–645

Jørgensen BB (2000) Bacteria and marine biogeochemistry. In: Schulz HD, Zabel M (eds) Marine geochemistry. Springer, New York, pp 173–207

Jørgensen BB, Boetius A (2007) Feast and famine – microbial life in the deep-sea bed. Nat Rev Microbiol 5:770–781

Kasten S, Jørgensen BB (2000) Sulfate reduction in marine sediments. In: Schulz HD, Zabel M (eds) Marine geochemistry. Springer, New York, pp 263–282

Kasten S, Freudenthal T, Gingele FX, Schulz HD (1998) Simultaneous formation of iron-rich layers at different redox boundaries in sediments of the Amazon deep-sea fan. Geochim Cosmochim Acta 62:2253–2264

Katz ME, Cramer BS, Mountain GS, Katz S, Miller KG (2001) Uncorking the bottle: what triggered the Paleocene/Eocene thermal maximum methane release? Paleoceanography 16:549–562

Keir RS, Greinert J, Rhein M, Petrick G, Sultenfuss J, Furhaupter K (2005) Methane and methane carbon isotope ratios in the Northeast Atlantic including the Mid-Atlantic Ridge (50 degrees N). Deep-Sea Res Part I Oceanogr Res Papers 52:1043–1070

Kelley DS, Karson JA, Blackman DK, Fruh-Green GL, Butterfield DA, Lilley MD, Olson EJ, Schrenk MO, Roe KK, Lebon GT, Rivizzigno P (2001) An off-axis hydrothermal vent field near the Mid-Atlantic Ridge at 30 degrees N. Nature 412:145–149

Kelley DS, Karson JA, Fruh-Green GL et al (2005) A erpentinite-hosted ecosystem: the lost city hydrothermal field. Science 307:1428–1434

Knab NJ, Dale AW, Lettmann K, Fossing H, Jørgensen BB (2008) Thermodynamic and kinetic control on anaerobic oxidation of methane in marine sediments. Geochim Cosmochim Acta 72:3746–3757

Kvenvolden KA (1988) Methane hydrate – a major reservoir of carbon in the shallow geosphere. Chem Geol 71:41–51

Kvenvolden KA (1998) A primer on the geological occurrence of gas hydrate. In: Henriet JP, Mienert J (eds) Gas hydrates relevance to world margin stability and climate change. Geological Society, London, pp 9–30

Larrasoana JC, Roberts AP, Musgrave RJ, Gracia E, Pinero E, Vega M, Martinez-Rulz F (2007) Diagenetic formation of greigite and pyrrhotite in gas hydrate marine sedimentary systems. Earth Planet Sci Lett 261:350–366

Levin LA, James DW, Martin CM, Rathburn AE, Harris LH, Michener RH (2000) Do methane seeps support distinct macrofaunal assemblages? Observations on community structure and nutrition from the northern California slope and shelf. Mar Ecol Prog Ser 208:21–39

Lide DR (2004) Handbook of chemistry and physics. CRC Press, Boca Raton, FL

Lippmann F (1973) Sedimentary carbonate minerals. Springer, New York

Lorenson TD, Kvenvolden KA, Hostettler FD, Rosenbauer RJ, Orange DL, Martin JB (2002) Hydrocarbon geochemistry of cold seeps in the Monterey Bay National Marine Sanctuary. Mar Geol 181:285–304

Lovley DR (2000) Fe(III) and Mn(IV) reduction. In: Lovley DR (ed) Environmental microbe-metal interactions. ASM Press, Washington, DC, pp 3–30

Lu H (2007) Mineralogical and geochemical studies on sediments from Dongsha Area, South China Sea: evidences for gas hydrate occurrence. PhD dissertation, Sun Yat-Sen University, Guangzhou, China

Luther GW (1991) Pyrite synthesis via polysulfide compounds. Geochim Cosmochim Acta 55:2839–2849

MacDonald GJ (1990) Role of methane clathrates in past and future climates. Clim Change 16:247–281

Makogon IF, Makogon YF (1997) Hydrates of hydrocarbons. Pennwell Books, Tulsa, Oklahoma, 482 p

Mazzullo SJ (2000) Organogenic dolomitization in peritidal to deep-sea sediments. J Sediment Res 70:10–23

Megonigal JP, Hines ME, Visscher PT (2003) Anaerobic metabolism: linkages to trace gases and aerobic processes. In: Schlesinger WH (ed) Tretise on geochemistry, volume 8: biogeochemistry. Elsevier, New York, pp 317–424

Michaelis W, Seifert R, Nauhaus K et al (2002) Microbial reefs in the Black Sea fueled by anaerobic oxidation of methane. Science 297:1013–1015

Murowchick JB, James B, Barnes HL (1986) Marcasite precipitation from hydrothermal solutions. Geochim Cosmochim Acta 50:2615–2629

Naehr TH, Eichhubl P, Orphan VJ, Hovland M, Paull CK, Ussler W, Lorenson TD, Greene HG (2007) Authigenic carbonate formation at hydrocarbon seeps in continental margin sediments: a comparative study. Deep-Sea Res Part II Top Stud Oceanogr 54:1268–1291

Niewöhner C, Hensen C, Kasten S, Zabel M, Schulz HD (1998) Deep sulfate reduction completely mediated by anaerobic methane oxidation in sediments of the upwelling area off Namibia. Geochim Cosmochim Acta 62:455–464

Orphan VJ, Hinrichs KU, Ussler W, Paull CK, Taylor LT, Sylva SP, Hayes JM, Delong EF (2001a) Comparative analysis of methane-oxidizing archaea and sulfate-reducing bacteria in anoxic marine sediments. Appl Environ Microbiol 67:1922–1934

Orphan VJ, House CH, Hinrichs KU, McKeegan KD, DeLong EF (2001b) Methane-consuming archaea revealed by directly coupled isotopic and phylogenetic analysis. Science 293:484–487

Paull CK, Dillon WP (2001) Natural gas hydrates: occurrence, distribution, and detection. American Geophysical Union, Washington, DC, 315 p

Peckmann J, Reimer A, Luth U, Luth C, Hansen BT, Heinicke C, Hoefs J, Reitner J (2001) Methane-derived carbonates and authigenic pyrite from the northwestern Black Sea. Mar Geol 177:129–150

Pohlman JW, Ruppel C, Hutchinson DR, Downer R, Coffin RB (2008) Assessing sulfate reduction and methane cycling in a high salinity pore water system in the northern Gulf of Mexico. Mar Petrol Geol 25:942–951

Reeburgh WS (2007) Oceanic methane biogeochemistry. Chem Rev 107:486–513

Rickard D (1997) Kinetics of pyrite formation by the H_2S oxidation of iron (II) monosulfide in aqueous solutions between 25 and 125 C: the rate equation. Geochim Cosmochim Acta 61:115–134

Rickard D, Luther GW (1997) H_2S oxidation of iron (II) monosulfide in aqueous solutions between 25 and 125 C: the mechanism. Geochim Cosmochim Acta 61:135–147

Sassen R, MacDonald IR, Guinasso NL, Joye S, Requejo AG, Sweet ST, Alcala-Herrera J, DeFreitas D, Schink DR (1998) Bacterial methane oxidation in sea-floor gas hydrate: significance to life in extreme environments. Geology 26:851–854

Sassen R, Roberts HH, Carney R, Milkov AV, DeFreitas DA, Lanoil B, Zhang CL (2004) Free hydrocarbon gas, gas hydrate, and authigenic minerals in chemosynthetic communities of the northern Gulf of Mexico continental slope: relation to microbial processes. Chem Geol 205:195–217

Schoonen MAA, Barnes HL (1991a) Mechanisms of pyrite and marcasite formation from solution. 3. Hydrothermal processes. Geochim Cosmochim Acta 55:3491–3504

Schoonen MAA, Barnes HL (1991b) Reactions forming pyrite and marcasite from solution. 1. Nucleation of FeS_2 below 100-degrees-C. Geochim Cosmochim Acta 55:1495–1504

Sloan ED (1998) Clathrate hydrates of natural gases. Marcel Dekker, New York, 705 p

Stadnitskaia A, Nadezhkin D, Abbas B, Blinova V, Ivanov MK, Damste JSS (2008) Carbonate formation by anaerobic oxidation of methane: evidence from lipid biomarker and fossil 16S rDNA. Geochim Cosmochim Acta 72:1824–1836

Stakes DS, Orange D, Paduan JB, Salamy KA, Maher N (1999) Cold-seeps and authigenic carbonate formation in Monterey Bay, California. Mar Geol 159:93–109

Stumm W, Morgan JJ (1996) Aquatic chemistry: chemical equilibria in natural waters. Wiley, New York

Suess E, Carson B, Ritger SD, Moore JC, Jones ML, Kulm LD, Cochrane GR (1985) Biological communities at vent sites along the subduction zone off Oregon. In: Jones ML (ed) The hydrothermal vents of the Eastern Pacific: an overview. Bulletin of the Biological Society of Washington, Washington, DC, pp 475–484

Takeuchi R, Matsumoto R, Ogihara S, Machlyama H (2007) Methane-induced dolomite "chimneys" on the Kuroshima Knoll, Ryukyu Islands, Japan. J Geochem Explor 95:16–28

Theberge SM, Luther GW (1997) Determination of the electrochemical properties of a soluble aqueous FeS species present in sulfidic solutions. Aquat Geochem 3:191–211

Trehu AM, Flemings PB, Bangs NL, Chevallier J, Gracia E, Johnson JE, Liu CS, Liu XL, Riedel M, Torres ME (2004) Feeding methane vents and gas hydrate deposits at south Hydrate Ridge. Geophys Resear Lett 31:L23310. doi:10.1029/2004GL021286

Treude T, Boetius A, Knittel K, Wallmann K, Jøorgensen BB (2003) Anaerobic oxidation of methane above gas hydrates at Hydrate Ridge, NE Pacific Ocean. Mar Ecol Prog Ser 264:1–14

Treude T, Orphan V, Knittel K, Gieseke A, House CH, Boetius A (2007) Consumption of methane and CO_2 by methanotrophic microbial mats from gas seeps of the anoxic Black Sea. Appl Environ Microbiol 73:2271–2283

Valentine DL, Reeburgh WS (2000) New perspectives on anaerobic methane oxidation. Environ Microbiol 2:477–484

Whitaker M (1994) Correlation of natural gases with their sources. In: Magoon LB, Dow WG (eds) The petroleum system – from source to trap. American Association of Petroleum Geologists, Tulsa, Oklahoma, pp 261–284

Zachos JC, Rohl U, Schellenberg SA et al (2005) Rapid acidification of the ocean during the Paleocene-Eocene thermal maximum. Science 308:1611–1615

Chapter 16
Bacterial Degradation of Polychlorinated Biphenyls

Martina Mackova, Ondrej Uhlik, Petra Lovecka, Jitka Viktorova,
Martina Novakova, Katerina Demnerova, Michel Sylvestre,
and Tomas Macek

Microbe bioremediation is the application of biological treatment to the cleanup of hazardous contaminants in soil and surface or subsurface waters. Normally microbe bioremediation treats organic contaminants. Most microbe bioremediation processes take advantage of indigenous microorganisms, although some rely on the introduction of bacterial or fungal strains. Bacterial digestion is the process of bacteria-consuming organic matter. The bacteria feed on the contamination, deriving nutrition for growth and reproduction. Undergoing complex chemical reactions, the waste is metabolized into the final metabolic waste products, water and carbon dioxide. This provides the bacteria with the energy they need to live.

Polychlorinated biphenyls (PCB) are synthetic chemical compounds that carry one to ten chlorine atoms on a biphenyl carbon skeleton. There are theoretically 209 possible congeners, and 20–60 congeners are present in commercial products. This versatile composition of PCB, with their high hydrophobicity and chemical stability, makes them persistent in the environment. PCB contamination results from commercial preparations composed of complex mixtures containing 60–90 congeners with commercial names such as Aroclor, Fenclor, Kanechlor, Delor, among others. Studies documenting PCB degradation frequently refer to Aroclor which

T. Macek (✉), M. Mackova, O. Uhlik, J. Viktorova, and M. Novakova
Department of Biochemistry and Microbiology, Faculty of Food and Biochemical Technology,
ICT Prague, Technicka 3, 16628 Prague, Czech Republic;

Institute of Organic Chemistry and Biochemistry CAS, Joint Laboratory of ICT Prague and
IOCB, Flemingovo n. 2, 16610 Prague, Czech Republic
e-mail: tom.macek@uochb.cas.cz

P. Lovecka and K. Demnerova
Department of Biochemistry and Microbiology, Faculty of Food and Biochemical Technology,
ICT Prague, Technicka 3, 16628 Prague, Czech Republic

M. Sylvestre
Institut National de la Recherche Scientifique, INRS-Institut Armand-Frappier, 531, boul. des
Prairies, Laval (Quebec) H7V 1B7, Canada

L.L. Barton et al. (eds.), *Geomicrobiology: Molecular and Environmental Perspective*,
DOI 10.1007/978-90-481-9204-5_16, © Springer Science+Business Media B.V. 2010

was produced in the United States. Commercial preparations of Aroclor are specified with a four-digit code. The first two numbers in the code refer to the parent structure (12 indicating biphenyl) and the second two digits refers to the weight percentage of chlorine. For example, Aroclor 1242, 1248, 1254 and 1260 refer to PCB mixtures with an average weight percentage of chlorine of 42%, 48%, 54% and 60%, respectively. Some studies have utilized single congeners, which will be abbreviated as CBp, DCBp, TCBp, TeCBp, PeCBp, HCBp, HeCBp, OCBp and NCBp for mono-, di-, tri-, tetra-, penta-, hexa-, hepta-, octa- and nonachlorobiphenyl, respectively (Field and Sierra-Alvarez 2008).

Polychlorinated biphenyls are included among the 12 worldwide priority persistent organic pollutants (or POPs) and are ranked fifth on the U.S. Environmental Protection Agency Superfund Priority List of Hazardous Compounds. PCB contaminate the sediments of many lakes, rivers, and harbors where they bioaccumulate and biomagnify in the food chain.

PCB have been used for hundreds of industrial and commercial purposes, such as in electrical applications, heat transfer, and hydraulic equipment; as plasticizers in paints, plastics and rubber products; in pigments, dyes and carbonless copy paper, and many other applications. More than 1.5 million tons of PCB were manufactured worldwide between 1927 and the early 1980s, a significant amount of which has been released to the environment. The first adverse health effects were recorded in the 1930s (Drinker et al. 1937). Since then, PCB have been shown to cause cancer in animals, and also to cause a number of serious effects on the immune, reproductive, nervous and endocrine systems, among others (Aoki 2001; Faroon et al. 2001).

PCB Biodegradation

In spite of the xenobiotic properties of PCB, various PCB-degrading microorganisms have been isolated. Studies have identified two distinct biological processes capable of biotransforming polychlorinated biphenyls: aerobic oxidative processes and anaerobic reductive processes. It is now known that these two complementary activities are occurring naturally in the environment (Barriault and Sylvestre 1993; Harkness et al. 1993; Dercova et al. 2008; Bedard et al. 2006; Mackova et al., 2006, 2009).

Anaerobic Degradation of PCB

Anaerobic reductive dechlorination of PCB is a natural process that provides a means of detoxification and, when coupled with aerobic degradation, completely destroys PCB (Bedard 2008). In anaerobic environments, higher chlorinated biphenyls can undergo reductive dehalogenation. *Meta-* and *para-*chlorines in PCB congeners are more susceptible to dechlorination than *ortho-*chlorines. Anaerobes catalyzing PCB dechlorination for a long time have not been isolated in pure culture but there was strong evidence from enrichment cultures that some *Dehalococcoides* spp. and other microorganisms within the *Chloroflexi* phylum can grow by linking the oxidation of H_2 to the reductive dechlorination of PCB (Field and Sierra-Alvarez 2008). The widespread anaerobic dechlorination of PCB that has been observed in many rivers and marine sediments results in reduction of both the potential risk from and potential exposure to PCB.

In general, this microbial process effects the preferential removal of *meta* and *para* chlorines, resulting in a depletion of highly chlorinated PCB congeners with corresponding increases in lower chlorinated, *ortho-*substituted PCB congeners. The altered congener distribution of residual PCB contamination observed in several aquatic sediments was the earliest evidence of the anaerobic dechlorination of PCB (Brown et al. 1984; Quensen et al. 1988; Häggblom et al. 2003). This same activity has been observed in the laboratory (Quensen et al. 1988, 1990; Abramowicz 1995), where the selective removal of *meta* and *para* chlorines was also noted. The preferential loss of *meta* and *para* chlorines catalyzed by anaerobic dechlorination results in dramatic reductions in the levels of coplanar, dioxin-like PCB congeners in the mixture (Quensen et al. 1990, 1992). These reductions in concentrations correlate with reductions in ethoxyresorufin-O-deethylase (EROD) induction potency and toxic equivalency factors for the mixture. Most importantly, these same extensive reductions are occurring in the environment (Quensen et al. 1992). The reduced carcinogenicity as a result of dechlorination is supported by the recent reanalysis of the original rat cancer studies.

Anaerobic reductive dechlorination involves the removal of the chlorine substituents as halogen ions and their replacement by hydrogen in the form of electrons and protons. Reductive dechlorination is an important process in biodegradation of various halogenated aliphatic and aromatic compounds (for reviews of reductive dechlorination, see Mohn and Tiedje 1992). This reductive dechlorination of highly chlorinated PCB decreases their toxicity and thus increases their degradability, e.g. by converting the 'co-planar', dioxin-like congeners into congeners with fewer chlorines (Mousa et al. 1996; Quensen et al. 1998, 1992). Experimental evidence indicates that environmental factors such as temperature, pH, partial pressure of H_2, and the presence or absence of utilizable carbon sources, electron donors and electron acceptors affect the dechlorination (Abramowicz 1995; Bedard and Quensen 1995). Anaerobic microbial communities in sediments dechlorinate Aroclor at rates of 3 μg Cl/g sediment per week. PCB dechlorination occurs at 12°C, a temperature relevant for remediation at temperate sites, and at concentrations of 100–1,000 ppm. The positions dechlorinated are usually *meta* > *para* > *ortho*. The biphenyl rings, and the mono-*ortho-* and diorthochlorobiphenyls were not degraded after a 1 year incubation. Hence subsequent aerobic treatment may be necessary to meet regulatory standards. The most important limitation to using PCB dechlorination as a remediation technology is the slower than desired dechlorination rates and no means yet discovered to substantially enhance these rates. Long term enrichments using PCB as the only electron acceptor resulted in an initial enhancement in dechlorination rate (Tiedje et al. 1993–1994).

In situ reductive dechlorination has been documented in anaerobic sediments at numerous locations (Table 16.1), and six distinct dechlorination patterns have been observed, giving rise to six recognizable profiles of congeners in the dechlorination products (Bedard and Quensen 1995). Aerobic microbial degradation of PCB by the biphenyl (or bph) pathway is well understood, but in situ degradation by this dioxygenase-based pathway is limited to PCB with three or fewer chlorines (Abramowicz 1995). In contrast, anaerobic microbial reductive dechlorination, preferentially attacks PCB with two to nine chlorines. Reductive dechlorination is the key process for PCB detoxification in aquatic sediments (Quensen et al. 1988; Abramowicz 1995; Bedard 2008). In situ microbial reductive dechlorination of Aroclors was first reported by Brown et al. (1987a, b), and has been documented in many contaminated rivers, harbors, and estuary sediments (Quensen et al. 1988; Abramowicz 1995; Bedard 2008), indicating that the organisms responsible for the dechlorination are widespread (see Table 16.1). Note that these organisms can be detected in a number of PCB-free (uncontaminated) environments upon the addition of PCB in the laboratory. This suggests that PCB-dechlorinating activity may be the result of a common reductive pathway present in many different anaerobic microorganisms located throughout the environment. Support for this hypothesis

Table 16.1 List of sites where PCB-dechlorinating microorganisms have been found in 1995 published by Abramowicz (1995), Dercova et al. (2008)

PCB contaminated sites	
Escambia Bay	Florida
Fox River/Green Bay	Wisconsin
Grass River	New York
Hoosic River	Massachusetts
Housatonic River	Massachusetts
Hudson River	New York
Kalamazoo River	Michigan
Lake Hartwell	South Carolina
Lake Ketelmeer	The Netherlands
Lake Shniji	Japan
Moreau Drag Strip	New York
New Bedford Barbor	Massachusetts
Otonabee River/Rice Lake	Canada
Rhine River	Germany, The Netherlands
Sheboygan River	Wisconsin
Silver Lake	Massachusetts
St. Lawrance River	New York
Waukegan Barbor	Illinois
Woods Pond	Massachusetts
Zemplinska Sirava	Slovakia, former Czechoslovakia
Strazsky kanal	Slovakia, former Czechoslovakia

comes from recent efforts demonstrating that several iron and cobalt heme cofactor systems are capable of reductively dechlorinating a wide variety of chlorinated organic compounds (Krone et al. 1989; Gantzer and Wackett 1991), including PCB. In general, environmental dechlorination is more extensive at higher PCB concentrations consistent with the faster dechlorination rates observed at higher PCB concentrations in the laboratory (Abramowicz et al. 1993).

The involvement of different microorganisms in the reductive dechlorination of PCB was first suggested by enrichment cultures with different behavior derived from the same sediment source. For example, cultures derived from Hudson River sediments after pasteurization were able to catalyze mainly *meta*-dechlorination, while the original untreated sediment culture maintained the capacity to dechlorinate both *meta*- and *para*-chlorine groups (Ye et al. 1992). The result clearly demonstrates the involvement of more than one population of dechlorinating bacteria. Similarly, organisms from Woods Pond sediment showed distinct patterns of dechlorination depending on temperature of cultivation (Wiegel and Wu 2000). Eight dechlorination processes have been identified based on careful examination of congener loss and product accumulation patterns in different sediment samples (Table 16.1) (Wiegel and Wu 2000). The documented patterns only include *meta*- and *para*-dechlorination, reflecting the relatively infrequent observation of *ortho*-dechlorination in many of the initial sediment studies (Brown et al. 1987a; Quensen et al. 1988, 1990; Nies and Vogel 1990; Alder et al. 1993; Ofjord et al. 1994) or later studies with enriched cultures developed from the same sediments (Bedard et al. 2005). Poor *ortho*-dechlorination is also observed in the field (Bedard et al. 2005). However, some studies with either enrichment cultures or sediment samples spiked with a defined congener have revealed many examples where *ortho*-dechlorination has taken place (Van Dort and Bedard 1991; Wu et al. 1997; Quensen et al. 1998; Wiegel and Wu 2000). As was mentioned above microbiologically mediated dechlorination of PCB typically removes *meta* and/or *para* chlorines to generate primarily *ortho*-substituted mono- through tetrachlorobiphenyls, though biphenyl has been observed in dechlorination of 4-4'-dichlorobiphenyl (4-4'-CB) (Mavoungou et al. 1991) 3,4,5-CB (Williams 1994), 3,4-3',4'-CB (Rhee et al. 1993), and 2,3,4,5,6-CB (Field and Sierra-Alvarez 2008).

Brown and co-workers (Brown et al. 1984, 1987a, b; Bedard and Quensen 1995) described various patterns of PCB dechlorination for both environmental and laboratory samples and gave each a letter designation. There are at least eight distinct microbial dechlorination processes in the environment and sediment microcosms known as processes named as M, Q, H', H, P, N, LP and T, see Table 16.2. These can be identified through careful comparison of the patterns of congener loss and product formation (Bedard and Quensen 1995; Wu et al. 1997; Bedard et al. 2005; Van Dort et al. 1997; Quensen et al. 1990; Mavoungou et al. 1991; Bedard 2008) (Table 16.2). Other processes can be explained as combinations of these eight.

Environmental Conditions Affecting Anaerobic Degradation of PCB. Environmental factors and conditions affect the growth and the variety of metabolic activities of different microorganisms differently and hence infuence divergently

Table 16.2 Positions of chlorines removed by each dechlorination process

Dechlorination process	Susceptible chlorines
M	Flanked and unflanked *meta*
Q	Flanked and unflanked *para, meta* of 2,3-group
H'	Flanked *para, meta* of 2,3- and 2,3,4- groups
	Susceptible chlorines
H	Flanked *para*, doubly flanked *meta*
P	Flanked *para*
N	Flanked *meta*
LP	Flanked and unflanked *para*
T	Flanked *meta* of 2,3,4,5-group, in hepta- and octachlorobiphenyls

Modified table from cited reference (Wiegel and Wu 2000)

the extent and rate of the various PCB-dechlorinating activities. Consequently, a better knowledge of whether and to what extent individual environmental factors can infuence PCB dechlorination is important (Wiegel and Wu 2000).

An important factor in the biotransformation and biodegradation plays the bioavailability of PCB, i.e. the adsorption and desorption processes. The rates of desorption are different for the different PCB congeners; the most hydrophobic congeners (with high Kow values) behaved roughly as predicted from models, whereas the less hydrophobic ones with low Kow values desorbed up to four magnitudes slower than models predicted. These problems also includes bioavailability connected with the process of aging of PCB contaminations, which leads over time to stronger adsorption and lower recovery rates of PCB.

Temperature has a signifficant effect on the growth and the physiological activity including uptake and enzymatic dehalogenation of PCB congeners. Effects include changes in the adsorption and desorption kinetics of PCB from soil particles (Jota and Hasset 1991) and thus both the hydrolytic (abiotic) dehalogenation and the availability of PCB for microbial transformations. However, these effects are probably minor in comparison with the effect of temperature on the growth of microorganisms and the catalytic activity of enzymes. In many environments, the temperature fluctuates not only on a seasonal basis, but also from day to night and as a result of strong rain or hot spells, although diurnal fluctuations are minimal in most water sediments covered with several feet of water. Presently, it is unknown how much or whether at all seasonal temperature changes the PCB dechlorination pathways after they have been induced in a given sediment. Wu et al. (2002) describe simulation of seasonal temperature changes to some extent by introducing temperature downshifts and upshifts and investigation of the effects of these temperature shifts on the dechlorination of 2,3,4,6-CB added as a primer and of residual PCB in Woods Pond sediment. These experiments showed that the temperature at the time of priming (initiation) of PCB dechlorination and not the subsequent incubation temperature was the primary determinant of which of the dechlorination processes occurred. Furthermore, these dechlorination processes were more or less retained during subsequent up- and downshifts of the incubation temperature. Changes in the incubation temperature, however, strongly influenced the rate and extent of dehalogenation. This phenomenon fully corresponded with the hypothesis that the observed dechlorination patterns were due to different populations which are established in response to the different temperatures at the time of priming and which change only slowly with subsequent changes in the incubation temperatures.

Other environmental factor influencing anaerobic PCB degradation is pH. Sediments are frequently well-buffered systems, but in contrast to strictly aerobic processes, anaerobic microbial processes may lead to an increase in acidic fermentation products and thus cause local changes in the pH. Furthermore, pH affects the equilibrium between PCB that are dissolved and those that are adsorbed to organic matter and thus influences the bioavailability of PCB in soil. The optimal pH for overall removal of chlorines was around 7.0–7.5. However, the stereospecificity of the dechlorination varied: e.g. for 2,3,4,6-CB, flanked *meta* dechlorination occurred at pH 5.0–8.0, unflanked *para* dechlorination at pH 6.0–8.0 and *ortho* dechlorination at pH 6.0–7.5. However, at pH 7.0 and 15°C, *ortho* dechlorination dominated whereas at 18°C and 25°C, unflanked *para* dehalogenation outpaced the other dehalogenation reactions (Wiegel and Wu 2000). These results indicate that the pH strongly influences not only the rate and extent of dechlorination but also the route of dechlorination.

Other factor studied in few cases was the addition of different carbon sources when a specific carbon source could enhance dechlorination. It was demonstrated that addition of fatty acids in carbon-limited sediments (Alder et al. 1993), also addition of pyruvate and malate increased the extent of PCB dechlorination in Aroclor 1242 (Klasson et al. 1996). However in further studies it was shown that the effect of malate was highly dependent on pH and temperature (Wu et al. 1999). In enrichment cultures addition of glucose, methanol, acetate or acetone enhanced the rate and extent of dechlorination (Nies and Vogel 1990). In other study the addition of fumarate, an electron acceptor potentially competing with PCB, also increased the extent of dehalogenation. These studies have led to the concept of priming the growth of reductive PCB dechlorinating microorganisms also with simple PCB congeners (Bedard et al. 1998; Klasson et al. 1996) or alternative substrates such as halogenated benzoates or chlorophenols (Deweerd and Bedard 1999) or bromobiphenyls (Bedard et al. 1998; Wu et al. 1999). Priming is based on the hypothesis that high concentrations of an appropriate substrate which is readily susceptible to microbial dehalogenation will promote the growth of microorganisms capable of utilizing analogous PCB compounds as electron acceptors. Other factor is the presence of an electron acceptor. Several investigations of PCB dechlorination were proceeded in presence of common electron acceptors for anaerobic microbes under methanogenic, sulfidogenic, iron (III) reducing and denitrifying conditions.

Various bacterial groups contain members capable of respiring halogenated aromatic compounds in a process known as (de)halorespiration (Adrian et al. 2009). Diversity of dehalorespiring bacteria was recently described by Hiarishi (2008). PCB-dechlorinating microorganisms are difficult to isolate and are generally a small portion of the total microbial community in natural sediments (Watts et al. 2001). The present knowledge of organisms involved in anaerobic PCB conversion has been obtained from either molecular ecology studies applied to PCB-dechlorinating enrichment cultures or from the use of pure cultures of halorespiring bacteria. Denaturing gradient gel electrophoresis (DGGE) analysis of 16S rRNA genes was utilized to identify microorganisms in an enrichment culture catalyzing the *ortho*-dechlorination of 2,3,5,6-TeCBp to 2,3,5-TriCBp and 3,5-DiCBp with acetate as electron donor (Cutter et al. 2001). One of the predominant bands in the DGGE gel was highly dependent on the presence of the PCB and was assigned to bacterium *o*-17, the first identified PCB-dechlorinating anaerobe. In a later study, bacterium *o*-17 was shown to dechlorinate eight PCB congeners including single-flanked *ortho*-PCB chlorines; however, double-flanked chlorines of PCB were preferentially dechlorinated (May et al. 2006). Nevertheless, the ability to dechlorinate more extensively chlorinated congeners was limited and some PCB congeners were shown to inhibit the process. Bacterium *o*-17 belongs to a deep branch of the phylum *Chloroflexi* and its 16S rRNA gene sequence is approximately 90% identical to that of the well known halorespiring bacterium, *Dehalococcoides ethenogens* (Cutter et al. 2001). Fennell and coworkers (2004) reported that another species within the *Chloroflexi*, *Dehalococcoides ethenogenes* 195, dechlorinated the PCB 2,3,4,5,6-pentachlorobiphenyl and other aromatic organochlorines when grown with perchloroethene. This microorganism was the first species in the *Dehalococcoides* group to be isolated and described. Although other *Dehalococcoides* spp. including strains VS (Cupples et al. 2003), FL2 (Löffler et al. 2002), BAV1 (He et al. 2003), CBDB1 (Bunge et al. 2003), and KB-1/VC-H$_2$ (Duhamel et al. 2004) use chlorinated ethenes and other chlorinated compounds as electron acceptors, no other species have been reported to reductively dechlorinate PCB. Considering that the isolated *Dehalococcoides* strains 195, FL2, BAV1, and CBDB1 contain 18, 14, 7 and 32 non-identical reductive dehalogenase genes, respectively (Holscher et al. 2004), there is a significant lack of knowledge on substrate specificity and distribution of such dehalogenases genes. *D. ethenogenes* sp. strain 195 has been shown to catalyze the reductive dechlorination of 2,3,4,5,6-PeCBp to 2,3,4,6-/2,3,5,6-TeCBp and 2,4,6-TCBp while utilizing perchloroethene (PCE) as its primary electron acceptor (Fennell et al. 2004), but apparently it cannot dechlorinate commercial PCB mixtures. Double-flanked chlorines could be dechlorinated by *D. ethenogenes* 195, as in the case of strain DF-1. *Dehalococcoides* have also been implicated in the removal of double-flanked *meta* and *para* chlorines from 2,3,4,5-tetrachlorobiphenyl to form 2,3,5-trichloro- and 2,4,5-trichlorobiphenyl, respectively, in different enrichment cultures (Yan et al. 2006a).

DGGE was also utilized to identify the PCB-dechlorinating organisms in a highly enriched culture capable of *para*-dechlorination of flanked chlorines from 2,3,4,5-TeCBp (Wu et al. 2002). The band associated with the presence of 2,3,4,5-TeCBp was sequenced and the corresponding dechlorinating microorganism, designated DF-1, was found to have a high sequence similarity to the *ortho*-PCB-dechlorinating bacterium *o*-17 (89% identity for 16S rRNA sequences) and thus it was described as *Dehalococcoides* as well.

Dehalococcoides strain CBDB1 was isolated for its ability to grow using chlorinated benzenes as electron acceptors (Adrian et al. 2000). This strain was particularly noted for its ability to dechlorinate chlorinated aromatics and dehalogenates all highly chlorinated benzenes, chlorinated phenols, and several chlorinated dioxins. Strain CBDB1 has been studied biochemically and the complete genome has been sequenced and annotated (Adrian et al. 2009). Congener-specific high-resolution gas chromatography revealed that CBDB1 extensively dechlorinated both Aroclor 1248 and Aroclor 1260 after 4 months of incubation (Adrian et al. 2009). For example, 16 congeners comprising 67.3% of the total PCB in Aroclor 1260 were decreased by 64%. From 43 different PCB congeners the most prominent dechlorination products were 2,3',5-chlorinated biphenyl (2,5-3'-CB) and 2,4-3'-CB from Aroclor 1248 and 2,3,5-2'5'-CB, 2,5-2'5'-CB, 2,4-2'5'-CB, and 2,3,5-2'3'6-CB from Aroclor 1260. Strain CBDB1 removed flanked *para* chlorines from 3,4-, 2,4,5-, and 3,4,5-chlorophenyl rings, primarily *para* chlorines from 2,3,4,5-chlorophenyl rings, primarily *meta* chlorines from 2,3,4- and 2,3,4,6-chlorophenyl rings, and either *meta* or *para* chlorines from 2,3,4,5,6-chlorophenyl rings. The site of attack on the 2,3,4-chorophenyl ring was heavily influenced by the chlorine configuration on the opposite ring and matched PCB process H dechlorination.

16S rRNA gene sequences closely related to the halorespiring bacterial group, *Dehalobacter*, were found in addition to those of *Dehalococcoides* in enrichment cultures that dechlorinated 2,3,4,5-TeCBp (Yan et al. 2006b). Analysis of an enrichment culture capable of *meta*- and *para*-dechlorination of PCB revealed the presence of nine *Clostridium* strains but did not provide proof that these bacteria were responsible for dechlorination (Hou and Dutta 2000). The occurrence of *Clostridium* species, which are spore-formers, would be consistent with the survival of PCB dechlorinating activity after pasteurization (Ye et al. 1992).

The history progresses from the characterization of the PCB in the sediment, to cultivation in sediment microcosms, to the identification of four distinct types of PCB dechlorination, to a successfull field test, to the cultivation in defined medium of the organisms responsible for extensive dechlorination of Aroclor 1260, and finally to the identification of a *Dehalococcoides* population that links its growth to the dechlorination of Aroclor 1260.

Aerobic Degradation of PCB

Aerobic degradation of PCB occurs by co-metabolism of biphenyl, which is not so alien to nature. Lunt and Evans (1970) were the first to isolate microorganisms capable of growing on biphenyl as a sole source of carbon and energy and since then several biphenyl-degrading organisms have been isolated. Ahmed and Focht (1973a, b) were the first to describe that biphenyl-degrading organisms have the capacity to transform several PCB congeners.

The microbial degradation of PCB is regarded as one of the most cost-effective and energy-efficient methods to remove PCB from the environment. Many isolates have already been reported, including Gram-negative strains, such as *Pseudomonas, Alcaligenes, Achromobacter, Janibacter, Burkholderia, Acinetobacter, Comamonas,*

Sphingomonas, and *Ralstonia, Enterobacter* and Gram-positive strains, such as *Corynebacterium, Rhodococcus, Bacillus, Paenibacillus,* and also *Arthrobacter* and *Micrococcus* (Bedard et al. 1986; Bopp 1986; Erickson and Mondello 1993; Ling-Yun et al. 2008). However, most of these strains are only capable of degrading PCB congeners that have five or fewer chlorines. Moreover, only a small number of strains have the ability to transform highly chlorinated congeners (five or more chlorines) and recalcitrant coplanar PCB, yet their abilities are still limited.

Figure 16.1 shows the upper biphenyl degradation pathway that is conserved among most degraders. In this pathway, the conversion of biphenyl to a tricarboxylic acid cycle intermediate and benzoate is catalyzed by a series of enzymes: BphA, BphB, BphC, BphD, BphE, BphF, and BphG. BphA is a biphenyl dioxygenase composed of four subunits, BphA1-A4, and converts biphenyl to dihydrodiol. Biphenyl 2,3-dioxygenases (BphA) usually belong to the toluene/biphenyl branch of Rieske non-heme iron oxygenases (Gibson and Parales 2000; Pieper and Seeger 2008) where a ferredoxin and a ferredoxin reductase act as an electron transport system to transfer electrons from NADH to the terminal oxygenase. The dioxygenation regiospecificity determines the sites of attack by the subsequent metabolic pathway while, on the other hand, their specificity determines the spectrum of PCB congeners that can be transformed. Rieske non-heme iron oxygenases are multicomponent enzyme complexes composed of a terminal oxygenase component (iron-sulfur protein [ISP]) and different electron transport proteins (a ferredoxin and a reductase or a combined ferredoxin-NADH-reductase) (Butler and Mason 1997). The catalytic iron-sulfur proteins are heteromultimers, containing a Rieske-type [2Fe-2S] cluster, a mononuclear non-heme iron oxygen activation center, and a substrate-binding site (Butler and Mason 1997; Furusawa et al. 2004) which is responsible for substrate specificity (Gibson and Parales 2000) (Fig. 16.1).

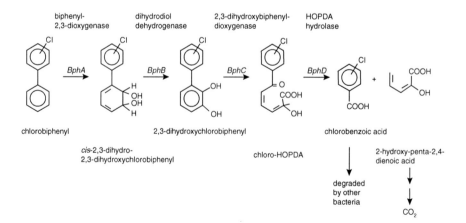

Fig. 16.1 Pathway of bacterial PCB degradation at aerobic conditions (Pieper 2005)

As an example, the biphenyl 2,3-dioxygenases of strains LB400 and KF707 differ in only 20 amino acid residues; however, the two enzymes differ dramatically in substrate specificity and regioselectivity. The LB400 dioxygenase shows broad substrate specificity and transforms PCB with up to six chlorine atoms (Mondello 1989). Congeners with chlorines at the 2- and 5-position on the ring were readily attacked, whereas activity was poor with congeners with chlorines at both *para*-positions (Mondello 1989). 2,5,2′,5′-Tetrachlorobiphenyl was subject to 3,4-dioxygenation (Haddock et al. 1995). A detailed analysis of the dioxygenation products of several di- to pentachlorinated biphenyls formed by this enzyme revealed a complex dependence of the regiospecificity and the yield of dioxygenation on the substitution patterns of both the oxidized and the nonoxidized rings (Seeger et al. 1999). The enzyme of strain KF707 has a much more restricted substrate range and is not capable of *ortho*-dechlorination, but catalyzes 5,6-dioxygenation of 2,2′-dichlorobiphenyl with low activity (Suenaga et al. 2002). However, it is superior in its activity in transforming 4,4′- dichlorobiphenyl (Erickson and Mondello 1993).

BphB is a dehydrogenase which converts dihydrodiol to 2,3-dihydroxybiphenyl. BphC is a ring-cleavage dioxygenase that converts 2,3-dihydroxybiphenyl to 2-hydroxy-6-oxo-6-phenyhexa-2,4-dienoic acid (HOPDA). BphD is a hydrolase and cleaves HOPDA into two compounds, benzoate and 2-hydroxypenta-2,4-dienoate. The latter compound is further converted to acetyl-CoA and pyruvate by 2-hydroxypenta-2,4-dienoate hydratase (BphE), 4-hydroxy-2-oxovalerate aldolase (BphF), and acetaldehyde dehydrogenase (BphG). All the Bph enzymes also catalyze the degradation of PCB congeners. However, the efficiency and reaction intermediates are different from those for biphenyl degradation and depend on the number and position of the chlorine atoms (Furukawa 1994; Furukawa and Fujihara 2008). On the basis of different studies, the following general concepts were drawn concerning the biodegradability of PCB (Bedard et al. 1986, 1987; Furukawa 1994; Pieper 2005; Pieper and Seeger 2008; Furukawa and Fujihara 2008):

1. Biodegradability decreases with increased number of chlorines.
2. PCB congeners with chlorines on only one biphenyl ring are degraded more easily than those with chlorines on both rings.
3. PCB congeners with chlorine at position 2,6- or 2,2′-(double *ortho*-substituted congeners) are poorly degraded.
4. Both the relative rates of PCB depletion and the choice of ring attacked depend on the bacterial strains.

For some PCB congeners, the reaction schemes of the initial dioxygenation and intermediate metabolites of PCB, such as chloro-substituted HOPDA or chlorobenzoate, have been identified (Seeger et al. 1995, 1999). Genes encoding enzymes of the biphenyl upper pathway (termed *bph*) were first cloned from *Pseudomonas pseudoalcaligenes* KF707 (Furukawa and Miyazaki 1986) and later from *Burkholderia* sp. LB400 (now *Burkholderia xenovorans* LB400) (Mondello 1989). A second type of *bph* gene cluster was observed in *Achromobacter georgiopolitanum* (formerly *Pseudomonas* sp.) strain KKS102 (Kimbara et al. 1989; Kikuchi et al. 1994) and *Wautersia oxalatica* A5 (Springael et al. 1993; Mouz et al. 1999).

In these gene clusters, the order of *bph* genes differs significantly from that found in LB400. A third type of gene cluster was observed in biphenyl-degrading *Rhodococcus* strains M5 (Péloquin and Greer 1993), RHA1 (Masai et al. 1995) and TA421 (Arai et al. 1998).

Most of the degrading bacteria only cometabolize chlorinated congeners, and thus different dead-end metabolites accumulate (Furukawa et al. 1979; Bedard and Haberl 1990), including dihydrodiols (Brühlmann and Chen 1999), dihydroxybiphenyls (including 3,4-dihydroxylated derivatives) (Triska et al. 2004), or chlorinated HOPDAs (Furukawa et al. 1979; Seeger et al. 1997). It was recently shown that dihydrodiols and dihydroxybiphenyls are very toxic metabolites for bacteria, affecting cell viability much more than (chloro)biphenyls (Cámara et al. 2004). It will be of critical importance to overcome dead-end reactions in the catabolic process to avoid accumulation of toxic metabolites. However, as different pathway bottlenecks for different PCB congeners operate in different organisms, it is expected that only diverse communities with complementary activities will be able to deal with complex PCB mixtures.

Some bacteria have the ability to grow by utilizing PCB congeners with one or two chlorines as sole sources of carbon and energy. Examples of growth on 4-chlorobiphenyl (4-CBp) are common (Furukawa et al. 1978; Furukawa and Miyazaki 1986; Barton and Crawford 1988; Ahmad et al. 1990); whereas examples on 2-CBp and 3-CBp are less frequent (Bedard et al. 1987; Hickey et al. 1992). *Burkholderia cepacia* P166 (formerly *Pseudomonas cepacia*) utilizes 4-CBp, 2-CBp and 3-CBp as growth substrates (Arensdorf and Focht 1994). However, growth on 2-CBp and 3-CBp is restricted due to accumulation of toxic intermediates. Unrestricted growth on 4-CBp is attributable to the ability of the strains to readily mineralize the 4-chlorobenzoate intermediate without formation of any toxic intermediates.

Some other PCB-degrading strains exhibiting novel capabilities have been isolated. A bacterial isolate, tentatively named SK-3, was able to grow on 2-chlorobiphenyl (2-CB), 3-chlorobiphenyl (3-CB), 4-chlorobiphenyl (4-CB), chloroacetate and 4-chlorobenzoic acid as sole carbon and energy sources (Kim and Picardal 2001). During the growth on 2,2'-dichlorobiphenyl (2,2'-CB) and 2,4'-dichlorobiphenyl (2,4'-CB), strain SK-4 produced stoichiometric amounts of 2-chlorobenzoate and 4-chlorobenzoate, respectively.

In recent years, exceptional bacterial strains have been discovered. A psychrotrophic bacterium, *Hydrogenophaga taeniospiralis* IA3-A, was isolated that co-metabolizes DCBp congeners and lower chlorinated PCB congeners in Aroclor 1221 at 5°C (Lambo and Patel 2006). A bacterium isolated from soil, *Paenibacillus* sp. KBC101, was shown to degrade highly chlorinated congeners; including NCBp congeners (Sakai et al. 2005). A marine bacterium lacking biphenyl dioxygenase genes, *Pseudomonas* CH07, was also isolated that degraded higher PCB congeners including HeCBp (De et al. 2006). A Gram-negative bacterium, named LY402, was isolated from contaminated soil. It was determined as *Enterobacter*. Degradation experiments showed that it had the ability to aerobically transform 79 of the 91 major congeners of Aroclor 1242, 1254, and 1260. Almost all the tri- and tetra-chlorobiphenyls (CBs), except for 3,4,3',4'-CB, were degraded in 3 days, whereas 73% of 3,4,3',4'-, 92% of the penta-, 76% of the hexa-, and 37% of the hepta-CBs were transformed after

6 days. In addition, among 12 octa-CBs, 2,2′,3,3′,5,5′,6,6′-CB was obviously degraded, and 2,2′,3,3′,4,5,6,6′- and 2,2′,3,3′,4,5,5′,6′-CB were slightly transformed.

Microbial degradation of PCB in the environment is influenced by various biological, chemical and physical factors as well as the survival of microorganisms in cases where bioaugmentation is the application of choice (Blumenroth and Wagner-Dobler 1998; Barriault et al. 1999), and by the fact that higher chlorinated PCB provide the organisms neither with energy nor with carbon.

It was speculated for successful bioremediation of PCB-contaminated soil that certain plant-derived compounds, such as flavonoids, may serve as growth substrates for PCB-degrading bacteria and may induce *bph* genes. Donelly et al. (1994) showed a high PCB turnover capacity of biphenyl-degrading organisms after the growth with certain flavonoids, suggesting that plant roots could serve as a natural injection system capable of inducing PCB degradation in indigenous microorganisms over long periods. Focht (1995) proposed that plant terpenes might stimulate PCB oxidation. Gilbert and Crowley 1997 used several terpenoid compounds, carvone, limonene and p-cymene to induce PCB transformation activity in a biphenyl-degrading *Arthrobacter* strain thus explaining why growth in a medium containing *Mentha spicata* (spearmint containing significant amounts of carvone) resulted in enhanced PCB metabolizing activity of that strain. Repeated application (Gilbert and Crowley 1998) of the carvone-induced bacterial strain *Arthrobacter* sp. B1B in soil enabled 27% degradation of PCB, while addition of carvone alone enabled 10% degradation of PCB. Surfactants were tested in combination with bioaugmentation for the aerobic bioremediation of an Aroclor 1242-polluted soil (Singer et al. 2000). The surfactant sorbitan trioleate was added with or without repeated application of the bacterial strains *Arthrobacter* sp. B1B and *Ralstonia eutropha* H850, which were induced by carvone and salicylic acid, respectively. Interestingly, the addition of orange peel, ivy leaves, pine needles or eucalyptus leaves containing terpenes to PCB-contaminated soil resulted in an enhanced PCB transformation activity and significantly enhanced levels of biphenyl degraders (Hernandez et al. 1997). Ionescu et al. (2009) and Mackova et al. (2009) confirmed that certain plant species may play an important role in direct or indirect removal of PCB. The findings of these studies strongly suggest that certain plant species are associated with increased numbers of PCB-metabolizing bacteria in the root zone, including organisms with broad congener specificities, following even short-term growth at a contaminated site. *Salix caprea* and *Armoracia rusticana* were identified as rhizoremediation candidates because of their association with increased numbers of active PCB degraders. Plants not only do accumulate and transform xenobiotics (Rezek et al. 2007, 2008), including PCB, but they do play an important role in supporting the nutrients and increasing of bioavailability of pollutants for rhizosphere microorganisms or in soil close to the roots of plants (Leigh et al. 2006, 2007).

As the majority of the microbial population found in the soil is associated with plant roots, the rhizosphere is the ideal site in which to modify microbial populations. *Pseudomonas fluorescens* F113, an important biocontrol strain for sugar beet, was chosen as a host strain for *bph* genes (Brazil et al. 1995). The modified organism colonized roots as effectively as the wild-type and *bph* genes were expressed in situ, indicating considerable potential for the manipulation of the rhizosphere as a useful strategy for bioremediation. Besides that, bacterial genes responsible for degradation have been successfully introduced into plants (Sylvestre et al. 2009).

The main strategy towards promoting the aerobic degradation of PCB has been through the addition of oxygen, co-substrates, surfactants, inducers and in some cases bioaugmentation of PCB-degrading bacteria (Abraham et al. 2002; Ohtsubo et al. 2004). Biphenyl, which is an important primary substrate supporting PCB co-metabolism, has successfully been utilized to stimulate aerobic degradation of PCB contaminated soil and sediments (Harkness et al. 1993). Brunner et al. (1985) studied the impact of adding biphenyl and a PCB-degrading *Acinetobacter* strain to soil on the mineralization of [^{14}C]Aroclor 1242. Biphenyl supplementation permitted up to 27% conversion of the label to $^{14}CO_2$ in 62 days, while treatments lacking biphenyl were mineralized by less than 1%. A combined treatment, containing O_2, biphenyl and inorganic nutrients, enhanced PCB biodegradation by 37–55% in 73 days. Bioaugmentation with the PCB-degrading strain *Alcaligenes eutrophus* H850 (*Ralstonia eutropha*) did not help, indicating that the natural microflora was sufficient to catalyze the aerobic degradation of the PCB. In a similar study, Anid et al. (1993) utilized H_2O_2 to promote the aerobic degradation of PCB previously dechlorinated under anaerobic conditions. PCB removals ranged from 55% to 59% with the combined bioaugmentation surfactant treatment or 30–36% with surfactant treatment only that relied on the natural microflora. Cyclodextrins, which are alternative compounds to aid in the solubilization of apolar compounds, also stimulate the aerobic degradation of PCB in soil (Fava et al. 1998). Up to 40% loss of PCB was observed during composting of contaminated soil mixed with yard trimmings (Michel et al. 2001).

PCB have also been successfully degraded in an aerobic packed-bed bioreactor (Fava et al. 1996). Tucker et al. (1975) studied the degradation of various Aroclor preparations in a laboratory activated sludge plant. The biodegradability was inversely correlated with the degree of chlorination (Borja et al. 2006).

Concluding Remarks

A substantial large body of information on bacterial potential to degrade PCB has been gathered together – especially during last 2 decades. This chapter summarizes experience of long-term research in many world laboratories which can help to speed up removal of organic pollutants from the environment. Efforts are being made to get better molecular insight into mobilization, uptake translocation and detoxication of PCB in bacteria including detailed structure of microbial communities in contaminated soil (material), their functional and taxonomic diversity and their changes in dependence to different environmental stimuli. New approaches should be conceived that will allow to better understand interactions of organisms in real environment and their impact. One of the most challenging tasks is exploitation of metagenomics and molecular biology methods (Uhlik et al. 2009a, b) which can reveal true composition of metabolising species within a particular environment and complement the exploited approaches.

Acknowledgements The authors thank to GACR 525/09/1058 and MSMT NPVII 2B06156, 2B08031, MSM 6046137305, Z 40550506, TANDEM FT-TA4/101.

References

Abraham WR, Nogales B, Golishin PN et al (2002) PCB-degrading communities in soils and sediments. Curr Opin Biotechnol 5:246–253

Abramowicz DA (1995) Aerobic and anaerobic PCB biodegradation in the environment. Environ Health Perspect 103(Suppl 5):97–99

Abramowicz DA, Brennan MJ, Van Dort HM et al (1993) Factors influencing the rate of PCB dechlorination in Hudson River sediments. Environ Sci Technol 27:1125–1131

Adrian L, Szewzyk U, Wecke J et al (2000) Bacterial dehalorespiration with 581 chlorinated benzenes. Nature 408:580–583

Adrian L, Dudkova V, Demnerova K et al (2009) "Dehalococcoides" sp. strain CBDB1 extensively dechlorinates the commercial PCB mixture Aroclor 1260. Appl Environ Microbiol 75:4516–4524

Ahmad D, Masse R, Sylvestre M (1990) Cloning and expression of genes involved in 4-chlorobiphenyl transformation by *Pseudomonas testosteroni* – homology to polychlorobiphenyl-degrading genes in other bacteria. Gene 86:53–61

Ahmed M, Focht DD (1973a) Degradation of PCB by two species of *Achromobacter*. Can J Microbiol 19:47–52

Ahmed M, Focht DD (1973b) Oxidation of PCB by *Achromobacter* pCB. Bull Environ Contam Toxicol 10:70–72

Alder AC, Haggblom MM, Oppenheimer S et al (1993) Reductive dechlorination of polychlorinated biphenyls in anaerobic sediments. Environ Sci Technol 27:530–538

Anid PJ, Alvarez PJJ, Vogel TM (1993) Biodegradation of monoaromatic hydrocarbons in aquifer columns amended with hydrogen peroxide and nitrate. Water Res 27:685–691

Aoki Y (2001) Polychlorinated biphenyls, polychlorinated dibenzo-*p*-dioxins, and polychlorinated dibenzofurans as endocrine disrupters–what we have learned from Yusho disease. Environ Res 86:2–11

Arai H, Kosono S, Taguchi K et al (1998) Two sets of biphenyl and PCB degradation genes on a linear plasmid in *Rhodococcus erythropolis* TA421. J Ferment Bioeng 86:595–599

Arensdorf JJ, Focht DD (1994) Formation of chlorocatechol *meta* cleavage products by a pseudomonad during metabolism of monochlorobiphenyls. Appl Environ Microbiol 60:2884–2889

Barriault D, Sylvestre M (1993) Factors affecting PCB degradation by an implanted bacterial strain in soil microcosms. Can J Microbiol 39:594–602

Barriault D, Vedadi M, Powlowski J et al (1999) Cis-2, 3-dihydro-2, 3-dihydroxybiphenyl dehydrogenase and cis-1, 2-dihydro-1, 2-dihydroxynaphathalene dehydrogenase catalyze dehydrogenation of the same range of substrates. Biochem Biophys Res Comm 260:181–187

Barton MR, Crawford RL (1988) Novel biotransformations of 4-chlorobiphenyl by a *Pseudomonas* sp. Appl Environ Microbiol 54:594–595

Bedard DL (2008) A case study for microbial biodegradation: anaerobic bacterial reductive dechlorination of polychlorinated biphenyls-from sediment to defined medium. Annu Rev Microbiol 62:253–70

Bedard DL, Haberl ML (1990) Influence of chlorine substitution pattern on the degradation of polychlorinated biphenyls by eight bacterial strains. Microbiol Ecol 20:87–102

Bedard DL, Quensen JF (1995) Microbial reductive dechlorination of polychlorinated biphenyls. In: Young LY, Cerniglia C (eds) Microbial transformation and degradation of toxic organic chemicals. Wiley-Liss Division/Wiley, New York, pp 127–216

Bedard DL, Unterman R, Bopp LH et al (1986) Rapid assay for screening and characterizing microorganisms for the ability to degrade polychlorinated biphenyls. Appl Environ Microbiol 51:761–768

Bedard DL, Wagner RE, Brennan MJ et al (1987) Extensive degradation of aroclors and environmentally transformed polychlorinated-biphenyls by *Alcaligenes eutrophus* H850. Appl Environ Microbiol 53:1094–1102

Bedard DL, Van Dort H, Deweerd KA (1998) Brominated biphenyls prime extensive microbial reductive dehalogenation of Aroclor 1260 in Housatonic River sediment. Appl Environ Microbiol 64:1786–1795

Bedard DL, Pohl EA, Bailey JJ et al (2005) Characterization of the PCB substrate range of microbial dechlorination process LP. Environ Sci Technol 39:6831–6838

Bedard DL, Bailey JJ, Reiss BL, Jerzak GV (2006) Development and characterization of stable sediment-free anaerobic bacterial enrichment cultures that dechlorinate Aroclor 1260. Appl Environ Microbiol 72:2460–2470

Blumenroth P, Wagner-Dobler I (1998) Survival of inoculants in polluted sediments: effect of strain origin and carbon composition. Microbiol Ecol 35:279–288

Bopp LH (1986) Degradation of highly chlorinated PCBs by *Pseudomonas* strain LB400. J Ind Microbiol Biotechnol 1:23–29

Borja JQ, Auresenia JL, Gallardo SM (2006) Biodegradation of polychlorinated biphenyls using biofilm grown with biphenyl as carbon source in fluidized bed reactor. Chemosphere 6:555–559

Brazil GM, Kenefick L, Callanan M et al (1995) Construction of a rhizosphere pseudomonad with potential to degrade polychlorinated-biphenyls and detection of *bph* gene-expression in the rhizosphere. Appl Environ Microbiol 61:1946–1952

Brown JF, Wagner RE, Bedard DL et al (1984) PCB transformations in upper Hudson sediments. Northeast Environ Sci 3:167–179

Brown JF, Bedard DL, Brennan MJ et al (1987a) PCB dechlorination in aquatic sediments. Science 236:709–712

Brown JF, Wagner RE, Feng H et al (1987b) Environmental dechlorination of PCBs. Environ Toxicol Chem 6:579–593

Brühlmann F, Chen W (1999) Tuning biphenyl dioxygenase for extended substrate specificity. Biotechnol Bioeng 63:544–551

Brunner W, Sutherland FH, Focht DD (1985) Enhanced biodegradation of polychlorinated-biphenyls in soil by analog enrichment and bacterial inoculation. J Environ Qual 14:324–328

Bunge M, Adrian L, Kraus A et al (2003) Reductive dehalogenation of chlorinated dioxins by an anaerobic bacterium. Nature 421:357–360

Butler CS, Mason J (1997) Structure-function analysisof the bacterial aromatic ring-hydroxylating dioxygenases. Adv Microb Physiol 38:47–84

Cámara B, Herrera C, González M et al (2004) From PCBs to highly toxic metabolites by the biphenyl pathway. Environ Microbiol 6:842–850

Cupples AM, Spormann AM, McCarty PL (2003) Growth of *Dehalococcoides*-like microorganism on vinyl chloride and *cis*-dichloroethene as electron acceptors as determined by competitive PCR. Appl Environ Microbiol 69:953–959

Cutter LA, Watts JEM, Sowers KR et al (2001) Identification of a microorganism that links its growth to the reductive dechlorination of 2,3,5,6-chlorobiphenyl. Environ Microbiol 3:699–709

De J, Ramaiah N, Sarkar A (2006) Aerobic degradation of highly chlorinated polychlorobiphenyls by a marine bacterium, *Pseudomonas* CH07. World J Microbiol Biotechnol 22:1321–1327

Dercova K, Cicmanova J, Lovecka P et al (2008) Isolation and identification of PCB-degrading microorganisms from contaminated sediments. Int Biodeterior Biodegr 62:219–225

Deweerd KA, Bedard DL (1999) Use of halogenated benzoates and other halogenated aromatic compounds to stimulate the microbial dechlorination of PCBs. Environ Sci Technol 33:2057–2063

Donelly PK, Hedge RS, Fletcher JS (1994) Growth of PCB-degrading bacteria on compounds from photo-synthetic plants. Chemosphere 28:984–988

Drinker C, Warren M, Bennet G (1937) The problem of possible systemic effects from certain chlorinated hydrocarbons. J Ind Hyg Toxicol 19:283–311

Duhamel M, Mo K, Edwards EA (2004) Characterization of a highly enriched *Dehalococcoides*-containing culture that grows on vinyl chloride and trichloroethene. Appl Environ Microbiol 70:5538–5545

Erickson BD, Mondello FJ (1993) Enhanced biodegradation of polychlorinated biphenyls after site-directed mutagenesis of a biphenyl dioxygenase gene. Appl Environ Microbiol 59:3858–3862

Faroon O, Jones D, de Rosa C (2001) Effects of polychlorinated biphenyls on the nervous system. Toxicol Ind Health 16:305–333

Fava F, Di Gioia D, Marchetti L (1996) Dichlorobiphenyl degradation by an uncharacterized Pseudomonas species, strain CPE1, in a fixed film bioreactor. Int Biodeterior Biodegr 37:53–59

Fava F, Di Gioia D, Marchetti L (1998) Cyclodextrins effects on the ex-situ bioremediation of a chronically polychlorobiphenyl-contaminated soil. Biotechnol Bioeng 58:345–355

Fennell DE, Nijenhuis I, Wilson SF et al (2004) *Dehalococcoides ethenogenes* strain 195 reductively dechlorinates diverse chlorinated aromatic pollutants. Environ Sci Technol 38:2075–2081

Field JA, Sierra-Alvarez R (2008) Microbial transformation and degradation of polychlorinated biphenyls. Environ Pollut 155:1–12

Focht DD (1995) Strategies for the improvement of aerobic metabolism of polychlorinated biphenyls. Curr Opi Biotechnol 6:341–346

Furukawa K (1994) Molecular genetics and evolutionary relationship of PCB-degrading bacteria. Biodegradation 5:289–300

Furukawa K, Fujihara H (2008) Microbial degradation of polychlorinated biphenyls: biochemical and molecular features. J Biosci Bioeng 105:433–449

Furukawa K, Miyazaki T (1986) Cloning of a gene-cluster encoding biphenyl and chlorobiphenyl degradation in *Pseudomonas pseudoalcaligenes*. J Bacteriol 166:392–398

Furukawa K, Matsumura F, Tonomura K (1978) *Alcaligenes* and *Acinetobacter* strains capable of degrading polychlorinated biphenyls. Agric Biol Chem 42:543–548

Furukawa K, Tomizuka N, Kamibayashi A (1979) Effect of chlorine substitution on the bacterial metabolism of various polychlorinated biphenyls. Appl Environ Microbiol 38:301–310

Furusawa Y, Nagarajan V, Tanokura M et al (2004) Crystal structure of the terminal oxygenase component of biphenyl dioxygenase derived from *Rhodococcus* sp strain RHA1. J Mol Biol 342:1041–1052

Gantzer CJ, Wackett LP (1991) Reductive dechlorination catalyzed by bacterial transition-metal coenzymes. Environ Sci Technol 25:715–722

Gibson DT, Parales RE (2000) Aromatic hydrocarbon dioxygenases in environmental biotechnology. Curr Opin Biotechnol 11:236–243

Gilbert ES, Crowley DE (1997) Plant compounds that induce polychlorinated biphenyl biodegradation by *Arthrobacter* sp. strain B1B. Appl Environ Microbiol 63:1933–1938

Gilbert ES, Crowley DE (1998) Repeated application of carvone-induced bacteria to enhance biodegradation of polychlorinated biphenyls in soil. Appl Microbiol Biotechnol 50:489–494

Haddock JD, Horton JR, Gibson DT (1995) Dihydroxylation and dechlorination of chlorinated biphenyls by purified biphenyl 2,3-dioxygenase from Pseudomonas sp. strain LB400. J Bacteriol 177:20–26

Häggblom MM, Ahn YB, Fennell DE et al (2003) Anaerobic dehalogenation of organohalide contaminants in the marine environment. Adv Appl Microbiol 53:61–84

Harkness MR, McDermott JB, Abramowicz DA (1993) In situ stimulation of aerobic PCB biodegradation in Hudson river sediments. Science 259:503–507

He J, Ritalahti KM, Yang KL et al (2003) Detoxification of vinyl chloride to ethene coupled to growth of an anaerobic bacterium. Nature 424:62–65

Hernandez BS, Koh S, Chial M et al (1997) Terpene utilizing isolates and their relevance to enhanced biotranformation of PCBs in soil. Biodegradation 8:153–158

Hiarishi A (2008) Biodiversity of dehalorespiring bacteria with special emphasis on polychlorinated biphenyl/dioxin dechlorinators. Microbes Environ 23:1–12

Hickey WJ, Brenner V, Focht DD (1992) Mineralization of 2-chloro-and 2,5-dichloro-biphenyl by Pseudomonas sp. strain UCR2. FEMS Microbiol Lett 98:175–180

Holscher T, Krajmalnik-Brown R, Ritalahti KM et al (2004) Multiple nonidentical reductive-dehalogenase-homologous genes are common in *Dehalococcoides*. Appl Environ Microbiol 70:5290–5297

Hou LH, Dutta SK (2000) Phylogenetic characterization of several *para*-and *meta*-PCB dechlorinating *Clostridium* species: 16S rDNA sequence analyses. Lett Appl Microbiol 30:238–243

Ionescu M, Beranova K, Dudkova V, Kochankova L, Demnerova K, Macek T, Mackova M (2009) Isolation and characterization of different plant associated bacteria and their potential to degrade polychlorinated biphenyls. Int Biodeterior Biodegrad 63:667–672

Jota MAT, Hassett JP (1991) Effects of environmental variables on binding of a PCB congener by dissolved humic substances. Environ Toxicol Chem 10:483–491

Kikuchi Y, Yasukochi Y, Nagata Y et al (1994) Nucleotide sequence and functional analysis of the *meta*-cleavage pathway involved in biphenyl and polychlorinated biphenyl degradation in *Pseudomonas* sp. strain KKS102. J Bacteriol 176:4269–4276

Kim S, Picardal F (2001) Microbial growth on dichlorobiphenyls chlorinated on both rings as a sole carbon and energy source. Appl Environ Microbiol 67:953–1955

Kimbara K, Hashimoto T, Fukuda M et al (1989) Cloning and sequencing of two tandem genes involved in degradation of 2,3-dihydroxybiphenyl to benzoic acid in the polychlorinated biphenyl-degrading soil bacterium Pseudomonas sp. strain KKS102. J Bacteriol 171:2740–2747

Klasson KT, Barton JW, Evans BS et al (1996) Reductive microbial dechlorination of indigenous polychlorinated biphenyls in soil using a sediment-free inoculum. Biotechnol Prog 12: 310–315

Krone UE, Thauer RK, Hogenkamp HPC (1989) Reductive dehalogenation of chlorinated C1-hydrocarbons mediated by corrinoids. Biochemistry 28:4908–4914

Lambo AJ, Patel TR (2006) Isolation and characterization of a biphenyl-utilizing psychrotrophic bacterium, *Hydrogenophaga taeniospiralis* IA3-A, that cometabolize dichlorobiphenyls and polychlorinated biphenyl congeners in Aroclor 1221. J Basic Microbiol 46:94–107

Leigh MB, Prouzova P, Mackova M, Macek T, Nagle DP, Fletcher JS (2006) Polychlorinated biphenyls (PCB)-degrading bacteria associated with the trees in a PCB-contaminated site. Appl Environ Microbiol 72:2331–2342

Leigh MB, Pellizari VH, Uhlik O, Sutka R, Rodrigues J, Ostrom NE, Zhou J, Tiedje JM (2007) Biphenyl-utilizing bacteria and their functional genes in a pine root zone contaminated with polychlorinated biphenyls (PCBs). ISME J 1:134–148

Ling-Yun J, Zheng AP, Xu L et al (2008) Isolation and characterization of comprehensive polychlorinated biphenyl-degrading bacterium, *Enterobacter* sp. LY402. J Microbiol Biotechnol 18:952–957

Löffler FE, Sun Q, Li J et al (2002) 16S rRNA gene-based detection of tetrachloroethene-dechlorinating *Desulfuromonas* and *Dehalococcoides* species. Appl Environ Microbiol 66:1369–1374

Lunt D, Evans WC (1970) The microbial metabolism of biphenyl. Biochem J 118:54–55

Mackova M, Dowling D, Macek T (eds) (2006) Phytoremediation and rhizoremediation, theoretical background. FOCUS Biotechnol 9A:300 pp (Springer)

Mackova M, Prouzova P, Stursa P, Ryslava E, Uhlik O, Beranova K, Rezek J, Kurzawova V, Demnerova K, Macek T (2009) Phyto/rhizoremediation studies using long-term PCB-contaminated soil. Environ Sci Pollut Res 16:817–829

Masai E, Yamada A, Healy JM et al (1995) Characterization of biphenyls catabolic genes of gram-positive polychlorinated biphenyls degrader *Rhodococcus* sp. strain RHA1. Appl Environ Microbiol 61:2079–2085

Mavoungou R, Masse R, Sylvestre M (1991) Microbial dehalogenation of 4, 4'-dichlorobiphenyl under anaerobic conditions. Sci Total Environ 101:263–268

May HD, Cutter LA, Miller GS et al (2006) Stimulatory and inhibitory effects of organohalides on the dehalogenating activities of PCB-dechlorinating bacterium *o*-17. Environ Sci Technol 40:5704–5709

Michel FC, Quensen J, Reddy CA (2001) Bioremediation of a PCB-contaminated soil via composting. Compost Sci Util 9:274–283

Mohn WW, Tiedje JM (1992) Microbial reductive dehalogenation. Microbiol Rev 56:482–507

Mondello FJ (1989) Cloning and expression in *Escherichia coli* of *Pseudomonas* strain LB400 genes encoding PCB degradation. J Bacteriol 171:1725–1732

Mousa MA, Quensen JF, Chou K et al (1996) Microbial dechlorination alleviates inhibitory effects of PCBs on mouse gamete fertilization *in vitro*. Environ Sci Technol 30:2087–2092

Mouz S, Merlin C, Springael D et al (1999) A GntR-like negative regulator of the biphenyl degradation genes of the transposon Tn4371. Mol Gen Genet 262:790–799

Nies L, Vogel TM (1990) Effects of organic substrates on dechlorination of Aroclor-1242 in anaerobic sediments. Appl Environ Microbiol 56:2612–2617

Ofjord GD, Puhakka JA, Ferguson JF (1994) Reductive dechlorination of aroclor 1254 by marine sediment cultures. Environ Sci Technol 28:2286–2294

Ohtsubo Y, Kudo T, Tsuda M et al (2004) Strategies for bioremediation of polychlorinated biphenyls. Appl Microbiol Biotechnol 65:250–258

Péloquin L, Greer CW (1993) Cloning and expression of the polychlorinated biphenyl-degradation gene cluster from Arthrobacter M5 and comparison to analogous genes from gram-negative bacteria. Gene 125:35–40

Pieper DH (2005) Aerobic degradation of polychlorinated biphenyls. Appl Microbiol Biotechnol 67:170–191

Pieper DH, Seeger M (2008) Bacterial metabolism of polychlorinated biphenyls. J Mol Microbiol Biotechnol 15:121–138

Quensen JF III, Boyd SA, Tiedje JM et al (1992) Expected dioxin-like toxicity reduction as a result of the dechlorination of Aroclors. In: General electric company research and development program for the destruction of PCBs, eleventh progress report. General Electric Corporate Research and Development, Schenectady, NY, pp 189–196

Quensen JF III, Tiedje JM, Boyd SA (1988) Reductive dechlorination of PCBs by anaerobic microorganisms from sediments. Science 242:752–754

Quensen JF III, Boyd SA, Tiedje JM (1990) Dechlorination of four commercial polychlorinated biphenyl mixtures (Aroclors) by anaerobic microorganisms from sediments. Appl Environ Microbiol 56:2360–2369

Quensen JF, Mueller SA, Jain MK et al (1998) Reductive dechlorination of DDE to DDMU in marine sediment microcosms. Science 280:722–724

Rezek J, Macek T, Mackova M, Triska J (2007) Plant metabolites of polychlorinated biphenyls in hairy root culture of black nightshade *Solanum nigrum* SNC-9O. Chemosphere 69:1221–1227

Rezek J, Macek T, Mackova M, Triska J, Ruzickova K (2008) Hydroxy-PCBs, methoxy-PCBs and hydroxy-methoxy-PCBs: metabolites of polychlorinated biphenyls formed *in vitro* by tobacco cells. Environ Sci Technol 47:5746–5751

Rhee GY, Sokol RC, Bethoney CM et al (1993) A long-term study of anaerobic dechlorination of PCB congeners by sediment microorganisms – pathways and mass-balance. Environ Toxicol Chem 12:1829–1834

Sakai M, Ezaki S, Suzuki N et al (2005) Isolation and characterization of a novel polychlorinated biphenyl-degrading bacterium, *Paenibacillus* sp. KBC101. Appl Microbiol Biotechnol 68:111–116

Seeger M, Timmis KN, Hofer B (1995) Degradation of chlorobiphenyls catalyzed by the bph-encoded biphenyl-2, 3-dioxygenase and biphenyl-2, 3-dihydrodiol-2, 3-dehydrogenase of Pseudomonas sp. LB400. FEMS Microbiol Lett 133:259–264

Seeger M, Timmis KN, Hofer B (1997) Bacterial pathways for the degradation of polychlorinated biphenyls. Mar Chem 58:327–333

Seeger M, Zielinski M, Timmis KN et al (1999) Regiospecificity of dioxygenation of di- to pentachlorobiphenyls and their degradation to chlorobenzoates by the bph-encoded catabolic pathway of Burkholderia sp. strain LB400. Appl Environ Microbiol 65:3614–3621

Singer AC, Gilbert ES, Luepromchai E et al (2000) Biodegradation of polychlorinated biphenyl-contaminated soil using carvone and surfactant-grown bacteria. Appl Environ Microbiol 54:838–843

Springael D, Kreps S, Mergeay M (1993) Identification of a catabolic transposon, Tn4371, carrying biphenyl and 4-chlorobiphenyl degradation genes in *Alcaligenes eutrophus* A5. J Bacteriol 175:1674–1681

Suenaga H, Watanabe T, Sato M et al (2002) Alteration of regiospecificity in biphenyl dioxygenase by active-site engineering. J Bacteriol 184:3682–3688

Sylvestre M, Macek T, Mackova M (2009) Transgenic plants to improve rhizoremediation of polychlorinated biphenyls (PCBs). Curr Opin Biotechnol 20:242–247

Tiedje JM, Quensen JF III, Chee-Sanford J et al (1993–1994) Microbial reductive dechlorination of PCBs. Biodegradation 4:231–240

Triska J, Kuncova G, Mackova M et al (2004) Isolation and identification of intermediates from biodegradation of low chlorinated biphenyls (Delor 103). Chemosphere 54:725–733

Tucker ES, Litschgi WJ, Mees WM (1975) Migration of polychlorinated biphenyls in soil induced by percolating water. Bull Environ Contam Toxicol 13:86–93

Uhlik O, Jecna K, Leigh MB, Mackova M, Macek T (2009a) DNA-based stable isotope probing: a link between community structure and function. Sci Total Environ 407:3611–3619

Uhlik O, Jecna K, Mackova M, Vlcek C, Hroudova M, Demnerova K, Paces V, Macek T (2009b) Biphenyl-metabolizing bacteria in the rhizosphere of horseradish and bulk soil contaminated by polychlorinated biphenyls as revealed by stable isotope probing. Appl Environ Microbiol 75:6471–6477

Van Dort HM, Bedard DL (1991) Reductive *ortho* and *meta* dechlorination of a polychlorinated biphenyl congener by anaerobic microorganisms. Appl Environ Microbiol 57:1576–1578

Van Dort HM, Smullen LA, May RJ et al (1997) Priming *meta*-dechlorination of PCB that have persisted in Housatonic River sediments for decades. Environ Sci Technol 31:3300–3307

Watts JEM, Wu Q, Schreier SB et al (2001) Comparative analysis of PCB-dechlorinating communities in enrichment cultures using three different molecular screening techniques. Environ Microbiol 3:710–719

Wiegel J, Wu QZ (2000) Microbial reductive dehalogenation of PCB. FEMS Microbiol Ecol 32:1–15

Williams WA (1994) Microbial reductive dechlorination of trichlorobiphenyls in anaerobic sediment slurries. Environ Sci Technol 28:630–635

Wu Q, Bedard DL, Wiegel J (1997) Temperature determines the pattern of anaerobic microbial dechlorination of Aroclor 1260 primed by 2,3,4,6-Tetrachlorobiphenyl in woods pond sediment. Appl Environ Microbiol 63:4818–4825

Wu Q, Bedard DL, Wiegel J (1999) 2, 6-Di bromobiphenyl primes extensive dechlorination of Aroclor 1260 in contaminated sediment at 8–30EC by stimulating growth of PCB-dehalogenating microorganisms. Environ Sci Technol 33:595–602

Wu Q, Joy EM, Watts K et al (2002) Identification of a bacterium that specifically catalyzes the reductive dechlorination of polychlorinated biphenyls with doubly flanked chlorines. Appl Environ Microbiol 68:807–812

Yan T, Lapara TM, Novak PJ (2006a) The reductive dechlorination of 2,3,4,5-tetrachlorobiphenyl in three different sediment cultures: evidence for the involvement of phylogenetically similar Dehalococcoides-like bacterial populations. FEMS Microbiol Ecol 55:248–261

Yan T, LaPara TM, Novak PJ (2006b) The impact of sediment characteristics on PCB-dechlorinating cultures: implications for bioaugmentation. Biorem J 10:143–151

Ye DE, Quensen JF, Tiedje JM et al (1992) Anaerobic dechlorination of polychlorobiphenyls (Aroclor-1242) by pasteurized and ethanol-treated microorganisms from sediments. Appl Environ Microbiol 58:1110–1114

Chapter 17
Role of Clay and Organic Matter in the Biodegradation of Organics in Soil

Laura E. McAllister and Kirk T. Semple

The fate of organic contaminants such as polycyclic aromatic hydrocarbons (PAHs), organochlorine pesticides, nitroaromatic compounds, chlorinated aliphatics, and other xenobiotics in soils is an issue of significant environmental and regulatory interest. Such chemicals are known to possess a variety of carcinogenic, mutagenic and toxic properties, making their destruction and/or removal from all environmental domains a highly desired end-point. In soil, the decomposition of organic contaminants occurs primarily through attack by the vast populations of microorganisms (Cerniglia 1992), an occurrence that is exploited for the purposes of land remediation. However, the biodegradation of organic compounds is a dynamic and complicated process. The viability and capability of a soil microbial population to degrade the target contaminant is essential. The physicochemical properties of the contaminant and their interaction with various soil components both influence a given contaminant's potential to be biodegraded, as well as the length of contaminant residence time (Semple et al. 2003). Soils are a heterogeneous assemblage of organic/inorganic constituents, organo-mineral complexes, ions, biota and other compounds (Fig. 17.1) which result in an array of mechanisms and reactions that all contribute to the fate and behaviour of organic contaminants (Reid et al. 2000, Stokes et al. 2005). Predicting the biodegradation end points in a given soil is therefore inherently difficult (Semple et al. 2007).

Clays are the smallest (<2 μm) class of inorganic materials present in soil and are composed of clay minerals, sesquioxides and amorphous minerals associated with soil organic matter (SOM) (Rowell 1994). Due to their small dimensions and ability to carry electrical charge, clay particles possess characteristics that allow interaction with other soil components. Of all soil inorganic constituents, clay minerals are

L.E. McAllister and K.T. Semple (✉)
Lancaster Environment Centre, Lancaster University, Lancaster LA1 4YQ, UK
e-mail: k.semple@lancaster.ac.uk

368 L.E. McAllister and K.T. Semple

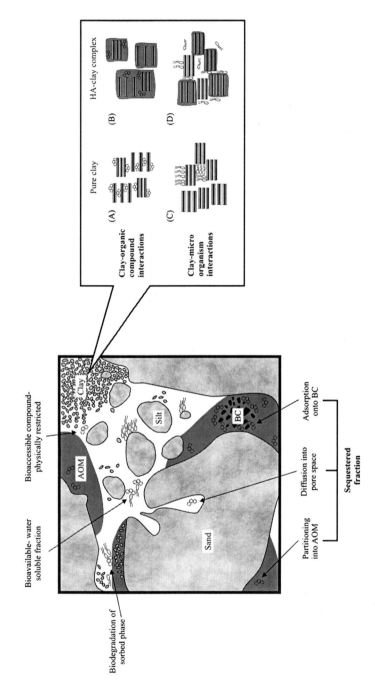

Fig. 17.1 Conceptual diagram illustrating a range of potential interactions between organic compounds (phenanthrene is used as a model pollutant), soil microorganisms and soil components, with particular reference to the clay fraction. Abbreviation: AOM – amorphous organic matter; BC – black carbon; HA – humic acid

empirically demonstrated to be the most reactive (Boyd et al. 2001), thus they are significant agents of reaction in soil. Understanding the nature of how clays influence biodegradation requires a composite knowledge of the factors which influence the availability of organic compounds in soil, the physicochemical properties of clays and their interactions with organics, consideration as to the impact of clay on microorganism activity in soil and finally, how these operate concomitantly in soil to influence organic compound biodegradation. Although this review focuses primarily on clay, SOM will be an integral part of discussions as clay does not exist in soils as a purified material but in conjunction with humic/fulvic substances and humin. As such, examining either material alone would likely be unrepresentative of true environmental conditions. The specific aims of this review are to (a) discuss the concepts of microbial bioavailability and bioaccessibility of organic contaminants in soil; (b) consider the physicochemical properties of the soil clay fraction; (c) explore the significance of clay-organic contaminant interactions upon compound mobility and sequestration; (d) consider the potential impact of clay upon soil microbial activity and finally, and (e) evaluate, using empirical data, the effects of the clay fraction on biodegradation. The review concludes by identifying some of the key processes/interactions involving clay that influence biodegradation.

Soil Contaminant Interactions – Bioavailability and Biodegradation

Organic compounds entering the soil environment may be lost or dissipated via manifold pathways including leaching, volatilisation or biotic/abiotic degradation (Semple et al. 2003; Stokes et al. 2005). In addition, intra-soil processing of these chemicals may lead to sequestration within the soil's organic/inorganic matrices. As soil-contaminant contact time lengthens, more intimate association with the soil solid phase leads to a reduction in chemical and biological availability, as well as toxicity, a process termed aging (Hatzinger and Alexander 1995). The interactions leading to aging are governed by several factors including soil characteristics and contaminant physicochemical properties (e.g. aqueous solubility, polarity, hydrophobicity, lipophilicity and molecular structure). For example, highly non-polar PAHs such as benzo[a]pyrene and anthracene will associate rapidly with soil solid components whereas more polar compounds including certain pesticides will remain in the aqueous phase for a longer duration. Sequestration processes are dynamic and can involve both the diffusion and entrapment into inaccessible micropores as well as sorption to the soil matrix (Hatzinger and Alexander 1995; Pignatello and Xing 1996; Reid et al. 2000). Many studies have been conducted to investigate the fate and behaviour of organic contaminants over time with particular interest directed at the bioavailability of contaminants in soil (e.g. Guthrie and Pfaender 1998; Hatzinger and Alexander 1995; Kelsey et al. 1997; White et al. 1997, 1999).

Bioavailability refers to the fraction of a chemical in soil that can be taken up or transformed by living organisms (Semple et al. 2004). Two prevailing factors determine the fraction of a chemical that is bioavailable: (1) the rate of compound mass transfer to the living cell and (2) the rate of uptake and metabolism (the intrinsic activity of the cell). Bioavailability therefore governs both rates of microbial degradation and soil ecotoxicity (Ehlers and Luthy 2003; Semple et al. 2004). However, Semple et al. (2004) stated that the use of 'availability' has an implied immediacy for the transfer of contaminants across cell membranes and excludes the fraction that may become available should any physical or temporal restraint upon the compound be removed (e.g. temporary sorption). The term 'bioaccessibility' was postulated as it encompasses both the bioavailable fraction and that which will/could become available. The distinction in terms is significant as either bioaccessibility or bioavailability may be more relevant depending on the perspective of the investigator. Bioavailability is a good indicator of biodegradation rates, whereas bioaccessibility has more bearing on final degradation end points (Semple et al. 2007). Mechanisms of interaction and the relative importance of soil constituents to bioavailability and bioaccessibility have been explored (Nam et al. 1998; Rhodes et al. 2008).

Clays and Clay Minerals

The clay fraction in soil is composed primarily of clay minerals; the majority of which are alumino(phyllo)silicates; these are articulated silicate or aluminosilicate layers in which individual sheets are built up from basic building blocks of either silicon tetrahedra or aluminium octahedra (Fig. 17.2). Clay minerals are classified on the basis of the ratio, in a unit layer, of these sheets, the spacing between unit layers and their interlayer components. When one octahedral sheet is linked to one tetrahedral sheet, a 1:1 layer is formed (e.g. kaolinite); when one octahedral sheet is enclosed between two tetrahedral sheets a 2:1 layer is produced (e.g. smectites). Table 17.1 provides a list of some of the most common clay mineral groups and

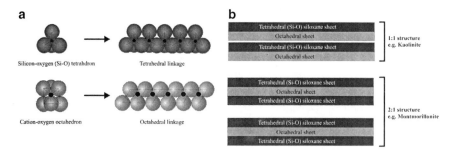

Fig. 17.2 Illustration and representation of the basic unit polyhedra that form clay minerals. Tetrahedra consist of a central silicon ion surrounded by four closed spaced coordinated oxygen ions. Aluminium, magnesium or ferrous ions form octahedra with six oxygen or hydroxyls. These units form sheets, which combine to form layers, by sharing oxygen atoms (Adapted from Singer and Munns 2002)

Table 17.1 The major clay mineral groups and selected key properties (Adapted from Rowell 1994; Newman 1987; Singer and Munns 2002; Velde 1996)

Clay mineral groupings	Layer type	CEC[a] (milliequivalents per 100 g)	SSA[b] (m^2 g^{-1})	Relative swelling (upon wetting)
Smectite	2:1 layers, 10 nm, variable thickness	80–120	40–800	High
Vermiculite	2:1 layers, variable thickness	100–150	760–800	High
Micas	2:1 layers, 10 Å	10–40	10–100	Low
Chlorites	2:1 layers, 14 Å	10–40	10–55	None
Kaolinite-serpentine	Uncharged 1:1 layers, 7–10 Å repeat units depending on the presence of interlayer water	3–15	5–40	Almost none
Palygorskites and sepiolites	2:1 type layers linked by their edges	3–20	40–180	None

[a]Cation exchange capacity at pH 7
[b]Specific surface area

selected properties of relevance. Internal chemical structure, size and crystal shapes of clay minerals can vary drastically and are dependant upon (1) the parent rock; (2) the environmental conditions of weathering and formation, including temperature and pH and (3) substitutions within the chemical formulae (Jackson 1957). Although 2:1 and 1:1 layers of clays are strongly bound internally, the bonds between layers are relatively weak, which can result in inter-stratifications of different structural units in mixed sequences. Common examples include illite-montmorillonite, vermiculite-chlorite and kaolinite-montmorillonite (Velde 1996). Such variation can vastly influence the physicochemical properties of clays in the environment. For a more detailed discussion of clay crystallography, mineralogy and physical properties, the reader is directed to reviews by Bailey (1980) and Brindley (1980).

Clay minerals possess important physicochemical properties, in particular, the ability to hold and retain water as well as possessing a net negative charge (Berkowitz et al. 2008). Water may be incorporated into the clay structure by a variety of mechanisms, including adsorption on to the surface of clay particles, entrapment in pore spaces, and in 2:1 structured swelling/expanding clays, water may occur in interlayer locations (Rowell 1994; Velde 1996). Adsorption of water and electrostatic charge are both related to the surface area of these minerals, which, given the relatively small dimension and the propensity of 2:1 clays to expand under hydrated conditions, swelling clays may have surface areas as high as $800 \text{ m}^2 \text{ g}^{-1}$ (Boyd et al. 2001). Therefore, expandable clays, such as montmorillinite, provide a large surface area consisting of both internal and external surfaces for interaction with organic compounds, in relation to non-expanding clays, such as kaolinite.

Due to isomorphic substitution of structural Si^{4+} and/or Al^{3+} cations by lower valent cations, clay minerals bear a permanent excess of negative charge (Haderlein et al. 1996). As a result certain cations are adsorbed and retained on surfaces, as well as in internal crystallographic sites of swelling clays; a property quantified by the cation exchange capacity (CEC), typically expressed in milliequivalents (meq) per 100 g. CECs can vary for a number of reasons, including clay particle size or the adsorbed ion composition; thus a range of CEC values is usually given for a defined mineral (Singer and Munns 2002) (see Table 17.1). Further, as soils are mixed-mineral systems, a wide range of CEC values will exist for a given soil (Rowell 1994; Velde 1996). The significance of this property for biodegradation is that positively charged organic molecules may replace inorganic exchangeable cations such as H^+, Ca^{2+}, Mg^{2+}, or K^+ on the surfaces on external surfaces and in interlayer positions (Alexander 1999). Therefore, ion exchange can represent an important process for the immobilisation and biodegradation of organic compounds.

Sorption and Desorption Phenomena of Clays and Clay Minerals

Sorption is a phase distribution process that accumulates solutes at surfaces and inter-phases (i.e. adsorption) or from one phase to another (i.e. partitioning) (Huang et al. 2003). Soils that possess a high capacity for sorption effectively reduce the size of the freely dissolved compound fraction, where it is more readily (but by no means exclusively) biodegraded (Semple et al. 2003; Stokes et al. 2005). Sorption is characterised by bi or tri-phasic kinetics, whereby a rapid phase of sorption is succeeded by slow and/or very slow sorption (Cornelissen et al. 1997, 1998a, 2000). A comparable profile occurs during desorption; an initial rapid phase is followed by slow and/or very slow compartments. Desorption has been empirically linked to biodegradation, particularly with well-aged soils; bioavailability is controlled primarily by the release (desorption) of these compounds, as well as mass movement (diffusion) and dissolution (Cornelissen et al. 1998a; White et al. 1999). Therefore, sorption is a governing process of biodegradation. A wealth of research has revealed that SOM exerts the greatest thermodynamic affinity for organic compounds in soil, predominantly for hydrophobic substances and thus is the principal determinant of sequestration (Nam et al. 1998, White et al. 1999, Reid et al. 2000, Stokes et al. 2005). Affinity for sorption often correlates well with the fraction of organic carbon (f_{oc}). SOM has been mechanistically split into two 'domains', (1) amorphous organic matter (AOM), which is defined by rapid, reversible and linear sorption and (2) black carbon (BC), sorption to BC is observed to be 100–1,000 times greater than AOM, in addition to exhibiting a highly non-linear and competitive nature (Cornelissen and Gustafsson 2005; Koelmans et al. 2006). Decreases in bioavailability and biodegradation with increased soil organic matter have been extensively studied (e.g. Nam et al. 1998; White et al. 1999).

The affinity of organic compounds to clays is generally considered to be of less significance in comparison to SOM due to the hydrophilicity of particle surfaces and crystalline nature of clay minerals (Cornelissen et al. 1998a). Despite this, sorption to clay minerals and other mineral fractions in soils is well documented (Ball et al. 1991a, b; Mader et al. 1997). Since the 1970s, considerable investigation has been focussed upon the interactions between clays and pesticides (Lagaly 2001); largely as a result of the large fractions of clays found in agricultural soils and the large surface areas available for interaction (Laird et al. 1992). High affinities have been documented for pesticides, many of which are polar compounds or possess polar moieties; for example metamitron (Cox et al. 1996, 1997), thiazofluoron (Cox et al. 1997), simazine (Sannino et al. 1999) metolachlor (Nennemann et al. 2001) and endrin (Peng et al. 2009). However, sorption to clay is also observed for PAHs, such as naphthalene (Ghosh and Keinath 1994), pyrene (Ogawa et al. 1995; Theng et al. 1998; Huang and Cutright 2003), phenanthrene (Hundal et al. 2001; Huang and Cutright 2003) as well as polychlorinated biphenyls (PCBs) and chlorobenzenes (Cornelissen et al. 1998a).

The mechanisms of sorption have been a central topic of investigation e.g. Haderlien et al. (1996); Boyd et al. (2001).Specific mechanisms of sorption include ion exchange, hydrogen bonding, van der Waals forces and chemisorption (Lagaly 2001). Intuitively, higher sorption capacities are associated with swelling clays. For example, Haderlien et al. (1996) investigated the association of nitroaromatic compounds (NACs) with clay minerals of varying interlayer ratios and swelling capabilities. The affinity and capacities of all NACs increased in the order kaolinite < illite < montmorillinite; total surface areas were 12, 70 and 820 m^2 g^{-1}, respectively. Montmorillinite, an expanding clay lattice, has a great capacity for adsorption, especially of cationic compounds that bind to the negatively charged magnesium-aluminosilicate sheets. In contrast, exchange sites on kaolinite are restricted to the external surface, with no interlayer sites made available as the lattice does not expand. In addition to increased ion exchange, large surface areas present an extensive pore network, which can provide steric hindrance to the molecular diffusion of organic compounds (White et al. 1999). Hundal et al. (2001) hypothesised that the quantified sorption of phenanthrene to smectites was explained by a model of capillary condensation of molecules into nanosites within clay quasi-crystals. Such a model was invoked as results did not reflect a chemical reaction between the PAH and mineral surfaces. Cornelissen et al. (1998a) quantified desorption of three chlorobenzenes and three PCBs from montmorillinite and zeolite in clean, aqueous systems, observing higher than expected sorbent-water distribution ratios. Similarly, the authors hypothesised that the interlayer spacing represented a network of nanopores in which diffusion would be retarded. Therefore, it seems evident that the physical inhibition of substrate movement rather than chemical interactions is important for highly non-polar, hydrophobic substances.

The chemical composition of the clays is influential on sorption kinetics of organic chemicals in soil. As soils are likely to contain a range of clay mineral types, a rich diversity of chemical surfaces and respective reactivities may be present. Pesticide adsorption is dramatically affected by the types of exchangeable cations present, commonly Ca^{2+}, Mg^{2+}, Al^{3+}, K^+, NH^{4+}, and Na^+ (Li et al. 2001). Clays (specifically smectites) saturated with K^+ ions have displayed some of the largest sorption capacities, as these ions modify the microenvironment within clay inter layers. Firstly, K^+ ions are weakly hydrated relative to other cations such as Ca^{2+} or Mg^{2+}, which limits the detrimental impact of associated water molecules upon the interactions of hydrophobic organic contaminants (HOCs) with uncharged, hydrophobic siloxane surfaces. Further, K^+ ions create basal spacings of ~12.5 Å, which is an optimal dimension to permit HOC intercalation, whilst inhibiting the interaction of water molecules (Li et al. 2001). In addition, Sannino et al. (1997) found that altering the $Al(OH)_x$ species coated on montmorillinite surfaces led to signficant variations in the adsorption capacity of the herbicide 2–4 dichlorophenoxyacetic. However, clay minerals may be easily modified to achieve the stabilisation and immobilisation of compounds (Lagaly 2001). For example, the generation of organoclays, specifically for this purpose, is an important market in remediation (Beall 2003). Organoclays are synthesised by replacing the native exchangeable inorganic cations with organic cations, typically quaternary ammonium cations. Organoclays have a greater propensity for the sorption of non-ionic organic compounds due to the combination of (1) adsorption to siloxane surfaces and (2) partitioning into the organic 'pseudophase' of the quaternary ammonium cations (Borisover et al. 2008). Hydrogen bonding between ammonium moieties and organic sorbates has also been reported (Cox et al. 2001). Critically, the presence of ammonium cations serves to increase the hydrophobicity of clay surfaces and internal environments by reducing the volume of complexed water at these localities (Borisover et al. 2008) Such is the efficacy of organoclays for adsorbing and inter-calating target sorbates that it is an established procedure for water treatment (Beall 2003). Further, considerable research has been extended into a dual sorption-deg-radation technology for the remediation of contaminated sub-soils and groundwater (Crocker et al. 1995; Witthuhn et al. 2005). When levels of contaminants are below toxic thresholds and sorption is rapid but easily desorbable (i.e. bioavailability is relatively high), the addition of a microbial consortia in situ could potentially lead to the effective degradation of pollution.

In terms of bioavailability, the mass transfer of contaminants to microbial populations, rather than catabolic activity, is often the limiting factor for biodegrada-tion. Therefore the strength of association and any subsequent desorption from clay minerals are likely to be pertinent parameters for consideration. In pure montmoril-linite clay suspensions, the desorption half-life of benzylamine was less than 1 min (Miller and Alexander 1991). Conditions in soil, particularly with well-aged con-taminants, are unlikely to reflect similar kinetics. For example, Chai et al. (2007) observed that the addition of 15% montmorillinite clay in artificial sediments resulted in an average reduction of 17% in the fast desorbing fraction of hexachlo-robenzene. Clay minerals led to a 64.8% decrease in the desorbed fraction of phenanthrene in soil when compared with the soil containing just SOM (Huang and Cutright 2003). Further, Ghosh and Keinath (1994) conducted equilibrium desorp-tion studies of naphthalene from montmorillonite and vermiculite using a surfactant solution and found that re-mobilization of the compound was negligible. In contrast, naphthalene sorbed to hexadecyltrimethylammonium (HDTMA)-modified smectite clay yielded very high desorption rates; the only limiting factor to naphthalene mass transfer was the impact of clay aggregates size on diffusion (Crocker et al. 1995). Talley et al. (2002) found that desorption of PAHs from the clay/silt fraction of dredged sediment was substantial; nearly 80% was readily desorbed within 2 months, Overall, the physicochemical nature of the clay fraction particles represents the decisive factor upon sorption/desorption strength and intensity.

Clays in environmental matrices such as sediments and soils are likely to exist in association with SOM; it has been reported that the majority of mineral surfaces are coated with SOM in the upper horizons of soil (Mayer and Xing 2001). The resultant formation of SOM-clay complexes may alter the sorptive properties of clay minerals; Bertsch and Seaman (1999) reported that SOM contents as low as 1% significantly influenced the surface properties of soil phyllosilicates. Several studies have reported the blocking of sorption sites by humic acids (e.g. Onken and Traina 1997; Shen and Jaffé 2000; Charles et al. 2006). Conversely, Murphy et al. (1990) found that overall sorption increased when humic acids were coated onto haematite and kaolinite and the most hydrophobic humic acid (peat) was the stron-gest sorbent. The presence of humic acid on K^+ smectites did not alter the sorption of two herbicides, but could suppress or facilitate sorption on Ca^{2+} clays as a func-tion of sorbate type (Li et al. 2003). Clay minerals may promote whole soil sorption by protecting SOM which may contain partitioned or adsorbed compounds (Pignatello 1990).

A question therefore exists; at which point does the contribution of the clay fraction become significant in soils to total sorption? Ambiguity over this issue is apparent in the literature. Hundal et al. (2001) compared several authors' findings by contrasting the sorption of HOCs and herbicides against a ratio of clay minerals to f_{oc}, with conflicting results. In some instances, above a threshold ratio (60:1 and 40:1) swelling clays made a marked contribution; other studies found sorption to be independent of both mineral type and content. Alternatively, other workers (e.g. Huang and Cutright [and references therein] 2003; Li et al. 2001) have shown that sorption of HOCs to SOM dominates under hydrated conditions, whereas clay minerals are the principal sorbents in dry soils as HOCs cannot compete for sorption site against water. Other researchers suggest that the significance of sorbent material type is largely controlled by the nature of the sorbate. Earlier discussion indicated that polar compounds may have a higher affinity for clay minerals than non-polar chemicals. Based upon sorption to clay minerals in aqueous suspension, sorption of pesticides (Laird et al. 1992), herbicides (Sheng et al. 2002) and NACs (Haderlein et al. 1996) by clay minerals may equal or exceed sorption to SOM. It has thus been concluded that characterising the behaviour of polar compounds by using f_{oc} -normalised sorption coefficients is unrepresentative (Li et al. 2001). However, it must be stressed that these summations are based largely upon investigations with model sorbents may not hold true in complex matrices such as soil. Indeed, Chen and Huang (2009) recognised this limitation and studied the sorption of NACs to three soils, with a portion of one soil treated with H_2O_2 to remove SOM. Sorption was an order of magnitude greater in untreated soil, suggesting that SOM was in fact an important sorption domain; a finding in direct contradiction to many purified sorbent studies in the literature. Interestingly, Huang and Cutright (2003) observed that natural soils containing only clay minerals sorbed 12.2% more phenanthrene, strictly a non-polar compound, than soils containing just SOM.

It is evident that clay minerals represent important sorbents for organic compounds and as decisive materials for phase distribution, have implications for (bio) availability and accessibility. Rather than a simple partition coefficient based upon f_{oc}, a model incorporating multiple sorption domains and mechanisms is likely required to predict compound mobility and sequestration. However, the processes are complex and dependant on many variables, including environmental conditions and the physicochemical properties of both clays and the target sorbate.

The Effect of Clays on the Mobility and Activity of Microorgansims in Soil

Thus far, discussion of the role that clays may impose on biodegradation has been directed at their role in immobilising contaminants. However, in order for biodegradation to occur, a viable population of microorganisms with the necessary catabolic capabilities must be present in soil. Crucially, contact between the microorganisms and their target substrate is necessary for degradation to occur. As pollutants are not homogenously distributed in soil and are often associated with different solid fractions, contact can be a limiting factor in biodegradation (Wick et al. 2007). Bacteria generally have low mobility in soil; for example, when applied to soil surfaces penetration is usually only a few centimetres (Lahlou et al. 2000 and references therein). In soils and other subsurface environments, microorganisms are predominantly attached or adhered to surfaces forming micro-colonies (Lahlou et al. 2000), rather than free-floating in the aqueous phase. As clay particles offer a large surface area for attachment and colony formation, large clay fractions could potentially influence the nature and location of microorganism populations in soil.

Despite this, chemotactic behaviour in microorganisms is a means by which the spatial distribution between substrate and bacteria may be reduced, thereby increasing biodegradation (Semple et al. 2007). Bacteria move up concentration gradients to access substrates, overcoming situations where mass transfer is low. Many physiological factors influence bacterial transport through soil, including cell size, shape, and the hydrophilic nature of the cell surfaces. Adaptations in the cell wall structure may facilitate attachment to geosorbents (which may already be associated with the target compound) or at least, alter the physicochemical characteristics of the bacteria, which dictate the phase distribution of the organism (Semple et al. 2007). Environmental conditions including chemical composition of the carrier solution (e.g. pH, surfactants) affect transport, as well as soil textural properties which permit preferential water flow through macropores such as the presence of large sand grains and soil aggregates. The importance of soil texture for the distribution of microbial degraders and organic substrates has been highlighted by Amellal et al. (2001). After spiking soil with a mixture of 8 PAHs, shifts in the localities of microorganisms were evident; larger phenanthrene degrading bacterial communities were found in the sand and clay fractions. As microorganisms rely on continuous water-filled pathways for translocation through soil (Wick et al. 2007), clay-rich soils (i.e. 60% clay) are known to have a propensity towards bacterial clogging (Niqui-Arroyo et al. 2006).

Lahlou et al. (2000) confirmed that clay is the main retarding agent of PAH-degrading bacteria in soil by calculating the adhesion efficiency (α_t) of nine different bacteria through packing columns containing either a clay-rich soil or the isolated constituents of montmorillinite clay, sand or soil humic acid (HA)-clay complexes. Despite the relatively low clay content of the soil (6.7%) all bacteria exhibited a strong affinity to clay. Coating of clay with HAs was found to sharply increase the transport of all bacteria tested. For example, the α_t values of *Mycobacterium* sp. LB501T dropped from 3.075 to 1.702 and 0.831 for the two types of HAs coating respectively. It was postulated that the organic matter coating interfered with the bacterial association to clay, facilitating greater movement through the column.

Despite behavioural adjustments for movement towards a contaminant, degradation may still be inhibited by the clay fraction if pollutant molecules are inaccessible. Earlier, it was noted that a clay interlayer basal spacing of ~12.5 Å is sufficient for PAHs to enter clay lattices. As bacterial cells are excluded from pores smaller than 0.2–0.8 μm (Johnson et al. 2005), which is up to two orders of magnitude greater than clay interlayers, bioaccessibility of compounds sorbed in this state remains impaired. The enzymes which oxidise PAHs are located within the cytoplasm of the bacteria, therefore compound uptake and diffusion across the bacterial membrane is essential (Theng et al. 2001). Some microorganisms synthesise extracellular enzymes to initiate degradation, which are not as impaired as bacteria in entering nanopores, in order to overcome access to physically restrained substrates (Semple et al. 2007). Evidence does exist for the sorption of extracellular enzymes onto clay fractions (e.g. Rao et al. 2000; Kelleher et al. 1997). Further, Huang et al. (2005) observed the preferential adsorption and low percentages of desorption of acid phosphatise (a hydrolase which facilitates the decomposition of phosphorus in soils) onto the clay-sized fraction in soil. Sorption of free enzymes may protect them from denaturation and preserve their activity, although in this instance, activity was reduced by a minimum of 27% in relation to activity of the free enzyme. Clay mineral surfaces may also retain other organic substances required by degraders to function, such as proteins and peptides. Clay rich soils are often nutrient and oxygen poor environments for microorganisms, owing to characteristically slow diffusion and low hydraulic conductivity (Niqui-Arroyo et al. 2006).

Impact of Clay on Biodegradation

Currently, it remains unclear whether clay minerals play an important role in organic compound biodegradation in soils. Despite the number of investigations determining the role of clays in sorption, undoubtedly a key process in ageing and sequestration, a paucity of studies exist that assess the direct impact of clay on bioavailability/bioaccessibility and biodegradation. In some instances, clay mineral type is the key characteristic in determining the impact; for example, the pre-adsorption of surfactants onto clays before amendment into soil was found to lead to reductions in their microbial utilisation, in the ranked order of (least to most) kaolinite < illite < montmorillinite (Knaebel et al. 1996). In aqueous systems, when the herbicide diquat was completely sorbed to montmorillinite clay, the compound was rendered completely resistant to microbial attack; however, the converse was observed when sorbed to kaolinite (Weber and Coble 1968). In the late 1990s, Alexander and co-workers began to investigate the effects of specific soil properties on biodegradation; however the results drawn were inconclusive. For example, Chung and Alexander (1999) compared the mineralisation and sequestration over time of phenanthrene and atrazine in 16 different soils with a range of physicochemical characteristics. Properties such as organic carbon (OC), log-OC and CEC correlated well with the percentages of both compounds that were not mineralised after sequestration was proceeding; however, the authors did identify that no single overriding factor determined availability. Interestingly, clay mineral type had a variable impact upon measured parameters; for example, the percentages of phenanthrene that was not mineralised after 200 days ageing were comparable between the Fuquay soil (62.0%), which contained 20.5% kaolinite and the Amor soil, which consisted of 28.4% smectite (57.1%). This contrasts with atrazine, where substantially more was mineralised in the Amor soil (51.3%) than in the Fuquay soil (34.5%). Clay type is more significant for polar compounds, such as atrazine, and the larger surface areas associated with swelling clays, such as smectite, appears to have reduced the bioavailability and subsequently biodegradability in this soil. However, caution must be exercised with such generalisations. Huang and Cutright (2003) were able to observe good agreement between the rates and extents of biodegradation of the non-polar PAH pyrene (65.0%, 78.3% and 81.8%) and the fraction of expandable clays present in three soils (4.4, 2.2 and 1.7 wt.% of vermiculite and smectite). Likewise, Ortega-Calvo et al. (1997) yielded a clear trend of decreasing ^{14}C-phenanthrene mineralisation with increasing clay content (5.6%, 13.45% and 22.4%), but comparable organic matter contents.

Sorption to clay minerals will sequester contaminants with respect to mobility and phase distribution; however, it does not necessarily render compounds non-biodegradable; sorbed contaminants can still be susceptible microbial attack (Guerin and Boyd 1992). Desorption of pyrene from three natural soils has been shown to be consistently lower than the fraction biodegraded; this observation was hypothesised to be a result of microbial movement and adhesion to clay particles, which enabled soil-phase biodegradation (Huang and Cutright 2003). Similar explanations have been developed to account for degradation in other systems with low aqueous concentrations and subsequent poor contaminant mass transfer (e.g. Rhodes et al. 2008). Such a theory is not weakened by earlier reported findings that clays obstruct microorganism movement, as it is likely the clay particles will be coated with OM in these natural soils, which may facilitate microbial transloca-tion. For example, Loar et al. (1999) documented a stimulation of phenanthrene biodegradation when phenanthrene was sorbed to a HA-clay complexes, compared with sorption and degradation associated with bare mineral surfaces. Whilst sorption capacity was enhanced by the presence of the HAs, it was postulated that mineral surfaces provided a platform for microbial adsorption in locations where they could directly degrade phenanthrene. However, this study was conducted in highly artificial, aqueous systems. In contrast, Ortega-Calvo et al. (1997) reported that phenanthrene sorbed to clay alone was more readily biodegraded than when associated with HA-clay complexes in soil. ^{14}C-Mineralization was more than 15% greater in soils containing HA-clay complexes than in clay exclusively. These findings were thought to be a result of retarded desorption and diffusion from HA-clay complexes to the autochthonous degraders (Ortega-Calvo et al. 1997).

The lack of agreement pertaining to the impact clay fractions present in soil to biodegradation owes largely to the range and complexity of clay-associated processes that influence bioavailability and bioaccessibility. While this contrasts with the general consensus HOC sequestration can be reconciled to the quantity and quality of f_{oc} in soil, there are still far-reaching implications for clay when assessing biodegradation. Elucidating exactly what these implications are can only be addressed by further laboratory and field based investigations.

Conclusions

The various interactions between clay-biota and clay-organic compounds, in conjunction with SOM fractions, are diverse and competing processes in terms of biodegradation. In Fig. 17.1 the various interactions that can impact upon clay bioavailability and biodegradation in soils have been conceptualised. Ultimately, the affect of clays upon organic chemical biodegradation can be inherently difficult to predict and difficult to unravel. Despite this, some tentative and competing conclusions can be drawn from the current pool of knowledge about processes that impact upon the long term bioaccessibility of organic compounds. (1) Differential sorption affinities are evident with sorbate polarity, although clays may represent significant sorption pools for HOCs under certain environmental conditions. (2) Swelling clay types, such as smectite, show a greater propensity towards sorption than non-swelling clays. (3) Clays are readily modified to increase sorption affinity and capacity by altering chemical composition of intercalated cations. (4) Clay-rich soils are hostile environments for bacterial translocation and biodegradation. (5) The presence of OM on clay surfaces may alter all the impacts that processes 1–4 can have upon bioaccessibility in comparison to OM-free mineral surfaces. (6) Processes 1–5 all interact at different levels to influence compound biodegradation and are affected by local environmental conditions.

The complexities of the processes that impact upon organic compound biodegradation warrant further systematic investigation, utilising a wide range of natural soils under environmentally relevant conditions (pH, temperature etc). Although experiments that only investigate one or two variables are not reflective or environmental scenarios, they still hold value as they aid our mechanistic understanding and facilitate the interpretation of results. Such research is necessary to further refine our understanding the fate and behaviour of organics in the environment. How these findings may be reconciled to allow prediction of biodegradation for the risk assessment and remediation of contaminated soils remains unresolved.

Acknowledgements The authors would like to thank the European Commision under FP7 – Environmental Technologies (ModelProbe, No 213161) for funding.

References

Alexander M (1999) Biodegradation and bioremediation, 2nd edn. Academic Press, London
Amellal N, Portal JM, Berthelin J (2001) Effect of soil structure on the bioavailability of polycyclic aromatic hydrocarbons within aggregates of a contaminated soil. Appl Geochem 16:1611–1619
Bailey SW (1980) Summary of recommendations of AIPEA Nomenclature Committee. Clay Miner 15:85–93
Ball WP, Roberts PV (1991a) Long-term sorption of halogenated organic chemicals by aquifer material 1. Equilibrium. Environ Sci Technol 25:1223–1237
Ball WP, Roberts PV (1991b) Long-term sorption of halogenated organic chemicals by aquifer material 2. Interparticle diffusion. Environ Sci Technol 25:1237–1249
Beall GW (2003) The use of organo-clays in water treatment. Appl Clay Sci 24:11–20

Berkowitz B, Dror I, Yaron B (2008) Contaminant geochemistry: interactions and transport in the subsurface environment. Springer, Berlin

Bertsch PM, Seaman JC (1999) Characterization of complex mineral assemblages: implication for contaminant transport and environmental remediation. Proc Natl Acad Sci USA 96: 3350–3357

Borisover M, Gerstl Z, Burshtein F, Yariv S, Mingelgrin U (2008) Organic sorbate-organoclay interactions in aqueous and hydrophobic environments: sorbate-water competition. Environ Sci Technol 42:7201–7206

Boyd SA, Sheng GO, Teppen BJ, Johnson CT (2001) Mechanisms for the adsorption of substituted nitrobenzenes by smectite clays. Environ Sci Technol 35:4227–4234

Brindley GW (1980) Crystal Structures of clay minerals and their X-ray identification. Mineralogical Society, London

Cerniglia CE (1992) Biodegradation of polycyclic aromatic hydrocarbons. Biodegradation 3:351–368

Chai Y, Qui X, Davis JW, Budinsky RA Jr, Bartels MJ, Saghir SA (2007) Effects of black carbon and montmorillinite clay on multiphasic hexachlorobenzene desorption from sediments. Chemosphere 69:1204–1212

Charles S, Teppen BJ, Li H, Laird DA, Boyd SA (2006) Exchangeable cation hydration properties strongly influence soil sorption of nitroaromatic compounds. Soil Sci Soc Am J 70:1470–1479

Chen BL, Huang WH (2009) Effect of background electrolytes on the adsorption of nitroaromatic compounds onto bentonite. J Environ Sci Chin 21:1044–1052

Cornelissen G, Gustafsson O (2005) Importance of unburned coal carbon, black carbon, and amorphous organic carbon to phenanthrene sorption in sediments. Environ Sci Technol 39:764–769

Cornelissen G, Van Noort PCM, Govers HAJ (1997) Desorption kinetics of chlorobenzenes, polycyclic aromatic hydrocarbons, and polychlorinated biphenyls: sediment extraction with tenax and effects of contact time and solute hydrophobicity. Environ Toxicol Chem 16:1351–1357

Cornelissen G, Rigterink H, Ferdinandy MMA, Van Noort PCM (1998a) Rapidly desorbing fractions of PAHs in contaminated sediments as a predictor of the extent of bioremediation. Environ Sci Technol 32:966–972

Cornelissen G, Rigterink H, van Noort PCM, Govers HAJ (2000) Slowly and very slowly desorbing organic compounds in sediments exhibit langmuir-type sorption. Environ Toxicol Chem 19:1532–1539

Cox L, Hermosin MC, Cornejo J (1996) Sorption of metamitron on soils with low organic matter content. Chemosphere 32:1391–1400

Cox L, Hermosin MC, Koskinen WC, Cornejo J (2001) Interactions of imidacloprid with organic- and inorganic-exchanged smectites. Clay Miner 36:267–274

Cox L, Hermosin MC, Celis R et al (1997) Sorption of two polar herbicides in soils and soil clays suspensions. Water Res 31:1309–1316

Crocker FH, Guerin WF, Boyd SA (1995) Bioavailability of naphthalene sorbed to cationic surfactant-modified smectite clay. Environ Sci Technol 29:2953–2958

Ehlers LJ, Luthy RG (2003) Contaminant bioavailability in soil and sediment. Improving risk assessment and remediation rests on better understanding bioavailability. Environ Sci Technol 37(15):295A–302A

Ghosh DR, Keinath TM (1994) Effect of clay minerals present in aquifer soils on the adsorption and desorption of hydrophobic organic compounds. Environ Prog 13:51–59

Guerin WF, Boyd SA (1992) Differential bioavailability of soil-sorbed naphthalene to two bacterial species. Appl Environ Microbiol 58:1142–1152

Guthrie EA, Pfaender FK (1998) Reduced pyrene bioavailability in microbially active soils. Environ Sci Technol 32:501–508

Haderlein SB, Weissmahr KW, Schwarzenbach RP (1996) Specific adsorption of nitroaromatic: Explosives and pesticides to clay minerals. Environ Sci Technol 30:612–622

Hatzinger PB, Alexander M (1995) Effect of aging of chemicals in soil on their biodegradability and extractability. Environ Sci Technol 29:537–545

Huang S, Cutright TJ (2003) Preliminary exploration of the relationships between soil characteristics and PAH desorption and biodegradation. Environ Int 29:887–894

Huang W, Peng P, Yu Z, Fu J (2003) Effects of the heterogeniety on sorption and desorption of organic contaminants by soils and sediments. Appl Geochem 18:955–972

Huang Q, Liang W, Cai P (2005) Adsorption, desorption and activities of acid phosphatise on various colloidal particles from an Ultisol. Colloid Surf B 45:209–214

Hundal LS, Thompson ML, Laird DA, Carmo AM (2001) Sorption of phenanthrene by reference smectites. Environ Sci Technol 35:3456–3461

Jackson ML (1957) Frequency distribution of clay minerals in major great soil groups as related to the factors of soil formation. Clay Clay Miner 6:133–143

Johnsen AR, Wick LY, Harms H (2005) Principles of microbial PAH-degradation in soil. Environ Pollut 133(1):71–84

Kelleher B, Simpson AJ, Willeford KO, Simpson MJ, Stout R, Rafferty A, Kelsey JW, Alexander M (1997) Declining bioavailability and inappropriate estimation of risk of persistent compounds. Environ Toxicol Chem 16:582–585

Kelsey JW, Kottler BD, Alexander M (1997) Selective chemical extractants to predict bioavailability of soil-aged organic chemicals. Environ Sci Technol 31:214–217

Knaebel DB, Federle TW, McAvoy DC, Vestal JB (1996) Microbial mineralisation or organic compounds in an acidic agricultural soil: effects of preadsorption to various soil constituents. Environ Toxicol Chem 15:1865–1875

Koelmans AA, Jonker MTO, Cornelissen G, Bucheli TD, Van Noort PCM, Gustafsson O (2006) Black carbon: the reverse of its dark side. Chemosphere 3:365–377

Lagaly G (2001) Pesticide–clay interactions and formulations. Appl Clay Sci 18:205–209

Lahlou M, Harms H, Springael D, Ortega-Calvo JJ (2000) Influence of soil components on the transport of polycyclic aromatic hydrocarbon-degrading bacteria through saturated porous media. Environ Sci Technol 34:3649–3656

Laird DA, Barriuso E, Dowdy RH, Koskinen WC (1992) Adsorption of atrazine on smectites. Soil Sci Soc Am J 56:62–67

Li H, Sheng G, Teppen BJ, Johnston CT, Boyd SA (2003) Sorption and desorption of pesticides by clay minerals and humic acid-clay complexes. J Soil Sci Soc Am 67:122–131

Li YR, Struger J, Fischer JD, Huang GH, Li YF (2001) Modeling of pesticide runoff losses from agricultural lands – A Canadian case study. Conference Information: 29th Annual Congress of the International-Association-of-Hydraulic-Engineering-and-Research (IAHR). Beijing Int Convent CTR, Beijing, Peoples Republic of China, 16–21 Sept 2001 (Source: Environmental hydraulics and eco-hydraulics, theme B, Proceedings – 21st century: The new era for hydraulic research and its applications, pp 616–626)

Loar Y, Strom PF, Farmer WJ (1999) Bioavailability of phenanthrene sorbed to mineral-associated humic acid. Water Res 33:1719–1729

Mader BT, Uwe-Goss K, Eisenreich SJ (1997) Sorption of nonionic, hydrophobic organic chemicals to mineral surfaces. Environ Sci Technol 31:1079–1086

Mayer LM, Xing B (2001) Organic matter-surface area relationships in acid soils. J Soil Sci Soc Am 65:250–258

Miller ME, Alexander M (1991) Kinetics of bacterial degradation of benzylamine in a montmorillonite suspension. Environ Sci Technol 25:240–245

Murphy EM, Zachara JM, Smith SC (1990) Influence of mineral-bound humic substances on the sorption of hydrophobic organic-compounds. Environ Sci Technol 24:1507–1516

Nam K, Chung N, Alexander M (1998) Relationship between organic matter content of soil and the sequestration of phenanthrene. Environ Sci Technol 32:3785–3788

Nennemann A, Mishael Y, Nir S, Rubin B, Polubesova T, Bergaya F, van Damme H, Lagaly G (2001) Clay-based formulations of metolachlor with reduced leaching. Appl Clay Sci 18:265–275

Newman ACD (ed) (1987) Chemistry of clays and clay minerals. Mineralogical Society Monograph No. 6, Longman Scientific and Technical

Niqui-Arroyo J, Bueno-montes M, Posada-Baquero R, Ortego-Calvo JJ (2006) Electrokinetic enhancement of phenanthrene biodegradation in creosote-polluted clay soil. Environ Pollut 142:326–332

Ogawa M, Wada T, Kuroda K (1995) Intercalation of pyrene into alkylammonium-exchanged swelling layered silicates: the effects of the arrangements of the interlayer alkylammonium ions on the states of adsorbate. Langmuir 11:4598–4600

Onken BM, Traina SJ (1997) The sorption of pyrene and anthracene to humic acid-mineral complexes: effect of fractional organic carbon content. J Environ Qual 26:126–132

Ortega-Calvo JJ, Lahlou M, Saiz-Jimenez C (1997) Effect of organic matter and clays on the biodegradation of phenanthrene in soil. Int Biodeterior Biodegrad 40:101–106

Peng X, Wang J, Fan B, Luan Z (2009) Sorption of endrin to montmorillonite and kaolinite clays. J Hazard Mater 168:210–214

Pignatello JJ (1990) Slowly reversible sorption of aliphatic hydrocarbons in soils. I. Formation of residual fractions. Environ Toxicol Chem 9:1107–1115

Pignatello JJ, Xing B (1996) Mechanisms of slow sorption of organic chemicals to natural particles. Environ Sci Technol 30:1–11

Rao MA, Violante A, Gianfreda L (2000) Interaction of acid phosphatase with clays, organic molecules and organo-mineral complexes: kinetics and stability. Soil Biol Biochem 32:1007–1014

Reid BJ, Jones KC, Semple KT (2000) Bioavailability of persistent organic pollutants in soils and sediments – a perspective on mechanisms, consequences and assessment. Environ Pollut 108:103–112

Rhodes AH, Carlin A, Semple KT (2008) Impact of black carbon in the extraction and mineralization of phenanthrene in soil. Environ Sci Technol 42:740–745

Rowell DL (1994) Soil science: methods and applications. Longman Group, UK

Sannino F, Violante A, Gianfreda L (1997) Adsorption-desorption of 2,4-D by hydroxy aluminium montmorillonite complexes. J Pest Sci 51:429–435

Sannino F, Filazzola MT, Violante A, Gianfreda L (1999) Adsorption-desorption of simazine on montmorillonite coated by hydroxy aluminum species. Environ Sci Technol 33:4221–4225

Semple KT, Doick KJ, Wick L, Harms H (2007) Microbial interactions with organic contaminants in soil: definitions, processes and measurement. Environ Pollut 150:166–176

Semple KT, Doick KJ, Burauel P, Craven A, Harms H, Jones KC (2004) Defining bioavailability and bioaccessibility of contaminated soil and sediment is complicated. Environ Sci Technol 38:228A–231A

Semple KT, Morriss AWJ, Paton GI (2003) Bioavailability of hydrophobic organic contaminants in soils: fundamental concepts and techniques for analysis. Eur J Soil Sci 564:1–10

Shen L, Jaffé R (2000) Interactions between dissolved petroleum hydrocarbons and pure and humic acid-coated mineral surfaces in artificial seawater. Mar Environ Res 49:217–231

Sheng G, Johnston CT, Teppen BJ, Boyd SA (2002) Adsorption of dinitrophenol herbicides from water by montmorillinites. Clay Clay Miner 50:25–34

Singer MJ, Munns DN (2002) Soils: an introduction, 5th edn. Prentice Hall, New Jersey

Stokes JD, Paton GI, Semple KT (2005) Behaviour and assessment of bioavailability of organic contaminants in soil: relevance for risk assessment and remediation. Soil Use Manage 21:475–486

Talley JW, Ghosh U, Tucker SG, Furey JS, Luthy RG (2002) Particle-scale understanding of the bioavailability of PAHs in sediment. Environ Sci Technol 36:477–483

Theng BKG, Aislabie J, Fraser R (2001) Bioavailability of phenanthrene intercalated into an alkylammonium montmorillonite clay. Soil Biol Biochem 33:845–848

Theng BKG, Newman RH, Whitton JS (1998) Characterization of an alkylammonium-montmorillonite-phenanthrene intercalation complex by carbon-13 nuclear magnetic resonance spectroscopy. Cly Miner 33:221–229

Velde B (1996) Origin and mineralogy of clays: clays and the environment. Springer, Germany

Weber JB, Coble HDJ (1968) Microbial decomposition of diquat adsorbed on montmorillonite and kaolinite clays. J Agric Food Chem 16:475–478

White JC, Kelsey JW, Hatzinger PB, Alexander M (1997) Factors affecting sequestration and bioavailability of phenanthrene in soils. Environ Toxicol Chem 16:2040–2045

White JC, Hunter M, Nam K, Pignatello JJ, Alexander M (1999) Correlation between biological and physical availabilities of phenanthrene in soils and soil humin in aging experiments. Environ Toxicol Chem 18:1720–1727

Wick LY, Remer R, Birgitwu RZ, Reichenbach RZ, Braun S, Chafer FS, Harms H (2007) Effect of fungal hyphae on the access of bacteria to phenanthrene in soil. Environ Sci Technol 41:500–505

Witthuhn B, Klauth P, Klumpp E, Narres H, Martinius H (2005) Sorption and biodegradation of 2,4-dichlorophenol in the presence of organoclays. Appl Clay Sci 28:55–66

Chapter 18
Electrodes as Electron Acceptors, and the Bacteria Who Love Them

Daniel R. Bond

When Faraday wanted to explain the forces of nature, he chose the example of a candle. The disappearance of paraffin and production of water was visible in any laboratory or lecture hall (Faraday 1988). However, when the topic turned to electricity, Faraday questioned his ability to provide a tangible example, saying, "I wonder whether we shall be too deep to-day or not". Electron movement clearly is a less obvious phenomenon, something that occurs invisibly between molecules, within the candle flame, or at the moment a glass rod is rubbed with wool (Faraday 2000).

Microbial metabolism also conceals from view the electrical flow that powers living systems. We measure the products of electron trafficking; gas evolution, accumulation of reduced metals, the generation of heat. But because electrons must be carefully passed tiny distances, between proteins and redox centers within the confines of an insulating membrane, gaining direct access to this flow seems unlikely. Despite these barriers, it has recently become common to place electrodes in anaerobic environments, and collect a current of electrons at the expense of microbial activity (Reimers et al. 2001; Tender et al. 2002, 2008; Liu et al. 2004; Logan 2005), or monitor electrons flowing out of bacteria using electrodes as respiratory electron acceptors (Bretschger et al. 2007; Marsili et al. 2008a, b; Richter et al. 2009). These demonstrations show that a window into life's electron flow exists.

The fact that some bacteria can direct electrons far beyond the cell surface elicits ideas for energy generation, bioremediation, and sensing, which all draw from the use of an electrode as one half of a bioelectrochemical reaction. The challenges related to harnessing bacteria for fuel-cell like devices have been detailed in multiple reviews (Logan and Regan 2006; Du et al. 2007; Rabaey et al. 2007; Schroder 2007; Lovley 2008a, b), and the field has even produced its first text (Logan 2008). But, at the heart of all of these technologies are the bacteria, who evolved this ability for other purposes, such as reduction of metals or redox-active compounds in

D.R. Bond (✉)
Department of Microbiology and Biotechnology Institute, University of Minnesota,
1479 Gortner Ave, St Paul, MN 55105, USA
e-mail: dbond@umn.edu

L.L. Barton et al. (eds.), *Geomicrobiology: Molecular and Environmental Perspective*,
DOI 10.1007/978-90-481-9204-5_18, © Springer Science+Business Media B.V. 2010

the environment. This chapter will discuss what respiration to a seemingly artificial electrode, and this tangible electron flow, can tell us about organisms active in biogeochemical cycling.

Is an Electrode a Defined Habitat?

Electrodes, especially as they are currently used in the cultivation of microorganisms, are not all the same. The surface chemistry of graphitic, glassy, or other carbon materials can differ widely, roughness can influence available surface area (McCreery 2008), and three dimensional structures (flat versus fibrous) can affect diffusion. For example, the electrodes used for microbial fuel cell research can be ammonia-treated carbon brushes (Cheng and Logan 2007; Logan et al. 2007), gold (Richter et al. 2008), carbon cloth (Nevin et al. 2008), stainless steel (Dumas et al. 2008), or carbon-coated titanium (Biffinger et al. 2008). The electrochemical potential of these surfaces to accept electrons can be precisely controlled (via a potentiostat), or allowed to drift (as in a fuel cell). These variables are somewhat analogous to the challenges in comparing data between laboratories studying metal oxide respiration, in terms of how surface charge, surface area, passivation-influenced reactivity, and accessibility of pore spaces can alter outcomes (Roden and Urrutia 2002; Roden 2006).

Another set of variables that affect measurements of microbial activity lie in the device used to house the electrode. As negative charge moves into the electrode and travels via a wire to a counter electrode (the cathode), positive charge must as quickly migrate this same distance, but through the biofilm and electrolyte (Torres et al. 2008a, b). This is again where porosity and three-dimensional effects can alter the environment, and the resistance imposed between the two electrodes can lead to incorrect interpretations of bacterial capability.

A demonstration of this effect was a set of comparisons by Liang et al. (2007), who inoculated reactors containing identical electrodes, but arranged in three different configurations that impacted charge equilibration between electrodes (commonly quantified as "internal resistance"). The rate these electrodes could collect current from bacteria varied more than 20-fold, yet the bacterial inoculum and conditions were otherwise identical. Dewan et al. (2008), illustrated this under even more controlled conditions. Simply changing the configuration of electrodes (such as the ratio of electrode surface areas), dramatically altered how a pure culture would perform. From these examples, one can see how confusion may arise, in terms of comparing bacterial abilities; there are cited examples of "power output" by bacteria in fuel cells that vary over 100-fold (as high as 5 W/m^2 of electrode to as low as 0.03 W/m^2) (Liang et al. 2007; Dewan et al. 2008; Zuo et al. 2008a; Yi et al. 2009). This likely reflects differences in internal resistance or electrode configuration of the devices, rather than isolation of bacteria capable of strikingly different electron transport rates.

Thus, the "electrode" is not a fixed or defined environment, but an electron acceptor that can vary widely in surface charge, porosity, and electron acceptor potential, which is incubated like any other electron acceptor in a medium controlled for salinity, microaerobic versus strictly anaerobic conditions, mixing, and other factors. The diversity of possible electrode-based experiments, largely conducted in fuel-cell like devices, has led to isolation of a wider variety of organisms known to direct electrons beyond their outer surface, compared to experiments with Fe(III) as the electron acceptor. In addition, as researchers have focused their attention on controlling the electrode environment more precisely, specific abilities related to extracellular electron transfer have become more apparent.

What Bacteria Can Use Electrodes as Electron Acceptors?

A long list of organisms have demonstrated a qualitative ability to produce electrical current, although the factors described above discourage direct comparison. Even *E. coli* (Zhang et al. 2008b), and yeast (Prasad et al. 2007), have been persuaded to produce some measure of current at electrodes, although these observations are typically linked to cell lysis (and release of redox-active compounds) or evolution under laboratory pressure. In fact, before metal-reducing bacteria were discovered, microbial-electrode research largely focused on fermentative growth of organisms such as *Proteus* (Kim et al. 2000), or *Bacillus* (Choi et al. 2001), which could divert a small percentage of their metabolism to reduction of soluble redox-active mediators, which could then be oxidized by electrodes. These observations do not necessarily mean that organisms such as *E. coli* (Zhang et al. 2008b), yeast (Prasad et al. 2007), *Clostridia* (Park et al. 2001; Prasad et al. 2006), and *Klebsiella* (Zhang et al. 2008a), are using extracellular electron acceptors such as Fe(III) in their environment, but it may indicate a competitive advantage exists from disposal of a small number of redox equivalents under anaerobic conditions.

A second way to approach this question is 16S-based rDNA surveys of electrodes used to enrich for bacteria under various conditions (Lee et al. 2003; Pham et al. 2003; Back et al. 2004; Holmes et al. 2004a, b; Kim et al. 2004; Phung et al. 2004; Logan et al. 2005; Jong et al. 2006; Kim et al. 2006, 2007; Jung and Regan 2007; Catal et al. 2008; Ishii et al. 2008a, b; Mathis et al. 2008; Park et al. 2008; Wrighton et al. 2008; Zuo et al. 2008b). It should be straightforward to ask who survives or is enriched in such devices, and use this to create a list of putative electrode-reducing bacteria. However, in many of these studies, oxygen leakage into the electrode chamber from the cathode can support growth of a subpopulation of aerobic heterotrophs, partial flux of electrons through the anaerobic food chain to methanogenesis may support a normal anaerobic community, and substrates used to enrich the bacteria may be partially fermentable (e.g. glucose, lactate, ethanol), supporting a mixed community of fermentative and respiratory organisms. Thus, it is not surprising that such surveys find a general enrichment of multiple low

G+C gram positive fermentative and syntrophic organisms, many *Proteobacteria* with known metal-reduction and respiratory abilities, and *Bacteriodes* commonly found in anaerobic habitats.

Sifting through these observations, however, a few trends can be noted. Sequences related to *Geobacteraceae* are commonly enriched when simple fatty acids (such as acetate) are used as the electron donor, especially when the medium is buffered with CO_2 (Bond et al. 2002; Holmes et al. 2004a, b; Jung and Regan 2007; Chae et al. 2008; Ha et al. 2008; Ishii et al. 2008c). Such findings are consistent with enrichments that repeatedly obtain *Geobacter* using acetate as the electron donor and Fe(III) as the electron acceptor (Lovley et al. 1993; Coates et al. 1996; Straub et al. 1998; Snoeyenbos-West et al. 2000; Bond et al. 2002; Kostka et al. 2002; Nevin et al. 2005), and with the fact that *Geobacter* requires CO_2 to build C3 metabolites from acetyl-CoA, and synthesize amino acids (Mahadevan et al. 2006; Tang et al. 2007; Sun et al. 2009). Thus, the metal-reduction machinery of *Geobacter* appears well-suited to also competing for electrodes as insoluble electron acceptors, especially when other strategies are not available.

Perhaps more interesting are cases where *Geobacteraceae* are not enriched on electrodes, even when similar surfaces and inocula have been used. Or, even when *Geobacter* is present, certain genera seem to be reported in multiple 16S rDNA libraries, indicating that they are able to take advantage of the electrode in some way. Examples include communities where *Pseudomonas* spp. (Rabaey et al. 2004), *Gammaproteobacteria* (Rabaey et al. 2004, 2005), *Rhizobiales* (Ishii et al. 2008b), *Azoarcus* and *Dechloromonas* (Kim et al. 2007), and Firmicutes (Jung and Regan 2007; Mathis et al. 2008; Wrighton et al. 2008) have either dominated or represented significant percentages of the electrode-attached population. In many of these cases, follow-up studies have isolated new pure cultures with electrode-reducing capabilities. These findings suggest that passing electrons to the outer surface may confer an advantage, even in nitrate-reducing or fermentative habitats, and that use of an electrode as bait allows us to catch organisms that would otherwise be missed when using metal reduction as the sole selective pressure.

Examples of isolates that support this idea include *Rhodopseudomonas palustris* DX-1 (Xing et al. 2008), *Dechlorospirillum* VDY (Thrash et al. 2007), *Ochrobactrum anthropi* YZ-1 (Zuo et al. 2008b) the prosthecae-like strain Mfc52 (Kodama and Watanabe 2008), and the thermophllic Gram-positive *Thermincola* sp. strain JR (Wrighton et al. 2008). Electron transfer sustained by these organisms to electrodes appears to be robust, both in terms of overall rates, and in terms of their ability to form biofilms. In addition, as most were obtained using the electrode as the enrichment tool, these isolates were truly competitive for the electrode as a primary source of energy generation. As most fundamental information regarding electron transfer to external acceptors is based on organisms such as *Geobacter* and *Shewanella*, further data showing the molecular basis and physiological role of extracellular electron transfer in these isolates may shed new light on the ecological role of this widespread ability.

What Does It Take to Use an Electrode?

When growth on an electrode is being described, it is important to consider the events taking place. Electrode reduction requires oxidation of a substrate (often in multiple steps), transfer of electrons across membranes, surface attachment, possible secretion of mediators, and, in the case of biofilms, cell-cell contact. Key questions that arise in the study of electron transfer to metals, or electrodes, involve which steps are rate-limiting, and how each step responds to changing in driving force (as either substrate concentration or voltage). Figure 18.1 shows a simplified view of these interactions.

For all cells, the rate of oxidation will be sensitive to the supply and concentration of donor, but only the first layer of electrode-interacting cells mediates electron transfer. If this basal layer of cells can pass electrons to the electrode directly, there will be a relationship between the electrode potential and the rate electrons traverse their final hop from proteins to the electrode. While the rate of enzymatic catalysis in response to substrate availability is typically visualized as a Michaelis-Menten system (reaction reaches half V_{max} at concentration $[K_m]$, and saturates at V_{max}), electron transfer rates respond according to Butler-Volmer theory, where the rate of the forward reaction increases exponentially with the difference in donor and acceptor potentials.

Thus, while an enzyme has a maximal turnover rate constant (k_{cat}), reached at high substrate concentrations, electron transfer is described by a rate constant achieved at

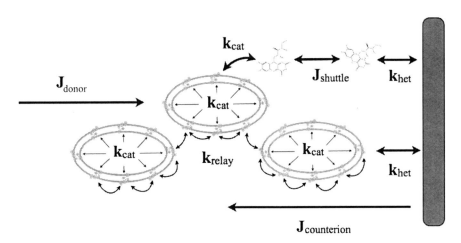

Fig. 18.1 Cartoon model of steps which may control the overall phenotype of current flux to electrodes. Electron donor flux into biofilm leads to oxidation of substrate, followed by electron transfer to electron shuttles or directly to the electrode surface. More distant cells rely on a relay between cells and electrodes, and all electron movement must be compensated by outward flux of ions. Fluxes (J) are limited by diffusion rates and concentration gradients, while enzymatic reaction rates (k_{cat}) are governed by substrate availability. Reaction rates at electrodes (k_{het}) respond to changes in electrode potential

the midpoint potential of the redox species (k_{het}), that can be increased nearly 50-fold simply by application of a small (~0.2 V) potential. This difference in responses usually means that an enzymatic reaction transferring electrons to an electrode can be easily brought to the point where enzyme turnover rate (k_{cat}) is rate-limiting, relative to the rate electrons can hop from the enzyme to the electrode, if sufficient potential is applied. In other words, if the electrical connection is good, one would expect application of a small potential to pull the system to a point where the maximum turnover rate of the cell's enzymatic machinery is reached. If the connection is poor, each increase in potential at the electrode would be met with an increase in current, as there remains an excess of enzymatic capacity waiting to be used.

If the organism is utilizing a soluble mediator to shuttle electrons to the electrode, the Butler-Volmer relationship will still control mediator-electrode reaction rates, and perhaps Michaelis-Menten kinetics will describe cell-mediator kinetics, but, the mediator needs to diffuse to the electrode. Unlike electron transfer, which responds to applied potential, diffusion is a function of the compound's mobility, and concentration. Thus, as with direct electron transfer, application of a small potential can easily accelerate surface oxidation rates, and leave delivery of electrons to the electrode (diffusion) as the rate-limiting step. In such a case, the primary mechanism of increasing flux to the electrode is to increase mediator concentrations.

If cells are to stack on top of each other, rates of current production can scale significantly. However, for this to occur, a mechanism of longer-range electron transfer is required. In addition, cells now attach to other cells, rather than the electrode. Thus, new factors must come into play; relay of electrons from cell to cell (perhaps via cytochrome-to-cytochrome collisions, mediators, or conductive macromolecules [Reguera et al. 2005; Gorby et al. 2006, 2008; Juarez et al. 2009]), and cell–cell aggregation rather than surface attachment. If long-range transfer is via soluble mediators, the relatively slow rates of diffusion restrict this strategy from being useful over distances greater than a few microns (Picioreanu et al. 2007, 2008). Basic diffusion calculations strongly support the conclusion that when organisms (such as *G. sulfurreducens*) form >20 μm thick biofilms, and the rate of respiration per unit biomass is increasing as cells are added to these biofilms, some non-soluble mediator based mechanism for relaying electrons to the electrode must be active. A TEM image of an actual *Geobacter* biofilm, showing the intimate cell–cell contact throughout the film, compared to the low percentage of cells that form the actual connection with electrodes, is shown in Fig. 18.2. This image shows how, when we think of bacteria "growing on electrodes", the majority of the bacteria are actually not in contact with the electrode, but rely on each other (or each other's outer surface) as conduits.

Finally, as cells grow thicker on the electrode, there is the question of flux out of the biofilm. Recent modeling (Torres et al. 2007, 2008a, c), and confocal microscopy experiments (Franks et al. 2009) have suggested that the flux of protons or positive charge out of the biofilm can ultimately become limiting for some organisms. In other words, even if the bacteria have solved the problem of attaching, bringing electrons to their outer surface, relaying these negative charges quickly

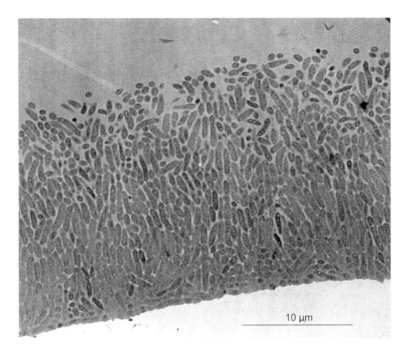

Fig. 18.2 TEM cross section of a *G. sulfurreducens* biofilm, grown using an electrode as the electron acceptor (3,400 × magnification). Cells at the *bottom* of the image are the only cells in contact with the electrode. The electron transfer rate per cell in this biofilm was consistent with each cell respiring at a similar rate, implying a mechanism for relaying electrons from more distant cells to the surface (Image credit; E. LaBelle and G. Ahlstrand)

between cells, and getting them to the electrode, they will eventually outpace the rate positive charge can escape the biofilm. Evidence for this issue can be seen in the recent evolution of a laboratory strain (KN400), which had an increased capacity for electrode reduction, while growing as thinner biofilms (Yi et al. 2009).

Some Examples of the Relationship Between Potential and Electron Transfer Rates

The mixture of enzymatic, diffusional, and electrochemical laws that govern electron transfer to electrodes can result in organisms capable of similar rates of electron transfer to electrodes, but at very different potentials. This is reminiscent of the classic descriptions of anaerobic microbial communities defined not by their *ability* to use hydrogen, but the threshold below which they cannot metabolize hydrogen (Lovley 1985; Lovley and Goodwin 1988). Some real and theoretical current-voltage curves are shown for illustration (Fig. 18.3).

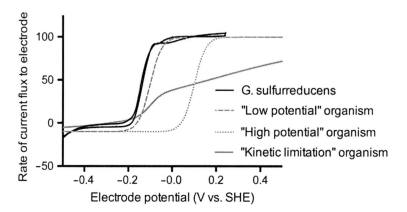

Fig. 18.3 Three theoretical slow scan rate current-potential curves, and representative data from *G. sulfurreducens* (*thick black line*). If the potential of the electrode is raised to a high enough potential, all organisms could be described as having a similar rate of electron transfer to electrodes. However, at lower potentials, some of these organisms are unable to transfer electrons, and show a different phenotype. For example, at −0.1 V, the low potential organism (*dashed line*) would respire faster than the high potential (*dotted line*) or kinetic limitation (*grey line*) organism

   To obtain this data, bacteria are typically grown on electrodes under conditions where electron donor is in excess (e.g., higher concentrations do not result in faster respiration rates), and the potential of the electrode is held sufficiently positive as to be non-limiting (e.g, higher potentials do not result in faster respiration rates). The potential is then swept from high to low potential slowly (1 mV/s), and this slow change in potential allows sampling of the rate of respiration across a range of applied potentials. Because the observed rates are being controlled by the electrode potential, they should be similar regardless of the direction of the potential sweep.
   In the examples shown in Fig. 18.3, there are two organisms ("high potential" and "low potential") who have identical maximum rates of current flux to the electrode. If the electrode were held at a sufficiently high potential (>0.2 V), then these would both appear similar, and would be expected to compete equally at an electrode. However, when the electrode potential is held at −0.1 V, the "high potential" organism cannot transfer electrons to the electrode, while the "low potential" organism could respire at half its maximal rate. Thus, in a device such as a fuel cell, where the potential is allowed to drift (and typically equilibrates to a low potential), the "low potential" organism would be highly favored, and would out compete other strains which have similar *rates* of electron transfer, but require a higher electron acceptor potential. Similarly, if these organisms were competing for an oxidized iron mineral with a low potential, the "low potential" organism would be favored.
   Both of these theoretical organisms respond similarly to changes in driving force. The switch from no electron transfer to a maximal rate occurs over a narrow window, which is defined by the Nernst equation. The flat plateau seen as the potential is raised higher suggests that electron transfer rate from terminal

cytochromes (which is accelerated by potential) is not rate-limiting, but more likely, a key step in substrate oxidation within the cell, or a key step in electron transfer within the membrane has become rate-limiting. This behavior, with a characteristic threshold near −0.2 V, and a half-maximal current around −0.15 V, is what is typically observed for *G. sulfurreducens* (Marsili et al. 2008b). This sigmoidal wave-like shape is what would be expected if the connection between bacteria and electrodes was facile, and some step in acetate oxidation or transfer of electrons to the outer membrane by cells represented the slowest step.

Such findings seem contrary to the notion that reduction of external metals is slow, an assumption based on slow growth rates of bacteria on insoluble metals versus chelated metals. These observations with electrodes suggest it is the process of finding metal particles, reducing them, and finding others, while fighting against the competing forces of passivation and limiting surface area – which cause slower growth rates, not slow rates of surface electron transfer per se. Thus, when an electrode is provided as a constant, unchanging electron acceptor, cells grow and respire as fast as when soluble acceptors are present.

A second possible type of response is also shown, where there is an initial rise in electron transfer rate, followed by a more general linear response. This could be caused by many factors affecting the kinetics of electron transfer. Proteins may be heterogeneously oriented at the electrode, with each requiring a different driving force for optimal electron transfer, or, electron transfer from proteins could have a slow rate constant, and require significant overpotential to drive the reaction. In any case, there appears to be some resistance to electron transfer, which can be overcome with potential, leading to a "kinetic limitation" in electron transfer. With enough driving force, even this organism could produce the same rate of electron transfer as the other examples.

This example again shows how a report of the overall rate of electron transfer can have less ecological meaning than the potential at which that rate is achieved. While the "kinetic limitation" organism looks, in the figure, to be rather sluggish, and to have a low rate of electron transfer versus the "high potential" organism, consider the case where the electrode is at −0.15 V. Under these conditions, the "kinetic limitation" and "low potential" organisms would be capable of similar electron transfer rates, and would out compete organisms requiring a higher potential. In the environment, the "high potential" organism would be expected to grow well in the presence of chelated metals, or high potential metal acceptors (perhaps Mn(IV)), but would not be thermodynamically capable of reducing lower potential metal oxides.

Similarities with the Thermodynamics of Metal Reduction

In the environment, Fe(III) minerals likely have a potential to accept electrons near 0 V or lower (Straub et al. 2001). Thus, to generate energy, an organism using donors near a potential of −0.3 V (typical of simple fatty acids) only has, at best, a small potential drop to use in energy generation. The fact that *Geobacter* can transfer electrons to external acceptors at potentials as low as −0.22 V, and reaches its maximal rate of electron transfer by nearly −0.1 V, shows how little energy this organism 'keeps for itself', while moving electrons through a proton-pumping electron transport chain, to reach the outer surface.

For the reduction of a −0.22 V acceptor to still be favorable, while linked to oxidation of a donor with a reduction potential of −0.28 V (the $E^{\circ\prime}$ of acetate), a ΔE of approximately 0.06 V per electron remains for proton pumping and ATP generation. In the case of acetate oxidation (an eight electron oxidation), this is equivalent to ($\Delta G = -nF\Delta E$), and predicts that only −46 kJ/acetate is available for ATP generation by *Geobacter*. This is consistent with models and yield data that estimate the ATP/acetate of *Geobacter* to be approximately 0.5 ATP/acetate (Mahadevan et al. 2006; Sun et al. 2009). Clearly, while higher potential acceptors exist (either as freshly precipitated ferrihydrite, or higher potential electrodes), this evidence suggests *G. sulfurreducens* is most competitive when *both* donors and acceptors are limiting, and near the thermodynamic threshold for growth.

Bacteria which couple very little of their net electron flow to ATP production should therefore dominate when the selective pressure is a low electrode potential. This ability of organisms to grow near thermodynamic equilibrium explains why anodes of fuel cells self-equilibrate to potentials near −0.2 V versus SHE, and why this can lead to powerful enrichment of *Geobacter*-like organisms. It also shows that there may be little that can be done in terms of discovering bacteria capable of creating stronger potentials, so long as the same donors are used as fuel. If the potential of the cathode, using oxygen as an acceptor, is fixed (e.g., at +0.4 to +0.5 V versus SHE), finding a bacterium that requires 0.05 V less overpotential for their own growth would increase the overall power output less than 10%.

Conclusions

Following a rush of initial findings showing the repeatability of electricity generation by bacteria in fuel cell-like devices, it has become clear that a wider diversity of bacteria are capable of extracellular respiration to electrodes than enrichments with Fe(III) may have suggested. Certainly, new isolates are yet to be discovered, especially from systems incubated at higher and lower temperature, pH, and salinities. As researchers have begun to recognize the factors that can influence measurements of electron transfer rates by these organisms, they have been able to dissect various adaptations which contribute to the overall phenotype of electron transfer to electrodes. Clear differences in strategy (direct versus mediators), complexity (thinner layers versus conductive biofilms), and potential-dependent behavior suggest that these abilities exist for competition in different niches, or to utilize different electron acceptors in the environment. The use of electrodes to study respiration offers a degree of control that should prove very useful in the future isolation and study of these bacteria.

Acknowledgements D. R. Bond is supported by grants from the Office of Naval Research (#N000140810162) and the National Science Foundation (#0702200). The author also acknowledges the College of Biological Sciences Imaging Center at the University of Minnesota for TEM imaging.

References

Back JH, Kim MS, Cho H, Chang IS, Lee JY, Kim KS, Kim BH, Park YI, Han YS (2004) Construction of bacterial artificial chromosome library from electrochemical microorganisms. FEMS Microbiol Lett 238:65–70

Biffinger J, Ribbens M, Ringeisen B, Pietron J, Finkel S, Nealson K (2008) Characterization of electrochemically active bacteria utilizing a high-throughput voltage-based screening assay. Biotechnol Bioeng 102:436–444

Bond DR, Holmes DE, Tender LM, Lovley DR (2002) Electrode-reducing microorganisms that harvest energy from marine sediments. Science 295:483–485

Bretschger O, Obraztsova A, Sturm CA, Chang IS, Gorby YA, Reed SB, Culley DE, Reardon CL, Barua S, Romine MF, Zhou J, Beliaev AS, Bouhenni R, Saffarini D, Mansfeld F, Kim BH, Fredrickson JK, Nealson KH (2007) Current production and metal oxide reduction by *Shewanella oneidensis* MR-1 wild type and mutants. Appl Environ Microbiol 73:7003–7012

Catal T, Xu S, Li K, Bermek H, Liu H (2008) Electricity generation from polyalcohols in single-chamber microbial fuel cells. Biosens Bioelectron 24:849–854

Chae KJ, Choi MJ, Lee J, Ajayi FF, Kim IS (2008) Biohydrogen production via biocatalyzed electrolysis in acetate-fed bioelectrochemical cells and microbial community analysis. Int J Hydrogen Energy 33:5184–5192

Cheng SA, Logan BE (2007) Ammonia treatment of carbon cloth anodes to enhance power generation of microbial fuel cells. Electrochem Comm 9:492–496

Choi Y, Song J, Jung S, Kim S (2001) Optimization of the performance of microbial fuel cells containing alkalophilic *Bacillus* sp. J Microbiol Biotechnol 11:863–869

Coates JD, Phillips EJ, Lonergan DJ, Jenter H, Lovley DR (1996) Isolation of *Geobacter* species from diverse sedimentary environments. Appl Environ Microbiol 62:1531–1536

Dewan A, Beyenal H, Lewandowski Z (2008) Scaling up microbial fuel cells. Environ Sci Technol 42:7643–8

Du Z, Li H, Gu T (2007) A state of the art review on microbial fuel cells: a promising technology for wastewater treatment and bioenergy. Biotechnol Adv 25:464–482

Dumas C, Mollica A, Feron D, Basseguy R, Etcheverry L, Bergel A (2008) Checking graphite and stainless anodes with an experimental model of marine microbial fuel cell. Bioresour Technol 99:8887–8894

Faraday M (1988) Faraday's chemical history of a candle. Chicago Review Press, Chicago, IL

Faraday M (2000) A course of six lectures on the various forces of matter, and their relations to each other. Scholarly Publishing Office, University of Michigan Library, Michigan

Franks AE, Nevin KP, Jia HF, Izallalen M, Woodard TL, Lovley DR (2009) Novel strategy for three-dimensional real-time imaging of microbial fuel cell communities: monitoring the inhibitory effects of proton accumulation within the anode biofilm. Energy Environ Sci 21:113–119

Gorby YA, Yanina S, McLean JS, Rosso KM, Moyles D, Dohnalkova A, Beveridge TJ, Chang IS, Kim BH, Kim KS, Culley DE, Reed SB, Romine MF, Saffarini DA, Hill EA, Shi L, Elias DA, Kennedy DE, Pinchuk G, Watanabe K, Ishii S, Logan S, Nealson KH, Fredrickson JK (2006) Electrically conductive bacterial nanowires produced by *Shewanella oneidensis* strain MR-1 and other microorganisms. Proc Natl Acad Sci USA 103:11358–11363

Gorby YA, McLean J, Korenevsky A, Rosso K, El-Naggar MY, Beveridge TJ (2008) Redox-reactive membrane vesicles produced by *Shewanella*. Geobiology 6:232–241

Ha PT, Tae B, Chang IS (2008) Performance and bacterial consortium of microbial fuel cell fed
    with formate. Energy Fuels 22:164–168
Holmes DE, Bond DR, O'Neil RA, Reimers CE, Tender LR, Lovley DR (2004a) Microbial com-
    munities associated with electrodes harvesting electricity from a variety of aquatic sediments.
    Microbiol Ecol 48:178–190
Holmes DE, Nicoll JS, Bond DR, Lovley DR (2004b) Potential role of a novel psychrotolerant
    member of the family *Geobacteraceae, Geopsychrobacter electrodiphilus gen. nov., sp. nov.*,
    in electricity production by a marine sediment fuel cell. Appl Environ Microbiol 70:6023–30
Ishii S, Hotta Y, Watanabe K (2008a) Methanogenesis versus electrogenesis: morphological
    and phylogenetic comparisons of microbial communities. Biosci Biotechnol Biochem
    72:286–94
Ishii S, Shimoyama T, Hotta Y, Watanabe K (2008b) Characterization of a filamentous biofilm
    community established in a cellulose-fed microbial fuel cell. BMC Microbiol 8:6
Ishii S, Watanabe K, Yabuki S, Logan BE, Sekiguchi Y (2008c) Comparison of electrode reduc-
    tion activities of *Geobacter sulfurreducens* and an enriched consortium in an air-cathode
    microbial fuel cell. Appl Environ Microbiol 74:7348–7355
Jong BC, Kim BH, Chang IS, Liew PWY, Choo YF, Kang GS (2006) Enrichment, performance,
    and microbial diversity of a thermophilic mediatorless microbial fuel cell. Environ Sci Technol
    40:6449–6454
Juarez K, Kim BC, Nevin K, Olvera L, Reguera G, Lovley DR, Methe BA (2009) PilR, a tran-
    scriptional regulator for pilin and other genes required for Fe(III) reduction in *Geobacter
    sulfurreducens*. J Mol Microbiol Biotechnol 16:146–58
Jung S, Regan JM (2007) Comparison of anode bacterial communities and performance in micro-
    bial fuel cells with different electron donors. Appl Microbiol Biotechnol 77:393–402
Kim N, Choi Y, Jung S, Kim S (2000) Effect of initial carbon sources on the performance of
    microbial fuel cells containing *Proteus vulgaris*. Biotechnol Bioeng 70:109–114
Kim BH, Park HS, Kim HJ, Kim GT, Chang IS, Lee J, Phung NT (2004) Enrichment of microbial
    community generating electricity using a fuel-cell-type electrochemical cell. Appl Microbiol
    Biotechnol 63:672–81
Kim GT, Webster G, Wimpenny JWT, Kim BH, Kim HJ, Weightman AJ (2006) Bacterial com-
    munity structure, compartmentalization and activity in a microbial fuel cell. J Appl Microbiol
    101:698–710
Kim JR, Jung SH, Regan JM, Logan BE (2007) Electricity generation and microbial community
    analysis of alcohol powered microbial fuel cells. Bioresour Technol 98:2568–2577
Kodama Y, Watanabe K (2008) An electricity-generating prosthecate bacterium strain Mfc52
    isolated from a microbial fuel cell. FEMS Microbiol Lett 288:55–61
Kostka JE, Dalton DD, Skelton H, Dollhopf S, Stucki JW (2002) Growth of Fe(III)-reducing
    bacteria on clay minerals as the sole electron acceptor and comparison of growth yields on a
    variety of oxidized iron forms. Appl Environ Microbiol 68:6256–6262
Lee JY, Phung NT, Chang IS, Kim BH, Sung HC (2003) Use of acetate for enrichment of electro-
    chemically active microorganisms and their 16S rDNA analyses. FEMS Microbiol Lett
    223:185–191
Liang P, Huang X, Fan MZ, Cao XX, Wang C (2007) Composition and distribution of internal
    resistance in three types of microbial fuel cells. Appl Microbiol Biotechnol 77:551–558
Liu H, Ramnarayanan R, Logan BE (2004) Production of electricity during wastewater treatment
    using a single chamber microbial fuel cell. Environ Sci Technol 38:2281–2285
Logan BE (2005) Simultaneous wastewater treatment and biological electricity generation. Water
    Sci Technol 52:31–37
Logan B (2008) Microbial fuel cells. Wiley, Hoboken, NJ
Logan BE, Regan JM (2006) Microbial challenges and applications. Environ Sci Technol
    40:5172–5180
Logan BE, Murano C, Scott K, Gray ND, Head IM (2005) Electricity generation from cysteine in
    a microbial fuel cell. Water Res 39:942–952

Logan B, Cheng S, Watson V, Estadt G (2007) Graphite fiber brush anodes for increased power production in air-cathode microbial fuel cells. Environ Sci Technol 41:3341–3346

Lovley DR (1985) Minimum threshold for hydrogen metabolism in methanogenic bacteria. Appl Environ Microbiol 49:1530–1531

Lovley DR (2008a) Extracellular electron transfer: wires, capacitors, iron lungs, and more. Geobiology 6:225–231

Lovley DR (2008b) The microbe electric: conversion of organic matter to electricity. Curr Opin Biotechnol 19:564–571

Lovley DR, Goodwin S (1988) Hydrogen concentrations as an indicator of the predominant terminal electron accepting reactions in aquatic sediments. Geochim Cosmochim Acta 52:2993–3003

Lovley DR, Giovannoni SJ, White DC, Champine JE, Phillips EJ, Gorby YA, Goodwin S (1993) *Geobacter metallireducens* gen. nov. sp. nov., a microorganism capable of coupling the complete oxidation of organic compounds to the reduction of iron and other metals. Arch Microbiol 159:336–344

Mahadevan R, Bond DR, Butler JE, Esteve-Nunez A, Coppi MV, Palsson BO, Schilling CH, Lovley DR (2006) Characterization of metabolism in the Fe(III)-reducing organism *Geobacter sulfurreducens* by constraint-based modeling. Appl Environ Microbiol 72:1558–1568

Marsili E, Baron DB, Shikhare ID, Coursolle D, Gralnick JA, Bond DR (2008a) *Shewanella* secretes flavins that mediate extracellular electron transfer. Proc Natl Acad Sci USA 105:3968–3973

Marsili E, Rollefson JB, Baron DB, Hozalski RM, Bond DR (2008b) Microbial biofilm voltammetry: direct electrochemical characterization of catalytic electrode-attached biofilms. Appl Environ Microbiol 74:7329–7337

Mathis BJ, Marshall CW, Milliken CE, Makkar RS, Creager SE, May HD (2008) Electricity generation by thermophilic microorganisms from marine sediment. Appl Microbiol Biotechnol 78:147–155

McCreery RL (2008) Advanced carbon electrode materials for molecular electrochemistry. Chem Rev 108:2646–2687

Nevin KP, Holmes DE, Woodard TL, Hinlein ES, Ostendorf DW, Lovley DR (2005) *Geobacter bemidjiensis sp. nov.* and *Geobacter psychrophilus sp. nov.*, two novel Fe(III)-reducing subsurface isolates. Int J Syst Evol Microbiol 55:1667–1674

Nevin KP, Richter H, Covalla SF, Johnson JP, Woodard TL, Orloff AL, Jia H, Zhang M, Lovley DR (2008) Power output and columbic efficiencies from biofilms of *Geobacter sulfurreducens* comparable to mixed community microbial fuel cells. Environ Microbiol 10:2505–2514

Park HS, Kim BH, Kim HS, Kim HJ, Kim GT, Kim M, Chang IS, Park YH, Chang HI (2001) A novel electrochemically active and Fe(III)-reducing bacterium phylogenetically related to *Clostridium butyricum* isolated from a microbial fuel cell. Anaerobe 7:297–306

Park HI, Sanchez D, Cho SK, Yun M (2008) Bacterial communities on electron-beam Pt-deposited electrodes in a mediator-less microbial fuel cell. Environ Sci Technol 42:6243–6249

Pham CA, Jung SJ, Phung NT, Lee J, Chang IS, Kim BH, Yi H, Chun J (2003) A novel electrochemically active and Fe(III)-reducing bacterium phylogenetically related to *Aeromonas hydrophila*, isolated from a microbial fuel cell. FEMS Microbiol Lett 223:129–134

Phung NT, Lee J, Kang KH, Chang IS, Gadd GM, Kim BH (2004) Analysis of microbial diversity in oligotrophic microbial fuel cells using 16S rDNA sequences. FEMS Microbiol Lett 233:77–82

Picioreanu C, Head IM, Katuri KP, van Loosdrecht MCM, Scott K (2007) A computational model for biofilm-based microbial fuel cells. Water Res 41:2921–2940

Picioreanu C, Katuri KP, Head IM, van Loosdrecht MCM, Scott K (2008) Mathematical model for microbial fuel cells with anodic biofilms and anaerobic digestion. Water Sci Technol 57:965–971

Prasad D, Sivaram TK, Berchmans S, Yegnaraman V (2006) Microbial fuel cell constructed with a micro-organism isolated from sugar industry effluent. J Power Sources 160:991–996

Prasad D, Arun S, Murugesan A, Padmanaban S, Satyanarayanan RS, Berchmans S, Yegnaraman V (2007) Direct electron transfer with yeast cells and construction of a mediatorless microbial fuel cell. Biosens Bioelectron 22:2604–2610

Rabaey K, Boon N, Siciliano SD, Verhaege M, Verstraete W (2004) Biofuel cells select for microbial consortia that self-mediate electron transfer. Appl Environ Microbiol 70:5373–5382

Rabaey K, Boon N, Hofte M, Verstraete W (2005) Microbial phenazine production enhances electron transfer in biofuel cells. Environ Sci Technol 39:3401–3408

Rabaey K, Rodriguez J, Blackall LL, Keller J, Gross P, Batstone D, Verstraete W, Nealson KH (2007) Microbial ecology meets electrochemistry: electricity-driven and driving communities. ISME J 1:9–18

Reguera G, McCarthy KD, Mehta T, Nicoll JS, Tuominen MT, Lovley DR (2005) Extracellular electron transfer via microbial nanowires. Nature 435:1098–1101

Reimers CE, Tender LM, Fertig S, Wang W (2001) Harvesting energy from the marine sediment-water interface. Environ Sci Technol 35:192–195

Richter H, McCarthy K, Nevin KP, Johnson JP, Rotello VM, Lovley DR (2008) Electricity generation by *Geobacter sulfurreducens* attached to gold electrodes. Langmuir 24:4376–4379

Richter H, Nevin KP, Jia HF, Lowy DA, Lovley DR, Tender LM (2009) Cyclic voltammetry of biofilms of wild type and mutant *Geobacter sulfurreducens* on fuel cell anodes indicates possible roles of OmcB, OmcZ, type IV pili, and protons in extracellular electron transfer. Energy Environ Sci 2:506–516

Roden EE (2006) Geochemical and microbiological controls on dissimilatory iron reduction. Comptes Rendus Geosci 338:456–467

Roden EE, Urrutia MM (2002) Influence of biogenic Fe(II) on bacterial crystalline Fe(III) oxide reduction. Geomicrobiol J 19:209–251

Schroder U (2007) Anodic electron transfer mechanisms in microbial fuel cells and their energy efficiency. Phys Chem Chem Phys 9:2619–2629

Snoeyenbos-West OL, Nevin KP, Anderson RT, Lovley DR (2000) Enrichment of *Geobacter* species in response to stimulation of Fe(III) reduction in sandy aquifer sediments. Microb Ecol 39:153–167

Straub KL, Hanzlik M, Buchholz-Cleven BE (1998) The use of biologically produced ferrihydrite for the isolation of novel iron-reducing bacteria. Syst Appl Microbiol 21:442–449

Straub KL, Benz M, Schink B (2001) Iron metabolism in anoxic environments at near neutral pH. FEMS Microbiol Ecol 34:181–186

Sun J, Sayyar B, Butler JE, Pharkya P, Fahland TR, Famili I, Schilling CH, Lovley DR, Mahadevan R (2009) Genome-scale constraint-based modeling of *Geobacter metallireducens*. BMC Syst Biol 3:15

Tang YJ, Chakraborty R, Martin HG, Chu J, Hazen TC, Keasling JD (2007) Flux analysis of central metabolic pathways in *Geobacter metallireducens* during reduction of soluble Fe(III)-nitrilotriacetic acid. Appl Environ Microbiol 73:3859–3864

Tender LM, Reimers CE, Stecher HA, Holmes DE, Bond DR, Lowy DA, Piloello K, Fertig SJ, Lovley DR (2002) Harnessing microbially generated power on the seafloor. Nat Biotechnol 20:821–825

Tender LM, Gray SA, Groveman E, Lowy DA, Kauffman P, Melhado J, Tyce RC, Flynn D, Petrecca R, Dobarro J (2008) The first demonstration of a microbial fuel cell as a viable power supply: powering a meteorological buoy. J Power Sources 179:571–575

Thrash JC, Van Trump JI, Weber KA, Miller E, Achenbach LA, Coates JD (2007) Electrochemical stimulation of microbial perchlorate reduction. Environ Sci Technol 41:1740–1746

Torres CI, Kato Marcus A, Rittmann BE (2007) Kinetics of consumption of fermentation products by anode-respiring bacteria. Appl Microbiol Biotechnol 77:689–697

Torres CI, Kato Marcus A, Rittmann BE (2008a) Proton transport inside the biofilm limits electrical current generation by anode-respiring bacteria. Biotechnol Bioeng 100:872–881

Torres CI, Lee H, Rittmann BR (2008b) Carbonate species as OH- carriers for decreasing the pH gradient between cathode and anode in biological fuel cells. Environ Sci Technol 42:8773–8777

Torres CI, Marcus AK, Parameswaran P, Rittmann BE (2008c) Kinetic experiments for evaluating the Nernst-Monod model for anode-respiring bacteria (ARB) in a biofilm anode. Environ Sci Technol 42:6593–6597

Wrighton KC, Agbo P, Warnecke F, Weber KA, Bodie EL, DeSantis TZ, Hugenholtz P, Andersen GL, Coates JD (2008) A novel ecological role of the Firmicutes identified in thermophilic microbial fuel cells. ISME J 2:1146–1156

Xing D, Zuo Y, Cheng S, Regan JM, Logan BE (2008) Electricity generation by *Rhodopseudomonas palustris* DX-1. Environ Sci Technol 42:4146–4151

Yi H, Nevin KP, Kim BC, Franks AE, Klimes A, Tender LM, Lovley DR (2009) Selection of a variant of *Geobacter sulfurreducens* with enhanced capacity for current production in microbial fuel cells. Biosens Bioelectron 24:3498–3503

Zhang LX, Zhou SG, Zhuang L, Li WS, Zhang JT, Lu N, Deng LF (2008a) Microbial fuel cell based on *Klebsiella pneumoniae* biofilm. Electrochem Comm 10:1641–1643

Zhang T, Cui C, Chen S, Yang H, Shen P (2008b) The direct electrocatalysis of Escherichia coli through electroactivated excretion in microbial fuel cell. Electrochem Comm 10:293–297

Zuo Y, Cheng S, Logan BE (2008a) Ion exchange membrane cathodes for scalable microbial fuel cells. Environ Sci Technol 42:6967–6972

Zuo Y, Xing DF, Regan JM, Logan BE (2008b) Isolation of the exoelectrogenic bacterium *Ochrobactrum anthropi* YZ-1 by using a U-tube microbial fuel cell. Appl Environ Microbiol 74:3130–3137

Chapter 19
The Biogeochemistry of Biomining

David Barrie Johnson

Biomining is a technology that harnesses the abilities of certain microorganisms to accelerate the dissolution of minerals, thereby facilitating the recovery of metals of value. In full-scale commercial operations, biomining currently mainly involves using consortia of acidophilic bacteria and archaea to bring about the oxidative dissolution of sulfide minerals. The major ore reserves of many base metals, such as copper and zinc, are sulfidic, and other valuable metals, such as gold and uranium, may also be associated with minerals such as pyrite (FeS_2). Where the destruction of minerals leads to the target metal being solubilised, the process of oxidative dissolution is referred to as bioleaching, whereas if the metal become more accessible to chemical extraction but remains in an insoluble form, the process is known more correctly as biooxidation (Fig. 19.1).

The general subject of biomining has been addressed in a number of review articles (e.g. Rawlings 2002; Rawlings et al. 2003; Olson et al. 2003) and in textbooks (Rawlings 1997; Rawlings and Johnson 2007a; Donati and Sand 2007), while articles by Brierley (2008a, b) provide insights into the emergence of this biotechnology and how this may develop in the future.

Principles and Practices of Biomining

The notion that microorganisms could be harnessed to extract metals from sulfidic ore arose from the discovery in the late 1940s of a bacterium that was able to generate ferric iron from ferrous in acidic liquors, and subsequent demonstrations that this microorganism could grow autotrophically on pyrite and other sulfide minerals, causing their dissolution. *Thiobacillus ferrooxidans*, as it was known at the time (it was renamed *Acidithiobacillus ferrooxidans* by Kelly and Wood (2000), as a result of assessment of phylogenetic data of the somewhat disparate bacteria then contained within the genus *Thiobacillus*) has been one of the most widely studied

D.B. Johnson (✉)
School of Biological Sciences, Bangor University, Deiniol Road, Bangor LL57 2UW, UK
e-mail: d.b.johnson@bangor.ac.uk

L.L. Barton et al. (eds.), *Geomicrobiology: Molecular and Environmental Perspective*, 401
DOI 10.1007/978-90-481-9204-5_19, © Springer Science+Business Media B.V. 2010

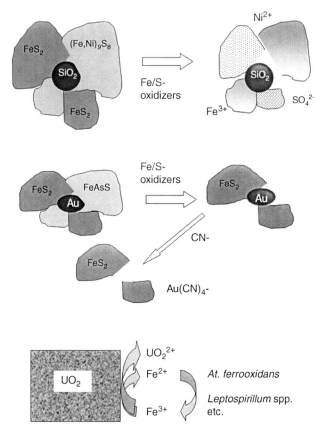

Fig. 19.1 Bioleaching of base metal sulfide ores (illustrated with a pyritic pentlandite ore; *top*), bio-oxidation of refractory gold ore (*middle*) and bio-solubilization of uraninite (*bottom*)

of all bacteria outside of those of medical importance. While other acidophilic bacteria and archaea can also degrade sulfide minerals in pure or mixed cultures (as described below), the early work with *At. ferrooxidans* remains an important foundation stone for the entire biomining industry.

Currently, there are two major engineering variants for biomining – irrigated systems and stirred tanks (Fig. 19.2). In the first of these, low grade sulfidic rocks (ranging from large boulders to ground and agglomerated ore) are piled into large dumps or heaps, which are then irrigated with an acidic liquor to stimulate the activities of microorganisms that either occur naturally with the ore or else are inoculated into the heaps. The percolating liquor becomes enriched with metals (and sulfate) released from the sulfide minerals and also other minerals (referred to as gangue minerals) that are unstable in acidic solutions. This "pregnant leach solution" (PLS) is collected from outflow channels from the base of the dump or heap, and the target metal(s) extracted. Dump leaching was the first practice of biomining to be adopted, and is still operated at many mines to extract copper from low-grade "run of mine"

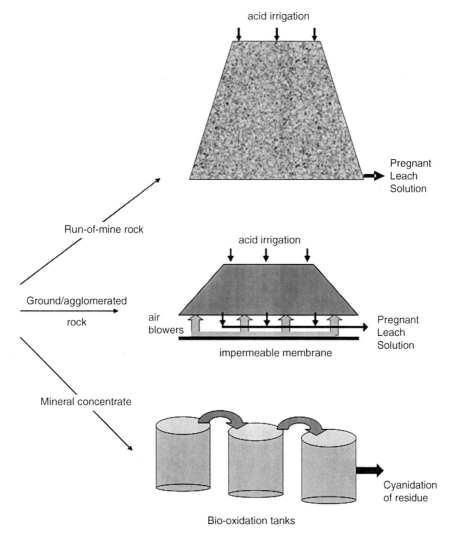

Fig. 19.2 Major process options used in biomining. *Top*, dump bioleaching; *middle*, heap bioleaching; *bottom*, biooxidation in stirred-tanks

ores that contain too little of the metal (typically <0.5% by weight) to merit conventional processing (grinding, concentrating and smelting). Rocks and large boulders that would otherwise be considered as waste, are assembled in dumps that may be extremely large (over 100 m in height) and the process of bioleaching may continue for several years or even decades. Dumps are seldom inoculated and, because the financial return in terms of copper recovery is often marginal, operational costs are kept to a minimum. In spite of this, dump leaching had been estimated to account (in 2008) for about 13% of global copper production (and about 65% of that obtained using biological processing of ores; Brierley 2008a).

Heap leaching is a more refined approach that again, in most cases, involves irrigation of constructed mounds of low-grade mineral ores (Fig. 19.2). The main differences between heap and dump leaching systems are: (a) ores are ground in the former to facilitate more rapid breakdown of sulfide minerals and, where fine-grain particles are produced, these may be agglomerated with sulfuric acid; (b) heaps are generally smaller in height (typically 6–10 m) than dumps (though their base areas may exceed those of dump operations) and several successive heaps (or "lifts") may be stacked upon each other as the operation develops; (c) heaps are usually aerated from beneath as well as irrigated from above, to provide the indigenous bacteria and archaea with both oxygen and carbon dioxide (the main key players are aerobic chemo-autotrophs, as described below); (d) heaps are constructed on pads that are lined with impermeable synthetic liners (generally high density polyethylene on which drain pipes, used to collect and convey PLS, are placed); (e) inoculation ponds, in which mineral-degrading bacteria and archaea can be grown to high cell densities and then introduced either in the irrigation liquor or else during the mineral agglomeration process, are commonly used in heap operations. The higher capital cost expenditure results in more rapid (heaps typically operate for about a year) and more extensive (up to 90%) metal extraction than is achieved in dump operations. Again, the main metal that is recovered in heap biomining operations is copper, though a polymetallic black schist ore (where the major metal in terms of its economic value is nickel) is currently processed via heap leaching in Finland, and heap leaching of a refractory gold ore has been operated for over a decade in a site in Nevada (Brierley 2008b).

Differentiating between dump and heap leaching systems is sometimes not clear cut. For example, in the first decade of the twenty first century the largest irrigation-based biomining operation was constructed at the Escondida copper mine in northern Chile. The operation is planned to cover a surface area of 10 km^2, and have a height of 126 m (comprising seven lifts, each 18 m high). Although it has many features of a conventional heap leaching operation (heaps are aerated and inoculated, constructed on base liners, and are insulated to reduce loss of irrigation liquors) the ore itself is not crushed, as in dump leaching.

Bioprocessing of sulfidic mineral concentrates in stirred tanks that are continually aerated and maintained at prescribed temperatures and pH values constitute the most efficient biomining systems currently in operation, in terms of both speed and efficiency of the operation. However, the greater costs involved in constructing and maintaining the vast bioreactors (up to 1,350 m^3 – the largest reactors used for any biological-based operation) means that these have been operated as full-scale commercial systems for recovering gold rather than base metals, with the single exception of a bioleaching plant in Uganda where cobalt is the target metal. Most commercial stirred tank biomining systems operate at pH of about 1.5 and temperatures between 40°C and 45°C. Since the oxidative dissolution of sulfide minerals is an exothermic reaction, cooling rather than heating of the tanks is required, constituting (along with stirring of the dense mineral slurries – 20% pulp density is typical) a significant part of the running costs of any single operations. Stirred tank configurations can vary from single to multiple tank systems (Fig. 19.2), into which mineral concentrate slurries are continuously fed. Nutrients (N and P) are added to promote growth of the mineral-oxidizing microflora, which may be added from enrichments prepared from the local ore source or else from another tank leaching operation that the company concerned is operating, possibly in a different global location.

A third engineering design for bioprocessing sulfidic ores is known as "in situ" biomining. This may involve allowing worked out deep mines to flood for several weeks (or longer) during which time accessible minerals are degraded by indigenous microorganisms, and then pumping out and recovering metals from the enriched water. This is essentially what occurred at the Mynydd Parys copper mine in Wales and the Rio Tinto copper mine in Spain, well before the role of acidophiles in mediated the process was recognized. In more recent times, this approach was used to extract uranium from worked out deep mines in Canada. To facilitate additional uranium extraction, the buried ore bodies were fractured by controlled blasting prior to being flooded. At one mine (the Denison mine), an estimated 300 t of uranium was extracted in 1 year using in situ biomining, and an estimated 4×10^6 t of otherwise inaccessible ore was processed using this method (Rawlings 2002). In an alternative approach to in situ biomining, a fresh ore body is disrupted and an acidic leach liquor containing a microbial inoculum is injected through the fracture planes (Fig. 19.3). The percolating liquor become increasing laden with dissolved metals, and is eventually pumped to the surface for processing. The major advantage to this approach is that it avoids the costly business of digging out or deep mining the ore body, though suitable geological strata are necessary for the process to work as lateral migration of the acidic metal-rich waters has the potential to cause serious pollution of adjacent groundwaters.

A number of innovative approaches for bioprocessing minerals have been suggested and demonstrated in pilot-scale systems. The Geocoat[R] process, developed by the American minerals biotechnology company GeoBiotics, involves coating support rock particles with mineral concentrate, and then stacking the material on lined pads for bio-oxidation in thin-layer heaps (Harvey and Bath 2007). This promotes more rapid mineral processing (typically around 60 days) and often higher recoveries of target metals (gold and base metal ores) than conventional heap leaching. In a radically different approach to minerals bioprocessing, the biological

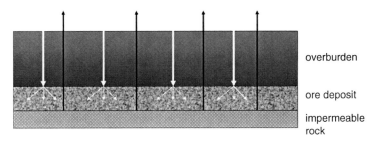

overburden

ore deposit

impermeable rock

Fig. 19.3 Idealized scenario for in situ bioleaching of a buried ore deposit. The *white arrows* indicate the flow path of the injection liquor, while the *black arrows* indicate the direction of flow of the metal-enriched liquid effluent

and mineralogical components are separated, allowing conditions to be optimized separately for each process. "Indirect" bioleaching of zinc sulfide concentrates has been demonstrated at pilot-scale by a European research consortium ("BioMinE"; Morin et al. 2008).

A Brief History of Biomining

Although biomining bacteria were not discovered until the mid twentieth century, humankind had been inadvertently making use of their activities for hundreds of years before this. Workers at copper mines in western Europe found that copper could be extracted by placing metallic iron in drainage streams. This simple electrochemical reaction ($Cu^{2+} + Fe^0 \rightarrow Cu^0 + Fe^{2+}$) was considered at the time to be an example of alchemy (the transmutation of one metal into another), and this process (referred to as "cementation" in the mining industry) is still used to recover soluble copper from PLS at some mines, though superior technologies (solvent extraction-electrowinning (SX-EW) and (bio)sulfide precipitation are now available). At the Mynydd Parys mines in north Wales, for example, copper production continued long after deep mining came to an end (in 1883) using, essentially, in situ biomining (allowing the mines to flood, pumping out the metal-enriched water and recovering soluble metal by cementation in large precipitation ponds).

The first recognized application of biomining was by the Kennecott Copper Corporation in the 1960s. They obtained a patent for using bacteria to extract copper from their waste rock dumps, and developed operations at the Bingham Canyon mine in Utah and later at the Chino mine in New Mexico. Since then, mine operations using dump bioleaching of low-grade copper ore have been established elsewhere in the world, including China and South America. Gradation from dump to heap leaching began with the use of "thin layer" engineering of mounds (crushed ore, stacked 2.5–6 m in height) at the Lo Aguirre mine, and later demonstration of the benefits of forced aeration of heaps at the Quebrada Blanca mine (both copper mines, located in Chile; Brierley 2008a). Heap bioreactors used to recover metals other than copper have been established in Nevada (since 1999; pre-treatment of refractory gold ore) and Finland (since 2008; bioleaching of a polymetallic black schist at the Talvivaara mine, extracting nickel, zinc and copper).

Stirred tank bioreactors processing sulfidic ores have been operating since 1986. Most of these are used to oxidize arsenopyrite and pyrite that are present in refractory gold ores, thereby liberating otherwise occluded fine particles of the precious metal (Fig. 19.1). The first plant to be commissioned, at the Fairview mine in Barberton, South Africa, was still in operation in 2009. Its design capacity (55 t/day) has far been exceeded by more recent plants established in Western Australia (Wiluna; 158 t/day), China (Jinfeng; 790 t/day), Ghana (Ashanti; 960 t/day) and elsewhere (Brierley 2008a). One of the more recent stirred tank bioleaching systems to have been commissioned (at Kokpatas, Uzbekistan) is designed to process over 160,000 t of refractory gold concentrate each day.

Geochemical Aspects

As mentioned previously, biomining, as currently practised, utilizes certain species of acidophilic bacteria and archaea that are able to catalyze the oxidative dissolution of sulfide minerals, either solubilising target metals (bioleaching) or making them accessible for chemical extraction (biooxidation). The most abundant of all sulfide minerals is pyrite (FeS_2). Other iron sulfide minerals include marcasite (FeS_2, which has an orthorhombic crystallinity rather than a cubic crystallinity, as is the case with pyrite), pyrrhotite ($Fe_{1-x}S$, where x = 0–0.2) and others that contain additional elements, such as arsenopyrite (FeAsS) and chalcopyrite (which is conventionally denoted as $CuFeS_2$, and which is the most abundant copper-containing mineral in the lithosphere). Due to their widespread distribution in sulfidic ore bodies, the biogeochemistries of iron and sulfur are of fundamental importance in biomining environments. Other sulfide minerals are listed in Table 19.1.

The oxidation state of iron in all sulfide minerals is +2 (i.e. ferrous iron) with the exception of greigite (Fe_3S_4) where it occurs in both ionic states (+2 and +3). Oxidative dissolution of these minerals, by mechanisms described below, results ultimately in iron being transformed to the +3 oxidation state (i.e. ferric iron). Besides zero-valent (metallic) iron, the only oxidation states in which this metal occurs are +2 and +3, which restricts the possible redox transformations to single

Table 19.1 Metal sulfide minerals of economic and environmental importance

Iron sulfides		
	Pyrite	FeS_2
	Marcasite	FeS_2
	Pyrrhotite	$Fe_{0.8}S - FeS$
	Greigite	Fe_3S_4
Other single metal/metalloid sulfides		
	Argentite	Ag_2S
	Chalcocite	Cu_2S
	Cinnabar	HgS
	Covellite	CuS
	Digenite	Cu_9S_5
	Galena	PbS
	Millerite	NiS
	Molybdenite	MoS_2
	Realgar	AsS
	Sphalerite	ZnS
	Stibnite	Sb_2S_3
Mixed metal/metalloid sulfides		
	Arsenopyrite	FeAsS
	Bornite	Cu_5FeS_4
	Chalcopyrite	$CuFeS_2$
	Cobaltite	CoAsS
	Pentlandite	$(Fe/Ni)_9S_8$

electron oxidation or reduction reactions. Although both ferrous and ferric iron can be complexed by a large number of organic ligands that can impact both the ferrous/ferric redox potentials and the solubilities of iron ions (particularly ferric), these tend to be of little or no importance in the organic carbon-poor liquors that are typical of biomining environments. The most important inorganic ligand of ferric iron in mineral leach liquors is the sulfate anion, which forms the soluble complexes $FeSO_4^+$ and $Fe(SO_4)_2^-$. The soluble hydroxylated species $Fe(OH)^{2+}$ and $Fe(OH)_2^+$ also occur at low pH (Welham et al. 2000).

The redox potential of the ferrous/ferric couple is pH-dependent, but in low pH liquors typical of biomining environments it is most often quoted as +770 mV, relative to a hydrogen reference electrode. A major implication of this is that, while ferrous iron is a potential electron donor for chemolithotrophic metabolism, the only feasible electron acceptor that can be coupled to it to generate energy is molecular oxygen, i.e. iron oxidation at low pH only occurs in aerobic environments. In addition, the amount of energy available from ferrous iron oxidation is very small compared to that from organic and other inorganic electron donors (Table 19.2). On the other hand, ferrous iron tends to be present in relatively high concentrations in sulfide mineral leachates, and relatively simple biochemical machinery is required to oxidize ferrous iron. Together these overriding factors probably account for the ability to utilize the energy from oxidizing iron being widespread among acidophilic bacteria (Table 19.3), and also help explain why some bacteria (notably *At. ferrooxidans*) preferably oxidize iron even when more "energy rich" substrates, such as reduced sulfur, are present.

Conversely, the relatively high redox potential of the ferrous/ferric couple together with the enhanced solubility of ferric iron in acidic liquors make ferric iron an attractive terminal electron acceptor from both thermodynamic and bioavailability

Table 19.2 Comparison of free energy changes associated with the oxidation of inorganic substrates and carbon dioxide fixation by chemolithotrophic acidophiles (Data from Kelly 1978)

Reaction	Free energy change ΔG° (kJ mole substrate^{-1})
Ferrous iron oxidation:	
$4FeSO_4 + O_2 + 2H_2SO_4 \rightarrow 2Fe_2(SO_4)_3 + 2H_2O$	−25 (pH 3.0)
	−30 (pH 2.0)
	−33 (pH 1.5)
Hydrogen oxidation:	
$H_2 + 0.5O_2 \rightarrow H_2O$	−237
Elemental sulfur oxidation:	
$S^\circ + 1.5O_2 + H_2O \rightarrow SO_4^{2-} + 2H^+$	−507
Hydrogen sulfide oxidation:	
$H_2S + 2O_2 \rightarrow H_2SO_4$	−714
Sulfur oxy-anion oxidation:	
(i) $S_2O_3^{2-} + 2O_2 + H_2O \rightarrow 2SO_4^{2-} + 2H+$	−739
(ii) $S_4O_6^{2-} + 3.5O_2 + 3H_2O \rightarrow 4SO_4^{2-} + 6H^+$	−1,225
Carbon dioxide fixation:	
$6CO_2 + 6H_2O \rightarrow C_6H_{12}O_6 + 6O_2$	+495

Table 19.3 Categorization of validated species and genera of extremely acidophilic iron and sulfur-oxidizing micro-organisms, based on growth temperature optima

Fe^{2+}-oxidizers	S-oxidizers	Fe^{2+}/S-oxidizers
(a) Mesophiles (temperature optima 20–40°C)		
L. ferrooxidans	At. thiooxidans	At. ferrooxidans
Fm. acidiphilum		At. ferrivorans
Fp. acidiphilum		
(b) Moderate thermophiles (temperature optima 40–60°C)		
L. ferriphilum	At. caldus	Sulfobacillus spp.
Am. ferrooxidans		Alicyclobacillus spp.[a]
Fx. thermotolerans		
Ap. cupricumulans		
(c) Extreme thermophiles (temperature optima >60°C)		
	H. acidophilum	S. metallicus
	Metallosphaera spp	S. tokodaii
	Sulfurococcus spp.	Ac. brierleyi
	Ac. infernus	
	Ac. ambivalens	

Genera abbreviations: *Ap.*, *Acidiplasma*; *At.*, *Acidithiobacillus*; *L.*, *Leptospirillum*; *Fm.*, *Ferrimicrobium*;, *Sb.*, *Sulfobacillus*; *Fp. Ferroplasma*; *Am.*, *Acidimicrobium*; *Fx.*, *Ferrithrix*; *H.*, *Hydrogenobaculum*; *S.*, *Sulfolobus*; *Ac.*, *Acidianus*

viewpoints. Many acidophiles that oxidize iron in aerobic waters can switch to using soluble ferric iron as an alternative electron acceptor when oxygen is absent or limiting, providing than an alternative electron donor (e.g. reduced sulfur or hydrogen) is available. In addition, many species of obligately heterotrophic acidophiles can also catalyse the dissimilatory reduction of ferric iron minerals (Johnson and McGinness 1991; Coupland and Johnson 2008). Table 19.4 lists bacteria and archaea that have been shown to reduce ferric iron in acidic liquors, though whether iron respiration can support growth of these microorganisms has not been ascertained in all cases. From a biomining perspective, biological reduction of ferric iron is detrimental to the objective of mineral dissolution, as iron(III) is the main oxidant of sulfides in acidic liquors, as described below. While it is often assumed that it is of no importance in highly aerated stirred tanks, dissimilatory iron reduction is probably more significant in dump and heap leaching operations.

In contrast to iron, redox transformations are sulfur are highly complex. Sulfur can exist in up to nine different oxidation states, ranging from −2 (as in hydrogen sulfide) to +6 (as in the sulfate anion; Steudel 2000). In sulfidic minerals, sulfur occurs either as −2 (as in chalcocite and sphalerite) or −1 (as in pyrite and arsenopyrite). Oxidative dissolution of sulfide minerals results, ultimately, in the production of sulfate, though intermediates such as elemental sulfur and various sulfur oxyanions (e.g. thiosulfate and trithionate) may occur either transiently or else accumulate during mineral decay (Schippers and Sand 1999). A large number of different species of acidophilic bacteria and archaea can utilize the energy from oxidation of reduced forms of sulfur to support growth. Indeed the first acidophile to be characterized (in 1921) was the sulfur-oxidizing chemo-autotroph (*Acidi*)

Table 19.4 Acidophilic bacteria and archaea that catalyze the dissimilatory reduction of ferric iron to ferrous

Species	Electron donor(s)
(a) Mesophiles (temperature optima 20–40°C)	
Acidthiobacillus ferrooxidans	Sulfur, hydrogen
Acidithiobacillus ferrivorans	Sulfur
Ferrimicrobium acidiphilum	Small MW[a] organic compounds
Acidiphilium cryptum	Small MW organic compounds
Acidiphlium acidophilum	Small MW organic compounds
Acidocella facilis	Small MW organic compounds
Acidobacterium capsulatum	Small MW organic compounds
Ferroplasma acidiphilum	Small MW organic compounds
(b) Moderate thermophiles (temperature optima 40–60°C)	
Sulfobacillus thermosulfidoxidans	Sulfur, small MW organic compounds
Sulfobacillus acidophilus	Sulfur, small MW organic compounds
Sulfobacillus benefaciens	Sulfur, small MW organic compounds
Alicyclobacillus tolerans	Small MW organic compounds
Acidimicrobium ferrooxidans	Small MW organic compounds
Ferrithrix thermotolerans	Small MW organic compounds
Acidicaldus organivorans	Small MW organic compounds
Acidiplasma cupricumulans	Small MW organic compounds

[a]Molecular weight

thiobacillus thiooxidans). In highly aerated biomining enviroments such as stirred tanks, sulfate, like ferric iron, is stable. In the absence of oxygen, however, dissimilatory reduction of sulfate, elemental sulfur and sulfur oxyanions is possible, though there have been relatively few reports of acidophilic microorganisms that have this ability. Sulfur reduction has been reported for the archaea *Acidianus* spp., *Stygiolobus azoricus, Sulfurisphaera ohwakuensis* and *Thermoplasma* spp. (described in Johnson and Hallberg 2008) and sulfur reduction genes have been found in the genome of the type strain of *At. ferrooxidans* (Valdes et al. 2008). Bacteria have been described that reduce sulfate to sulfide at pH values of 3 and above, though currently no extremely acidophilic sulfate-reducing bacteria have been described (reviewed in Koschorreck 2008). Figure 19.4 shows reactions in the sulfur cycle that have been identified in acidic environments, including biomining operations and acid mine drainage waters.

Other metals and metalloids that have variable oxidation states and that occur in, or may be associated with, sulfide ores include arsenic, copper and uranium. Arsenic exists in an oxidation state of −1 in arsenopyrite (FeAsS). This mineral is particularly abundant in many refractory gold ores, and biological oxidation results in the production of soluble forms of the metalloid. Initially, the species produced is As(III) (as undissociated arsenious acid at the pH of mineral leach liquors) but this is a transient species, and is oxidized to the less biotoxic form As (V) ($H_2AsO_4^-$). In biomining environments, ferric iron catalyzes this oxidation though the reaction proceeds relatively slowly except in the presence of a suitable reaction surface (pyrite can fulfil this role). Some species of *Thiomonas* are known to derive

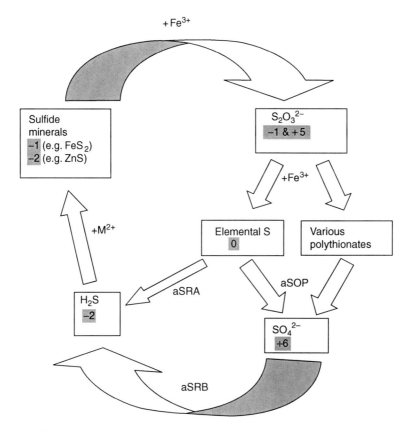

Fig. 19.4 The sulfur cycle in biomining environments. Highlighted numbers in *bold* show the various oxidation states of sulfur. aSOP, acidophilic sulfur-oxidizing prokaryotes; aSRA, acidophilic sulfur-reducing archaea; aSRB, acid-tolerant sulfate-reducing bacteria. M^{2+} is a generic representation of a chalcophilic divalent metal (e.g. Fe^{2+}, Zn^{2+} etc.)

energy from oxidizing As(III) to As(V), but these are moderate acidophiles and, while they are widespread in acid mine drainage waters, they are rarely found in biomining operations. The thermo-acidophilic archaeon *Sulfolobus metallicus* (strain BC, incorrectly referred to at the time as *Sulfolobus acidocaldarius*) has also been shown to oxidize As(III) to As(V) (Sehlin and Lindstrom 1992). While reduction of As(V) to As(III) occurs within bacterial cells, including some acidophiles, as part of their As resistance mechanism, there has been no convincing evidence of biological dissimilatory reduction of As (V) at low pH. Copper exists as Cu(I) in chalcopyrite and chalcocite, and as both Cu(I) and Cu(II) in covellite (Table 19.1) but is rapidly oxidized to Cu(II) by ferric iron in leach liquors. Copper (I) may be readily detected however, when these minerals are leaching at low redox potentials. Uranium does not form a sulfide phase, but the mineral uraninite (UO_2) frequently occurs in association with sulfide minerals such as pyrite. Oxidation of uraninite by ferric iron results the transformation of insoluble U(IV) to soluble U(VI) (UO_2^{2+};

Fig. 19.1). While the iron-and sulfur-oxidizing acidophile *At. ferrooxidans* has been reported to catalyze the dissimilatory oxidation of both Cu(I) and U(VI), the extent to which this occurs, compared to abiotic oxidation, in iron-rich biomining environments is probably very minor.

Mechanisms of Mineral Dissolution

The mechanism(s) by which microorganisms catalyze the oxidative dissolution of sulfide minerals has been the subject of considerable debate for many years. Minerals, such as pyrite, are stable in environments that are totally dry or from which oxygen is excluded. At circum-neutral pH values, in moist aerobic environments, molecular oxygen is the main oxidant of pyrite and other sulfides, as illustrated in equation (19.1):

$$FeS_2 + 1.75O_2 + 2.5H_2O \rightarrow Fe(OH)_3 + S_2O_3^{2-} + 2H \qquad (19.1)$$

Ferrous iron released from pyrite undergoes spontaneous oxidation at neutral pH and the ferric iron produced hydrolyses (reacts with water) to form insoluble ferric iron hydroxide (as shown) and hydrous oxide minerals, such as ferrihydrite ($Fe_2O_3.9H_2O$). The reaction generates protons, and in the presence of sulfur-oxidizing bacteria, addition acidity is generated by the oxidation of thiosulfate to sulfuric acid (equation (19.2)):

$$S_2O_3^{2-} + 1.5O_2 + 2H_2O \rightarrow 4H^+ + 2SO_4^{2-} \qquad (19.2)$$

Depending on the nature of the environment in which these reactions are occurring, the acidity generated may be neutralized by basic minerals (chiefly carbonates) that are present, or else the site of mineral oxidation becomes increasingly acidic. An important consequence of the latter is that the pH may fall low enough (<2.5) to restrict the hydrolysis of ferric iron. Soluble ferric iron is a more powerful oxidant of sulfide minerals than is molecular oxygen, and therefore the rate of mineral dissolution can increase dramatically at low pH. This is entirely dependent on a continuous supply of oxidized iron, as ferric iron is reduced when it reacts with pyrite (equation (19.3)):

$$FeS_2 + 6Fe^{3+} + 3H_2O \rightarrow 7Fe^{2+} + S_2O_3^{2-} + 6H^+ \qquad (19.3)$$

Ferric iron can originate either from an external supply (e.g. in percolating liquors) or may be re-generated in situ. Although, as illustrated in equation (19.3), oxidation of pyrite by ferric iron can occur in anoxic environments, regeneration of ferric iron is an oxygen-requiring reaction (equation (19.4)):

$$Fe^{2+} + 0.25O_2 + H^+ \rightarrow Fe^{3+} + 0.5H_2O \qquad (19.4)$$

One other requirement for ferrous iron oxidation at low pH is a suitable (iron-oxidizing) bacterium or archaeon, since abiotic rates of iron oxidation are extremely slow at pH < 3.5 (Stumm and Morgan 1981). These two important requirements explain why biomining operations are carried out at low pH (to ensure ferric iron-enhanced mineral dissolution) and require suitable microorganisms (for replenishment of the oxidant).

Schippers and Sand (1999) categorized sulfide minerals into two types – those that are acid-soluble and others that are acid-insoluble. The former includes sphalerite, chalcopyrite, arsenopyrite and galena, while pyrite, molybdenite and tungstenite (WS_2) are examples of sulfides that are not susceptible to proton attack. In acidic liquors, the latter are oxidized by soluble ferric iron, and a total of six successive one-electron oxidation steps are required to break the sulfur-metal bonds. Since the initial sulfur product released from the degrading mineral is thiosulfate, this form of oxidative sulfide mineral dissolution has been described as the "thiosulfate mechanism" (Rohwerder et al. 2003). Thiosulfate is unstable in acidic liquors (more so when ferric iron is present) and oxidizes via tetrathionate and other sulfur oxy-anions, ultimately to sulfate. Due to the requirement of ferric iron, only iron-oxidizing acidophilic bacteria and archaea are able to accelerate the dissolution of acid-insoluble sulfides. In contrast, metal-sulfur bonds in acid-soluble sulfides can be disrupted by protons alone, liberating hydrogen sulfide, though more commonly, in the presence of ferric iron, the first free sulfur compound is thought to be an unstable sulfide cation (H_2S^+) which then dimerizes (to H_2S_2) and then oxidizes via various polysulfides, to elemental sulfur (S^0), a process that has been referred to as the "polysulfide mechanism" (Schippers and Sand 1999). In theory, bacteria and archaea that can generate protons (i.e. sulfur-oxidizers) can degrade acid-soluble sulfide minerals, though rates of dissolution tend to be far greater when iron-oxidizing acidophiles are also present.

Many acidophilic bacteria display a propensity to attach to sulfide minerals such as pyrite (Rohwerder et al. 2003) though there are significant differences between species and even strains within a species in how rapidly and permanently they attach to surfaces (Ghauri et al. 2007). Historically, two scenarios were described to account for mineral dissolution either catalyzed by bacteria that attached to sulfide minerals (the "direct" mechanism) or planktonic-phase cells that generate the oxidant (ferric iron) in solution, which subsequently diffuses to the mineral where it induces dissolution (the "indirect" mechanism). It is now recognized that ferric iron mediates mineral oxidation in both cases (ferric iron is tightly associated with the glycocalyx of attached cells), and the terms "contact" and "non-contact" leaching, proposed by Hallberg and Johnson (2001) are now used widely to describe that carried out by attached and free-swimming bacteria and archaea, respectively (Fig. 19.5).

The means by which acidophiles attach to mineral surfaces has been the subject of a considerable body of research in recent years. In the case of the iron- and sulfur-oxidizer, At. ferrooxidans, ferric iron is complexed by two uronic acid residues in the bacterial glycocalyx, and the resulting net positive charge facilitates attachment to negatively charged pyrite (Rohwerder et al. 2003). Once attached,

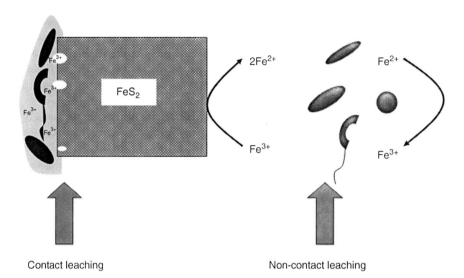

Contact leaching   Non-contact leaching

Fig. 19.5 Dissolution of pyrite (FeS$_2$) by iron-oxidizing acidophilic bacteria and archaea attached to the mineral surface ("contact leaching") and by unattached (free-swimming) microorganisms ("non-contact leaching"). Ferric iron mediates mineral attack in both cases, either as soluble iron or that contained in bacterial glycocalyx within the biofilms of attached acidophiles

further production of exopolymeric substances (EPS) by the bacteria enhances the development of biofilms on the mineral surfaces, within which conditions (pH, dissolved oxygen concentrations) may differ from the bulk solution phase and be more conducive for mineral dissolution.

Tributsch and Rojas-Chapana (2007) described contrasting strategies for pyrite dissolution by *At. ferrooxidans* and *Leptospirillum ferrooxidans*, two species of mesophilic mineral-oxidizers that are considered to be of major importance in heap leaching operations and in the genesis of acid mine drainage. *L.ferrooxidans* is known only to use ferrous iron as an electron donor. When it is grown on pyrite, the exopolymeric capsule it produces is dotted with small particles, identified as colloidal, microcrystalline pyrite, and the oxidation of these microscopic grains appears to support the growth of this acidophile. The pitting caused by this process can develop into deep holes within the minerals, a process described as "electrochemical machining". *At. ferrooxidans* can oxidize both ferrous iron and reduced forms of sulfur, but tends to generate redox potentials that are some 200 mV lower that those generated by *L. ferrooxidans* when oxidizing pyrite (Rawlings et al. 1999). Colloidal sulfur has been detected in the EPS of *At. ferrooxidans* attached to pyrite. Tributsch and Rojas-Chapana (2007) have provided evidence for the role of thiol groups being involved in the break up of interfacial chemical complexes at these lower redox potentials, and have suggested that cysteine might be the organic compound involved in this process in vivo.

Diversity and Interactions of Biomining Microorganisms

The microbiology of biomining has been the subject of numerous reviews, including Rawlings (2005), Schippers (2007), Norris (2007) and Rawlings and Johnson (2007b). Acidophiles that can have either a direct or indirect process in the dissolution of sulfide minerals can be categorized as primary, secondary and tertiary microorganisms (Johnson and Hallberg 2008). Primary bacteria and archaea are those chemo-autotrophs that can catalyze ferrous iron oxidation, and thereby initiate mineral oxidation; secondary microorganisms are sulfur-oxidizers that generate the acidity required for the acidophilic consortia, while tertiary acidophiles are those that degrade organic compounds excreted from living bacteria (most of which in biomining operations are autotrophic) or lysed from dead cells, and thereby help maintain suitable conditions for the more "organic sensitive" lithotrophs, such as *Leptospirillum* spp. In reality, there is considerable overlap between these categories. For example, some *Acidithiobacillus* spp. (*At. ferrooxidans* and *At. ferrivorans*) can oxidize both ferrous iron and reduced sulfur as can all *Sulfobacillus* spp., and some heterotrophic acidophiles (e.g. *Ferrimicrobium acidiphilum* and *Ferroplasma* spp.) can also oxidize iron as can some mixotrophic acidophiles (e.g. *Acidimicrobium ferrooxidans*). Primary producers in stirred tank operations are exclusively chemo-litho-autotrophic bacteria, though acidophilic algae may play a minor role in net CO_2-assimilation on the surfaces of heap reactors and inoculation ponds (Johnson 2009). Another important characteristic that can be use to differentiate acidophilic bacteria and archaea is their response to temperature. A pragmatic and much-used distinction has been made between (extreme) thermophiles, that have temperature optima above 60°C, moderate thermophiles (40–60°C) and mesophiles (20–40°C). Archaea account for almost all of the first group, most moderate acidophiles are Gram-positive bacteria, and Gram-negative bacteria make up the majority of mesophilic species. Some more recently described acidophiles have tended to buck this trend, such as the Gram-negative thermophilic sulfur-oxidizing heterotroph *Acidicaldus organivorans* (Johnson et al. 2006), and the mesophilic archaeon *Ferroplasma acidiphilum* (Golyshina et al. 2000). Relatively little attention has been given to cold-adapted mineral-oxidizing acidophiles. However, the recently described species *Acidithiobacillus ferrivorans* can grow at temperatures of 4°C and above (Hallberg et al. 2009) and one strain (SS3) has been demonstrate to accelerate the oxidation of pyrite at 5°C (Dopson et al. 2006).

Although oxidative dissolution of simple and complex sulfide ores and concentrates may be mediated by pure cultures of iron-oxidizing acidophiles, as has often been described in laboratory studies, axenic cultures are never found in actual biomining operations. Consortia of microorganisms with synergistic (and sometimes complimentary) metabolic physiologies, have been identified in all commercial-scale systems that have been examined. Laboratory studies using defined populations of acidophiles have demonstrated that mixed cultures are more robust and are also frequently superior to pure cultures in terms of leaching kinetics. Ways in which acidophiles interact in a sulfide mineral leaching environment have been described by Johnson (2001). Among the most important of these is the generation of sulfur oxy-anions and elemental sulfur by ferric iron-mediated attack of pyrite by iron-oxidizing acidophiles. If the iron-oxidizer can also utilize sulfur as an electron donor (e.g. *Sulfobacillus* spp. and some *Acidithiobacillus* spp.) these liberated energy sources can be utilized by the same bacteria, though it appears that cells oxidize either iron or sulfur, but not both simultaneously. However, if the iron-oxidizer cannot utilize reduced sulfur (e.g. *Leptospirillum* spp. and *Fm. acidiphilum*) then other acidophiles, such as *At. caldus* and *At. thiooxidans*, can exploit this situation. Oxidation of elemental sulfur and sulfur oxy-anions generates acidity, helping to maintain conditions that favour the growth of the iron-oxidizers, so the interaction can be considered to be mutualistic. Interactions based on carbon transfer are also important, particularly in stirred tanks. Both primary and secondary acidophiles in sulfide mineral leaching operations are autotrophic, and their numbers can reach ~10^9 mL^{-1} of leachate liquor, and consequentially amounts of dissolved organic carbon (DOC) released from these primary producers will be significant. Some of the organic compounds that have been identified (notably small molecular weight aliphatic acids) are toxic at low pH, with some acidophiles, such as the important primary bacteria, *Leptospirillum* spp., being particularly sensitive. Associations between chemo-autotrophic acidophiles and others than can metabolize these organic compounds can again be considered as a mutualistic interaction. It is also likely that release of CO_2 by heterotrophs and mixotrophs benefits the autotrophs. The solubility of CO_2 in acidic liquors is very small, and transfer of the gas between bacteria within biofilms may well be important to carbon assimilation by autotrophic acidophiles.

An example of a synergistic interaction between two bacteria in a mineral leaching context was described by Bacelar-Nicolau and Johnson (1999). *Fm. acidiphilum* is an obligately heterotrophic iron-oxidizing acidophilic bacterium. It can oxidize pyrite in pure culture only if suitable organic carbon (e.g. yeast extract) is provided. *At. thiooxidans* is an obligately sulfur-oxidizing autotroph that cannot oxidize pyrite. Mixed cultures of the two bacteria can, however, successfully accelerate the dissolution of pyrite in the absence of added organic carbon. This is initiated by trace amounts of ferric iron (generated by *Fm. acidiphilum*) attacking the mineral and releasing reduced sulfur which *At. thiooxidans* oxidizes and couples to the fixation of CO_2. DOC released by the autotroph is used as a carbon source by *Fm. acidiphilum*, further stimulating growth and iron oxidation, allowing the cycle to continue (Fig. 19.6).

Bioleaching Microbial Communities

There were relatively few reports on the microbial communities that establish in biomining operations published in the scientific literature for the first 30 years or so after the technology was established. However, this situation has improved since

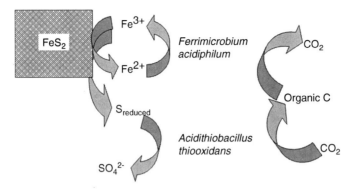

Fig. 19.6 Synergistic interactions between the obligately heterotrophic iron-oxidizer *Ferrimicrobium acidiphilum* and the obligately autotrophic sulfur-oxidizer *Acidithiobacillus thioooxidans* that allow mixed cultures to grow on pyrite as sole energy source

the start of the new millennium, mostly due to the application of biomolecular (DNA- and RNA-based) techniques and improvements in cultivation-based approaches for isolating biomining bacteria and archaea (reviewed in Johnson and Hallberg 2007).

Biomining processes operate as highly specialized environments (extremely acidic and enriched with soluble metals and other solutes) which severely restrict the biodiversity of indigenous life-forms that can survive, let alone be metabolically active, within them. All biomining systems operate under necessarily non-sterile conditions. The ore/concentrate feed material inevitably hosts its own microbial communities, in addition to those already present in the tank or heap, and those introduced as an inoculum.

Stirred tanks operate as continuous flow systems and are pH- and temperature-controlled, and the homogeneous conditions that develop within any single tank tend to limit the indigenous microbial diversity to relatively few (3–4) species (Rawlings and Johnson 2007b). Where tanks are operated in sequence, changes in liquor chemistry (increasing concentrations of metals, sulfate and DOC as mineral dissolution progresses) can greatly impact the composition of the microbial communities (Okibe et al. 2003). One further pressure on microorganisms that establish in stirred tank reactors is the rate of flow of the concentrate slurry through the system, which selects for acidophiles that can grow relatively fast under the prevailing physico-chemical conditions. Experience has shown that, following commissioning of a new stirred tank system, there is a gradual improvement in the performance of the bioleaching consortium over several years (resulting in shorter mean residence times) beyond which further improvements are marginal (Rawlings and Silver 1995). Most stirred tank systems operate at about 40–45°C and are used to process refractory gold concentrates, and many use commercial cultures (e.g. Biox[R] and BacTech[R] cultures). Analysis of microbial populations in stirred tanks processing pyrite/arsenopyrite, cobaltiferous pyrite and polymetallic (copper, zinc and iron) sulfides has revealed remarkable similarity between systems operated in different

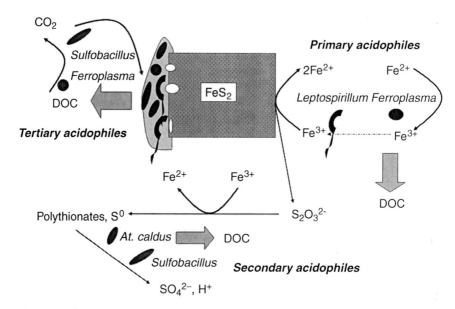

Fig. 19.7 Microbial consortium identified in stirred tank bioreactors bio-oxidizing refractory gold ore or bioleaching cobaltiferous pyrite. Iron-oxidizing autotrophs constitute the primary mineral-degraders, secondary sulfuric acid-generating bacteria the secondary-level subgroup, and organic carbon (DOC)-degrading acidophiles the tertiary-level subgroup

geographical locations, with the dominant (or exclusive) primary iron-oxidizer being *L. ferriphilum*, the dominant secondary (sulfur-oxidizer) identified as *At. caldus*, and a *Sulfobacillus* sp. (often *Sb. benefaciens*) and an archaeon (*Ferroplasma* sp.) operating as tertiary-level microorganisms (Fig. 19.7). In terms of their relative numbers in single tank systems and the first tank in an in-line series operation, the autotrophs generally outnumber the tertiary level mixotrophs/heterotrophs by at least one order of magnitude, though the situation can be reversed (in particular, large increases in relative numbers of *Ferroplasma*) in downstream stirred tanks due to increasing inputs of DOC (Okibe et al. 2003). In the single example to date where a near full-size stirred tank has been operated at an elevated temperature (78°C) specifically to bioleach the recalcitrant copper mineral chalcopyrite, the population was identified (as would be expected) as being comprised entirely of thermo-acidophilic archaea (Mikkelsen et al. 2006). Some were identified as being related to known species, while others probably represented novel organisms, though their composite roles (iron oxidation, sulfur oxidation, and carbon assimilation and metabolism) were thought to be the same as those observed in lower temperature systems.

Heap bioreactors present very different challenges to mineral-leaching microorganisms. On the one hand, the far longer time-scale over which heaps operate means that the pressure to select for fast-growing strains and species is no longer an issue, but rather the ability to attach to mineral surfaces (to prevent washout in the percolating leach liquor) is more important. In contrast to stirred tanks, heaps are highly heterogeneous systems, both spatially and temporally, particularly with regard to temperature which may vary from ambient to >80°C. Besides this, aeration conditions can vary even in aerated heaps, as can pH, redox potentials and solute compositions of the leach liquor. Different micro-environments can develop within heaps, presenting opportunities for colonization by bacteria and archaea with very different responses to pH, temperature etc. It would therefore be anticipated that a far wider diversity of acidophiles may be encountered in mineral heaps than in stirred tanks, and this has been borne out by the limited number of microbiological analyses that have been published. Ensuring sufficient microbial biodiversity in heaps (e.g. via inoculation) is therefore more of a priority issue to ensure leaching efficiency. Pregnant leach solutions (PLS) have often been used to assess the diversity of mineral heap microflora, though care must be taken in this case as bacteria and archaea that attach strongly to mineral phases are likely to be underestimated in these drainage waters. What is apparent is that, while many of the bacteria and archaea found in stirred tank systems have also been identified in heap leachates, the latter generally contain relatively more *Firmicutes* (low GC Gram-positive bacteria that form endospores, which probably gives them a competitive advantage in fluctuating conditions within heaps) and also other microorganisms that are either novel (e.g. *Acidiplasma cupricumulans* (formerly *Ferroplasma cupriculums*) or are unknown. Mesophilic autotrophic and heterotrophic acidophiles (e.g. *At. ferrooxidans, At. ferrivorans, At. thiooxidans, Fm. acidiphilum, Acidiphilium* and *Acidisphaera* spp.) have also been found in relatively large numbers in heap leaching operations (reviewed in Rawlings and Johnson 2007b). Two interesting recent findings are, (a) even in relatively cool conditions, thermotolerant *L. ferriphilum* rather than mesophilic *L. ferrooxidans* is one of the main primary bacteria present in copper heaps, which is probably related to its superior tolerance of copper (Galleguillos et al. 2009), and (b) the dominant iron/sulfur bacterium present is often *At. ferrivorans,* rather than *At. ferrooxidans*.

Laboratory simulations of biomining operations usually use either pH- and temperature-controlled bioreactors (to simulate stirred tanks) or columns (to simulate heap reactors). These can give useful insights into how microbial communities develop and interact and also how microbial consortia can be designed to optimize leaching of different ores and concentrates. Okibe and Johnson (2004) examined relative rates of pyrite dissolution by various combinations of seven species of acidophilic eubacteria (autotrophic and mixotrophic iron- and sulfur-oxidizers, and one obligate heterotroph) and one archaeon (a heterotrophic iron-oxidizing *Ferroplasma* sp.) and followed changes in microbial populations using a combined cultivation-based (plating on selective solid media) and cultivation-independent (fluorescent in situ hybridization; FISH) approach. Concentrations of DOC accumulated to >100 mg L^{-1} in cultures containing only autotrophic bacteria, and the most effective cultures contained both carbon-fixing and carbon-metabolizing acidophiles. A consortium containing *L. ferriphilum, At. caldus* and *Ferroplasma* (i.e. similar to that found in some commercial applications) was the most effective of those examined. Column reactors may be set up as flooded and aerated systems, percolated systems using recycled or single-throughput liquor, or air-lift systems, and can vary in capacities from tens of grams to many kilograms of mineral ore. Examination of mineral dissolution and microbial kinetics of a complex polymetallic black schist ore subjected to bioleaching in columns at 37°C showed that microbial communities displayed major changes over a period of a year, which appeared to be determined at least in part by changing leachate chemistries (Wakeman et al. 2008). *At. ferrooxidans* was the dominant bacterium present in the early stages of the leaching process, but was replaced by *L. ferriphilum* and, on a more protracted basis, by *At. ferrivorans*. While some other bacteria introduced in the inoculum were also detected occasionally, iron-oxidizing *Alicyclobacillus*-like *Firmicutes* that were present (presumably as spores) on the dried but non-sterilized ore, were more frequently found in large numbers in leach liquors, and were considered to have important roles during the early stages of mineral dissolution.

Metal Recovery Technologies

Oxidative dissolution of sulfide minerals is the first stage in metal recovery in biomining operations. Subsequently, the target metals need either to be extracted from the leach liquors (in bioleaching) or from the solid residues (in bio-oxidation and some bioleaching operations; Fig. 19.1). Copper, the most extensively biomined base metal, is most effectively recovered by extracting the metals from PLS using a specific solvent, followed by electrolytic recovery (electrowinning) of the metals on large plate cathodes (SX-EW; Fig. 19.8). Silver (which may be present in significant concentrations in some copper sulfide ores and concentrates) can be extracted from bioleached residues by leaching with chloride (a mixture of HCl and NaCl), while fine-grain gold liberated from enshrouding pyrite and arsenopyrite is extracted with cyanide (forming a soluble gold-cyanide complex, $Au(CN)_4^-$).

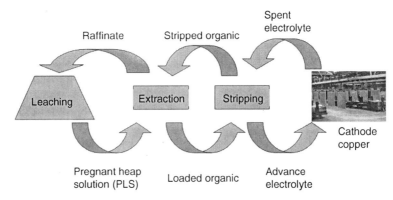

Fig. 19.8 Stages in the recovery of bioleached copper using solvent extraction-electrowinning (SX-EW)

Biology does not play a role in any of these downstream processing techniques. However, biosulfidogenesis (biological generation of hydrogen sulfide, often by sulfate-reducing bacteria; SRB) can be used to precipitate and recover metals in leach liquors. At least two commercial-scale biosulfidogenic systems are in operation for recovering metals from waste streams and process waters. The THIOTEQ[R] process developed by the Dutch company Paques bv, uses SRB to produce hydrogen sulfide in an off-line bioreactor using sulfate or elemental sulfur as an electron acceptor, and ethanol or hydrogen gas as electron donor. The sulfide is transported to a separate reactor where it contacts the metal-rich waste stream or process water, forming insoluble metal sulfides. Any excess sulfide is oxidized to elemental sulfur, which can be re-used for the process itself or sold as a product. Paques have also developed a related waste water treatment technology (BIOMETEQ[R]) that uses SRB housed in self-cleaning sand filters. The Canadian company BioteQ Environmental Technologies Inc. also has developed two technologies for recovering metals from polluted and process waters by precipitation as sulfides, though only the BioSulfide[R] process uses biologically-generated sulfide. Elemental sulfur rather than sulfate is used to generate sulfide in the BioSulfide[R] process, and by control of pH in contacting reactors it is possible to facilitate the selective recovery of metals.

Challenges and Opportunities

While biomining has emerged as a viable niche industry over the past 40 or so years, there remain both challenges that, if solved, and opportunities that, if taken, could lead to major expansions in this area of biotechnology in future years. Some of these are described below.

Bioleaching Minerals Using Saline and Brackish Waters

Quality and availability of water are global issues, particularly with concerns of how these might be impacted by climate change. Already, some of the largest biomining operations are carried out in and around desert areas. In northern Chile, for example, massive heap bioreactors are operating in areas adjacent to the Atacama desert. At the Escondida mine, water is pumped from the South Pacific over a distance of 170 km and an elevation of 3,200 m to the mine site, where it is desalinated. Here and in other locations (e.g. parts of Australia) there would be a considerable advantage in being able to use acidified brackish water or seawater as irrigation liquors in heap reactors. A major problem though is in the well-documented sensitivity of biomining microorganisms to elevated concentrations of chloride. Many acidophiles are sensitive to anions (organic and inorganic) other than sulfate, for reasons discussed by Norris and Ingledew (1992). There are exceptions, however, such as the iron-oxidizer *Thiobacillus prosperus* (Davis-Belmar et al. 2008). The problem is that none of the halotolerant acidophiles that have been described appear to be adept at leaching sulfide minerals. The challenge is therefore to find new strains of bacteria and/or archaea (and more pertinently, stable consortia) that are effective at accelerating the oxidative dissolution of sulfide minerals in a saline milieu.

Bioleaching Chalcopyrite

Chalcopyrite is the most abundant copper sulfide in the lithosphere but it is notoriously difficult to extract more than about 20% of copper from this mineral using conventional biomining technology (Watling 2006). Bioleaching chalcopyrite at high temperatures (~80°C) using thermo-acidophilic archaea has been demonstrated to recover >90% of copper from a chalcopyrite concentrate in two primary (50 and 300 m^3) stirred tank reactors (Batty and Rorke 2006). However, there were additional costs incurred in construction (the need to line the tanks with ceramic tiles to cope with the highly corrosive hot acidic leach liquors) and operation (aerating with oxygen-enriched gas to counter the problems of mass transfer of oxygen) of the plant compared lower temperature processing of mineral concentrates. An alternative approach to promoting effective bioprocessing of chalcopyrite is to leach at relatively low redox potentials. This has been demonstrated successfully in abiotic systems (Hiroyoshi et al. 2000). Conditions within stirred tank reactors can be geared to ensure low redox potentials (e.g. by controlling rates of aeration) but controlling redox potentials in heap reactors and dumps, which account for the bulk of copper ore bioleaching, is far more challenging.

Complex Ores and Mine Wastes

Polymetallic ores (i.e. those containing significant concentrations of more than one target metal) are becoming increasingly important as primary source materials for the mining industry. In some cases, producing concentrates from such ores is not pragmatic or economically viable, such as the black schist ore at the Talvivaara mine (Finland) where the high graphite content of the ore precluded manufacture of mineral concentrate. Bioleaching of polymetallic ores in heap reactors or dumps is a viable alternative. Sulfide minerals vary in their reactivities, which is reflected in different rates of dissolution and release of metals during bioleaching. Downstream processing can then be used to selectively remove soluble metals (e.g. as sulfides) from PLS.

Dwindling resources of metal-rich ores means that waste rocks and tailings from older mining operations (where technologies for extracting were not so well advanced) are now often being reconsidered as resources. Reprocessing of mine wastes represents an ideal opportunity to utilize biomining, as this can be far more cost-effective than pyrometallurgy and other competing technologies where metal contents are low. Often, reprocessing wastes from old mines also has an environmental bonus. One such example is a tailings stockpile at the former Kilembe mine in Uganda (estimated at about 900,000 t of waste, containing ~80% pyrite and 1.38% cobalt) which was causing a severe environmental threat to a nearby lake and national park. The decision to bioprocess these tailings to recover cobalt (in stirred tanks) has essentially removed this threat as well as being a commercial success in terms of the metal product (Morin and d'Hugues 2007).

Oxidized Ores

As discussed, commercial biomining focuses exclusively on recovering metals from reduced (sulfidic) ores. However, there are major ores where target metals are contained within or associated with oxidized (ferric iron) minerals. An example of these are "nickel laterites", which derive from the weathering of ultra-mafic rocks in tropical conditions, and which often contain significant amounts of cobalt and manganese as well as nickel (Coto et al. 2008). Large reserves of nickel laterites are located in countries such as Cuba, Western Australia and New Caledonia. Nickel and cobalt are conventionally extracted from these ores by acid leaching at either high pressures and temperatures (ca. 250°C) or at atmospheric pressures in heaps, but generate residues that still contain significant concentrations of metals (~0.25% nickel; Coto et al. 2008). Bioleaching of nickel laterites under aerobic conditions has been investigated, using either biogenic sulfuric acid produced by acidophilic sulfur-oxidizing bacteria such as *At. thiooxidans*, or metal-complexing organic acids (oxalic, citric etc.) produced by fungi such as *Aspergillus niger*. Alternatively, anaerobic bio-processing of nickel laterites may be feasible. Much of the non-ferrous metal in these ores is associated with ferric iron minerals such as goethite ($\alpha FeO.OH$), and it has been suggested that, by inducing the reductive dissolution of these minerals, it might be possible to significantly enhance nickel extraction. Many species of acidophilic bacteria and archaea are able to catalyze the dissimilatory reduction of ferric iron to ferrous (Table 19.4), and the reductive dissolution of several ferric iron minerals (including goethite) has been reported for the obligately heterotrophic acidophile *Acidiphilium* (Bridge and Johnson 2000), and the moderately thermophilic iron-oxidizing bacteria *Sulfobacillus* and *Acidimicrobium* (Bridge and Johnson 1998). Possibilities exist, therefore, of developing new areas of biomining ores and concentrates using reductive as well as oxidative processing of minerals.

Acknowledgement The author is grateful to the Royal Society (U.K.) for the provision of an Industrial Fellowship.

References

Bacelar-Nicolau P, Johnson DB (1999) Leaching of pyrite by acidophilic heterotrophic iron-oxidizing bacteria in pure and mixed cultures. Appl Environ Microbiol 65:585–590

Batty JD, Rorke GV (2006) Development and commercial demonstration of the BioCOP™ thermophile process. Hydrometallurgy 83:83–89

Bridge TAM, Johnson DB (1998) Reduction of soluble iron and reductive dissolution of ferric iron-containing minerals by moderately thermophilic iron-oxidizing bacteria. Appl Environ Microbiol 64:2181–2186

Bridge TAM, Johnson DB (2000) Reductive dissolution of ferric iron minerals by *Acidiphilium* SJH. Geomicrobiol J 17:193–206

Brierley CL (2008a) How will biomining be applied in future? Trans Nonferrous Met Soc China 18:1302–1310

Brierley JA (2008b) A perspective on developments in biohydrometallurgy. Hydrometallurgy 94:2–7

Coto O, Galizia F, Hernandez I, Marrero J, Donati E (2008) Cobalt and nickel recoveries from lateritic tailings by organic and inorganic bio-acids. Hydrometallurgy 94:18–22

Coupland K, Johnson DB (2008) Evidence that the potential for dissimilatory ferric iron reduction is widespread among acidophilic heterotrophic bacteria. FEMS Microbiol Lett 279:30–35

Davis-Belmar CS, Nicolle JLC, Norris PR (2008) Ferrous iron oxidation and leaching of copper ore with halotolerant bacteria in ore columns. Hydrometallurgy 94:144–147

Donati ER, Sand W (eds) (2007) Microbial processing of metal sulfides. Springer, Dordrecht, The Netherlands

Dopson M, Halinen A-K, Rahunen N, Ozkaya B, Sahinkaya E, Kaksonen AH, Lindstrom EB, Puhakka JA (2006) Mineral and iron oxidation at low temperatures by pure and mixed cultures of acidophilic microorganisms. Biotechnol Bioeng 97:1205–1215

Galleguillos P, Hallberg KB, Johnson DB (2009) Microbial diversity and genetic response to stress conditions of extremophilic bacteria isolated from the Escondida copper mine. Adv Mater Res 71–73:55–58

Ghauri MA, Okibe N, Johnson DB (2007) Attachment of acidophilic bacteria to solid surfaces: the significance of species and strain variations. Hydrometallurgy 85:72–80

Golyshina OV, Pivovarova TA, Karavaiko GI, Kondrat'eva TF, Moore ERB, Abraham WR, Lunsdorf H, Timmis KN, Yakimov MM, Golyshin PN (2000) *Ferroplasma acidiphilum* gen. nov., sp. nov., an acidophilic, autotrophic, ferrous-iron-oxidizing, cell-wall-lacking, mesophilic member of the *Ferroplasmaceae* fam. nov., comprising a distinct lineage of the *Archaea*. Int J Syst Evol Microbiol 50:997–1006

Hallberg KB, Johnson DB (2001) Biodiversity of acidophilic prokaryotes. Adv Appl Microbiol 49:37–84

Hallberg KB, González-Toril E, Johnson DB (2010) *Acidithiobacillus ferrivorans*, sp. nov., facultatively anaerobic, psychrotolerant iron- and sulfur-oxidizing acidophiles isolated from metal mine-impacted environments. Extremophiles 14:9–19

Harvey TJ, Bath M (2007) The GeoBiotics GEOCOAT^R technology – progress and challenges. In: Rawlings DE, Johnson DB (eds) Biomining. Springer, Heidelberg, pp 97–112

Hiroyoshi N, Miki H, Hirajima T, Tsunekawa M (2000) A model for ferrous-promoted chalcopyrite leaching. Hydrometallurgy 57:31–38

Johnson DB (2001) Importance of microbial ecology in the development of new mineral technologies. Hydrometallurgy 59:147–158

Johnson DB (2009) Extremophiles: acid environments. In: Schaechter M (ed) Encyclopaedia of microbiology. Elsevier, Oxford, pp 107–126

Johnson DB, Hallberg KB (2007) Techniques for detecting and identifying acidophilic mineral-oxidising microorganisms. In: Rawlings DE, Johnson DB (eds) Biomining. Springer, Heidelberg, pp 237–262

Johnson DB, Hallberg KB (2008) Carbon, iron and sulfur metabolism in acidophilic microorganisms. Adv Microbiol Physiol 54:202–256

Johnson DB, McGinness S (1991) Ferric iron reduction by acidophilic heterotrophic bacteria. Appl Environ Microbiol 57:207–211

Johnson DB, Stallwood B, Kimura S, Hallberg KB (2006) Isolation and characterization of *Acidicaldus organovorus*, gen. nov., sp. nov.; a novel sulfur-oxidizing, ferric iron-reducing thermo-acidophilic heterotrophic *Proteobacterium*. Arch Microbiol 185:212–221

Kelly DP (1978) Bioenergetics of chemolithotrophic bacteria. In: Bull AT, Meadows PM (eds) Companion to microbiology: selected topics for further discussion. Longman, London, pp 363–386

Kelly DP, Wood AP (2000) Reclassification of some species of *Thiobacillus* to the newly designated genera *Acidithiobacillus* gen. nov., *Halothiobacillus* gen. nov., and *Thermothiobacillus* gen. nov. Int J Syst Evol Microbiol 50:511–516

Koschorreck M (2008) Microbial sulphate reduction at a low pH. FEMS Microbiol Ecol 64:329–342

Mikkelsen D, Kappler U, McEwan AG, Sly LI (2006) Archaeal diversity in two thermophilic chalcopyrite bioleaching reactors. Environ Microbiol 8:2050–2055

Morin DHR, d'Hugues P (2007) Bioleching of a cobalt-containing pyrite in stirred reactors: a case study from the laboratory scale to industrial application. In: Rawlings DE, Johnson DB (eds) Biomining. Springer, Heidelberg, pp 35–56

Morin D, Pinches T, Huisman J, Frias C, Norberg A, Forssberg E (2008) Progress after three years of BioMinE – research and technical development project for a global assessment of biohydrometallurgical processes applied to European non-ferrous metal resources. Hydrometallurgy 94:58–68

Norris PR (2007) Acidophilic diversity in mineral sulfide oxidation. In: Rawlings DE, Johnson DB (eds) Biomining. Springer, Heidelberg, pp 199–216

Norris PR, Ingledew WJ (1992) Acidophilic bacteria: adaptations and applications. In: Herbert RA, Sharp RJ (eds) Molecular biology and biotechnology of extremophiles. Blackie, Glasgow, pp 115–142

Okibe N, Johnson DB (2004) Biooxidation of pyrite by defined mixed cultures of moderately thermophilic acidophiles in pH-controlled bioreactors: the significance of microbial interactions. Biotechnol Bioeng 87:574–583

Okibe N, Gericke M, Hallberg KB, Johnson DB (2003) Enumeration and characterization of acidophilic microorganisms isolated from a pilot plant stirred tank bioleaching operation. Appl Environ Microbiol 69:1936–1943

Olson GJ, Brierley JA, Brierley CL (2003) Bioleaching review part B: progress in bioleaching: applications of microbial processes by the minerals industry. Appl Microbiol Biotechnol 63:249–257

Rawlings DE (ed) (1997) Biomining: theory, microbes and industrial processes. Springer/Landes Biosciences, Georgetown, TX

Rawlings DE (2002) Heavy metal mining using microbes. Annu Rev Microbiol 56:65–91

Rawlings DE (2005) Characteristics and adaptability of iron- and sulfur-oxidizing microorganisms used for the recovery of metals from minerals and their concentrates. Microb Cell Fact 4:13

Rawlings DE, Johnson DB (eds) (2007a) Biomining. Springer, Heidelberg

Rawlings DE, Johnson DB (2007b) The microbiology of biomining: development and optimization of mineral-oxidizing microbial consortia. Microbiology 153:315–324

Rawlings DE, Silver S (1995) Mining with microbes. Biotechnology 13:773–778

Rawlings DE, Tributsch H, Hansford GS (1999) Reasons why '*Leptospirillum*'-like species rather than *Thiobacillus ferrooxidans* are the dominant iron-oxidizing bacteria in many commercial processes for the biooxidation of pyrite and related ores. Microbiology 145:5–13

Rawlings DE, Dew D, du Plessis C (2003) Biomineralization of metal-containing ores and concentrates. Trends Biotechnol 21:38–44

Rohwerder T, Gehrke T, Kinzler K, Sand W (2003) Bioleaching review part A: progress in bioleaching: fundamentals and mechanisms of bacterial metal sufide oxidation. Appl Microbiol Biotechnol 63:239–248

Schippers A (2007) Microorganisms involved in bioleaching and nucleic acid-based molecular methods for their identification and quantification. In: Donati ER, Sand W (eds) Microbial processing of metal sulfides. Springer, Dordrecht, The Netherlands, pp 3–33

Schippers A, Sand W (1999) Bacterial leaching of metal sulfides proceeds by two indirect mechanisms via thiosulfate or via polysulfides and sulfur. Appl Environ Microbiol 65:319–321

Sehlin HM, Lindstrom EB (1992) Oxidation and reduction of arsenic by *Sulfolobus acidocaldarius* strain BC. FEMS Microbiol Lett 93:8–92

Steudel R (2000) The chemical sulfur cycle. In: Lens P, Hulshoff Pol L (eds) Environmental technologies to treat sulfur pollution: principles and engineering. International Association on Water Quality, London, pp 1–31

Stumm W, Morgan JJ (1981) Aquatic chemistry: an introduction emphasizing chemical equilibria in natural waters. Wiley, New York

Tributsch H, Rojas-Chapana J (2007) Biological strategies for obtaining energy by degrading sulfide minerals. In: Rwalings DE, Johnson DB (eds) Biomining. Springer, Heidelberg, pp 263–280

Valdes J, Pedroso I, Quatrini R, Holmes DS (2008) Comparative genome analysis of *Acidithiobacillus ferrooxidans, A. thiooxidans* and *A. caldus*: insights into their metabolism and ecophysiology. Hydrometallurgy 94:180–184

Wakeman K, Auvinen H, Johnson DB (2008) Microbiological and geochemical dynamics in simulated heap leaching of a polymetallic sulfide ore. Biotechnol Bioeng 101:739–750

Watling HR (2006) The bioleaching of sulphide minerals with emphasis on copper sulphides – a review. Hydrometallurgy 84:81–108

Welham NJ, Malatt KA, Vukcevic S (2000) The effect of solution speciation on iron-sulphur-arsenic-chloride systems at 298°K. Hydrometallurgy 57:209–223

Index

Breinigsville, PA USA
30 August 2010
244437BV00004B/21/P